S0-CAH-143

Gas Phase Inorganic Chemistry

MODERN INORGANIC CHEMISTRY

Series Editor: John P. Fackler, Jr.
Texas A&M University

CARBON-FUNCTIONAL ORGANOSILICON COMPOUNDS
Edited by Václav Chvalovský and Jon M. Bellama

GAS PHASE INORGANIC CHEMISTRY
Edited by David H. Russell

HOMOGENEOUS CATALYSIS
WITH METAL PHOSPHINE COMPLEXES
Edited by Louis H. Pignolet

THE JAHN-TELLER EFFECT AND
VIBRONIC INTERACTIONS IN MODERN CHEMISTRY
I. B. Bersuker

METAL INTERACTIONS WITH BORON CLUSTERS
Edited by Russell N. Grimes

MÖSSBAUER SPECTROSCOPY
APPLIED TO INORGANIC CHEMISTRY, *Volumes 1 and 2*
Edited by Gary J. Long

MÖSSBAUER SPECTROSCOPY
APPLIED TO INORGANIC CHEMISTRY, *Volume 3*
Edited by Gary J. Long and Fernande Grandjean

A Continuation Order Plan is available for this series. A continuation order will bring delivery of each new volume immediately upon publication. Volumes are billed only upon actual shipment. For further information please contact the publisher.

Gas Phase Inorganic Chemistry

Edited by
David H. Russell

Texas A&M University
College Station, Texas

Plenum Press • New York and London

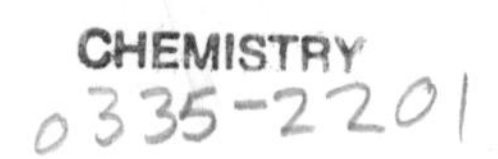

Library of Congress Cataloging in Publication Data

Gas phase inorganic chemistry / edited by David H. Russell
p. cm.—(Modern inorganic chemistry)
Includes bibliographies and index.
ISBN 0-306-42972-1
1. Chemical reaction, Conditions and laws of. 2. Gases, Ionized. 3. Metal ions. I. Russell, David H. II. Series.
QD501.G3238 1989 88-30269
541.3′9—dc19 CIP

A Division of Plenum Publishing Corporation
233 Spring Street, New York, N.Y. 10013

Printed in the United States of America

Contributors

Peter B. Armentrout • Department of Chemistry, University of Utah, Salt Lake City, Utah 84112

Steven W. Buckner • Department of Chemistry, Purdue University, West Lafayette, Indiana 47907

Robert C. Dunbar • Department of Chemistry, Case Western Reserve University, Cleveland, Ohio 44106

Donnajean Anderson Fredeen • Department of Chemistry, Texas A&M University, College Station, Texas 77843

Ben S. Freiser • Department of Chemistry, Purdue University, West Lafayette, Indiana 47907

Edward R. Grant • Department of Chemistry, Purdue University, West Lafayette, Indiana 47907

Michael L. Gross • Department of Chemistry, University of Nebraska, Lincoln, Nebraska 68588

Martin F. Jarrold • AT&T Bell Laboratories, Murray Hill, New Jersey 07974

Glen Eugene Kellogg • Department of Chemistry, University of Arizona, Tucson, Arizona 85721

Kelley R. Lane • Department of Chemistry, Purdue University, West Lafayette, Indiana, 47907

Dennis L. Lichtenberger • Department of Chemistry, University of Arizona, Tucson, Arizona 85721

Denise K. MacMillan • Department of Chemistry, University of Nebraska, Lincoln, Nebraska 68588

Wilma K. Meckstroth • Department of Chemistry, Ohio State University, Newark, Ohio 43055

Douglas P. Ridge • Department of Chemistry, University of Delaware, Newark, Delaware 19716

David H. Russell • Department of Chemistry, Texas A&M University, College Station, Texas 77843

Kenneth E. Schriver • Department of Chemistry and Biochemistry, University of California, Los Angeles, California 90024

Robert R. Squires • Department of Chemistry, Purdue University, West Lafayette, Indiana 47907

Ronald E. Tecklenburg • Department of Chemistry, Texas A&M University, College Station, Texas 77843

Veronica Vaida • Department of Chemistry and Biochemistry, University of Colorado, Boulder, Colorado 80309

Bruce H. Weiller • Department of Chemistry, Purdue University, West Lafayette, Indiana 47907

Robert L. Whetten • Department of Chemistry and Biochemistry, University of California, Los Angeles, California 90024

Preface

The field of gas phase inorganic ion chemistry is relatively new; the early studies date back approximately twenty years, but there has been intense interest and development in the field in the last ten years. As with much of modern chemistry, the growth in gas phase inorganic ion chemistry can be traced to the development of instrumentation and new experimental methods. Studies in this area require sophisticated instruments and sample introduction/ionization methods, and often these processes are complicated by the need for state-selecting (or collisionally stabilizing) the reactive species in order to assign the chemistry unequivocally. At the present level of experimental development, a wide range of experiments on diverse ionic systems are possible and many detailed aspects of the chemistry can be studied.

Gas Phase Inorganic Chemistry focuses on the reactions of metal ions and metal clusters, and on the study of these species using the available modern spectroscopic methods. Three of the twelve chapters cover the chemistry of ionic monometal transition metal ions and the chemistry of these species with small diatomics and model organics. Two of the chapters focus on the studies of the chemical and physical properties of (primarily) transition metal clusters, and these chapters review experimental methods and capabilities. Two chapters also deal with the chemistry of transition metal carbonyl clusters, and these chapters address issues important to cluster growth and activation as well as the characterization of such species. Two chapters deal with the photochemistry (single-photon and multiphoton) of transition metal species, and one chapter covers time-resolved vibrational spectroscopy of organometallic species. One chapter covers collisional activation studies of metal ion complexes formed by ion–molecule reactions, and one chapter deals with photoelectron spectroscopy of complex metals.

As with any text of this type there are obvious omissions. One such case is a detailed chapter on studies of metal clusters by supersonic nozzle methods. These studies have progressed at a rapid rate, but several volumes have already been published that highlight this area. Additional studies on the reaction dynamics of ionic metal species could obviously have been included; however, at the time of preparation of this volume it was unclear that a complete understanding of these processes was available. Certainly, it is now clear that the chemistry of ionic metal species is more complex than that of organic ions, owing to the abundance of long-lived states. This is certainly an area in which much additional work is required—quite possibly suggesting a future topic for a similar volume.

Collectively, this volume encompasses many of the important areas of gas phase inorganic chemistry. The volume is relatively evenly split between chemical studies and probes of physical properties.

The authors were chosen because of their stature in the field. The topics and organization reflect their individual styles. I am grateful to the authors for their willingness to contribute, and for the quality and promptness of their contributions.

David H. Russell

College Station, Texas

Contents

Chapter 1

Reactions of Atomic Metal Ions with H_2, CH_2, and C_2H_6: Electronic Requirements for H—H, C—H, and C—C Bond Activation

Peter B. Armentrout

Chapter 2

Nucleophilic Addition Reactions of Negative Ions with Organometallic Complexes in the Gas Phase

Robert R. Squires and Kelley R. Lane

Chapter 3

Reactions in Ionized Metal Carbonyls: Clustering and Oxidative Addition

Douglas P. Ridge and Wilma K. Meckstroth

Chapter 4

Structure–Reactivity Relationships for Ionic Transition Metal Carbonyl Cluster Fragments

David H. Russell, Donnajean Anderson Fredeen, and Ronald E. Tecklenburg

Chapter 5

Metal and Semiconductor Cluster Ions

Martin F. Jarrold

Chapter 6

Atomic Clusters in the Gas Phase

Robert L. Whetten and Kenneth E. Schriver

Chapter 7

Time-Resolved Kinetics of Organometallic Reactions in the Gas Phase by Transient Infrared Absorption Spectrometry

Bruce H. Weiller and Edward R. Grant

Chapter 8

Characterization of Metal Complex Positive Ions in the Gas Phase by Photoelectron Spectroscopy

Dennis L. Lichtenberger and Glen Eugene Kellogg

Chapter 9

Chemistry and Photochemistry of Bare Metal Cluster Ions in the Gas Phase

Steven W. Buckner and Ben S. Freiser

1

Reactions of Atomic Metal Ions with H_2, CH_4, and C_2H_6: Electronic Requirements for H—H, C—H, and C—C Bond Activation

Peter B. Armentrout

1. INTRODUCTION

1.1. Relation between Gas Phase and Condensed Phase

One of the obligations (and difficulties) faced by gas phase inorganic chemists is relating their work to the general field of condensed phase inorganic chemistry. One specific area of common interest to both gas phase and condensed phase inorganic chemists is the activation of C—H bonds by metal centers. Research into homogeneous systems which activate alkanes has seen a flurry of activity in recent years.[1] Similarly, ever since the first observation by Allison, Freas, and Ridge[2] in 1979 that atomic transition metal ions can activate C—H *and* C—C bonds, a tremendous amount of experimental work has centered on the gas phase reactions of M^+ with alkanes. Unfortunately, the connection between these gas phase and condensed phase systems has been elusive. This relationship is critical if the data of the gas phase chemist are to be most meaningful.

The general advantages of examining inorganic chemistry in the gas phase are fairly obvious. Complications due to solvation are avoided in the gas phase. Since ion chemists always are dealing with reactive species, a full (or even nearly full) coordination sphere around the metal is not required. Thus, metal

---

Peter B. Armentrout • Department of Chemistry, University of Utah, Salt Lake City, Utah 84112.

centers with active sites (coordinatively unsaturated species) can be prepared and reacted under well-defined conditions. This allows the gas phase ion chemist to study exactly the same reaction while varying only the identity of the metal, i.e., to examine the periodic trends in reactivity.

Our work in this area involves the use of ion beam techniques to make detailed studies of the periodic trends in the reactions of atomic metal ions with small alkanes, particularly methane and ethane,[3-7] and with dihydrogen,[8-15] a closely related system. With few exceptions, atomic metal ions and these small hydrocarbons do not react exothermically. These systems are therefore ideally suited to beam studies in which hyperthermal endothermic processes can be examined. Analysis of these endothermic reactions provides thermodynamic data which yield a quantitative view of the alkane activation process. This thermochemistry is examined in section 3. Further analysis of the beam results also provides a detailed mechanistic view of the reactions as well as an understanding of the electronic character on the metal center which promotes efficient reaction. This is discussed in Sections 4 to 6.

1.2. Electronic Requirements for Alkane Activation

In a recent review, Crabtree[1] outlined the requirements for alkane activation by a metal–ligand system.

1. The system must be coordinatively unsaturated, i.e., there must be an empty metal orbital capable of accepting electron density from the σ orbital of the bond to be broken.
2. There should be a filled metal orbital having the correct symmetry to donate electrons into an antibonding orbital of the bond to be broken.
3. There must be strong M—C and M—H bonds.
4. The system must be sterically uncongested.
5. Cyclometalation must not be a competitive reaction.
6. Precoordination of the alkane may be desirable.

Now consider atomic metal ions as the template for alkane activation. It is fairly obvious that such species obey most of the rules given above. They are coordinatively unsaturated, sterically uncongested, and, having no ligands, do not have cyclometalation as a competitive reaction path. As we shall see in Section 3, atomic metal ions have strong M—C and M—H bonds. Because the metal system is ionic, a long-range ion–induced dipole attraction exists which provides precoordination of the alkane. Clearly then, atomic metal ions bypass the rules related to solvation and ligation effects. It is not obvious, however, that all atomic metal ions have an empty acceptor orbital to fulfill rule 1 and an occupied donor orbital to fulfill rule 2. Both of these rules relate to the electronic requirements for efficient alkane activation. The interesting point here is that these rules can be tested experimentally by varying the electronic configuration of the atomic metal ions being studied and relating this to the observed reactivity. The isolation of this purely electronic effect is impossible in condensed phase systems because of the restrictions of ligation and solvation.

1.3. State-Specific Chemistry

The variation of the electron configuration can be achieved by examining the reactions of specific states of the metal ions. Unfortunately, the availability of d orbitals which lie close in energy to the valence s orbital means that there are numerous closely spaced electronic levels. This, of course, is one of the underlying reasons for the versatile reactivity of transition metal species, but experimentally, this means that the elucidation of the state-specific chemistry of transition metal ions is a formidable task. The close spacing of the states makes simple ion generation schemes unworkable. Also, all low-lying states have only s and d electrons such that transitions between these states are parity forbidden. Therefore, the lifetimes of the excited states are expected to be on the order of seconds although no experimental information is yet available.[16] This precludes optical pumping schemes as efficient modes of state preparation.

In this chapter (Sections 4 to 6), we summarize our recent results concerning the differences in reactivity exhibited by different electronic states of atomic ions of the first-row transition metals.[3–15] In our work to date, we have used a combination of ion sources to produce ions with varying populations of electronic states. While the true distribution of these states is not always known unambiguously, the reactivity of various states can be assessed fairly reliably. In future work, we plan to use an experimental method which appears to be a promising means for generation of state-specific metal ions, resonantly enhanced multiphoton ionization. This is under development in our labs[17] and by Weisshaar and co-workers.[18]

2. ION BEAM TECHNIQUES

2.1. Ion Sources

We have routinely used three types of ion sources to produce atomic metal ions in our experiments: an electron impact ionization source (EI),[19] a surface ionization source (SI),[4] and an EI source combined with a high-pressure drift cell (DC).[20] In the first of these sources (EI), ions are formed by high-energy electron impact on a volatile organometallic compound. Variation of the electron energy (E_e) above the appearance potential (AP) of the atomic metal ion allows rough control over the extent of electronic excitation. Generally, we have found that the amount of excitation does not appear to vary substantially once the electron energy is 10–20 eV above the AP. Also, the states which are formed most abundantly often appear to be the lowest-energy state having a particular spin. This source is the most intense ($>10^7$ ions/s) if the source is operated with an electron energy that is well above the AP.

The second source (SI) is a more controlled means of producing atomic metal ions. Here, the vapor of an organometallic compound or a metal salt is exposed to a rhenium filament heated to ~2200 K. Decomposition occurs and atoms with low ionization potentials desorb from the filament surface. Intensities range from 10^5 to 10^7 ions/s. Since the energy available is on the order of kT (~ 0.2 eV at 2200 K), the only states which can be populated appreciably are low-lying electronic levels (Table 1). Although no absolute measure of the state population is available, we have independent indications that the ionization process is statistical such that the desorbing ions have reached equilibrium at the filament temperature. These indications include determining the dependence of certain reactions on the filament temperature[4,7] and comparison of experimental cross section magnitudes for specific electronic states to those calculated by phase space theory[9,12] or to

Table 1. Electronic States of First-Row Transition Metal Ions

Ion	State	Configuration	Energy[a]	SI populated[b]
Sc^+	^3D	$4s3d$	0.01	0.886
	^1D	$4s3d$	0.32	0.060
	^3F	$3d^2$	0.61	0.054
Ti^+	^4F	$4s3d^2$	0.03	0.626
	^4F	$3d^3$	0.13	0.356
	^2F	$4s3d^2$	0.59	0.016
	^2D	$4s3d^2$	1.08	<0.001
	^2G	$3d^3$	1.12	0.001
V^+	^5D	$3d^4$	0.03	0.806
	^5F	$4s3d^3$	0.36	0.191
	^3F	$4s3d^3$	1.10	0.002
	^3P	$3d^4$	1.45	<0.001
Cr^+	^6S	$3d^5$	0.00	0.998
	^6D	$4s3d^4$	1.52	<0.002
	^4D	$4s3d^4$	2.46	
	^4G	$3d^5$	2.54	
Mn^+	^7S	$4s3d^5$	0.00	0.999
	^5S	$4s3d^5$	1.17	0.001
	^5D	$3d^6$	1.81	<0.001
Fe^+	^6D	$4s3d^6$	0.05	0.793
	^4F	$3d^7$	0.30	0.204
	^4D	$4s3d^6$	1.03	0.003
Co^+	^3F	$3d^8$	0.09	0.851
	^5F	$4s3d^7$	0.52	0.148
	^3F	$4s3d^7$	1.30	0.001
Ni^+	^2D	$3d^9$	0.08	0.990
	^4F	$4s3d^8$	1.16	0.009
	^2F	$4s3d^8$	1.76	<0.001
Cu^+	^1S	$3d^{10}$	0.00	1.000
	^3D	$4s3d^9$	2.81	<0.001
Zn^+	^2S	$4s3d^{10}$	0.00	1.000

[a] Statistically weighted averages over the J levels. Values are from Reference 64.
[b] Maxwell-Boltzmann distribution at 2200 K.

those measured using the EI source[8-13] or to those measured using the DC source.[11-13]

The third source (DC) takes ions produced via EI and focuses them into a 2-cm-long cell filled with a bath gas (Ar or CH_4). Drift rings establish a field which draws the ions to the exit of the cell. Under typical conditions, ions undergo several thousand collisions with the bath gas at ambient temperatures (300 K) and pressures of 0.1–0.3 torr. These collisions attenuate the beam intensity to 10^4–10^6 ions/s, but most importantly, they quench excited states such that the ions emerging from the DC source are primarily in their ground electronic state.[7,11,20] Again, no absolute measure of the state population is available; however, the DC conditions are maintained at a level such that additional collisions result in no further changes in the reactivity of the ions. This suggests that the quenching process is complete. An alternative use for this source is to introduce a gas in the cell that the injected ions react with. This enables the production of new ionic species or the depletion of a specific state of the injected ion beam.[20,21]

2.2. Experimental Considerations

Once formed, the ions are extracted from the source, focused into a magnetic sector for mass analysis, decelerated to a specific kinetic energy, and focused into an octopole ion beam guide.[19,22] This device utilizes rf potentials to trap ions in the radial direction without influencing the ion energy along the axis of the octopole. The octopole passes through a gas cell containing the neutral reactant at pressures which are sufficiently low that single-collision conditions prevail. Product ions and unreacted metal ions are trapped by the octopole until they are extracted and focused into a quadrupole mass filter for analysis. The ions are detected using a secondary electron scintillation ion counter for high sensitivity. A computer sweeps the kinetic energy of the reactant ions, monitors all reactant and product ions, and allows extensive signal averaging. Ion intensities are converted to absolute reaction cross sections as described elsewhere.[19] Absolute cross sections are estimated to be accurate to ±20% while relative cross sections are probably accurate to ±5%. It is difficult to determine whether there are systematic errors in the absolute cross sections measured with this apparatus since these are among the most precise measurements ever made. Comparison with calculated cross sections[20a] and experimental rate constants at thermal energies[19,20b,21] suggests that the accuracy is very good.

The absolute energy scale is measured by sweeping the ion energy through its nominal zero, the difference in potential between the ion source and the dc bias on the octopole. The derivative of the cutoff curve is nearly Gaussian with a peak taken to be the energy scale origin. Because the energy analysis region and the reaction zone are physically the same, ambiguities in this energy analysis due to contact potentials, space charge effects, and focusing aberrations are minimized. We conservatively quote the absolute uncertainty in our laboratory energy scale as 0.1 eV. The energy dependence of ion–molecule reaction cross sections provides a more severe test of the accuracy of our kinetic energy and indicates that our energy scale is accurate to better than 0.05 eV in the laboratory frame.[20,21,23] The ion energy as measured in the laboratory frame is converted to relative kinetic energy [center-of-mass frame (CM)] using the stationary target assumption, $E(\text{CM}) = E(\text{lab}) \times m/(M + m)$, where m and M are the masses of the neutral and ionic reactants, respectively. The CM energy is the kinetic energy available for chemical reaction since the remainder is conserved by the motion of the center of mass of the reacting system.

The distribution of the ion beam kinetic energy typically has a FWHM of 0.7 eV (lab frame) for the EI and SI sources and 0.4 eV (lab frame) for the DC source. An additional uncertainty in the true reaction energy is introduced by the thermal motion of the reactant gas. These molecules have a Maxwell-Boltzmann distribution of velocities at 305 K, the gas cell temperature. The effects of this motion can be severe and will obscure sharp features in the true excitation functions, particularly at threshold. This so-called Doppler broadening and the ion beam energy distribution are both taken into account when analyzing the experimental results.[19,24]

2.3. Energy Behavior of Ion–Molecule Reactions

2.3.1. Exothermic Reactions

Exothermic ion–molecule reactions are often observed to proceed at the capture rate. For molecules with dipoles, the calculation of the capture rate is an ongoing area of discussion. For polarizable molecules without dipole moments, the energy-dependent capture cross sections for a reaction, $\sigma(E)$, is given fairly accurately by the Langevin–Gioumousis–Stevenson (LGS) equation,[25]

$$\sigma_{\text{LGS}}(E) = \pi e(2\alpha/E)^{1/2} \tag{1}$$

where e is the electron charge, α is the polarizability of the neutral molecule, and E is the relative kinetic energy of the reactants. Since the long-range ion–induced dipole potential is often sufficiently strong to overcome small barriers,[26] exothermic ion–molecule reactions are often observed to proceed without an activation energy. Nevertheless, deviations from this behavior abound, as discussed by Henchman[27] and Armentrout.[28]

2.3.2. Endothermic Reactions: Thresholds

One reason that deviations occur is that the reaction may be endothermic. Procedures for determining the threshold behavior of ion–molecule reactions (or any reactions for that matter) include trajectory calculations, statistical theories, and empirical theories. Trajectory calculations potentially provide the most direct comparison between theory and experiment. They are not generally useful, however, because they require a very good potential energy surface, which is rarely available. More useful is the application of statistical theories, e.g., phase space theory (PST)[29] or transition state theory (TST).[30] These theories require the molecular constants for all species in every reaction channel and, in the case of TST, for the intermediates as well. Dynamics are not explicitly included although conservation of angular momentum and energy is required. In our work we have made extensive use of PST and found it to be fairly useful in describing the behavior of several atom–diatom ion–molecule systems.[8,9,11–15,31,32] These comparisons make it clear that angular momentum effects are very important for a proper description of bimolecular reactions.

The final method of describing the threshold behavior of endothermic reactions is to develop empirical theories (or more accurately, models) of this behavior. Such models can be quite general. A well-crafted model is particularly useful because it can often lead to physical insight which is easily lost in more thorough and thus complex theories. In our work, the general model used is

$$\sigma(E) = \sigma_0(E - E_T)^n / E^m \tag{2}$$

where σ_0 is a scaling factor and E_T is the reaction threshold. Most theoretical predictions of the threshold behavior can be expressed in this form.[4] Unfortunately, for even the simple atom–diatom case, there are many theoretically predicted values of n and m. No one form emerges as the threshold dependence, which is as it should be since experimentally no one set of values for n and m works for all systems. Hence, in applying Equation 2 to cross section data, the values of σ_0, n, m, and, if unknown, E_T are treated as adjustable parameters. This procedure has proven to be quite adequate for reproducing the data accurately over broad energy ranges. In cases where the threshold energies are unknown, the values of E_T determined by this procedure have uncertainties which reflect variations in different data sets, variations in the different models, and the uncertainty of the absolute energy scale. Overall, the uncertainty of our determinations varies between 1 and 5 kcal/mol depending on the nature of the reaction being studied.

2.3.3. High-Energy Behavior

At high kinetic energies, the consequences of energy disposal among the products need to be accounted for. In the general atom–diatom reaction

$$\mathrm{A + BC \rightarrow AB + C} \tag{3}$$

the excess energy available to the products must be in translation or in internal modes of AB. At sufficiently high reactant kinetic energies, the internal energy of AB can exceed the dissociation energy of this diatom such that the process

$$\mathrm{A + BC \rightarrow A + B + C} \tag{4}$$

occurs. This so-called indirect collision-induced dissociation has a thermodynamic threshold equal to $D^{\circ}(BC)$, the bond energy of BC. Cross sections for formation of AB in the process in Equation 3 are often observed to decline rapidly above $D^{\circ}(BC)$ due to the onset of the process in Equation 4. This decline can be delayed in energy if the reaction dynamics tend to place much of the excess energy in product translation.

In more complex systems (where A, B, or C represents chemical groups instead of atoms), this decline in the cross section for the process in Equation 3 can also be delayed if the product C carries away energy in internal modes. This is often the result if C is a large fragment with low-frequency vibrations and AB is a small species with only high-frequency vibrations (such as a diatomic metal hydride). Complex systems also exhibit cross sections which decline *before* the process in Equation 4 is thermodynamically accessible. This interesting result indicates that one of the intermediates formed during production of AB is being depleted by an alternate reaction channel. In such circumstances, the peak in the cross section for AB usually correlates with the onset of some other reaction channel.

This situation is somewhat more complex when the reaction involves an electronically excited state of the metal ion, A*. If the electronic state is conserved in the reaction in Equation 4, then the thermodynamic threshold for this process is still equal to $D^{\circ}(BC)$. However, a new reaction channel is the process

$$A^* + BC \rightarrow A + B + C \tag{5}$$

in which the excited state is quenched to the ground state. If this occurs, the peak in the cross section can be lowered by the electronic energy. This is, in fact, the most common observation.

2.4. Comparison to ICR (FTMS) Techniques

The other technique which has been used extensively to study transition metal ion chemistry in the gas phase is ion cyclotron resonance (ICR) mass spectrometry and its close cousin, Fourier transform mass spectrometry (FTMS).[33] While these methods and the ion beam techniques do yield some similar information (for instance, quantitative rate constants at 300 K), they are largely complementary, each having particular strengths. For example, the ion beam method is used to best advantage to study reactions under single-collision conditions. It can be used to examine multiple-reaction sequences but only under poorly characterized conditions. ICR, on the other hand, is ideally suited to the study of sequential reactions, and this ability has been used extensively. Another example is the recent study of Reents *et al.*[34] of the quenching rate of excited Cr^+, which could not easily be accomplished using ion beam methods.

The most obvious capability of the ion beam method which is not shared by ICR technology is the ability to alter the ion energy and thereby examine endothermic reactions. This provides a direct measurement of the thermochemistry of these systems and also can yield information related to the mechanisms of the reactions. Ordinarily, ICR technology can only measure thermal energy reactions although there have been some studies in which ions in an ICR spectrometer are translationally excited by rf irradiation.[35] Unfortunately, this technique results in very broad ion energy distributions which are not easily characterized. The extraction of quantitative thermochemical data is therefore difficult and suspect.

Another advantage of the beam technology is the physical separation of the ion source from the reaction region. This allows much more versatility in the ion sources used. This has been a key ability in deciphering the state-specific behavior of the atomic metal ion reactions. With one favorable exception,[34] no such studies have been performed using ICR or FTMS techniques.

3. THERMOCHEMISTRY

3.1. From Endothermicities to Bond Energies

The threshold value, E_T, for the reaction in Equation 3 can be related to the bond dissociation energy (BDE) of the species produced by using Equation 6,

$$D°(\mathrm{A—B}) = \Delta H_f(\mathrm{B}) + \Delta H_f(\mathrm{C}) - \Delta H_f(\mathrm{BC}) - E_T - E_{\mathrm{el}} \tag{6}$$

where E_{el} is the electronic excitation of the metal ion reactant. In simple bond exchange processes, Equation 6 can be simplified by noting that

$$D°(\mathrm{B—C}) = \Delta H_f(\mathrm{B}) + \Delta H_f(\mathrm{C}) - \Delta H_f(\mathrm{BC}) \tag{7}$$

For reactions with dihydrogen, we also correct for the internal energy of the neutral reactant and derive bond energies at 0 K. For reactions with alkanes, no such corrections are made and the thermochemistry obtained is characteristic of 298 K.

For Equation 6 to be valid, there must be no activation energy in excess of the reaction endothermicity. This is, of course, equivalent to there also being a barrier to reaction in the exothermic direction. As noted above, the long-range ion–induced dipole potential is often sufficiently strong to overcome small barriers[26] such that exothermic ion–molecule reactions are often (though not always) observed to proceed without an activation energy. The converse is also apparently true. Endothermic ion–molecule reactions are often (though not guaranteed) to proceed once the available energy exceeds the thermodynamic threshold.

This assumption is one which we have tested directly for a number of reactions.[9,20a,23,31,36] However, no direct check is available for transition metal-containing species since there are no alternate experimental determinations having better accuracy. However, high-quality *ab initio* calculations on the diatomic metal hydrides have been performed recently.[37-39] The calculated values average 0 ± 3 kcal/mol[39] and 3 ± 3 kcal/mol lower[37] than the experimental values derived in these studies. The latter is a typical error for such calculations. Note that if an activation barrier were present, the true bond energies would be larger, making the agreement with theory even worse.

3.2. Ionic Metal Hydrides

The diatomic metal hydride ion bond dissociation energies (Table 2) show a strong variation with the identity of the metal. This has been discussed

Table 2. Bond Dissociation Energies (kcal/mol) at 298 K[a]

M	$D°(M^+—H)$	$D°(M^+—CH)$	$D°(M^+—CH_2)$	$D°(M^+—CH_3)$	$D°(M—CH_3)$
Sc	56.2 (2.3)[b]		≥93 (3)[b]	59 (3)[b]	32 (7)
Ti	54.2 (2.5)[c]	121 (4)	93 (4)	58 (3)	46 (7)
V	48.2 (1.4)[d]	114 (2)	76 (2)	50 (2)[e]	37 (9)
Cr	32.5 (2.0)[f]	75 (8)	54 (2)	30 (3)	41 (7)
Mn	48.4 (3.4)[g]		94 (7)[h]	51 (5)[i]	<30, ≈12[i]
Fe	49.8 (1.4)[j]	101 (5)[k]	82 (5)[k]	58 (2)	37 (7)
Co	46.6 (1.4)[l]	100 (5)[k]	84 (4)[k,h]	51 (4)	45 (5)
Ni	39.5 (1.8)[l]		86 (6)[h]	47 (5)	54 (4)
Cu	22.1 (3.0)[l]			30 (3)	55 (4)
Zn	60[m], 55 (3)[c]			71 (3)[n]	19 (3)[n]
Y	62 (2)[c]				
Zr	55 (3)[q]				
Nb	54 (3)[q]	145 (8)[o]	109 (7)[o]		
Mo	42 (3)[q]				
Ru	41 (3)[p]			54 (5)[p]	
Rh	36 (3)	102 (7)[o]	91 (5)[o]	47 (5)[p]	
Pd	47 (3)			59 (5)[p]	
Ag	16 (3)[q]				

[a] Unless otherwise noted values are from Reference 7. Uncertainties are in parentheses.
[b] Reference 6.
[c] Reference 15.
[d] Reference 8.
[e] Reference 4.
[f] Reference 13.
[g] Reference 9.
[h] Reference 41.
[i] Armentrout, P.B. In *Laser Applications in Chemistry and Biophysics*; El-Sayed, M.A., Ed.; *Proc. SPIE* 1986, *620*, 38–45.
[j] Reference 11.
[k] Hettich, R.L.; Freiser, B.S. *J. Am. Chem. Soc.* 1986, *108*, 2537–2540.
[l] Reference 12.
[m] Po, P.L.; Radus, T.P.; Porter, R.F. *J. Phys. Chem.* 1980, *82*, 520–526.
[n] Reference 5.
[o] Reference 48.
[p] Reference 40.
[q] Reference 10.

several times in the past.[10,37,40,41] The first-row metals on the left side of the periodic table, Sc^+ and Ti^+, have the strongest BDEs while Cr^+ and Cu^+ have the weakest. These weak bond energies are easily understood by noting that the ground state configuration of Cr^+ is the very stable half-filled shell, $3d^5$, while Cu^+ has a filled-shell ground state, $3d^{10}$. To form a covalent bond to either of these ions, one of the electrons must be decoupled from the others. This concept can be formalized in terms of a promotion energy, E_p, defined for first-row metals as the energy necessary to take a metal ion in its ground state to an electron configuration where only one electron is in the 4*s* orbital and it is spin decoupled from the 3*d* electrons. This energy is just the mean energy of the electronic states which have high-spin and low-spin $4s3d^{n-1}$ configurations.[10,40] The MH^+ BDEs are plotted versus this promotion energy in Figure 1. Obviously, the correlation between $D°(M^+—H)$ for the first-row metals and E_p is excellent. This implies that the dominant binding orbital on the metal is the 4*s* orbital. *Ab initio* calculations[37,39] support this idea although they establish that there is significant 3*d* character in the bonds.

In contrast to those of the first row, the second-row bond energies do not correlate well with the promotion energy, Pd being the most striking deviation. Since the electron configuration of Pd^+ is $4d^9$, this deviation can be explained by postulating that Pd^+ utilizes a 4*d* orbital to bond with hydrogen.[10,40] This requires no promotion since the 4*d* bonding electron is already decoupled from all other electrons. *Ab initio* calculations[39] verify this hypothesis and suggest that the other second-row metals, especially those to the right, also utilize a 4*d* orbital in their bonding.

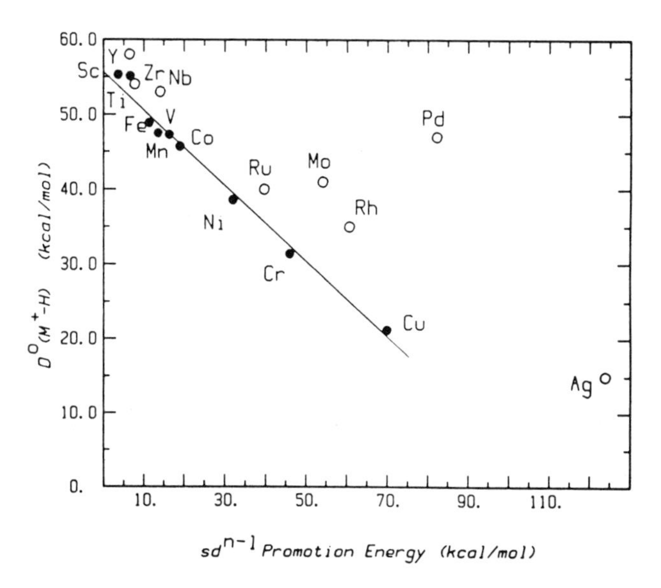

Figure 1. First-row (●) and second-row (○) transition metal hydride ion bond energies versus atomic metal ion promotion energy to an sd^{n-1} spin-decoupled state; see text. The line is a linear regression fit to the first-row data.

The maximum metal hydride ion bond energy occurs at $E_p = 0$ for both the first and the second row. We have argued previously[10] that the BDE at this point, ~56 kcal/mol, is a good estimate of the intrinsic metal-hydrogen bond energy. In other words, this is the BDE expected for a hydrogen atom bound to a metal with a directional and sterically unhindered orbital containing a single electron that is electronically decoupled from the other electrons. An example of such a species is $ScH^+(^2\Delta)$. We have recently measured that $D°(HSc^+—H) = 58.3 \pm 3.7$ kcal/mol,[6] in good agreement with our expectations.

3.3. Ionic Metal Methyls

One of the earliest thermochemical results of ion beam methods as applied to organometallic chemistry was the determination that $D°(M^+—CH_3)$ exceeded $D°(M^+—H)$ by almost 10 kcal/mol.[42] This was a surprising result to most condensed phase inorganic chemists since the limited data available for metal-carbon bond strengths indicated that they were almost half as strong as metal-hydrogen bond strengths.[43] Nevertheless, isolobal arguments suggest that a methyl group and a hydrogen atom should bond similarly. It therefore seems reasonable that they both form strong bonds with transition metals *if* the metal center is sterically uncrowded. Halpern has recently contended that steric effects are largely responsible for the weak condensed phase M—C bond strengths, using these gas phase data as evidence.[43] In the gas phase, the metal ion-methyl BDEs can be stronger due to the increased polarizability of the methyl group compared to a hydrogen atom.[40,42]

Further investigation of the gas phase metal-methyl BDEs finds a somewhat more complex picture. For Fe, Co, and Ni, the $M^+—CH_3$ BDEs are stronger than the $M^+—H$ BDEs by ~11 kcal/mol (Table 2). For Sc, Ti, V, Cr, and Mn, however, the metal-methyl bond energies are within 3 kcal/mol of the $M^+—H$ bond energies. Within these groups, the relative values of $D°(M^+—CH_3)$ parallel those of $D°(M^+—H)$, which indicates that the periodic trends and promotion energy ideas discussed above continue to be of value for the $M^+—CH_3$ species. The dichotomy between the metal—methyl BDEs of left-side and right-side metals remains to be explained.

One interesting suggestion is that agostic[44] M···H—C interactions are present.[45] If we again take the view that the principal binding orbital on the atomic metal ion is the 4*s*, then the 3*d* orbitals can be viewed as largely nonbonding. An agostic M···H—C interaction can occur by donation of the C—H bonding electron pair into an empty metal orbital, the $3d\pi$ in the M—H—C plane. This interaction is favored by small M—C—H bonds although this type of motion also reduces the overlap between the 4*s* and the radical orbital on the methyl group.[46] When applied to the $M^+—CH_3$ species, these ideas suggest that an agostic interaction will be strongest when there is an empty 3*d* orbital. This clearly can occur for Sc^+, Ti^+, V^+, and Cr^+, which have one electron involved in bonding the methyl and one, two, three and four nonbonding electrons, respectively. From Mn^+ to Cu^+, there are no low-lying

Table 3. Second Bond Dissociation Energies (kcal/mol) at 298 K[a]

M	$D^\circ(M^+—H)$	$D^\circ(M^+—CH_3)$
Sc	56 (2)[b]	59 (3)[b]
ScH	59 (4)[b]	64 (5)
$ScCH_3$	61 (5)	60 (5)
V	48 (1)[c]	50 (2)[d]
VH		45 (6)
VCH_3	43 (6)	48 (5)
Zn		71 (3)[e]
$ZnCH_3$		27 (3)[e]

[a] Unless otherwise noted values are from Reference 7. Uncertainties are in parentheses.
[b] Reference 6.
[c] Reference 8.
[d] Reference 4.
[e] Reference 5.

states which have an empty $3d$ orbital, and therefore agostic interactions seem less likely. The conclusion drawn from these ideas is the proposition that agostic interactions *weaken* the metal-methyl bond energy. The dominant effect here may be the reduced overlap between the methyl radical orbital and the metal $4s$.

3.4. Ionic MR_2: $R = H, CH_3$

To understand oxidative addition and reductive elimination processes, it is clearly important to know the second metal-ligand BDE as well as the first. Unfortunately, the experimental determination of this second bond energy is more difficult than determining the first for radical ligands like hydrogen atoms and alkyls. Nevertheless, we have now determined this type of information for several systems. The results of this work are listed in Table 3. It can be seen that the first and second BDEs are quite similar for Sc^+ and V^+ but very different for Zn^+. Promotion energy ideas can again qualitatively account for these results. The promotion energy for Sc^+ is small (3.5 kcal/mol)[10] and for ScR^+ is ~0 since this species has only one unpaired electron. Similarly, the promotion energies of V^+ and VR^+ are expected to be comparable (~16 kcal/mol).[10,37,40] For Zn^+, the first bond is strong since Zn^+ has a $4s3d^{10}$ configuration ($E_p = 0$). Forming a second covalent bond to Zn^+, however, requires disruption of the filled $3d^{10}$ shell, leading to a weak bond.

3.5. Neutral Metal Methyls

In a limited number of cases, we have also measured the neutral metal-methyl bond energies (Table 2). In most cases, the neutral BDEs lie below the ionic counterparts. Also note that isoelectronic neutrals and ions do not have similar bond energies. These results can again be qualitatively understood using the promotion energy ideas. Now, however, the ground state configuration of the neutral atomic metals is generally $4s^2 3d^{n-2}$. This requires that a $4s$ electron be promoted out of a stable filled subshell to reach the $4s3d^{n-1}$ electron configuration preferred for bonding. Thus, the promotion energies will ordinarily be higher for the neutral atoms than for the ions. An exception is Cr, which has a $4s3d^5$ ground state configuration compared to the $3d^5$ configuration for Cr^+. Here, E_p is less for spin decoupling the $4s$ electron of Cr than for promoting and spin decoupling a $3d$ electron of Cr^+. A plot similar to Figure 1 does not yield as good a correlation for MCH_3 as for MH^+. This may be because the $4p$ electrons on the metal are involved in the bonding to a much greater extent in the neutral species.[47]

3.6. Ionic Metal Methylidenes and Methylidynes

In an early paper,[3] we pointed out that the V^+—CH_3, V^+—CH_2, and V^+—CH bond strengths increase roughly as the organic analogues, CH_3—CH_3, CH_2—CH_2, and CH—CH, respectively. This simple correlation between the bond energy and the bond order is intuitively reasonable but had never been documented for transition metal species. Our work since then shows that Sc^+, Ti^+, and Cr^+ also follow this type of correlation,[6,7] and Hettich and Freiser have similarly documented it for Rh^+, Nb^+, and La^+ species.[48] *Ab initio* calculations by Harrison and co-workers[38b,49] have also seen this trend for M^+—CH_3, M^+—CH_2, and M^+—CH where M = Ti, V, and Cr.

To the degree that this correlation is meaningful, it implies that these species have certain spins. For example, Cr^+ has five electrons in a high-spin configuration, leading to a sextet state. $CrCH_3^+$ uses one of these to bind the methyl group, leading to a quintet state. If the Cr—C bond in $CrCH_2^+$ is a double bond, two metal electrons are used for bonding such that the ground state is presumably a quartet. Likewise, the ground state of triply bonded $CrCH^+$ should be a triplet. In this case, these assignments have been verified by *ab initio* calculations.[38b] Clearly, such calculations will always be useful in helping to identify the electronic character of these species from their thermochemistry.

While this simple correlation is qualitatively useful, it clearly misses the extensive reorganization and electron decoupling necessary for the metal to bond to these ligands. Its utility therefore should not be overinterpreted. This is illustrated by the recent calculations of Carter and Goddard,[50] who find that the ground state of $RuCH_2^+$ is a metal methylidene having a covalent M—C double bond (and hence is a doublet). However, there are low-lying quartet excited states which are metal carbenes, i.e., they have a donor–acceptor bond between Ru^+ and $CH_2(^1A_1)$. These states have comparable bond dissociation energies to the doublet ground state. Thus, the thermochemistry of these species is an imprecise tool for evaluating their electronic structure.

4. REACTIONS WITH DIHYDROGEN

The reaction system which we have documented most carefully is

$$M^+ + H_2 \rightarrow MH^+ + H \qquad (8)$$

and its isotopic analogues with HD and D_2.[8-15] This is an ideal system for the investigation of the detailed nature of electronic effects in metal ion reactions for several reasons. First, it is simple. Reaction 8 is the only process energetically possible at low energies. Second, it is endothermic. The H_2 bond energy of 104 kcal/mol is considerably stronger than the metal-hydride ion bond energies, which vary from 20 to 60 kcal/mol (Table 2). Because of this, the presence of electronically excited metal ions may be observed as a decrease in the apparent threshold for reaction 8. This provides an experimental tool for evaluating the effects of electronic excitation. Third, examination of reactions with HD and D_2 provides details concerning the reaction dynamics. Fourth, the diatomic metal hydrides provide an ideal interface for *ab initio* calculations and experiment, as discussed above. Fifth, the activation of molecular hydrogen by transition metals is important to understanding a variety of homogeneous and heterogeneous catalytic processes. Reaction 8 is a model system for understanding details of the electronic interactions of this activation step. By analogy, it is also a model system for activation of any covalent hydrocarbon single bond, i.e., C—H and C—C bonds.

We have published a number of papers detailing our results for the reactions of atomic metal ions with dihydrogen[8-13] and recently reviewed these results.[14] These studies indicate that the differences in reactivity can be understood by considering simple molecular orbital ideas and the conservation of spin throughout the reaction.

4.1. Molecular Orbital Considerations

Detailed discussions of our view of the molecular orbitals (MOs) of $M^+ + H_2$ interactions have been given before.[8,9,14] For simplicity, the metal *p* orbitals are ignored since they are too high in energy to be very influential in the bonding.[14,37] Most simply put, the ability of the atomic metal ions to react efficiently with dihydrogen can be described by the interactions displayed in Figure 2. Since the 4*s* is the outermost valence orbital as well as the principal binding orbital in the diatomic metal hydride ions (see above), the interaction

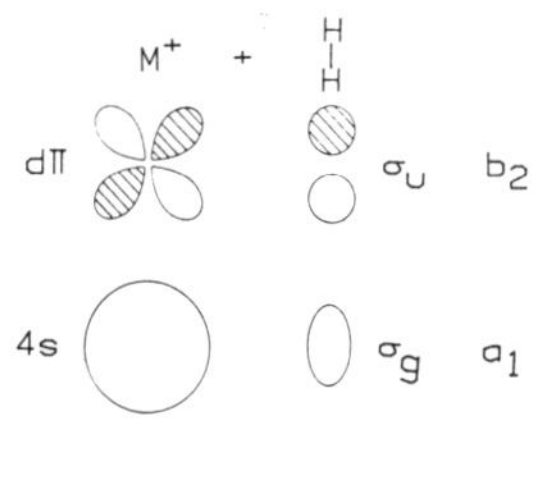

Figure 2. Orbital interaction for the *s* and $d\pi$ orbitals of M^+ with the molecular orbitals of H_2 in C_{2v} symmetry. Symmetry groups are indicated on the right.

of this orbital with H_2 is expected to be dominant at long range. In C_{2v} symmetry, the $4s$ and σ_g orbitals both have a_1 symmetry and thus combine into bonding and antibonding MOs. Since the σ_g orbital is fully occupied, occupation of the $4s$ orbital will lead to occupation of the antibonding a_1^* MO. The resulting interaction is repulsive overall. In $C_{\infty v}$ symmetry, this repulsion is relieved since the $4s$ orbital can interact with both the σ_g and σ_u MOs of H_2. A bonding, a nonbonding, and an antibonding MO are formed. Occupation of the $4s$ orbital now leads to occupation of the nonbonding MO. This is less repulsive than the interaction resulting from occupation of the antibonding a_1^* MO in C_{2v} symmetry but is still less attractive than if the $4s$ orbital were empty.

If the $4s$ orbital is unoccupied, interactions with $3d$ orbitals become important. In C_{2v} symmetry, the $3d\sigma$ orbital has a_1 symmetry and therefore acts like the $4s$. Indeed, mixing of the $4s$ and $3d\sigma$ orbitals is inevitable in these systems. The in-plane $3d\pi$ orbital has b_2 symmetry and therefore mixes with the σ_u orbital of H_2 to form bonding and antibonding MOs (Figure 2). Since the σ_u orbital is always empty, occupation of this $3d\pi$ orbital leads to occupation of the bonding MO and an attractive interaction. The other $3d$ orbitals are largely nonbonding in all symmetries and need not be considered.

This molecular orbital picture is easily related to the rules for C—H activation outlined by Crabtree[1] (see Section 1.2). These say that there must be orbitals on the metal which accept electron density from the σ bond to be activated (rule 1) and which donate electron density into the σ^* orbital of the activated bond (rule 2). For atomic metals, these orbitals are the $4s$ and $3d\pi$, respectively. Note that merely having a coordinatively unsaturated species is insufficient for efficient bond activation. The electronic requirements are much more explicit.

4.2. Spin Considerations

Another consideration in understanding the reactivity of the first-row metal ions is the spin of the reactant state. Not surprisingly, all reactions observed between M^+ and H_2 are spin allowed, i.e., they all conform to the reaction

$$M^+(s \pm \tfrac{1}{2}) + H_2(0) \rightarrow MH^+(s) + H(\tfrac{1}{2}) \tag{9}$$

where the numbers in parentheses are the spin quantum numbers of the species. For all metal ions, efficient reaction of a low-spin $(s - \frac{1}{2})$ state is observed. This corresponds to the ground states of Co^+, Ni^+, and Cu^+ and excited states of Ti^+, V^+, Cr^+, Mn^+, and Fe^+. High-spin $(s + \frac{1}{2})$ states are fairly reactive for Sc^+, Ti^+, and V^+ but are less reactive as one moves across the periodic table. When the ground state is high spin, as for Sc^+, Ti^+, V^+, Cr^+, Mn^+, and Fe^+, the reactions of both high-spin and low-spin states are observed. When the ground state is low spin, as for Co^+, Ni^+, and Cu^+, however, reactions of high-spin excited states are observed only with difficulty.

This systematic correspondence of reactivity with spin is probably related to the stability of the MH_2^+ reaction intermediate. In all cases, the spin of ground state MH_2^+ should be the same as that of the low spin metal ion states. This presumes that ground state MH_2^+ can be characterized as having two covalent metal–hydrogen bonds, a reasonable proposition.[38] Therefore, high-spin metal ion states do not have access to the ground state intermediates (except via spin–orbit transitions). This suggests that, in general, the potential energy surfaces of the high-spin states are more repulsive than surfaces of low-spin states.

The role of spin in controlling reactivity is particularly clear-cut in a comparison of the reactivity of low-spin and high-spin $4s3d^{n-1}$ configurations. As we shall see, the former is much more reactive than the latter, a result not accounted for by the MO considerations discussed above. The difference must lie in the interactions between the $3d$ electrons and H_2. One way to consider this is to envision the reverse reaction, i.e., approach of MH^+ and H. The H($1s$) electron can either be low-spin or high-spin coupled with the nonbonding $3d$ electrons of MH^+. If low-spin coupled, there can be favorable bonding interactions between the H atom and the metal. These are most effective if the reaction geometry differs from $C_{\infty v}$. If high-spin coupled, a node must exist between the incoming H($1s$) and the metal $3d$ electrons. This leads to a repulsive interaction which is weakest for strict $C_{\infty v}$ symmetry. Thus, the difference in reactivity between low- and high-spin $M^+(4s3d^{n-1}) + H_2$ interactions is explained as a change in the repulsiveness of the potential energy surface as the reaction deviates from a collinear geometry. For high-spin $4s3d^{n-1}$ states, the repulsiveness increases as this deviation occurs. For low-spin $4s3d^{n-1}$ states, the repulsion is mediated by bonding interactions between the trailing hydrogen atom and the metal $3d$ electrons.

4.3. Periodic Trends in Reactivity

Having discussed the fundamental aspects of the interactions of atomic metal ions with hydrogen, it is reasonably straightforward to understand the variations in reactivity across the periodic table. As discussed above, the states with high-spin $4s3d^{n-1}$ configurations should be unreactive. Not only is the $4s$ orbital occupied, leading to repulsion, but the high spin prevents attractive bonding interactions with the metal $3d$ electrons as well. Examples of such states include $Mn^+(^7S, 4s3d^5)$,[9] $Fe^+(^6D, 4s3d^6)$,[11] and $Ni^+(^4F, 4s3d^8)$.[12] $Zn^+(^2S, 4s3d^{10})$[15b] also falls into this category since the filled $3d$ shell also prevents attractive $3d$ bonding interactions. All of these states are observed to react via an impulsive reaction mechanism which indicates a largely repulsive potential energy surface. In addition, reactions of the species $Cr^+(^6D, 4s3d^4)$,[13] $Co^+(^5F, 4s3d^7)$,[12] and $Cu^+(^3D, 4s3d^9)$[12] are not evident, a result which we attribute to their relative inertness.

Now consider the high-spin $4s3d^{n-1}$ states for the metals on the left side of the periodic table, $Sc^+(^3D, 4s3d)$,[15c] $Ti^+(a^4F, 4s3d^2)$,[15a] and $V^+(^5F, 4s3d^3)$.[8] While experimental results concerning $V^+(^5F)$ are ambiguous since it is a low-lying excited state, the other two are both ground states and appear to react efficiently via a metal dihydride ion intermediate. These results are in obvious contrast to the reactivity exhibited by the high-spin $4s3d^{n-1}$ states discussed above. We believe that the potential energy surfaces evolving from the Sc^+ and Ti^+ ground states are repulsive also; however, unlike the situation for the states discussed above, there are low-lying excited states of Sc^+, the $^3F(3d^2)$, and Ti^+, the $b^4F(3d^3)$, which are also high spin (Table 1). The surfaces evolving from these excited states are attractive (since the $4s$ orbital is unoccupied) and will therefore cross with the repulsive surfaces from $Sc^+(^3D)$ and $Ti^+(a^4F)$. Since these diabatic surfaces (i.e., pertaining to a particular electron configuration) have the same spin, the crossings are avoided to form adiabatic surfaces. Another way to think about this crossing is that the electron in the $4s$ orbital moves into an empty $3d$ orbital of the same symmetry to avoid the repulsive interactions. To the right of chromium ($n > 5$), there are no empty $3d$ orbitals in the high-spin $4s3d^{n-1}$ states.

These ideas lead to the conclusion that the adiabatic surfaces correlating to $Sc^+(^3D)$ and $Ti^+(a^4F)$ have an electronic character of $3d^2$ and $3d^3$, respectively, in the MH_2^+ intermediate. This explanation is consistent with the observation that the behavior of Ti^+ is similar to that of $V^+(^5D, 3d^4)$. Sc^+ behaves somewhat differently, perhaps because the mixing of these states is less efficient than for Ti^+. In the case of Sc^+, the $^3D(4s3d)$ state and $^3F(3d^2)$ states are separated by 0.6 eV; while for Ti^+, the two ^4F states are separated by only 0.10 eV (Table 1).

The two ions $Ti^+(^4F)$ and $V^+(^5D)$ are the only metal states which give a statistical isotope effect which implies that a metal dihydride intermediate is formed.[8,15] This behavior is consistent with the MO arguments above. These configurations have an unoccupied $4s$ orbital and thus avoid repulsive interactions. In addition, for at least one of the surfaces evolving from these states, the $3d\sigma$ orbital is also empty, again avoiding repulsive interactions, and the inplane $3d\pi$ orbital is occupied, leading to an attractive interaction. This type of surface is apparently attractive enough to exhibit statistical behavior even though access to the ground state of MH_2^+ is spin forbidden. While not necessary to explain the experimental results, it is possible that spin-orbit interactions are sufficient to permit crossing to the ground state MH_2^+ surface.

The importance of the $3d\sigma$ orbital occupation can be explored further by comparing the reactivity of $Cr^+(^6S, 3d^5)$[13] and $V^+(^5D, 3d^4)$.[8] Rather than being statistically behaved, Cr^+ displays a strong component of impulsive reactivity. The only difference between the high-spin $3d^4$ and $3d^5$ configurations is that the $3d\sigma$ orbital must be occupied in the latter. This apparently prevents $Cr^+(^6S)$ from inserting into hydrogen. The high spin then leads to a relatively repulsive surface similar to that for high-spin $4s3d^{n-1}$ configurations.

The next ions in the $3d^n$ series ($n > 5$) differ from those with $n \leq 5$ because they are all low spin with one or more of the $3d$ orbitals doubly occupied. While the reactivity of $Mn^+(^5D, 3d^6)$[9] has not been cleanly isolated, results for $Fe^+(^4F, 3d^7)$,[11] $Co^+(^3F, 3d^8)$, $Ni^+(^2D, 3d^9)$, and $Cu^+(^1S, 3d^{10})$[12] are available. They all behave very similarly via a direct mechanism. Unlike the $3d^n$ configurations where $n < 5$, these states cannot have an empty $3d\sigma$ orbital. As noted for $Cr^+(^6S)$, occupation of this orbital is apparently sufficient that these reactions cannot occur via insertion. However, unlike the high-spin $Cr^+(^6S)$, these low-spin configurations provide attractive bonding interactions between the $3d$ orbitals and the trailing hydrogen. In C_{2v} geometry, this corresponds to the fact that it is spin allowed to doubly occupy the b_2 bonding orbital (Figure 2). Thus, these species react efficiently via a direct mechanism.

Because MH^+ bonding is largely $4s$–$1s$ in character, it may appear counterintuitive that $3d^n$ configurations react efficiently to form ground state MH^+ while $4s3d^{n-1}$ configurations do not. Indeed, diabatically, the $4s$–$1s$ bound ground state of MH^+ can only be formed from $M^+(4s3d^{n-1}) + H_2$ reactants. However, the surfaces evolving from a low-spin $M^+(4s3d^{n-1})$ configuration cross with those evolving from the $M^+(3d^n)$ ground state which are also low spin. These crossings are avoided to form adiabatic surfaces which connect ground state $M^+(3d^n)$ reactants with ground state MH^+ products.

This argument helps explain the somewhat surprising result that the closed-shell $Cu^+(^1S, 3d^{10})$ reacts readily at its thermodynamic threshold.[12] The molecular orbital ideas show that the primary interaction is the donation of electron density from the doubly occupied $\sigma_g(H_2)$ orbital to the empty $4s$(M) orbital. The CuH_2^+ intermediate formed in this process should be the ground state species. An interesting question for future spectroscopic or theoretical characterization is whether this species is best viewed as a metal dihydride with two covalent Cu—H bonds (b_2 MO is strongly bonding) or as a two-electron three-center bond (b_2 MO is nonbonding). Similar questions exist for other MH_2^+ species. In the exit channel, there is a crossing between the MO evolving to the $1s$(H) orbital and one of the $3d$(M) orbitals in all symmetries. The surface crossing between the $3d^n$ and $4s3d^{n-1}$ configurations discussed above is the result of this MO crossing.

The final type of electron configuration examined in these studies is low-spin $4s3d^{n-1}$. Because this configuration is always an excited state, there is no cleanly isolated example of this type of state. The best examples are found in EI data where the reactivity is dominated by such states: e.g., $V^+(^3F, 4s3d^3)$,[8] $Cr^+(^4D, 4s3d^4)$,[13] and $Mn^+(^5S, 4s3d^5)$.[9] Preliminary results indicate that $Ti^+(^2F, 4s3d^2)$ is probably another example.[15a] Each of these species exhibits behavior indicative of a direct reaction. Occupation of the $4s$ orbital is expected to be repulsive in C_{2v} symmetry such that insertion into H_2 is not expected. However, this repulsion is relieved in $C_{\infty v}$ symmetry and the low spin can lead to attractive bonding interactions. These factors are consistent with the efficient reaction of these states (in contrast to the inertness of the high-spin $4s3d^{n-1}$ configuration) but via a direct mechanism. This behavior is similar to that noted above for the $3d^n$ ($n > 5$) configurations. This is consistent with the fact that these species build in low-spin $4s3d^{n-1}$ character as the MH^+ products are formed. The low-spin $4s3d^{n-1}$ states of Ti^+, V^+, Cr^+, and Mn^+ probably do not undergo such a crossing since the low-spin $3d^n$ states of these metal ions are higher in energy (Table 1).

5. REACTIONS WITH METHANE

Having developed a reasonably detailed picture for the activation of molecular hydrogen by atomic metal ions, we wish to see whether similar concepts can also characterize the mechanisms and electronic requirements for C—H activation. We start by studying methane since it is the simplest alkane system and is also of technological importance. Investigations of state-specific chemistry again should permit us to test the general electronic requirements outlined by Crabtree.[1] Superficially, dihydrogen and methane have similar energetics since $D°(H—H) = 104$ kcal/mol and $D°(H—CH_3) = 105$ kcal/mol. However, as we shall see, dehydrogenation of methane to form a metal methylidene ion, MCH_2^+, and H_2 is a low-energy reaction channel which has no equal in the dihydrogen system. Other differences between H_2 and CH_4 include a large polarizability (0.8 Å^3 to 2.6 Å^3)[51] and more degrees of freedom (3 to 12). These changes will tend to increase the lifetime of the reaction intermediate and may enhance or alter the interactions between electronic surfaces, resulting in different kinds of reactivity.

5.1. $Sc^+ + CH_4$

We start our examination of methane reactivity with Sc^+. This system has the advantage of being the simplest transition metal ion system to study since Sc^+ has only two electrons. This necessarily reduces the number of covalent bonds it can form to two and thereby simplifies the types of intermediates which need to be postulated. The difficulty with this system is that the effects of electronic states have not been examined in any detail yet. Our ideas concerning the electronic excitation are derived from results for other first-row metals.

Results[7e] for the reaction of Sc^+ (generated by SI and hence largely ground state ^3D) with methane are shown in Figure 3. The three products observed correspond to the reactions:

$$Sc^+ + CH_4 \rightarrow ScH^+ + CH_3 \tag{10}$$

$$\rightarrow ScCH_3^+ + H \tag{11}$$

$$\rightarrow ScCH_2^+ + H_2 \tag{12}$$

All three reactions begin at their thermodynamic threshold. The dominant reaction at high energies is the process given by Equation 10. Reaction 11 has an apparent threshold comparable to that for reaction 10, consistent with the fact that the measured bond energies are similar (Table 2). Given this, it seems odd that the probability of forming $ScCH_3^+$ is much less than that of forming ScH^+. The dominant reason for this lies in the relative ability to conserve orbital angular momentum in reactions 10 and 11. This argument has been

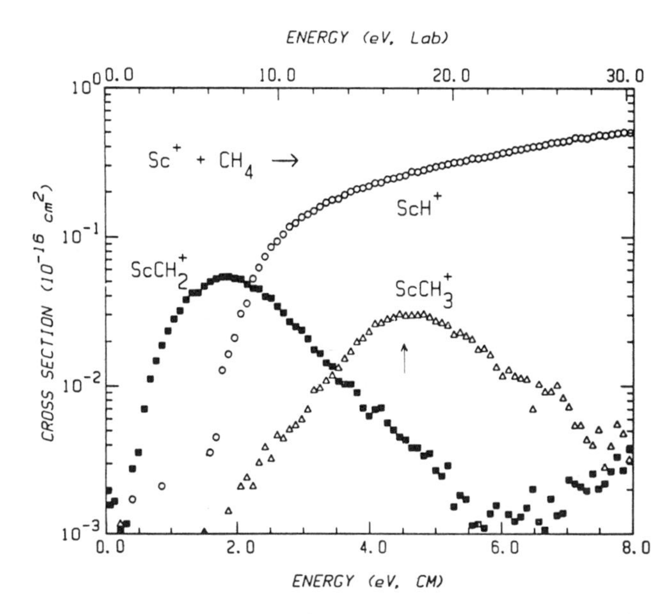

Figure 3. Cross sections for reaction of Sc^+ (produced by surface ionization) with CH_4 as a function of relative kinetic energy (lower scale) and laboratory ion energy (upper scale). The vertical arrow marks $D°(H—CH_3)$.

detailed elsewhere[4,6] but can be briefly summarized as follows. The reduced mass of the products in reaction 10 is similar to that of the reactants (11.3 amu versus 11.8 amu). In contrast, the reduced mass of the products in reaction 11 is much less (1.0 amu). As a consequence, it is much more difficult for the ($ScCH_3^+ + H$) products to conserve orbital angular momentum than it is for the ($ScH^+ + CH_3$) products since these are endothermic reactions. This results in a reduction in the cross section for reaction 11 relative to that for reaction 10. If this were the only effect, then formation of ScH^+ should be favored by a factor of about 38 compared to the $ScCH_3^+$ product. The fact that the observed ratio is less than this (about a factor of 8) is consistent with a more statistical process where the internal density of states favors formation of $ScCH_3^+$.

The cross section for reaction 11 reaches a peak at about 4.5 eV. This corresponds to decomposition of $ScCH_3^+$ in the reaction

$$Sc^+ + CH_4 \rightarrow ScCH_3^+ + H \rightarrow Sc^+ + CH_3 + H \quad (13)$$

which has a thermodynamic threshold of 4.5 eV, corresponding to $D°(CH_3—H)$. Interestingly, the ScH^+ product, which can also begin decomposing at this energy, shows no decline up to the 8-eV energy examined. This means that the ScH^+ product carries away little of the excess energy in internal modes. Most of this energy must reside in translation or in internal modes of the CH_3 neutral product. A secondary decomposition pathway for $ScCH_3^+$ is loss of a hydrogen atom in the following reaction:

$$Sc^+ + CH_4 \rightarrow ScCH_3^+ + H \rightarrow ScCH_2^+ + H + H \quad (14)$$

This channel can be observed as the rise in the $ScCH_2^+$ cross section beginning at about 6 eV. Note that the fact that the primary decomposition pathway is the process in Equation 13 is evidence that $ScCH_3^+$ is truly a metal methyl ion rather than some other structure such as a hydrido metal carbene, H—Sc^+=CH_2.

At low energies, the dominant reaction is the process in Equation 12, elimination of molecular hydrogen. At higher energies, the $ScCH_2^+$ cross section reaches a peak and declines in magnitude. As seen for $ScCH_3^+$, this type of behavior ordinarily occurs becaue the product begins to decompose. For $ScCH_2^+$, the possible decomposition products are $Sc^+ + CH_2$, $ScC^+ + H_2$, and $ScCH^+ + H$. The latter two possibilities are discounted since the ionic decomposition products are not observed in these studies. (They are observed in other transition metal ion–methane systems, however, verifying that they can be observed if present.) The first possibility cannot be detected explicitly since the decomposition product is identical to the reactant. Overall, this reaction corresponds to the process

$$Sc^+ + CH_4 \rightarrow Sc^+ + CH_2 + H_2 \quad (15)$$

which is endothermic by 4.8 eV. Therefore, it cannot account for the decline in the $ScCH_2^+$ cross section either. The only remaining rationale for the observed behavior is that the intermediate involved in reaction 12 is being depleted by another reaction (or reactions). Reactions 10 and 11, forming ScH^+ and $ScCH_3^+$, respectively, are the only possible candidates, and both have thresholds which correspond nicely to the peak in the $ScCH_2^+$ cross section. This shows that *these three products all have a common intermediate.* Since reactions 10 and 11 are thermodynamically less favorable, these processes must dominate reaction 12 because they are kinetically favored, i.e., they have larger preexponential factors (or, equivalent, larger phase space). This is direct evidence that reaction 12 involves a tighter transition state than either reaction 10 or 11.

5.2. Reaction Mechanism

A mechanism and potential energy surface for the interaction of Sc^+ with methane consistent with all this information is shown in Figure 4. Sc^+ inserts into the C—H bond of methane to form H—Sc^+—CH_3 (**I**). At low energies, a four-center elimination of H_2 occurs. While ordinarily symmetry forbidden, this process is allowed in this reaction because the metal–ligand bonds have substantial 3*d* character in them.[52] At higher energies, **I** can decompose to form ScH^+ or $ScCH_3^+$. These simple bond fissions should have loose transition states and consequently are much more facile than the constrained four-center elimination. At still higher energies, $ScCH_3^+$ decomposes to $Sc^+ + CH_3$ and $ScCH_2^+ + H$.

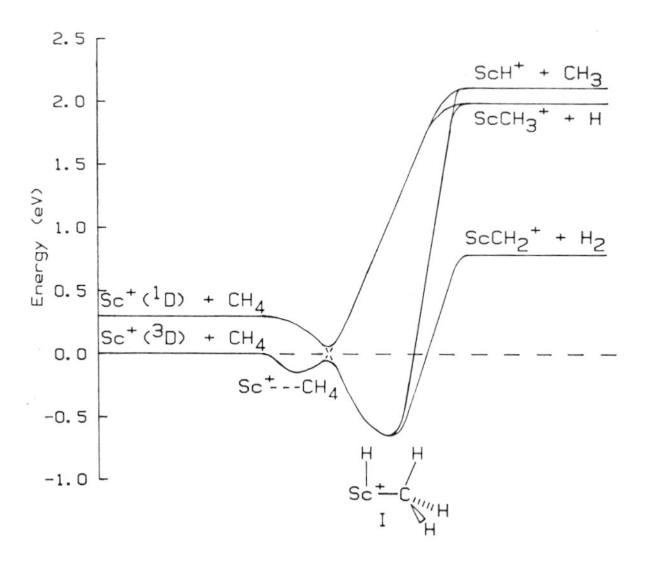

Figure 4. Proposed potential energy surface for the reaction of Sc^+ with methane. Full lines represent adiabatic surfaces and the dashed lines represent diabatic surfaces.

Other possible intermediates could include species like **II**:

$$H_2Sc^+{=}CH_2$$

II

however, since Sc^+ has only two valence electrons, it cannot form the four covalent bonds shown in **II**. Another possible structure is similar to **II** but replaces the $Sc^+{=}C$ bond with a dative interaction where singlet CH_2 donates its electron pair into an empty Sc^+ orbital, $H_2Sc^+ \leftarrow CH_2$ (**III**). However, such a species is inconsistent with the thermochemistry of $ScCH_4^+$ in Table 3 since this requires that the strength of this dative bond be 115 ± 7 kcal/mol. This value exceeds the BDE of the covalent double bond in $ScCH_2^+$ (Table 2).

The next consideration in the mechanism is the spin of the reactants, intermediates, and products. Our studies indicate that the $Sc^+(^3D)$ ground state is the dominant state reacting although contributions from excited states, $Sc^+(^1D)$ and $Sc^+(^3F)$, cannot be eliminated. The products of reactions 10 and 11 are all doublets in their ground states such that these reactions are spin allowed from both singlet and triplet metal ion reactants. The ground states of the products of reaction 12 however, are both singlet species and therefore their production from the $Sc^+(^3D)$ reactant is spin forbidden. Note also that the intermediate **I** must have a singlet spin if the Sc—H and Sc—C bonds are covalent as indicated by the thermochemistry. Intermediate **I** therefore correlates with the $Sc^+(^1D)$ excited state, 0.3 eV above the ground state reactants (Figure 4).

Since there seems to be little ambiguity that the reaction proceeds via **I**, it apparently involves a transition between the triplet surface of the ground state reactants and a singlet surface correlating with excited state reactants (Figure 4). Reaction 12 is evidence that spin-forbidden transitions can take place in these systems. This differs from any observations made in the reaction with H_2, which has no comparable reaction channel.

5.3. Ti^+, V^+ + CH_4

Both Ti^+ and V^+ show comparable behavior to Sc^+ in their interactions with methane when in their ground states, $Ti^+(a^4F)$ and $V^+(^5D)$.[7b,c] The primary product at low energies is MCH_2^+. The thermochemical evidence (see Section 3.6) suggests that both $TiCH_2^+$ and VCH_2^+ are double-bonded species. Thus, these species are presumably in doublet and triplet states, respectively, such that these reactions are spin forbidden. At high energies, the dominant product is MH^+. MCH_3^+ is formed with a cross section about one-fifth that of MH^+. Ti^+ and V^+ both exhibit small amounts of MCH^+ products at elevated energies. These species are from a minor decomposition channel for MCH_3^+ involving loss of dihydrogen.

The mechanisms of these reactions are presumably identical to that of Sc^+ and involve the formation of $HMCH_3^+$. For Ti^+, this species must be a doublet and therefore correlates with $Ti^+(^2F) + CH_4$, 0.6 eV above the ground state reactants. We estimate that $HTiCH_3^+$ lies ~16 kcal/mol below $Ti^+(^2F)$ + CH_4 (~2 kcal/mol below ground state reactants).[7c] For V^+, the intermediate is a triplet which correlates with $V^+(^3F)$, 1.1 eV above ground state reactants. From the thermochemistry determined for $HVCH_3^+$ (Table 3), this species lies 12 kcal/mol above ground state reactants but is in a 14-kcal/mol potential energy well compared with the reactive $V^+(^3F)$ state. Thus, the potential energy surfaces for the Ti^+ and V^+ reactions are qualitatively similar to that shown in Figure 4.

When the reactivities of excited states of Ti^+ and V^+ are probed, it is found that the production of MCH_2^+ and MH^+ is enhanced. In both cases, the states primarily responsible for the effect are $Ti^+(^2F)$ and $V^+(^3F)$, the lowest-energy low-spin states. This observation is consistent with potential energy surfaces analogous to that proposed for the case of Sc^+ (Figure 4). From these low-spin states, the $HMCH_3^+$ intermediate can be formed in a spin-allowed process which requires no spin-orbit interactions. These intermediates can then decompose to form MH^+, MCH_3^+, *and* MCH_2^+ in spin-allowed reactions. The enhanced reactivities of these low-spin states are hinted at in the H_2 system but are more dramatically displayed here because of the sensitivity of the dehydrogenation reaction to the metal spin state.

5.4. Cr^+ + CH_4

This system has been studied several times using both ICR[34] and beam techniques.[7d,53] Qualitatively, our results[7d] show that the reaction of ground state $Cr^+(^6S, 3d^5)$ with methane is quite similar to that observed for Sc^+. Formation of $CrCH_2^+$ is the lowest-energy channel and has a threshold consistent with the bond energy in Table 2. Based on the arguments in Section 3.6., this indicates an M—C double bond. *Ab initio* calculations[38b,54] obtain a similar bond energy for the double-bonded ground state, $CrCH_2^+(^4B_1)$. The cross section for $CrCH_2^+$ peaks at the onset for production of CrH^+ and $CrCH_3^+$, again indicating a common intermediate. The thresholds of these two channels also correspond to the Cr—H and Cr—C bond energies in Table 2. The total cross section in this system is smaller than for the Sc^+ system by about an order of magnitude, mainly because of the less favorable product energetics.

When Cr^+ is formed by electron impact,[7d,53] the threshold for CrH^+ formation shifts by about 2.5 eV, indicating the presence of the ^4D and ^4G excited states (Table 1). The dehydrogenation channel becomes exothermic and efficient.[7d,34,53] As in the case of H_2, no hint of reaction due to the first excited state, $Cr^+(^6D)$, is observed. By a detailed examination of the kinetics of reaction, Ridge and co-workers[34] also determined that the excited states react to form $CrCH_2^+$ on only 1 in every 17 collisions. They further determined that 80% of the collisions quench the excited states into a nonreactive state.

The reaction mechanism for this system is presumably similar to that proposed for Sc^+ based on the similarity in reaction features. We again postulate an intermediate like **I** and discount ones like **II** and **III**. While Cr^+ has enough electrons to form the four covalent bonds in a species like **II**, neither it nor an analogue to **III** appear to be energetically feasible. (We estimate that they are 50 kcal/mol higher in energy.[7d]) The semiquantitative potential energy surface for this reaction is shown in Figure 5. The intermediate **IV** must have a quartet spin and therefore correlates with excited state reactants $Cr^+(^4D, ^4G) + CH_4(^1A_1)$ and the dehydrogenation products $CrCH_2^+(^4B_1) + H_2(^1\Sigma_g^+)$. Formation of CrH^+ and $CrCH_3^+$ is spin allowed from either the sextet

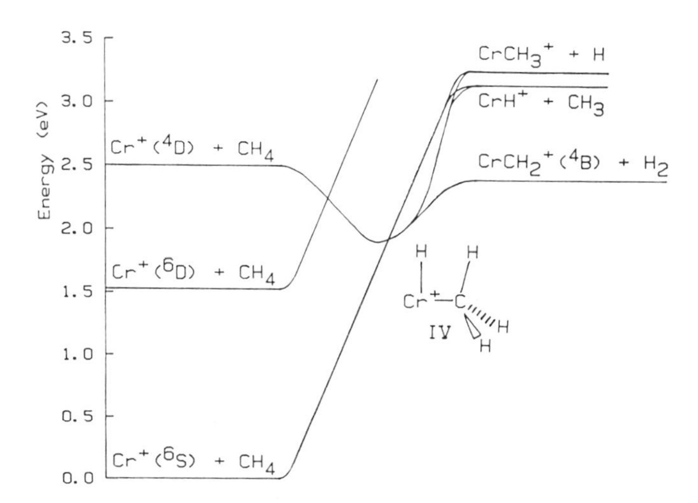

Figure 5. Proposed potential energy surface for the reaction of Cr^+ with methane.

or quartet metal ion reactants. Unlike the Sc^+ case, **IV** has an energy well in excess of ground state reactants. Note, however, that compared with the reactant state of the same spin, the well depths are comparable, ~15 kcal/mol in the Cr^+ case and 22 kcal/mol in the Sc^+ case. It seems evident that in order to form ground state $CrCH_2^+(^4B_1) + H_2(^1\Sigma_g^+)$ from $Cr^+(^6S) + CH_4(^1A_1)$, spin-orbit interactions must allow the reactants to form **IV**. This proposal is made more plausible by noting that this same potential energy surface also provides a mechanism for the quenching reaction observed by Ridge and co-workers.[34] Namely, $Cr^+(^4D, ^4G)$ reacts with methane to form **IV** which then crosses to the repulsive sextet surface. The fact that this spin-forbidden quenching process competes favorably with the spin-allowed dehydrogenation reaction demonstrates that this spin-orbit transition is reasonably efficient.

5.5. $Fe^+ + CH_4$

Ground state $Fe^+(^6D, 4s3d^6)$ reacts with methane to form only FeH^+.[7a] The cross section reaches its maximum at a much higher energy than observed for Sc^+, Ti^+, V^+, or Cr^+. No $FeCH_3^+$ or $FeCH_2^+$ is observed. These results clearly indicate that this state reacts with methane in an impulsive manner without formation of an $HFeCH_3^+$ intermediate. This is similar to the reactivity observed with dihydrogen. We again attribute this to the repulsive nature of the potential energy surface due to the occupation of the 4*s* orbital and the high spin of this state.

Results[7a] for the reaction of excited state $Fe^+(^4F, 3d^7)$ with methane are compared with those for ground state $Fe^+(^6D)$ in Figure 6. A very small

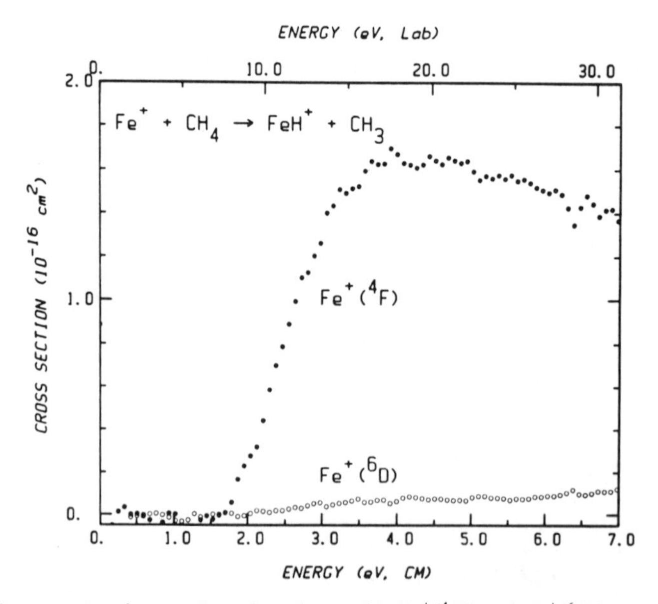

Figure 6. Cross section for reaction of methane with $Fe^+(^4F)$ and $Fe^+(^6D)$ as a function of relative kinetic energy (lower scale) and laboratory ion energy (upper scale).

amount of $FeCH_3^+$ (~1/40 that of FeH^+) is also produced in the reaction of $Fe^+(^4F)$, but no $FeCH_2^+$ is observed. As in the H_2 system, $Fe^+(^4F)$ is much more reactive than $Fe^+(^6D)$. The failure to observe $FeCH_2^+$ suggests that reaction of $Fe^+(^4F)$ may not occur via an insertion intermediate akin to **I**. This is also indicated by several other pieces of information. Reaction of this state with hydrogen is direct. The ratio of MCH_3^+ to MH^+ is less in the case of Fe^+ than in the case of Sc^+, Ti^+, V^+, or Cr^+. Indeed, it is about a factor of 40, close to the ratio of 42 predicted by the conservation of orbital angular momentum argument outlined above. Finally, our best estimate of the stability of $HFeCH_3^+$ places it ~19 kcal/mol above ground state reactants and ~12 kcal/mol above $Fe^+(^4F)$, the state to which it presumably correlates. Thus, unlike the situation for Sc^+, Ti^+, V^+, and Cr^+, the $HMCH_3^+$ intermediate for Fe^+ apparently does not lie in a potential energy well.

The case against the formation of a stable $HFeCH_3^+$ is furthered by the observation of Jacobson and Freiser[55] that $FeCH_2^+$ does not undergo H/D exchange with D_2. Neither does $FeCH_2^+$ react with H_2 to form $Fe^+ + CH_4$, a process which is exothermic by 29 ± 5 kcal/mol. These observations are taken as evidence that addition of hydrogen to $FeCH_2^+$ has a barrier to formation of $HFeCH_3^+$.[55] Similar observations were made for $CoCH_2^+$.[55] The origins of this barrier are as yet unclear.

6. REACTIONS WITH ETHANE

Examination of the reactions of metal ions with ethane again allows an exploration of the activation of C—H bonds and opens the possibility of activating a C—C bond as well. The question of whether similar electronic requirements exist for C—C activation as for C—H activation is of considerable interest since the former appears to be much more difficult in condensed phase chemistry. The energetics of this system are such that the C—H bond is weaker than that in methane, $D^\circ(H—C_2H_5) = 101$ kcal/mol. The C—C bond is weaker still, $D^\circ(H_3C—CH_3) = 90$ kcal/mol. Dehydrogenation is an even lower energy reaction channel for ethane than for methane. This can be seen by noting that in the absence of the metal ion, dehydrogenation of methane costs 111 kcal/mol but dehydrogenation of ethane requires only 33 kcal/mol. Thus, dehydrogenation of ethane by M^+ is exothermic if $D^\circ(M^+—C_2H_4) > 33$ kcal/mol, which is probably true for most M^+. Compared with the methane system, the reaction of metal ions with ethane continues the trend of increasing polarizability (4.4 Å^3),[51] degrees of freedom (21), and thus intermediate lifetimes.

6.1. $V^+ + C_2H_6$

We start by discussing our results[4] for the reaction of V^+ (produced by SI) with ethane, shown in Figure 7. The dominant channel is endothermic cleavage of the C—C bond to form VCH_3^+:

$$V^+ + C_2H_6 \rightarrow VCH_3^+ + CH_3 \tag{16}$$

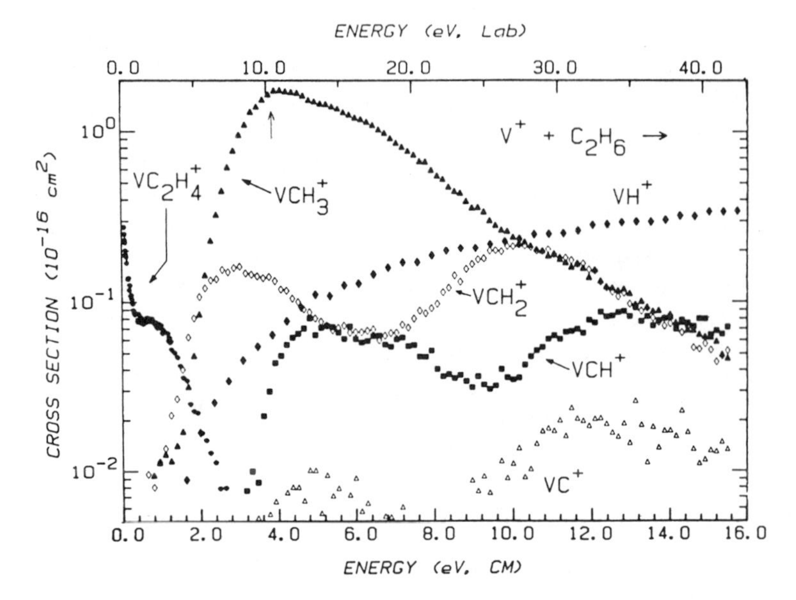

Figure 7. Cross sections for reaction of V^+ (produced by surface ionization) with ethane as a function of relative kinetic energy (lower scale) and laboratory ion energy (upper scale). The vertical arrow marks $D^\circ(H_3C—CH_3)$.

This cross section reaches a peak at 4 eV due to decomposition of this product to $V^+ + CH_3$. This is in agreement with the thermodynamic onset equal to $D^\circ(CH_3—CH_3) = 3.9$ eV. Formation of VCH_2^+ can be seen to have two features. The lower-energy feature corresponds to the reaction

$$V^+ + C_2H_6 \rightarrow VCH_2^+ + CH_4 \tag{17}$$

The peak in the cross section for the process in Equation 17 occurs below any threshold for decomposition of VCH_2^+. This indicates that reaction 16, which has an onset at an appropriate energy, is depleting a common intermediate, presumably **V**.

$$\begin{array}{cc} V^+ — & CH_2 \\ | & | \\ H_3C & H \end{array}$$

V

This can be formed by insertion of V^+ into the C—C bond of ethane. At low energies, **V** decomposes via a four-center elimination in the process given by Equation 17. At high energies, the bond fission process of Equation 16 dominates the reaction probability. The second peak in the VCH_2^+ cross section relates to the following reaction:

$$V^+ + C_2H_6 \rightarrow VCH_2^+ + H + CH_3 \tag{18}$$

this process requires 4.5 eV [$=D^\circ(H—CH_3)$] more energy than that in Equation 17. Reaction 18 is a secondary decomposition of VCH_3^+ which can also be seen as a break in the VCH_3^+ cross section at about 6.5 eV. The other high-energy products, VCH^+ and VC^+, come from secondary decompositions of VCH_3^+ and VCH_2^+ by loss of H_2 and H.

Formation of VH^+ begins at the thermodynamic limit for the process

$$V^+ + C_2H_6 \rightarrow VH^+ + C_2H_5 \tag{19}$$

Competing with this is dehydrogenation of ethane:

$$V^+ + C_2H_6 \rightarrow VC_2H_4^+ + H_2 \tag{20}$$

Deuterium labeling studies with CH_3CD_3 demonstrate that this product is a vanadium-ethene ion complex.[4] The odd part about this exothermic cross section is that while it follows the $E^{-1/2}$ energy dependence of Equation 1, its magnitude is too small by a factor of ~500. The reasons for this become apparent by varying the temperature of the SI filament from 1800 to 2200 K. This changes the population of the ^5D and ^5F states very little (Table 1), but the ^3F population varies by over a factor of 2. The magnitude of the exothermic part of the process in Equation 20 tracks with this variation. This demonstrates that this process is due primarily to the ^3F excited state; however, at about 0.4 eV, the quintet states also begin to form $VC_2H_4^+$.

In retrospect, this result may not be so surprising. Based on the mechanisms discussed above, dehydrogenation should occur via insertion into a C—H bond of ethane to form $H—V^+—C_2H_5$ (**VI**). Since two of the four electrons on the metal are involved in binding the H and ethyl radical ligands, only two unpaired electrons remain on the metal ion. Thus, the ground state of **VI** (and **V** as well) must have triplet spin. Therefore, these intermediates correlate not with ground state reactants, $V^+(^5D) + C_2H_6(^1A_1)$, but with the lowest triplet state, $V^+(^3F)$. Apparently there is a ~0.4 eV barrier to formation of **VI** from the $V^+(^5D)$ ground state. We hypothesize that this corresponds to the curve crossing between the surfaces correlating with ground state $V^+(^5D)$ and $V^+(^3F)$.[4] For the endothermic reactions, Equations 16–19, the effects of such a curve crossing are not easily observed because the endothermicity exceeds the height of the curve-crossing barrier.

6.2. $Sc^+ + C_2H_6$

Figure 8 shows the results for reaction of Sc^+ (produced by SI) with ethane.[6] Variation of the SI filament temperature indicates that the reactivity is due primarily to ground state $Sc^+(^3D)$. The cross sections for the C—C cleavage products, $ScCH_3^+$ and $ScCH_2^+$, are similar in shape and only somewhat reduced in magnitude compared to those for V^+ (Figure 7). We presume that a similar reaction mechanism is involved. Interestingly, production of $ScCH_2^+$ is not observed until 14 to 21 kcal/mol above the thermodynamic threshold.

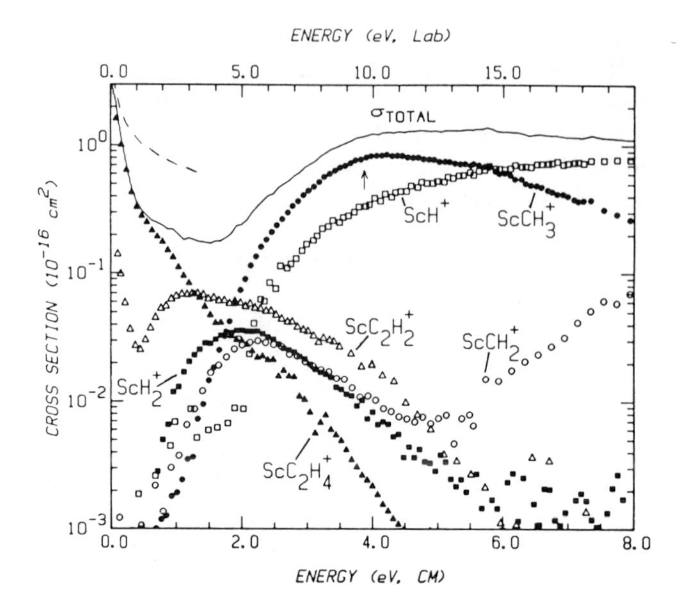

Figure 8. Cross sections for reaction of Sc^+ (produced by surface ionization) with ethane as a function of relative kinetic energy (lower scale) and laboratory ion energy (upper scale). The dashed line shows $\sigma_{LGS}/50$. The vertical arrow marks $D^\circ(H_3C—CH_3)$.

Thus, elimination of methane from $ScC_2H_6^+$ (presumably having a dimethyl structure like **V**) exhibits a barrier while elimination of H_2 from $ScCH_4^+$ (Section 5.1) does not. This type of barrier has also been noted by *ab initio* calculations for reductive elimination from $HPdCH_3$ (barrier height = 10 kcal/mol) and HPdH (2 kcal/mol).[56] The reason for the increased barrier in the former case is that the sp^3 orbital on the methyl radical is directional. It therefore cannot easily form bonds to both the metal and the H atom necessary for the transition state. In contrast, the $1s$ orbital of the hydrogen atom is spherical and can form bonds simultaneously in all directions.

The main differences between the results shown here and those for V^+ lie in the C—H bond cleavage products. Most evident is that two new products are observed: $ScC_2H_2^+$ and ScH_2^+. In addition, the C—H bond cleavage products, $ScC_2H_4^+$ and ScH^+, are larger than for V^+ although production of the former still occurs at a cross section which is well below the collision limit, about $\sigma_{LGS}/60$. Further insight into the C—H cleavage process can be obtained by isotopic labeling studies. In reaction of Sc^+ with CH_3CD_3, we find that the major process at low energies is the reaction

$$Sc^+ + CH_3CD_3 \rightarrow ScC_2H_2D_2^+ + HD \tag{21}$$

As for V^+, this shows that dehydrogenation mainly occurs across the C—C bond and implies that the product is a Sc^+-ethene complex. At the very lowest energies, the reactions

$$Sc^+ + CH_3CD_3 \longrightarrow ScC_2HD_3^+ + H_2 \quad (22)$$

$$\qquad\qquad\quad \longrightarrow ScC_2H_3D^+ + D_2 \quad (23)$$

are also observed with equal intensities that are about a factor of 5 less than reaction 21. These two reaction products decline as the energy is raised, indicating that scrambling is occurring at these low energies (corresponding to the longest intermediate lifetimes). Purely random dehydrogenation favors reaction 21 by a factor of 4 to 1 over reactions 22 and 23. These results are consistent with a mechanism which involves C—H bond activation to form **VII** followed by β-H transfer to form **VIII** (Figure 9). Scrambling can occur via an equilibrium between **VII** and **VIII**. Reductive elimination of H_2 from **VIII** yields the major product, $Sc^+—C_2H_4$, while loss of ethene from **VIII** yields ScH_2^+. This metal dihydride ion is not observed for reaction of ethane with any other first-row transition metal ion although it is observed for Y^+ and La^+, which are isovalent with Sc^+.[7e] This is presumably because these three metal ions have only two electrons, both of which are involved in bonding the two H atoms in **VIII.** Thus, the $H_2Sc^+—C_2H_4$ bond is a dative bond involving only donation of the ethene π electrons to the metal. Metal ions with more electrons can backbond to the ethene ligand, making elimination of C_2H_4 energetically less favorable and therefore less likely.

At higher energies, reactions 22 and 23 become more probable than reaction 21. We attribute this to a 1,1-dihydrogen elimination to form a

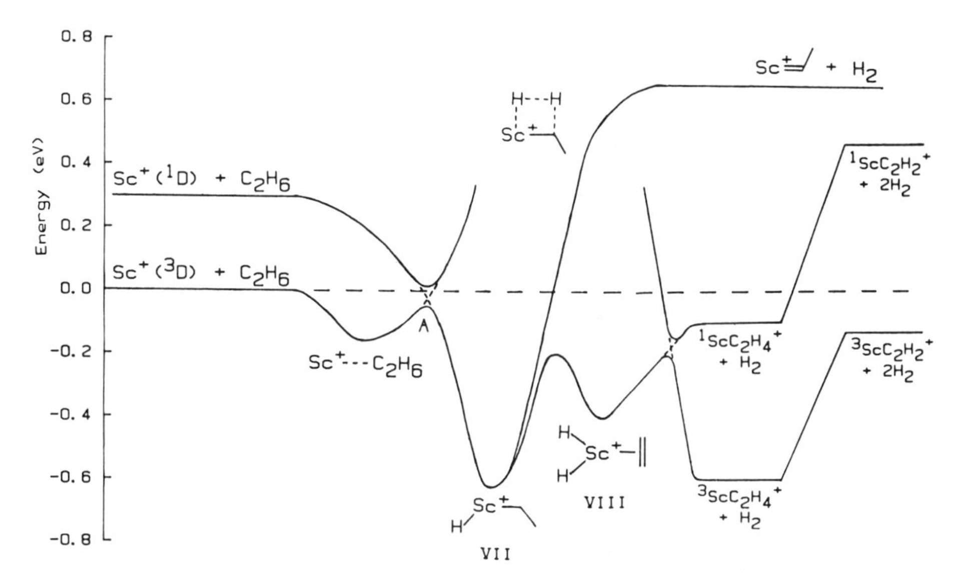

Figure 9. Proposed potential energy surface for the dehydrogenation and double dehydrogenation of ethane by Sc^+. Full lines represent adiabatic surfaces and dashed lines represent diabatic surfaces.

scandium ethylidene ion, $Sc^+{=}CHCH_3$ (Figure 9). This 1,1-H_2 elimination is analogous to reaction 12 in the methane system and begins at a comparable kinetic energy.

Another effect of isotopic substitution is observed in the total cross sections for dihydrogen elimination from d_0-, 1,1,1-d_3-, and d_6-ethane. A strong effect is seen such that the efficiency of dehydrogenation decreases as the degree of deuterium substitution increases. This is interpreted in terms of the need for a singlet–triplet surface crossing in the formation of **VII.** Note that this intermediate must be in a singlet state while the Sc^+ ground state is a triplet. We hypothesize that the potential energy surface (PES) looks something like Figure 9. The efficiency of reaction decreases because the position of the crossing, point A, on the PES depends on the relative strength of the C—H versus the C—D bond. Since the latter is stronger, point A is higher for d_6-ethane and the reaction is less efficient.

As in the case of V^+, it seems clear that a spin-forbidden surface crossing must occur if insertion to form **VII** in its ground state is to take place. It is the need for this crossing that we believe explains the inefficiency of the reaction between Sc^+ and ethane even though the reaction is exothermic with no activation barrier. Note that this differs from the V^+ case which *does* exhibit a barrier for ground state reaction. The reason for this probably lies in the larger splitting between the $V^+(^5D)$ and $V^+(^3F)$ states, 1.1 eV, compared to that for $Sc^+(^3D)$ and $Sc^+(^1D)$, 0.3 eV. The curve crossing thus occurs at a much lower energy in the Sc^+ system.

One final point of interest in this system is the dual nature of the $ScC_2H_2^+$ product. As can be seen in Figure 8, this product is formed both at the lowest energies in an exothermic reaction with no barrier and also at higher energies with a barrier of about 0.5 eV. This process also shows[6] a very strong

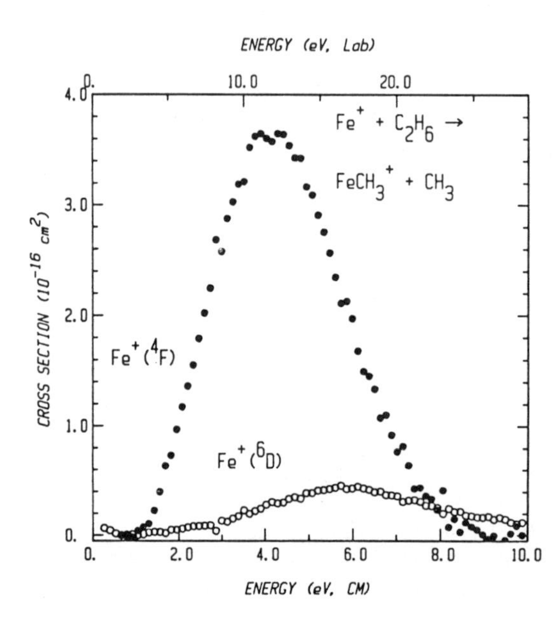

Figure 10. Cross sections for reaction of $Fe^+(^4F)$ and $Fe^+(^6D)$ with ethane to form $FeCH_3^+$ as a function of relative kinetic energy (lower scale) and laboratory ion energy (upper scale).

dependence on the degree of deuterium substitution such that the low-energy feature disappears for d_6-ethane and is enhanced for d_3-ethane. While there are several possible explanations for this complex behavior,[6] one possibility involves a second surface crossing (Figure 9). This presumes that $ScC_2H_4^+$ can be formed either in a triplet or a singlet state corresponding to triplet or singlet Sc^+. Since **VIII** is a singlet, formation of singlet $ScC_2H_4^+$ is spin allowed but formation of triplet $ScC_2H_4^+$ requires a singlet–triplet surface crossing. Similarly, there are both triplet and singlet $ScC_2H_2^+$ which evolve from triplet and singlet $ScC_2H_4^+$. If the triplet state of $ScC_2H_4^+$ is formed, it can further decompose by loss of a second H_2 molecule in an exothermic reaction. If the singlet state is formed, production of $ScC_2H_2^+$ is endothermic by about 0.4 eV. Zero-point energy differences between H_2, HD, and D_2 can then explain the severe isotope dependences observed.

6.3. $Fe^+ + C_2H_6$

$Fe^+(^4F)$ reacts with ethane as follows[7a]:

$$Fe^+ + C_2H_6 \rightarrow FeCH_3^+ + CH_3 \quad (24)$$

$$\rightarrow FeC_2H_4^+ + H_2 \quad (25)$$

$$\rightarrow FeH^+ + C_2H_5 \quad (26)$$

$$\rightarrow FeC_2H_5^+ + H \quad (27)$$

$$\rightarrow FeCH_2^+ + CH_3 + H \quad (28)$$

The behavior of reactions 24 and 26 is comparable to that for V^+ and Sc^+. The cross section for reaction 27 has a threshold comparable to that of reaction 26 but is smaller by a factor of about 100 (angular momentum constraints predict a factor of 86). Reaction 25 exhibits a threshold of several tenths of an eV and peaks at the onset for reaction 26. While reaction 28, analogous to the process in Equation 18, is observed, no reaction like the process in Equation 17—loss of methane to form $FeCH_2^+$—is observed at low energies. This may imply that the C–C activation in this system occurs via a direct mechanism as postulated for reaction with H_2 and CH_4.

The reaction of $Fe^+(^6D)$ with ethane is quite distinct from that of $Fe^+(^4F)$. The cross sections for the processes in Equations 24 and 26 decrease by factors of about 2 and 14, respectively, at 4 eV. Reaction 27 also decreases such that it is no longer observed. This behavior is quite similar to the changes in reactivity observed for H_2 and CH_4. This is evident from the state-specific reactivity for reaction 24 shown in Figure 10. Again the reactivity of this state is characteristic of a largely impulsive mechanism.

The most intriguing observation is that $Fe^+(^6D)$ reacts to form $FeC_2H_4^+$ exothermically, although with an efficiency of about 0.002. Again the observation of this dehydrogenation channel suggests that C—H bond activation may be occurring via an intermediate analogous to **VI** and **VII**. However, this insertion intermediate should have a quartet spin. Yet, $Fe^+(^4F)$ exhibits an activation barrier and $Fe^+(^6D)$ does not. This is in stark contrast to the observations made in the Sc^+ and V^+ systems. The explanation for this effect becomes more obvious in our examination of the reaction of Fe^+ with propane.

6.4. $Fe^+ + C_3H_8$

Fe^+ reacts exothermically with propane ($\alpha = 6.2$ Å^3)[51] via loss of methane to form $FeC_2H_4^+$ and loss of H_2 to form $FeC_3H_6^+$.[7a,57,58] The efficiency of these reactions is about one-tenth of the collision rate (Equation 1). The branching ratio at low energies is 3:1 in good agreement with ICR[58] and previous beam results.[57] At higher energies, the C—C bond cleavage products $FeCH_3^+$ and $FeC_2H_5^+$ and the C—H bond cleavage products FeH^+ and $C_3H_7^+$ are formed.[7a] When the state-specific reactions are examined,[7a] the cross sections for all endothermic reactions behave similarly to the reactions with ethane, methane, and hydrogen, namely, $Fe^+(^4F)$ is much more reactive than $Fe^+(^6D)$.

Surprisingly, the exothermic reactions do not conform to this general reactivity, just as for reaction 25, the exothermic reaction in ethane. As shown in Figure 11, the $Fe^+(^6D)$ ground state reacts exothermically with propane while $Fe^+(^4F)$ does not begin to react until about 0.5 eV. We believe that the explanation for this behavior again lies in the idea of a surface crossing. Presume that formation of $FeC_2H_4^+$ proceeds via the C-C bond activation

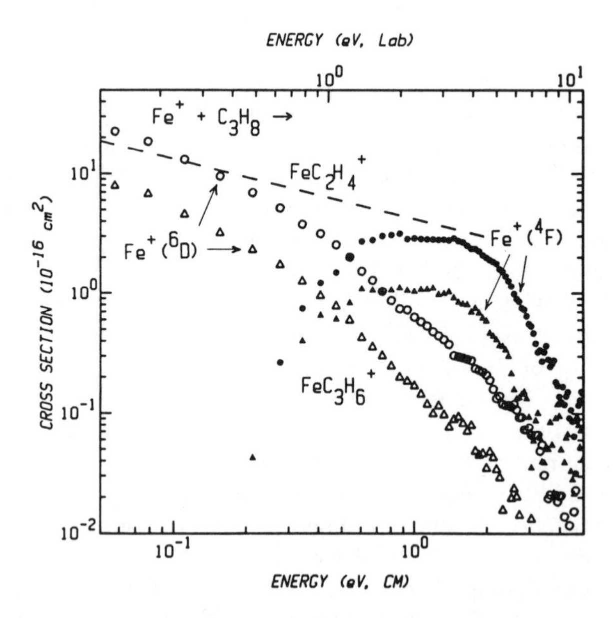

Figure 11. Cross sections for reaction of $Fe^+(^6D)$ (open symbols) and $Fe^+(^4F)$ (closed symbols) with propane to form $FeC_2H_4^+$ (circles) and $FeC_3H_6^+$ (triangles) as a function of relative kinetic energy (lower scale) and laboratory ion energy (upper scale). The dashed line shows $\sigma_{LGS}/10$.

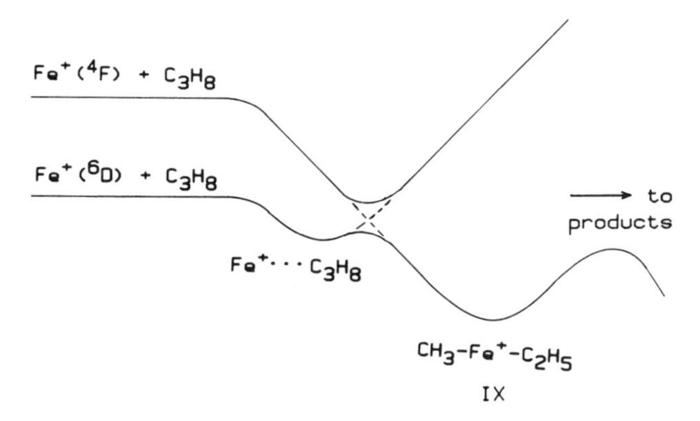

Figure 12. Potential energy surfaces for C—C insertion of Fe^+ in propane. Full lines represent adiabatic surfaces and dashed lines represent diabatic surfaces.

intermediate $H_3C—Fe^+—C_2H_5$ (**IX**). A parallel argument can be used for dehydrogenation. In order to have two covalent M—C bonds, **IX** must be in a state having quartet spin and thus correlates diabatically with the $Fe^+(^4F)$ state of reactants. The diabatic correlation rules established for reaction with H_2 suggest that $Fe^+(^6D)$ and alkanes should have largely repulsive interactions. Thus, the diabatic potential energy surfaces shown in Figure 12 can be drawn. If spin-orbit interactions are strong, the crossing between these surfaces can be avoided to produce two new adiabatic potential energy surfaces. At low kinetic energies, the reactants follow these adiabatic curves such that ground state reactants form ground state products while excited state reactants have no exothermic channels available. As the kinetic energy is raised, the reactants begin to behave diabatically, i.e., there is a transition between the adiabatic surfaces. A "threshold" is thus observed for reaction of $Fe^+(^4F)$, corresponding to the onset of this diabatic behavior, i.e., conservation of spin. By the time the endothermic products are energetically accessible, the diabatic behavior dominates the reactivity as for the smaller systems (H_2, CH_4, and C_2H_6). For alkanes larger than propane (for which state-specific experiments are incomplete), these considerations suggest that the exothermic channels will also be due to reaction of $Fe^+(^6D)$. Indeed, it is known that the efficiency of the reactions is near unity.[57,58]

6.5. $Zn^+ + C_2H_6$

$Zn^+(^2S)$ exhibits no low-energy reactions with ethane.[5] The dominant reaction channels are the following processes:

$$Zn^+ + C_2H_6 \longrightarrow ZnCH_3^+ + CH_3 \quad (29)$$

$$ \longrightarrow ZnH^+ + C_2H_5 \quad (30)$$

The behavior of these reactions indicates that they occur via an impulsive mechanism. This is consistent with the $4s3d^{10}$ electron configuration of Zn^+. Interestingly, the following reactions are also observed:

$$Zn^+ + C_2H_6 \rightarrow ZnCH_3 + CH_3^+ \quad (31)$$

$$\rightarrow ZnH + C_2H_5^+ \quad (32)$$

The fact that they are not observed in other metal systems is partly because of the high ionization potential of Zn compared with the other transition metals and partly due to the lack of competing processes. We have also studied the reactions

$$Zn^+ + RCH_3 \rightarrow ZnCH_3^+ + R \quad (33)$$

$$\rightarrow ZnCH_3 + R^+ \quad (34)$$

for propane ($R = C_2H_5$), isobutane [$R = CH(CH_3)_2$], and neopentane [$R = C(CH_3)_3$].[5] In this series of experiments, reaction 34 becomes progressively more favorable, primarily because the ionization potential of the alkyl radical decreases. Further, we have shown that the molecular orbital ideas discussed above lead to the conclusion that this reaction is the diabatically favored process in all these systems. This explains why this process is observed to be efficient once enough energy is available while reaction 33 is inefficient even if it is much more favorable energetically.

7. SUMMARY

7.1. Thermochemistry

The ion beam method is clearly a powerful tool for measuring thermodynamic data on a wide variety of small transition metal-containing species, both ionic and neutral. While the species amenable to study by this technique are far removed from condensed phase species, careful analysis of the data can lead to trends which should prove to be useful for both the gas and condensed phases. This type of information already includes an experimental determination of the intrinsic metal-hydrogen bond energies for both first- and second-row transition metals; a comparison of M—H and M—alkyl bond energies; and trends in single, double, and triple metal-carbon bond energies. Other data which are still at an early state of accumulation include direct comparisons of ionic and neutral M—H and M—C bond energies; first and second M—H and M—C bond energies; and possibly an evaluation of the energetics of agostic interactions. Some of this data is discussed in somewhat more detail elsewhere.[10] The application of these techniques to other types of metal-ligand systems is also a limitless area of expansion in which work is already under way.

7.2. Reactions with Dihydrogen

The experimental results and theoretical considerations discussed in Section 4 lead to a reasonably comprehensive view of the reactions of atomic transition metal ions with molecular hydrogen. Three categories of reactivity appear to exist based on the electron configuration and spin state of the metal:

1. If the $4s$ and $3d\sigma$ orbitals are unoccupied, the systems react efficiently. Overall, the behavior is near statistical. MO concepts indicate that these states should be able to insert into H_2 to form a metal dihydride intermediate.
2. If either the $4s$ or $3d\sigma$ orbital is occupied, the systems can react efficiently via a direct mechanism if they have low spin (i.e., the same spin as the ground state of MH_2^+). MO concepts indicate that these states may prefer a collinear reaction geometry but that other geometries are not unfavorable.
3. If either the $4s$ or $3d\sigma$ orbital is occupied and the ion has a high spin, the systems react inefficiently at the thermodynamic threshold and via an impulsive mechanism at elevated energies. MO concepts indicate that these states should have repulsive surfaces which strongly favor a collinear reaction geometry.

Note that these rules for $M^+ + H_2$ reactions are formulated for the *diabatic* character of the reaction surfaces, i.e., they pertain to specific electron configurations of the metal ions. The fact that these rules successfully explain our experimental observations demonstrates what kind of metal orbital character is necessary for efficient activation of molecular hydrogen. The ideas forwarded here are consistent with several theoretical discussions of the interactions of H_2 with metal atoms,[59] metal complexes,[60,61] and metal surfaces.[61,62] These studies are the first experimental tests of these ideas. Further, these rules constitute a specific example of the more general electronic requirements outlined by Crabtree (Section 1.2).

Despite the utility of these rules, they clearly can be broken. For instance, the reactivities of the ground states of Sc^+ and Ti^+ do not fall neatly into these categories. This is postulated to be due to mixing of the potential energy surfaces evolving from these states with those from more reactive excited states. Thus, the diabatic rules can fail when adiabatic interactions become important. Note however that the general electronic requirements of donor–acceptor orbitals remain in such cases. Further studies of such surface interactions may be accomplished by comparison of first-, second-, and third-row transition metals[10,15,40,63] since this progression changes the relative energies of the interacting states. Other effects that need to be considered in these investigations are the orbital size, the population of f orbitals, and increased spin–orbit interactions.

7.3. Reactions with Methane

While detailed state-specific results for reactions with methane at the level performed for reactions with dihydrogen have not yet been completed for all metal ions, the results of Section 5 indicate that the molecular orbital ideas discussed above for the activation of H_2 appear to be valid for activation of C—H bonds of methane. The same three categories of reactivity outlined above can be differentiated:

1. Statistical behavior resulting from formation of a hydrido metal methyl ion. Sc^+, Ti^+, V^+, and Cr^+ all behave in this manner.
2. Direct reaction strongly favoring formation of MH^+ over MCH_3^+ with no production of MCH_2^+. $Fe^+(^4F)$ is the prototypical species.
3. Impulsive behavior for high-spin ions with occupied $4s$ or $3d\sigma$ orbitals, e.g., $Fe^+(^6D)$. Only MH^+ is formed.

While the general electronic requirements for H_2 and CH_4 activation are qualitatively similar, there are important differences. The reaction with methane is substantially more sensitive to the spin state of the ion because the dehydrogenation reaction *requires* a low-spin configuration to be allowed. A comparable reaction channel does not exist in the case of the reactions with dihydrogen. Thus, interactions between surfaces of different spin which were hinted at in the results for hydrogen become clear in the results for reactions of methane, especially with Sc^+, Ti^+, V^+, and Cr^+. Note that these interactions involve the breakdown in the diabatic rules of reactivity formulated for the H_2 reactions due to spin–orbit interactions. The more general electronic requirements for C—H activation [i.e., the concept of donor-acceptor orbitals (Section 1.2)] are still upheld.

Another observation of mechanistic importance is that dehydrogenation of a hydrido metal methyl intermediate, $HMCH_3^+$, almost certainly occurs via a four-center transition state. In addition, this process appears to be facile for metals on the left side of the periodic table, i.e., Sc^+, Ti^+, V^+, and Cr^+, but may have a barrier for metals on the right, i.e., Fe^+, Co^+, and probably others. *Ab initio* calculations are of obvious interest in elucidating the detailed electronic character of these dehydrogenation reactions, but this difference in reactivity probably involves the nonbonding d electrons on the metal. On the left side of the periodic table, there are empty d orbitals, while on the right side, all d orbitals are occupied. It is possible then that agostic interactions (as discussed in Section 3.3 for MCH_3^+ species) withdraw electron density from the C—H bond being broken and thus facilitate the dehydrogenation reaction. A related explanation comes from a theoretical study[52] of $2+2$ isotope exchange reactions between MH and D_2. This work notes that the nonbonding d electrons on the metal must remain orthogonal to the bonds being made and broken, a requirement which is energy destabilizing.

7.4. Reactions with Ethane

The results of Section 6 indicate that the molecular orbital ideas again appear to be valid for activation of the C—H *and* the C—C bonds of ethane. The same three categories of reactivity observed for H_2 and CH_4 activation are observed: statistical behavior for Sc^+ and V^+; direct reaction for $Fe^+(^4F)$;

and impulsive behavior for $Fe^+(^6D)$ and $Zn^+(^2S)$. As for CH_4 activation, reaction with ethane is sensitive to the spin state of the ion. This is again because the low-energy reaction channels, loss of H_2 and CH_4, require a low-spin electron configuration on the metal ion to be spin allowed. Reactions of metal ions with high-spin configurations are still observed but with greatly reduced probabilities. We propose that this is due to the need for spin-orbit-induced transitions to the low-spin potential energy surface. While these spin-orbit interactions lead to a breakdown in the diabatic rules of reactivity observed for H_2 chemistry, the donor-acceptor concept outlined by Crabtree[1] still retains its validity.

At first glance, the state-specific reactions of Fe^+ with propane do not appear to be consistent with the molecular orbital ideas developed for H_2, CH_4, and C_2H_6 reactions since $Fe^+(^6D)$ is more reactive than $Fe^+(^4F)$ at low kinetic energies. Above about 0.5 eV, however, the reactivity for both exothermic and endothermic reactions returns to that observed in these smaller systems. Our rationale for this behavior invokes potential energy surfaces (Figure 12) not unlike those shown in Figures 4, 5, and 9. The very dramatic effect observed in the exothermic channels of the $Fe^+ + C_3H_8$ system appears to be due to the sensitivity of the nonadiabatic transition probability to the kinetic energy of the reactants. At low kinetic energies, adiabatic behavior is most likely. These transition probabilities should also depend on the reaction system for two reasons: (1) because the energy of the surface crossing is determined by the depth of the well corresponding to the Fe^+-alkane complex, which in turn relies on the polarizability of the neutral, and (2) because the lifetime of the complex, which depends on the number of degrees of freedom, will affect the access to the crossing point (the transition state). Thus, larger alkanes have lower crossing points and longer-lived complexes than smaller reactants. It is therefore not surprising that the adiabatic behavior observed for propane at low energies is not observed for smaller reactant species.

7.5. Outlook

The relationship between the gas phase ion molecule chemistry of transition metal ions and condensed phase organometallic chemistry is still evolving. The elimination of both ligand and solvent effects makes direct comparisons difficult. Nevertheless, when viewed judiciously, the quantitative information available from gas phase studies on well-characterized transition metal species can lend considerable insight into the reaction mechanisms and energetics of complex organometallic transformations. Further, the ability to examine the state-specific chemistry of atomic transition metal ions has permitted a reasonably definitive test of the electronic requirements for covalent bond activation. We find that simple molecular orbital ideas can explain this chemistry in a straightforward fashion for H—H, C—H, and C—C bond activation. These ideas are easily related to the electronic requirements outlined in the literature for C—H bond activation.[1] These studies verify that the active metal site must be capable of accepting electron density from the bond to be broken and preferably can also donate electron density into the antibonding orbitals of this same bond. Further work concerning the state-specific reactivity should enable a quantitative evaluation of the role of spin-orbit interactions in transition metal chemistry.

Future work in this field will see a more complete evaluation of the requirements for C—H bond activation by extending these studies to more complex systems. By systematically ligating the metal, the contribution of steric congestion and competitive reactions (like cyclometalation) can be ascertained. Such ligation will also have direct effects on the electronic environment around the metal, which may be assessed by comparing metal systems and by alteration of the type of ligand. For instance, one should be able to enhance the activating power of a metal ion by ligating with a π-donating or σ-accepting species. Conversely, addition of a π-accepting or σ-donating ligand to the metal ion should poison its reactivity. Variations on this theme are endless and promise to be extremely rewarding directions for future research.

ACKNOWLEDGMENTS

The work described in this article was performed by my co-workers over the past several years, Dr. Natasha Aristov, Dr. Jerry L. Elkind, Dr. Kent M. Ervin, Steve K. Loh, Rosina Georgiadis, David A. Hales, Lee Sunderlin, Mary Ellen Weber, Joel Burley, Richard Schultz, and Dr. Bong Hyun Boo. I am very grateful for their dedication, hard work, and insight. Early phases of this work were funded by the ACS PRF, Research Corp., and the Dreyfus Foundation. Continuing support has been provided by the National Science Foundation and the NSF Presidential Young Investigator program. Contributions from Monsanto Corp., Exxon Corp., and Chevron Corp. are also greatly appreciated.

REFERENCES

1. Crabtree, R.H. *Chem. Rev.* 1985, *85*, 245–269.
2. Allison, J.; Freas, R.B.; Ridge, D.P. *J. Am. Chem. Soc.* 1979, *101*, 1332–1333.
3. Aristov, N.; Armentrout, P.B. *J. Am. Chem. Soc.* 1984, *106*, 4065–4066.
4. Aristov, N.; Armentrout, P.B. *J. Am. Chem. Soc.* 1986, *108*, 1806–1819.
5. Georgiadis, R.; Armentrout, P.B. *J. Am. Chem. Soc.* 1986, *108*, 2119–2126.
6. Sunderlin, L.; Aristov, N.; Armentrout, P.B. *J. Am. Chem. Soc.* 1987, *109*, 78–89.
7. (a) Schultz, R.H.; Armentrout, P.B. *J. Phys. Chem.* 1987, *91*, 4433–4435. Schultz, R.H.; Elkind, J.L.; Armentrout, P.B. *J. Am. Chem. Soc.* 1988, *110*, 411–423. (b) Aristov, N.; Armentrout, P.B. *J. Phys. Chem.* 1987, *91*, 6178–6188. (c) Sunderlin, L.; Armentrout, P.B. *J. Phys. Chem.* 1988, *92*, 1209–1219. (d) Georgiadis, R.; Armentrout, P.B. *J. Phys. Chem.* accepted for publication. (e) Aristov, N.; Fisher, E.R.; Georgiadis, R.; Schultz, R.H.; Sunderlin, L.S.; Armentrout, P.B., work in progress.
8. Elkind, J.L.; Armentrout, P.B. *J. Phys. Chem.* 1985, *89*, 5626–5636.
9. Elkind, J.L.; Armentrout, P.B. *J. Chem. Phys.* 1986, *84*, 4862–4871.
10. Elkind, J.L.; Armentrout, P.B. *Inorg. Chem.* 1986, *25*, 1078–1080. Armentrout, P.B.; Georgiadis, R. *Polyhedron* 1988, in press.

11. Elkind, J.L.; Armentrout, P.B. *J. Am. Chem. Soc.* 1986, *108*, 2765-2767; *J. Phys. Chem.* 1986, *90*, 5736-5745.
12. Elkind, J.L.; Armentrout, P.B. *J. Phys. Chem.* 1986, *90*, 6576-6586.
13. Elkind, J.L.; Armentrout, P.B. *J. Chem. Phys.* 1987, *86*, 1868-1877.
14. Elkind, J.L.; Armentrout, P.B. *J. Phys. Chem.* 1987, *91*, 2037-2045.
15. (a) Elkind, J.L.; Armentrout, P.B. *Int. J. Mass Spectrom. Ion Proc.* 1988, *83*, 259-284. (b) Georgiadis, R.; Armentrout, P.B., *J. Phys. Chem.*, accepted for publication. (c) Elkind, J.L.; Sunderlin, L.S.; Armentrout, P.B., work in progress.
16. Garstang, R.H. *Mon. Not. R. Astron. Soc.* 1962, *124*, 321; private communication.
17. Georgiadis, R.; Armentrout, P.B. *Chem. Phys. Lett.* 1987, *137*, 144-148.
18. Sanders, L.; Sappy, A.D.; Weisshaar, J. *J. Chem. Phys.* 1986, *85*, 6952-6963; Sanders, L.; Hanton, S.; Weisshaar, J.C. *J. Phys. Chem.* 1987, *91*, 5145-5148.
19. Ervin, K.; Armentrout, P.B. *J. Chem. Phys.* 1985, *83*, 166-189.
20. (a) Ervin, K.M.; Armentrout, P.B. *J. Chem. Phys.* 1986, *84*, 6738-6749; (b) Ervin, K.M.; Armentrout, P.B. *J. Chem. Phys.* 1986, *85*, 6380-6395.
21. Burley, J.D.; Ervin, K.M.; Armentrout, P.B. *J. Chem. Phys.* 1987, *86*, 1944-1953. *Int. J. Mass Spectrum. Ion Proc.* 1987, *80*, 153-175.
22. Teloy, E.; Gerlich, D. *Chem. Phys.* 1974, *4*, 417-427; Frobin, W.; Schlier, Ch.; Strein, K.; Teloy, E. *J. Chem. Phys.* 1977, *67*, 5505-5516.
23. Ervin, K.M.; Armentrout, P.B. *J. Chem. Phys.* 1987, *86*, 2659-2673.
24. Chantry, P.J. *J. Chem. Phys.* 1971, *55*, 2746-2759; Lifshitz, C.; Wu, R.L.C.; Tiernan, T.O.; Terwilliger, D.T. *J. Chem. Phys.* 1978, *68*, 247-259.
25. Gioumousis, G.; Stevenson, D.P. *J. Chem. Phys.* 1958, *29*, 294.
26. Talrose, V.L.; Vinogradov, P.S.; Larin, I.K. In *Gas Phase Ion Chemistry*; Bowers, M.T., Ed.; Academic: New York, 1979; Vol. 1, pp. 305-347.
27. Henchman, M. In *Ion-Molecule Reactions*; Franklin, J.L., Ed.; Plenum: New York, 1972; Vol. 1, pp. 101-259.
28. Armentrout, P.B. In *Structure, Reactivity and Thermochemistry of Ions*; Ausloos, P.; Lias, S.G.; Eds.; Reidel: Dordrect, 1987; 97-164.
29. Light, J.C. *J. Chem. Phys.* 1964, *40*, 3221; Pechukas, P.; Light, J.C. *J. Chem. Phys.* 1965, *42*, 3281; Nikitin, E.E. *Teor. Eksp. Khim.* 1965, *1*, 135, 144, 248 [*Theor. Exp. Chem.* (*Engl. Trans.*) 1975, *1*, 83, 90, 275]; Chesnavich, W.J.; Bowers, M.T. *J. Chem. Phys.* 1977, *66*, 2306-2315; *ibid.* 1978, *68*, 900-905; Webb, D.A.; Chesnavich, W.J. *J. Phys. Chem.* 1983, *87*, 3791-3798.
30. Bates, D.R. *Proc. R. Soc. London, Ser. A* 1978, *360*, 1.
31. Weber, M.E.; Elkind, J.L.; Armentrout, P.B. *J. Chem. Phys.* 1986, *84*, 1521-1529.
32. Ervin, K.; Armentrout, P.B. *J. Chem. Phys.* 1986, *84*, 6750-6760.
33. See, for example, Chapters by B.S. Freiser, D.P. Ridge, and D.H. Russell in this volume.
34. Reents, W.D., Jr.; Strobel, F.; Freas, R.B.; Wronka, J.; Ridge, D.P. *J. Phys. Chem.* 1985, *89*, 5666-5670.
35. Bensimon, M.; Houriet, R. *Int. J. Mass Spectrom. Ion Proc.* 1986, *72*, 93-98.
36. Elkind, J.L.; Armentrout, P.B. *J. Phys. Chem.* 1984, *88*, 5454-5456.
37. Schilling, J.B.; Goddard, W.A.; Beauchamp, J.L. *J. Am. Chem. Soc.* 1986, *108*, 582-584.
38. (a) Vincent, M.A.; Yoshioka, Y.; Schaefer, H.F. *J. Phys. Chem.* 1982, *86*, 3905-3906, (b) Alvarado-Swaisgood, A.E.; Allison, J.; Harrison, J.F. *J. Phys. Chem.* 1985, *89*, 2517-2525; (c) Alvarado-Swaisgood, A.E.; Harrison, J.F. *J. Phys. Chem.* 1985, *89*, 5198-5202; (d) Rappe, A.K.; Upton, T.H. *J. Chem. Phys.* 1986, *85*, 4400-4410; (e) Dupuis, M.; Hammond, B.L.; Lester, W.A. Private communication.
39. Pettersson, L.G.M.; Bauschlicher, C.W.; Langhoff, S.R.; Partridge, H. *J. Chem. Phys.* 1987, *87*, 481.
40. Mandich, M.L.; Halle, L.F.; Beauchamp, J.L. *J. Am. Chem. Soc.* 1984, *106*, 4403-4411.
41. Armentrout, P. B.; Halle, L.F.; Beauchamp, J.L. *J. Am. Chem. Soc.* 1981, *103*, 6501-6502.
42. Armentrout, P.B.; Beauchamp, J.L. *J. Am. Chem. Soc.* 1981, *103*, 784-791.
43. Halpern, J. *Acc. Chem. Res.* 1982, *15*, 238-244; *Inorg. Chim. Acta* 1985, *100*, 41-48.
44. Brookhart, M.; Green, M.L.H. *J. Organomet. Chem.* 1983, *250*, 395-408.
45. Calhorda, M.J.; Simoes, J.A.M. Private communication.

46. Eisenstein, O.; Jean, Y. *J. Am. Chem. Soc.* 1985, *107*, 1177–1186.
47. Demuynck, J.; Schaefer, H.F. *J. Chem. Phys.* 1980, *72*, 311–315; Walch, S.P.; Bauschlicher, C.W. *J. Chem. Phys.* 1983, *78*, 4597–4605.
48. Hettich, R.L.; Freiser, B.C. *J. Am. Chem. Soc.* 1986, *108*, 5086.
49. Alvarado-Swaisgood, A.E.; Allison, J.; Harrison, J.F. *J. Phys. Chem.* 1985, *89*, 2517–2525; Harrison, J.F. Private communication.
50. Carter, E.A.; Goddard, W.A. *J. Am. Chem. Soc.* 1986, *108*, 2180–2191.
51. Rothe, E.W.; Bernstein, R.B. *J. Chem. Phys.* 1959, *31*, 1619–1627.
52. Steigerwald, M.L.; Goddard, W.A. *J. Am. Chem. Soc.* 1984, *106*, 308–311.
53. Halle, L.F.; Armentrout, P.B.; Beauchamp, J.L. *J. Am. Chem. Soc.* 1981, *103*, 962–963.
54. Carter, E.A.; Goddard, W.A. *J. Phys. Chem.* 1984, *88*, 1485–1490.
55. Jacobson, D.B.; Freiser, B.S. *J. Am. Chem. Soc.* 1985, *107*, 5870–5883.
56. Low, J.J.; Goddard, W.A. *J. Am. Chem. Soc.* 1984, *106*, 8321–8322.
57. Halle, L.F.; Armentrout, P.B.; Beauchamp, J.L. *Organometallics* 1982, *1*, 963–968; Houriet, R.; Halle, L.F.; Beauchamp, J.L. *Organometallics* 1983, *2*, 1818–1829.
58. Byrd, G.D.; Burnier, R.C.; Freiser, B.S. *J. Am. Chem. Soc.* 1982, *104*, 3565–3569; Jacobson, D.B.; Freiser, B.S. *J. Am. Chem. Soc.* 1983, *105*, 5197–5206.
59. Ruiz, M.E.; Garcia-Prieto, J.; Novaro, O. *J. Chem. Phys.* 1984, *80*, 1529–1534; Siegbahn, P.E.M.; Blomberg, M.R.A.; Bauschlicher, C.W. *J. Chem. Phys.* 1984, *81*, 1373–1382; Sevin, A.; Chaquin, P. *Nouv. J. Chim.* 1983, *7*, 353–360.
60. See, for example, Brothers, P.J. *Prog. Inorg. Chem.* 1981, *28*, 1; Noell, J.O.; Hay, P.J. *J. Am. Chem. Soc.* 1982, *104*, 4578–4584.
61. Saillard, J.; Hoffman, R. *J. Am. Chem. Soc.* 1984, *106*, 2006–2026.
62. Shustotovich, E.; Baetzold, R.C.; Muetterties, E.L. *J. Phys. Chem.* 1983, *87*, 1100–1113; Siegbahn, P.E.M.; Blomberg, M.R.A.; Bauschlicher, C.W. *J. Chem. Phys.* 1984, *81*, 2103–2111.
63. Tolbert, M.A.; Beauchamp, J.L. *J. Am. Chem. Soc.* 1986, *108*, 5675–5683.
64. Moore, C.E. *Atomic Energy Levels*; National Bureau of Standards: Washington, DC, 1949; Sugar, J.; Corliss, C. *J. Phys. Chem. Ref. Data* 1977, *6*, 317–383, 1253–1329, 1978, *7*, 1191–1262; 1979 *8*, 1–62; 1980, *9*, 473–511; 1981, *10*, 197–289, 1097–1174; 1982, *11*, 135–241.

2

Nucleophilic Addition Reactions of Negative Ions with Organometallic Complexes in the Gas Phase

Robert R. Squires and Kelley R. Lane

1. INTRODUCTION

1.1. Nucleophilic Addition in Organometallic Chemistry

Nucleophilic addition and substitution reactions involving metal compounds are among the most widely applied and extensively documented reactions in organometallic chemistry. From the familiar and time-honored Grignard syntheses[1] to the most recent methods for preparing organolanthanides[2] and organoactinides,[3] nucleophilic reactions are found in innumerable synthetic procedures for forming carbon–metal bonds.[4] Nucleophilic addition continues to be the method of choice for preparing anionic transition metal acyls, formyls, hydrides, η^3-allyls, η^5-cyclohexadienyl complexes, and other low-oxidation-state intermediates for both stoichiometric and catalytic metal-mediated organic synthesis.[5,6] Current efforts to activate CO, CO_2, and other small molecules using catalytic cycles in homogeneous solution frequently employ nucleophilic addition strategies in conjunction with transition metal complexes as catalysts.[7,8] For example, the industrially important water–gas shift reaction has been shown to exhibit catalysis in aqueous alkaline solutions containing a wide variety of mono- and polynuclear metal carbonyls.[9-11] Furthermore, the Reppe reaction, the Wacker oxidation, and most homogeneous models for the

---

Robert R. Squires and Kelley R. Lane • Department of Chemistry, Purdue University, West Lafayette, Indiana 47907.

Fischer–Tropsch process all involve nucleophilic attack on a metal-coordinated ligand during one or more of the key steps in the proposed mechanisms.[5,12-14]

As a consequence of the widespread significance of nucleophilic addition reactions, numerous mechanistic, kinetic, thermochemical, and product-isolation studies have been carried out over the last few decades, and several extensive reviews of this chemistry have become available.[5,15-18] Two general classes of behavior have been recognized for nucleophilic addition reactions involving organometallic substrates: those proceeding by direct attack of the nucleophile on the metal center, and those occurring by nucleophilic addition to a coordinated ligand. Moreover, the selection among the possible sites of attack and the ultimate fate of the addition complexes that result are known to be sensitive functions of the nucleophile type, the nature of the metal and its ligands, and the specific solvents and counterions employed in these reactions.[19-21] The latter factor can be a particularly important influence on the outcome of nucleophilic addition reactions involving ionic reactants.[20]

It has been repeatedly demonstrated in the last 20 years that experimental studies of gas phase ion–molecule reactions can provide useful new information about the intrinsic behavior of ionic reactions that is uncomplicated by possible medium effects and counterion influences.[23] This has been particularly so for anionic nucleophilic addition and displacement reactions involving *organic* compounds such as methyl halides and carbonyl derivatives.[24-27] Recognizing that a similar potential exists for gas phase ion studies to provide new physical and mechanistic insights for organometallic nucleophilic addition reactions, we began a program in 1982 using the flowing afterglow method to study the reactions between negative ions and transition metal organometallic complexes. Our aim was to delineate the major mechanistic features, kinetics, and thermochemistry of bimolecular reactions between simple, volatile transition metal substrates and various anionic nucleophiles that had familiar analogues in solution. At the same time, we wished to explore the potential of the flowing afterglow technique for permitting the synthesis and characterization of reactive organometallic species of the type that were commonly proposed as transient intermediates in certain nucleophilic addition-based catalytic cycles, but were seldom directly observed.

This chapter presents an overview of our past and continuing work in this area. The flowing afterglow method is particularly well suited for the study of negative metal ion chemistry, and its basic operation and important advantages in this regard are summarized in Section 2. The reactions of bare and partially solvated negative ions with mononuclear metal carbonyls are described in Section 3, with particular emphasis on our more extensive work involving $Fe(CO)_5$. Anionic nucleophilic addition reactions involving transition metal complexes possessing arene, cyclopentadienyl, and isomeric C_4H_6 ligands are presented in Section 4. Finally, in Section 5 we describe the formation of certain metal anion complexes by gas phase nucleophilic addition reactions which are believed to be of importance in condensed phase catalytic cycles.

1.2. Previous Gas Phase Studies

Earlier reports of negative ion–molecule reactions with neutral transition metal complexes are few in number and are largely concerned with the condensation-type clustering reactions that take place with simple metal carbonyl complexes under relatively high pressure ion source conditions ($P \geq 10^{-6}$ torr). For example, Foster and Beauchamp reported the occurrence of reaction 1 in an ion cyclotron resonance (ICR) spectrometer operating with an $Fe(CO)_5$ pressure of 10^{-6} torr.[28]

$$Fe(CO)_3^- + Fe(CO)_5 \rightarrow Fe_2(CO)_6^- + 2CO \tag{1}$$

The $Fe(CO)_3^-$ reactant in this case was the minor fragment ion resulting from 70-eV electron impact on $Fe(CO)_5$. The major fragment ion, $Fe(CO)_4^-$, was initially reported to be unreactive; however, more recent studies by Wronka and Ridge have shown that it does react, albeit slowly[29]:

$$Fe(CO)_4^- + Fe(CO)_5 \rightarrow Fe_2(CO)_8^- + CO \tag{2}$$

Related anionic condensation reactions have also been reported for $Cr(CO)_6$, $Ni(CO)_4$, and $CpCo(CO)_2$, all under ICR conditions.

The first and, until recently, the only deliberate attempt to examine the gas phase reactions of independently generated negative ions with a neutral transition metal complex was reported by Foster and Beauchamp in 1971.[30] The negative ions F^-, $CH_3CH_2O^-$, CN^-, and Cl^- were allowed to react with $Fe(CO)_5$ at relatively low pressure in an ICR. Only F^- and $CH_3CH_2O^-$ were found to react:

$$F^- + Fe(CO)_5 \rightarrow (CO)_3FeF^- + 2CO \tag{3}$$

$$C_2H_5O^- + Fe(CO)_5 \rightarrow (CO)_3FeOC_2H_5^- + 2CO \tag{4}$$

It was noted at the time that this was probably a reflection of the fact that CN^- and Cl^- are much weaker bases. The metal ion products of Equations 3 and 4 were presumed to be four-coordinate, 16-electron complexes. Similar conclusions were reported later by Corderman and Beauchamp in an ICR investigation of the negative ion chemistry of $CpCo(CO)_2$.[31] Of the six negative ions examined in this study, CD_3O^-, F^-, CN^-, NO_2^-, Cl^-, and I^-, again only the stronger bases F^- and CD_3O^- were observed to react:

$$F^- + CpCo(CO)_2 \begin{cases} \xrightarrow{68\%} CpCo(CO)F^- + CO \\ \xrightarrow{32\%} CpCoF^- + 2CO \end{cases} \tag{5}$$

$$CD_3O^- + CpCo(CO)_2 \longrightarrow \begin{cases} \xrightarrow{84\%} CpCo(CO)OCD_3^- + CO \\ \xrightarrow{8\%} CpCoOCD_3^- + 2CO \\ \xrightarrow{8\%} CpCo(CO)_2OCD_3^- \end{cases} \quad (6)$$

Substituted, even-electron cyclopentadienyl cobalt structures, $CpCo(CO)_nX^-$, $n = 0, 1, 2,$ were suggested for the products, corresponding to 16- and 18-electron complexes for $n = 0$ and $n = 1$, respectively.

Kahn *et al.* recently measured the gas phase acidities of mixed chromium arenes $ArCr(C_6F_6)$, $Ar = C_6H_5CH_3$, m-$C_6H_4(CH_3)_2$, and p-$C_6H_4(CH_3)_2$.[32] Proton transfer equilibria were established in an ICR between the chromium arenes and a series of amines, thiols, and their conjugate base anions:

$$H_3C{-}C_6H_4{-}Cr(C_6F_6) + nPrS^- \rightleftharpoons H_2\bar{C}{-}C_6H_4{-}Cr(C_6F_6) + nPrSH \quad (7)$$

No reactions other than proton transfer were reported, and benzylic proton abstraction was assumed. The measured acidity enhancements in the chromium arenes relative to the uncomplexed hydrocarbons (22–23 kcal/mol) were found to be comparable to the acidity-enhancing effects accompanying *para*-NO_2 and *para*-CN substitution in toluene. An MO analysis of the charge distribution in $(CO)_3Cr(C_6H_5CH_3)$ and its deprotonated form showed that very little of the excess negative charge in the conjugate base is delocalized into the $Cr(CO)_3$ fragment, suggesting that the metal complexes incorporate true benzylic anions.

Sidorov and co-workers have made extensive use of Knudsen cell mass spectrometry to determine transition metal–fluoride bond energies, electron affinities, and other thermochemical data from analysis of negative ion-molecule equilibria at high temperatures.[33,34] For example, the saturated vapor above a binary mixture of RhF_3 and FeF_3 at 995 to 1064 K was found to contain RhF_4^- and FeF_4^- formed by reversible F^- transfer reactions.[35] Analysis of the temperature dependence of the equilibrium concentrations of the ions and neutral species along with the previously established thermochemical data for the FeF_3/FeF_4^- system[36] allowed determination of the heterolytic and homolytic Rh—F bond strengths and the electron affinity of RhF_4. A compilation of recent results from this group may be found in a recent review.[37]

Finally, Dillow and Gregor have recently discussed the analytical utility of Cl^- attachment to organometallic complexes in chemical ionization (NCI) sources.[38] A series of zinc(II) β-keto-enolates were examined under NCI conditions with a CF_2Cl_2 reagent gas, and abundant $M + Cl^-$ ions were observed. Attachment of the chloride nucleophile was assumed to occur directly at the metal. Considering the increasing number of NCI reagent ions which are now finding useful analytical applications (e.g., OH^-, O_2^-, F^-, NH_2^-), the potential for future developments in the analyses of organometallic compounds in this way appears excellent.[39]

2. THE FLOWING AFTERGLOW METHOD[40]

A simplified schematic diagram of the flowing afterglow apparatus used in our gas phase studies is shown in Figure 1. The system consists of an electron impact ion source coupled to the upstream end of a 100 cm × 7 cm i.d. stainless-steel flow reactor. Helium buffer gas enters the instrument near the electron-emitting filament through a dispersive inlet and flows at a high velocity through the tube as maintained by a Roots mechanical blower and mass-flow controller. Typical operating pressures and flow velocities range from 0.1 to 1.0 torr and 8000 to 9000 cm/s, respectively. Primary reactant anions are generated by electron capture and dissociative electron capture processes with any of a variety of precursor gases added near the filament. The ions are entrained in the flowing helium and carried out of the source region through the reactor. At these relatively high helium pressures, the ions are rapidly thermalized by ca. 10^7 collisions/s, and a room temperature Boltzmann distribution is achieved within 10 to 15 cm of the source. Ion–molecule reactions are initiated by adding volatile liquids, sublimable solids, or pure gases through a movable inlet or fixed-position inlets located along the length of the tube. The steady-state ion composition in the flow reactor is monitored by a quadrupole mass spectrometer located behind a sampling orifice in a nosecone at the downstream end of the tube.

Rate coefficients for bimolecular ion–molecule reactions are measured under pseudo-first-order conditions by monitoring the intensity of a reactive ion as a function of the neutral reagent concentration (flow rate) or the total reaction time. Since the bulk flow velocity in the reactor is maintained constant, the reaction time is directly proportional to the distance between the point of reagent entry into the tube and the sampling orifice. For most of the studies involving $Fe(CO)_5$, a movable injector was employed for rate measurements. The $Fe(CO)_5$ flow rate is measured by momentarily diverting the flow into a calibrated volume and measuring the pressure increase with time. Rate measurements using these methods are estimated to be accurate to within 20%, with a typical ±5% precision. Primary product distributions are determined by extrapolating the observed product ratios to zero reaction distance or zero neutral reactant flow rate, or from the slope of plots of normalized product ion abundances versus percent reactant ion conversion. Since branching ratios are quite sensitive to the mass discrimination which attends quadrupoles, we normally operate at as low a resolution as is permissible for a given experiment. The reported product ratios are estimated to be accurate to ±10%.

The flowing afterglow method possesses several important advantages for investigating the reactions of preformed negative ions with neutral organometallic compounds. Primarily, the flow reactor supports a variety of

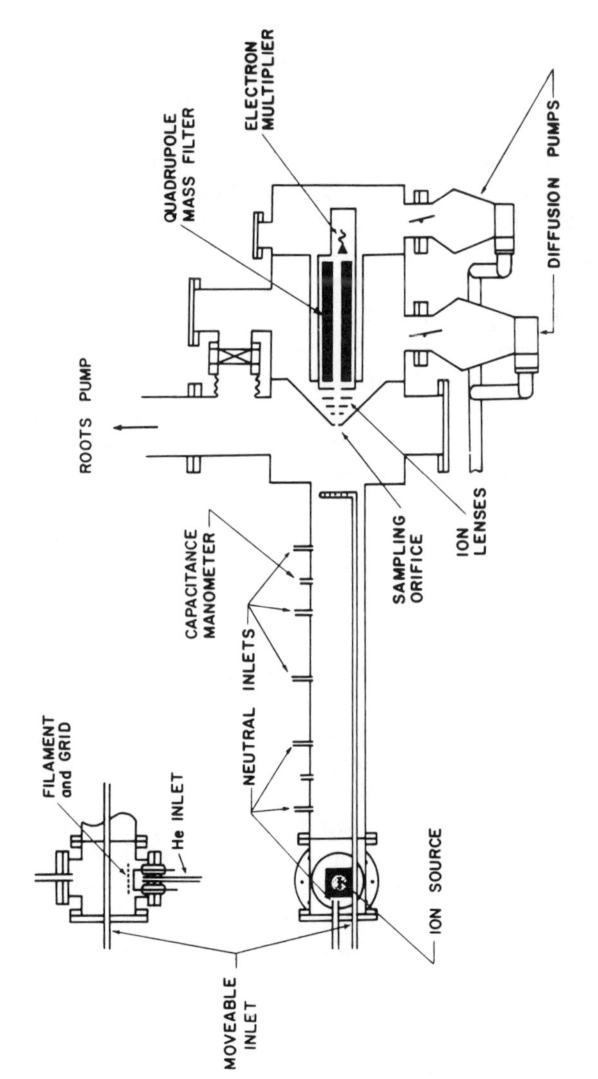

Figure 1. Simplified schematic diagram of the flowing afterglow apparatus used in these studies.

stepwise synthetic strategies for generating a desired reactant ion.[40] For example, NH_2^- and OH^-, the most commonly used primary reagent ions, can be conveniently formed in the source region by dissociative electron capture by NH_3 and by electron impact on an N_2O/CH_4 mixture, respectively. These strongly basic anions may then be employed as proton abstraction agents with hydrocarbons, alcohols, thiols, or any of a variety of acidic precursors.[41] Since several reagent inlet ports are available along the length of the tube, additional synthetic steps may be performed to produce the final reactant ion prior to the point of addition of the organometallic substrate. The advantage here is that the metal compound is added to the system at a point which is remote from the electron impact source, thereby removing the possibility of its ionization or pyrolysis on hot filaments. This feature facilitates identification of binary electron transfer reactions with negative ions which might otherwise be obscured by large signals due to the organometallic parent ion or its fragments. The relatively high pressure configuration of the flow reactor also provides important advantages. Not only do the reactant ions possess a well-defined Maxwell–Boltzmann energy distribution, but direct (termolecular) nucleophilic addition reactions may take place since collisional stabilization of long-lived adduct species in the bath gas can be rapid at 0.1–1.0 torr.[42] Normally, only bimolecular reactions can be observed in low-pressure experiments such as ICR, although addition reactions have been reported in a few instances.[31] The flowing afterglow is not without some disadvantages for the study of metal ion chemistry. Neutral organometallic reactants with relatively high vapor pressures (>0.1 torr) must be used. This problem can be overcome to a certain extent with use of a heated helium-flow inlet, but reliable measurements of neutral flow rates into the instrument are precluded. Furthermore, the quadrupole-based detector is limited to low resolution (ca. 1200 at 150 amu).

Complete details of most of the experiments described in the following sections may be found in the original papers. In general, the flow tube pressure, bulk flow velocity, and flow rate were maintained at 0.4 torr, 9000 cm/s, and 180 STP cm^3/s, respectively, corresponding to ion residence times in the flow tube of 10–70 ms.

3. MONONUCLEAR TRANSITION METAL CARBONYLS

3.1. $Fe(CO)_5$

3.1.1. General Reactions

Simple mononuclear metal carbonyls have served as prototype substrates for our initial studies of nucleophilic addition reactions in the gas phase, just as they have since the earliest studies of these reactions in solution. Because of its high volatility, relative stability, and ease of handling, $Fe(CO)_5$ has been the focus of our most extensive work, and much of our thinking about gas phase nucleophilic addition reactions directly stems from our experiences with this compound. Compared to the results from the earlier study of negative ion reactions with $Fe(CO)_5$ in an ICR, an enhanced reactivity is evident under flowing afterglow conditions.[43] Examination of a large series of negative ion reactions at 0.4 torr shows that, in addition to the double CO loss channels reported by Foster and Beauchamp[30] and illustrated by Equation 8a, negative ions may react by direct addition (Equation 8c), addition with loss of a single CO (Equation 8b), and a variety of other ion or electron transfer processes.

$$X^- + Fe(CO)_5 \longrightarrow \begin{cases} \xrightarrow{(a)} (CO)_3FeX^- + 2CO \\ \xrightarrow{(b)} (CO)_4FeX^- + CO \\ \xrightarrow{(c)} (CO)_4FeC(O)X^- \end{cases} \tag{8}$$

Table 1 presents a sampling of a few of these reactions, and a brief highlighting of some of the more interesting examples follows.

Strongly basic, localized nucleophiles such as OH^-, NH_2^-, and most alkoxide ions, RO^-, react mainly or exclusively via the path in Equation 8a

Table 1. Gas Phase Negative Ion Reactions of $Fe(CO)_5$ at 0.4 torr[a]

Anion	PA[b]	Primary CO expulsion products[c]			Other products
		$(CO)_3FeX^-$	$(CO)_4FeX^-$	$(CO)_4FeC(O)X^-$	
NH_2^-	403.6	+			
H^-	400.4	+	+		
$C_6H_5^-$	398.8	+	+		
$CH_2{=}CHCH_2^-$	391.3		+	+	$(CO)_4Fe^-$
OH^-	390.7	+			
CH_3O^-	381.4	+			$(CO)_4FeCHO^-$
$PhCH_2^-$	379.2	+	+	+	
$CH_3CH_2O^-$	376.1	+			$(CO)_4FeCHO^-$
HC_2^-	375.4	+	+	+	
CH_2CN^-	372.1	+	+	+	
F^-	371.3	+	+	+	
$CH_3C(O)CH_2^-$	368.8			+	
$CF_3CH_2O^-$	364.4	+			
HS^-	353.5			+	
HCO_2^-	345.2			+	$(CO)_4FeH^-$, $(CO)_4FeCHO^-$
N_3^-	344.6				$(CO)_2FeNCO^-$ $(CO)_3FeNCO^-$
Cl^-	333.4			+	

[a] Taken from Reference 43.
[b] Proton affinity in units of kcal/mol; from Reference 44.
[c] Plus sign indicates appearance of metal ion as a primary product.

to yield tricarbonylferrate product ions:

$$NH_2^- + Fe(CO)_5 \rightarrow (CO)_3FeNH_2^- + 2CO \tag{9}$$

The measured rates of reaction with these ions are generally quite high, approaching the computed ion–molecule collision rate limits (ca. 3×10^{-9} cm/s). Other strongly basic anions that do not possess a heteroatom at the nucleophilic site also react rapidly, but with a lesser extent of CO expulsion from the metal ion products. For example, hydride ion [proton affinity (PA) = 400.4 kcal/mol][44] and phenide ion (PA = 399 kcal/mol)[44] both react to produce tricarbonyl- and tetracarbonylferrate products,

$$C_6H_5^- + Fe(CO)_5 \longrightarrow \begin{cases} (CO)_3Fe(C_6H_5)^- + 2CO \\ (CO)_4Fe(C_6H_5)^- + CO \end{cases} \tag{10}$$

with the observed branching ratios exhibiting pressure dependence (Section 3.1.2). Allyl anion (PA = 391.3 kcal/mol)[45] yields an adduct, a monodecarbonylation product, and the dissociative charge-transfer product, $Fe(CO)_4^-$:

$$CH_2{=}CH{-}CH_2^- + Fe(CO)_5 \longrightarrow \begin{cases} Fe(CO)_4^- + CO + C_3H_5 \\ (CO)_4FeC(O)C_3H_5^- \\ (CO)_4Fe(C_3H_5)^- + CO \end{cases} \tag{11}$$

while benzyl anion, $PhCH_2^-$, and acetonitrile anion, CH_2CN^-, produce all three products. In most all instances where tricarbonylferrate ions appear as primary products, they are observed to undergo secondary clustering reactions with the excess $Fe(CO)_5$ in the reactor to yield diiron carbonyl ions from which zero, one, or two CO ligands are expelled:

$$X^- \xrightarrow[-2CO]{Fe(CO)_5} (CO)_3FeX^- \xrightarrow{Fe(CO)_5} Fe_2(CO)_n(X)^- \qquad n = 8, 7, 6 \tag{12}$$

In contrast, the tetracarbonyl product ions, $(CO)_4FeX^-$, and the adducts, $(CO)_4FeC(O)X^-$, are found to be completely inert towards secondary reactions with $Fe(CO)_5$, and towards further reactions with added ligands such as CO, PF_3, and NO. This reactivity pattern is consistent with an assignment of tetracoordinate, 16-electron iron structures for the tricarbonyl complexes (**I**) and 18-electron structures for the tetracarbonyl ions (**II**) and addition products (**III**).

The more weakly basic nucleophiles such as the enolates, halide ions, thiolates, and carboxylates react somewhat more slowly with $Fe(CO)_5$ by

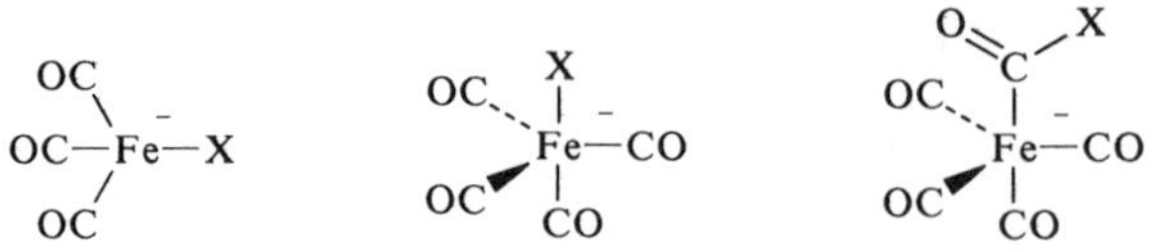

$$\underset{\mathbf{I}}{(OC)_3Fe^{-}\!-\!X} \qquad \underset{\mathbf{II}}{(OC)_4Fe^{-}\!-\!X} \qquad \underset{\mathbf{III}}{(OC)_4Fe^{-}\!-\!C(=O)X}$$

exclusive (termolecular) addition. This explains, in part, why the weaker-base anions (Cl^-, CN^-) examined by Foster and Beauchamp were found to be unreactive,[30] since at the reduced pressures in an ICR cell (10^{-6} torr), termolecular addition reactions are usually too slow to be observed. The addition products are idealized as tetracarbonylacylferrates (**III**) by analogy with nucleophilic addition products isolated from $Fe(CO)_5$ in solution.[46,47] Interestingly, the formate ion and certain other carboxylate ions that we have examined exhibit anion-transfer reactions in addition to direct association:

$$^{-}O\!-\!C(=O)\!-\!X + Fe(CO)_5 \rightarrow (CO)_4\bar{F}e\!-\!C(=O)\!-\!X + CO_2 \qquad X = H, CH_3O, HO \tag{13}$$

Azide ion exhibits surprising reactivity for such a weakly basic nucleophile:

$$N_3^- + Fe(CO)_5 \longrightarrow \begin{cases} (CO)_2FeNCO^- + N_2 + 2CO \\ (CO)_3FeNCO^- + N_2 + CO \end{cases} \tag{14}$$

While we cannot establish unequivocally that the observed product ions are not the isobaric azidoferrates [$(CO)_nFeN_3^-$, $n = 2, 3$] without benefit of labeling experiments, we prefer the cyanate ligand structures shown since there are well-established precedents for their formation from azide ion and metal carbonyls in solution,[48,49] and the loss of three CO ligands that is required to form $(CO)_2FeN_3^-$ seems unlikely. The mechanism is believed to involve initial addition of N_3^- to a CO ligand followed by intramolecular attack of the nucleophilic nitrogen on the carbon–metal bond with concomitant loss of N_2 (Scheme 1).[48] Subsequent expulsion of one or two CO ligands accompanies formation of the iron–nitrogen bond and conversion of the cyanate to a 3-electron donor ligand, viz., the final products are viewed as 16- and 18-electron complexes. The $(CO)_2FeNCO^-$ ion also undergoes secondary clustering with $Fe(CO)_5$ to form $Fe_2(CO)_6NCO^-$, while the $(CO)_3FeNCO^-$ ion appears to be unreactive. These observations provide additional support for the proposed ion structures.

3.1.2. Pressure Dependence and Mechanism

The observation of pressure-dependent product distributions for certain of the nucleophilic addition–decarbonylation reactions provides important

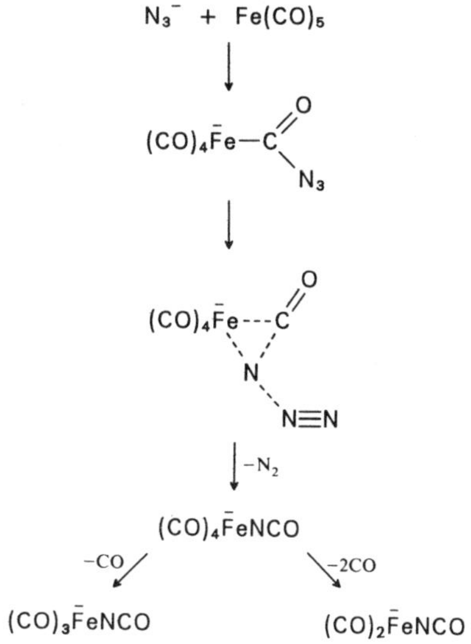

$$N_3^- + Fe(CO)_5 \rightarrow (CO)_4\bar{Fe}{-}C({=}O)N_3 \rightarrow (CO)_4\bar{Fe}{\cdots}C({=}O){\cdots}N{\cdots}N{\equiv}N \xrightarrow{-N_2} (CO)_4\bar{Fe}NCO$$

$$(CO)_4\bar{Fe}NCO \xrightarrow{-CO} (CO)_3\bar{Fe}NCO \qquad (CO)_4\bar{Fe}NCO \xrightarrow{-2CO} (CO)_2\bar{Fe}NCO$$

Scheme 1

clues about the general mechanism for these reactions.[43] Changes in the product ion ratios over a range of flow tube pressures from 0.1 to 1.0 torr are observed for most of the reactant ions that yield all three products, and for both the H^- and $C_6H_5^-$ reactants that produce tri- and tetracarbonylferrate products. No new products appear in either the OH^- or NH_2^- reactions at the highest reactor pressures that we can achieve. Figure 2 illustrates the observed behavior for $C_6H_5^-$ and CH_2CN^-. It can be seen that the extent of CO loss accompanying nucleophilic addition generally decreases with increasing flow tube pressure.

The simplest conclusions we can draw from the above behavior are that (1) a sequential fragmentation mechanism is operative, as opposed to a concerted displacement mechanism, and (2) the fragmenting intermediates possess CO-dissociation lifetimes near the metal ion-helium collision interval at 0.1-1.0 torr (ca. 10^{-7} s). A stepwise mechanism for nucleophilic addition/decarbonylation reactions which is consistent with these requirements is shown in Scheme 2. Initial nucleophilic attack occurs at a coordinated CO ligand to produce a tetracarbonylacylferrate anion complex **IV**[50] This intermediate possesses an amount of excess internal energy equal in magnitude to the negative ion binding energy of $Fe(CO)_5$, $D[(CO)_4FeCO{-}X^-]$ (Section 3.1.3). If this amount of energy is sufficiently large, then fragmentation of **IV** occurs by loss of a CO ligand from the metal to produce an initially coordinatively unsaturated tricarbonylacylferrate ion, **V**. Deactivation of the excited complex **IV** by collisions with helium takes place in competition with its fragmentation, with complete trapping of the adduct occurring if enough excess energy is removed by the bath gas. While an alternative structure for intermediate **V** as an 18-electron η^2-acyl complex (**Va**) is also conceivable,[5,51,52] both forms would be expected ultimately to succumb to a migratory-deinsertion-type

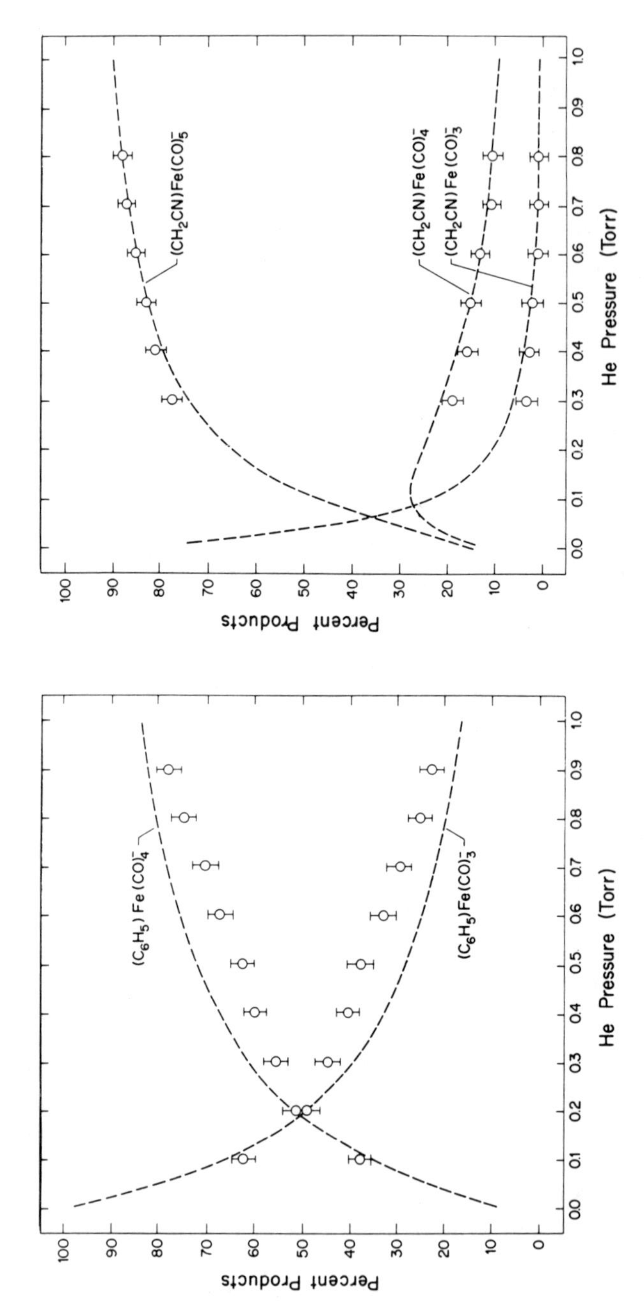

Figure 2. Plot of primary product ion yields versus helium pressure for the reaction between $Fe(CO)_5$ and (A) $C_6H_5^-$, (B) CH_2CN^-. The dashed lines represent the yield curves calculated using the model depicted in Scheme 3.

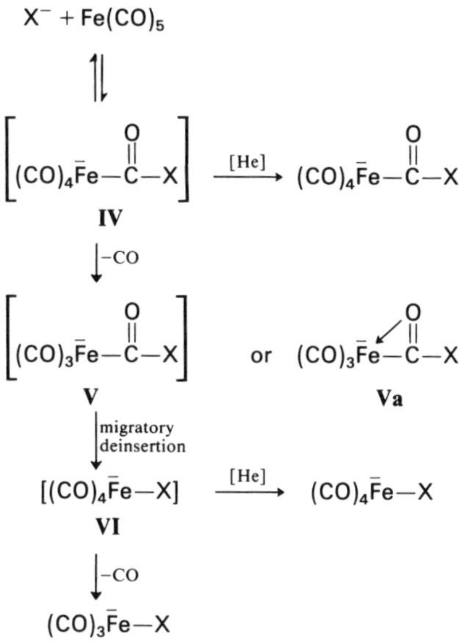

$$X^- + Fe(CO)_5$$

$$\rightleftharpoons$$

$$\left[(CO)_4\bar{F}e{-}\overset{\overset{\displaystyle O}{\|}}{C}{-}X\right] \xrightarrow{[He]} (CO)_4\bar{F}e{-}\overset{\overset{\displaystyle O}{\|}}{C}{-}X$$

IV

$$\downarrow -CO$$

$$\left[(CO)_3\bar{F}e{-}\overset{\overset{\displaystyle O}{\|}}{C}{-}X\right] \quad \text{or} \quad (CO)_3\bar{F}e{-}\overset{\overset{\displaystyle O}{\|}}{C}{-}X$$

V **Va**

$$\downarrow \text{migratory deinsertion}$$

$$[(CO)_4\bar{F}e{-}X] \xrightarrow{[He]} (CO)_4\bar{F}e{-}X$$

VI

$$\downarrow -CO$$

$$(CO)_3\bar{F}e{-}X$$

Scheme 2

rearrangement of the carbon-bonded group X to produce the tetracarbonylferrate complex **VI**.[53] If sufficient excess energy still remains in **VI** following decarbonylation and rearrangement, then a second CO dissociation may occur, again in competition with collisional deactivation. A key feature of this mechanism is that the $(CO)_4FeX^-$ product must arise from a termolecular channel, i.e., collisional stabilization of the metastable intermediate **VI** is obligatory for its formation. Good support for this hypothesis comes from the computed energy profiles for these reactions (Section 3.1.3), which show that if intermediate **VI** is ever formed by fragmentation and rearrangement from the initial nucleophilic addition, then it will necessarily possess sufficient internal energy to expel the second CO.

We can recast Scheme 2 in terms of the microscopic rate processes involved and derive estimates for the relative lifetimes of the fragmenting intermediates. Scheme 3 is a modified version of the general mechanism in which we have ignored intermediate **V**. This is done for economy and is justified if we make the reasonable assumption that all such species rapidly rearrange to complex **VI** and, therefore, do not enter into the kinetic expressions for product formation. In Scheme 3, $k_1(E)$ and $k_2(E)$ represent the energy-dependent unimolecular rate coefficients for loss of the first and second CO ligands, k_b is the rate of back-dissociation of **IV** to reactants, and k_L and k_s are the capture rate constants for collisions between the initial ion/molecule pair and subsequent collisions between He atoms and intermediates **IV** and **VI**, respectively. These rate constants may be reasonably estimated from Langevin theory.[54] The two β factors are inversely proportional to the number of He collisions required to quench CO dissociation, and they will generally differ since the excess internal energies of complexes **IV** and **VI** will differ. Thus, according to this

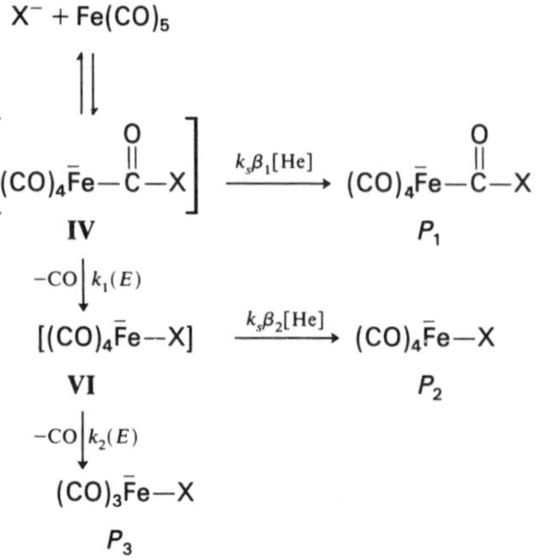

$$
\begin{array}{lcl}
X^- + Fe(CO)_5 & & \\
\updownarrow & & \\
\left[(CO)_4\bar{F}e{-}C(=O){-}X\right] \ \mathbf{IV} & \xrightarrow{k_s\beta_1[He]} & (CO)_4\bar{F}e{-}C(=O){-}X \ \ P_1 \\
\downarrow -CO,\ k_1(E) & & \\
\left[(CO)_4\bar{F}e{-}X\right] \ \mathbf{VI} & \xrightarrow{k_s\beta_2[He]} & (CO)_4\bar{F}e{-}X \ \ P_2 \\
\downarrow -CO,\ k_2(E) & & \\
(CO)_3\bar{F}e{-}X \ \ P_3 & &
\end{array}
$$

Scheme 3

kinetic model the relative yields of $(CO)_4FeC(O)X^-$ (P_1), $(CO)_4FeX^-$ (P_2), and $(CO)_3FeX^-$ (P_3) will depend upon the balance of the unimolecule rate coefficients k_1 and k_2, the magnitude of the excess internal energies of the intermediates, and the helium pressure.

Application of the steady-state approximation to Scheme 3 permits derivation of the following expressions for the individual product yields:

$$P_1/\Sigma P = k_s[\mathrm{He}]/(k_1/\beta_1 + k_s[\mathrm{He}]) \tag{15}$$

$$P_2/\Sigma P = (k_1\beta_1)k_s[\mathrm{He}]/(k_1/\beta_1 + k_s[\mathrm{He}])(k_2/\beta_2 + k_s[\mathrm{He}]) \tag{16}$$

$$P_3/\Sigma P = (k_1/\beta_1)(k_2/\beta_2)/(k_1/\beta_1 + k_s[\mathrm{He}])(k_2/\beta_2 + k_s[\mathrm{He}]) \tag{17}$$

It can be seen from these expressions that in the low-pressure limit ($[\mathrm{He}] = 0$), the yields of P_1 and P_2 are zero and only P_3 is formed, while in the high-pressure limit the P_1 yield approaches unity while those for P_2 and P_3 rapidly drop off. This is consistent with the observed behavior depicted in Figure 2, and with the differences noted between the present study and the earlier ICR work.[30] The yield equations may be used in conjunction with estimates for the individual rate constants derived from the experimental data in order to obtain limits on the dissociative lifetimes of the intermediates in Scheme 3. The reader is referred to the original paper for details concerning the estimation procedures.[43] In general, it is found that the initial adducts (**IV**) formed in the reactions with strongly basic nucleophiles such as H^-, $C_6H_5^-$, OH^-, and NH_2^- possess subnanosecond lifetimes towards CO loss (<50 ps), while for the weaker bases such as $PhCH_2^-$, CH_2CN^-, and F^-, intermediate **IV** persists nearly a microsecond. The estimated lower limits to the lifetimes for the excited tetracarbonyl intermediates vary from 0.1 μs to 2 μs, depending upon the particular nucleophile involved, with the F^- and CH_2CN^- complexes representing the shortest- and longest-lived intermediates, respectively.

3.1.3. Energetics

Evaluation of the relative energies for the separate steps in the proposed mechanism provides further insight into the origins of the variable CO-loss distributions accompanying nucleophilic attack. It is convenient to discuss the energetics in terms of the four reactions shown in Scheme 4, where the overall process given by their sum corresponds to Equation 8a. Table 2 provides a summary listing of selected values for certain of these quantities.

The energy released from attachment of a negative ion to $Fe(CO)_5$, ΔH_1, is a key parameter since it represents the amount of excess energy in intermediate **IV** that is initially available to drive the first CO loss. We have developed

$$X^- + Fe(CO)_5 \xrightarrow{\Delta H_1} (CO)_4FeC(O)X^-$$

$$(CO)_4FeC(O)X^- \xrightarrow{\Delta H_2} (CO)_3FeC(O)X^- + CO$$

$$(CO)_3FeC(O)X^- \xrightarrow{\Delta H_3} (CO)_4FeX^-$$

$$(CO)_4FeX^- \xrightarrow{\Delta H_4} (CO)_3FeX^- + CO$$

$$X^- + Fe(CO)_5 \xrightarrow{\Delta H_{TOT}} (CO)_3FeX^- + 2CO$$

Scheme 4

Table 2. Energetics for Nucleophilic Addition of Negative Ions with $Fe(CO)_5^a$

Anion	PA	$-\Delta H_1$ [b]	$D[(CO)_4Fe{-}X^-]$ [c]	$\Delta H(27)$ [d]	$\Delta H(28)$ [e]
NH_2^-	403.6	60.4	110.3	−8.4	−68.8
H^-	400.4	58.2	107.7	−8.0	−66.2
$C_6H_5^-$	398.8	57.2	106.4	−7.7	−64.9
$CH_2{=}CHCH_2^-$	391.3	52.4	100.6	−6.7	−59.1
OH^-	390.7	51.9	100.0	−6.6	−58.5
CH_3O^-	381.4	45.7	92.6	−5.4	−51.1
$PhCH_2^-$	379.2	44.4	91.0	−5.1	−49.5
$CH_3CH_2O^-$	376.1	42.2	88.4	−4.7	−46.9
HC_2^-	375.4	41.8	87.8	−4.5	−46.3
CH_2CN^-	372.1	39.6	85.2	−4.1	−43.7
F^-	371.3	39.0	84.5	−4.0	−43.0
$CH_3C(O)CH_2^-$	368.8	37.4	82.6	−3.7	−47.1
$CF_3CH_2O^-$	364.4	34.5	79.1	−3.1	−38.6
HS^-	353.5	27.3	74.4	−1.6	−28.9
HCO_2^-	345.2	21.8	63.8	−0.4	−22.2
N_3^-	344.6	21.0	62.8	−0.3	−21.3
Cl^-	333.4	14.0	54.4	1.1	−12.9

[a] All data in kcal/mol.
[b] Calculated with Equation 25.
[c] Calculated with Equation 26.
[d] CO-deinsertion enthalpy for tetracarbonylacylferrate ions given by $(\Delta H_1 + \Delta H_2)$ as defined in Scheme 4.
[e] Enthalpy of reaction producing $(CO)_4FeX^- + CO$ from $X^- + Fe(CO)_5$ given by $(\Delta H_1 + \Delta H_2 + \Delta H_3)$ as defined in Scheme 4.

means for experimentally determining certain of these binding energies and, in so doing, have discovered a useful thermochemical correlation that enables us to estimate values for a large number of other negative ions.[55] Binding energies ("anion affinities") of $Fe(CO)_5$, as defined in the following equation:

$$(CO)_4FeC(O)X^- \xrightarrow{D[(CO)_4FeCO-X^-]} Fe(CO)_5 + X^- \qquad (18)$$

may be determined in a number of different ways, including direct equilibrium methods, bracketing techniques, and thermodynamic cycles based on observed ion-molecule reactions. For instance, the fluoride and chloride affinities of $Fe(CO)_5$ can be readily bracketed by noting the direction of F^- and Cl^- transfer in a series of ion-molecule reactions involving reference fluoride or chloride donors and acceptors possessing established affinities. These reference data are available from previous studies of Cl^- and F^- binding to numerous Brønsted and Lewis acids by Larson and McMahon[56] and others.[57-59] As an example, $Fe(CO)_5$ is found to have a fluoride affinity close to that of PF_3; close enough, in fact, to permit direct assay of the equilibrium reaction shown below:

$$PF_4 + Fe(CO)_5 \underset{k_r}{\overset{k_f}{\rightleftharpoons}} PF_3 + (CO)_4FeC(O)F^- \qquad (19)$$

Measurements of the forward and reverse rate coefficients provide a value of $K(19)$ equal to 21.3 ± 1.0 and a corresponding free energy change of -1.8 kcal/mol. Using an estimate for the entropy change, $\Delta S(19)$, of 3.8 eu yields a value for the enthalpy of F^- transfer from PF_4^-, $\Delta H(19)$, of -0.7 kcal/mol. This may be combined with the absolute fluoride affinity of PF_3 (40.2 kcal/mol)[56] to give a final value for $D[(CO)_4FeCO—F^-]$ of 40.9 ± 3.0 kcal/mol. The H^- binding energy of $Fe(CO)_5$ is similarly bracketed to be 56 ± 4 kcal/mol from the observation of reactions involving reference hydride donor anions, as described in greater detail in Section 5.1.

Anion transfer to $Fe(CO)_5$ from negative ion clusters, $X(HX)_n^-$, also may be used to set limits on $(CO)_4FeCO—X^-$ bond strengths, provided the cluster ion possesses known or estimable thermochemistry. For example, the occurrence of reaction 20 for $n = 1$-3, reaction 21 for $n = 4$, and reaction 22 for $n = 0$

$$CH_3O(CH_3OH)_n^- + Fe(CO)_5 \rightarrow (CO)_4FeC(O)OCH_3^- + nCH_3OH \qquad (20)$$

$$OH(H_2O)_n^- + Fe(CO)_5 \rightarrow (CO)_4FeC(O)OH^- + nH_2O \qquad (21)$$

$$CH_3CO_2(CH_3CO_2H)_n^- + Fe(CO)_5 \rightarrow (CO)_4FeC(O)O_2CCH_3^- + nCH_3CO_2H \qquad (22)$$

but not for $n = 1$ sets the limits $D[(CO)_4FeCO—OCH_3^-] \geq 39.1$ kcal/mol, $D[(CO)_4FeCO—OH^-] \geq 53.1$ kcal/mol, and $D[(CO)_4FeC(O)—OAc^-] \leq 27.0$ kcal/mol.[55] Finally, limiting values of the binding energies for OH^-, HS^-, CH_3S^-, and other alkoxide ions can be derived from the occurrence or nonoccurrence of the protolytic exchange reaction involving Brõnsted acids and the $(CO)_4FeC(O)F^-$ ion shown in Equation 23.

$$(CO)_4FeC(O)F^- + HX \rightarrow (CO)_4FeC(O)X^- + HF \qquad X = OH, OR, SH, SR \tag{23}$$

This reaction appears to be general, occurring rapidly and completely when exothermic. Its utility for estimating $D[(CO)_4FeCO—X^-]$ values relies on the previously established value for X = F and a knowledge of the gas phase acidities of HX and HF.[44] The reaction in Equation 23 is observed to occur with HX = H_2S, CH_3SH, and CF_3CH_2OH, but not with simple aliphatic alcohols nor with water. Using Equation 23,

$$D[(CO)_4FeCO—X^-] \underset{\geq}{\overset{\leq}{}} \Delta H_{acid}(HX) - 330.4 \text{ kcal/mol} \tag{24}$$

(derived from an appropriate thermochemical cycle)in conjunction with the above results provides lower limits for $D[(CO)_4FeCO—X^-]$, $X^- = HS^-$, CH_3S^-, $CF_3CH_2O^-$, of 23.1, 28.6, and 34.0 kcal/mol, respectively. Upper limits are similarly inferred for the nonreactive systems.

Table 3 presents a summary listing of the experimentally measured $Fe(CO)_5$ anion affinities. The negative ions are listed in order of decreasing basicity (proton affinity), and a general correlation is evident. Figure 3 illustrates this relationship further in the form of a plot of $D[(CO)_4FeCO—X^-]$ versus $PA(X^-)$. The equation of the best straight line drawn through the bounded and half-bounded data is shown below:

$$D[(CO)_4FeCO—X^-] = 0.66 \times PA(X^-) - 206 \text{ kcal/mol} \tag{25}$$

Table 3. Experimentally Determined Negative Ion Binding Energies of $Fe(CO)_5$ [a]

X^-	$D[CO)_4FeCO—X^-]$ (kcal/mol)
H^-	56 ± 4
OH^-	$53.1 \leq x \leq 60.3$
MeO^-	$39.1 \leq x \leq 51.0$
EtO^-	≤ 45.7
n-PrO^-	≤ 44.3
n-BuO^-	≤ 43.6
t-BuO^-	≤ 42.9
F^-	40.9 ± 3.0
$CF_3CH_2O^-$	≥ 34.0
MeS^-	≥ 28.6
HS^-	≥ 23.1
$CH_3CO_2^-$	≤ 29.5
Cl^-	13.9 ± 3.0

[a] Taken from Reference 55.

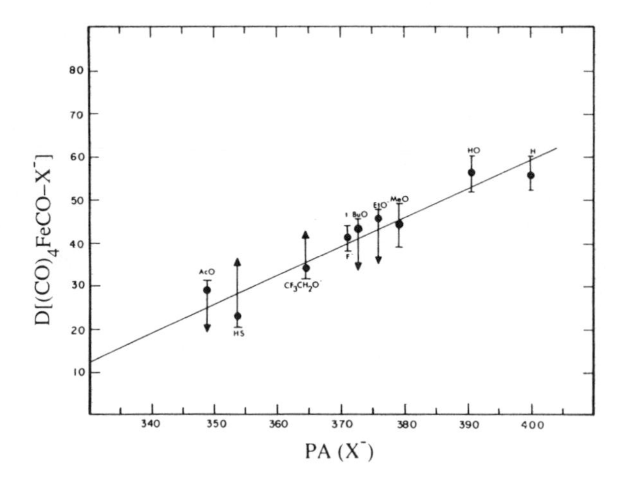

Figure 3. Plot of measured $Fe(CO)_5$ binding energies for negative ions (kcal/mol) versus proton affinities of the ions.

while the correlation is not perfect, the general trend is quite clear. Thus, within the accuracy of this relationship (±5 kcal/mol), the binding energy of a negative ion to $Fe(CO)_5$ can be predicted from its thermodynamic basicity. Selected values derived in this way are listed in the third column of Table 2.

Comparison of an expanded series of $Fe(CO)_5$ anion affinities with the corresponding nucleophilic addition product distributions shows that most all negative ions which bind to $Fe(CO)_5$ with an energy greater than ca. 34 kcal/mol always produce a product in which decarbonylation has occurred, while those with binding energies less than 34 kcal/mol yield only adduct. Therefore, we can use 34 kcal/mol as an empirical estimate of the average decarbonylation activation energy for $(CO)_4FeC(O)X^-$ ions. Furthermore, if we assume that CO loss from the acylferrate produces a $(CO)_3FeC(O)X^-$ species (or the η^2-acyl form) lying close in energy to the transition state for the reaction, then 34 kcal/mol represents an estimate for the average decarbonylation energy, ΔH_2. An estimate of this magnitude is consistent with the known Fe—CO bond energy in $Fe(CO)_5$ (41.2 kcal/mol),[60] and the known CO-labilizing influence of an η^1-acyl ligand.[61-64]

Evaluation of the CO-deinsertion energies for coordinatively unsaturated acylferrate complexes, ΔH_3, requires a knowledge of the heats of formation for $(CO)_4FeX^-$ species. We can use a similar approach to that adopted for the acyl ions by noting that a limited series of heterolytic bond dissociation energies are available for tetracarbonylferrate ions, $D[(CO)_4Fe—X]^-$, and that they too appear to be roughly correlated with the proton affinities of X^-.[43] Figure 4 shows a plot of a few $D[(CO)_4Fe—X^-]$ values taken from the literature[65] versus $PA(X^-)$, and Equation 26 gives the equation of the best straight line which fits these data.

$$D[(CO)_4Fe\text{-}X^-] = 0.796 \times PA(X^-) - 211 \text{ kcal/mol} \tag{26}$$

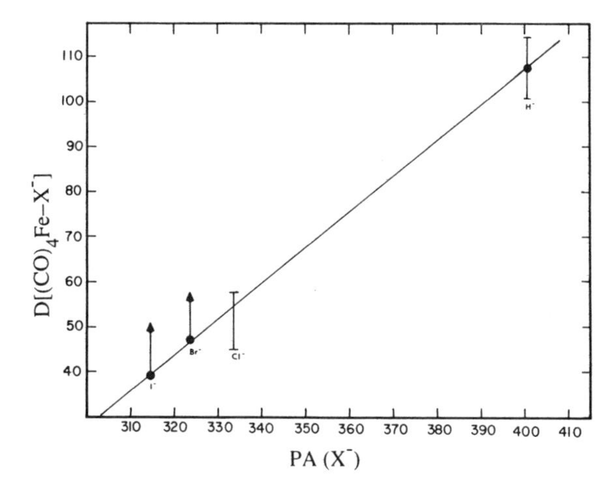

Figure 4. Plot of $Fe(CO)_4$ binding energies for negative ions (kcal/mol) versus proton affinities of the ions.

Thus, as with Equation 25, one may estimate the binding energy of a negative ion to a metal complex [in this case $Fe(CO)_4$] from its thermodynamic basicity. Representative bond energies derived in this way are given in the fourth column of Table 2. By combining these values with $\Delta H_f[\mathrm{Fe(CO)_4, g}] = -105.6$ kcal/mol[60] and corresponding $\Delta H_f[\mathrm{X^-, g}]$ values from gas phase acidity measurements,[44] we can obtain estimates of $\Delta H_f[\mathrm{(CO)_4FeX^-, g}]$ for any ion X^- which has a known proton affinity. With these heats of formation in hand, we may now compute values for ΔH_2 and, more importantly, the summation terms $(\Delta H_1 + \Delta H_2)$ and $(\Delta H_1 + \Delta H_2 + \Delta H_3)$. The former sum represents the CO-deinsertion enthalpy for the tetracarbonylacylferrate ions (Equation 27), and the latter gives the overall energy of displacement of one CO ligand by the X^- nucleophile (Equation 28).

$$\mathrm{(CO)_4\bar{F}e{-}\overset{\overset{O}{\|}}{C}{-}X} \xrightarrow{\Delta H_1+\Delta H_2} \mathrm{(CO)_4FeX^- + CO} \tag{27}$$

$$\mathrm{X^- + Fe(CO)_5} \xrightarrow{\Delta H_1+\Delta H_2+\Delta H_3} \mathrm{(CO)_4FeX^- + CO} \tag{28}$$

These quantities are tabulated in the fifth and sixth columns of Table 2.

The data needed to evaluate the energetics for the last step in Scheme 4, ΔH_4, are not presently available. However, we can make some qualitative conclusions about the Fe—CO bond strengths in $(CO)_4FeX^-$ ions from a consideration of the reactivity trends illustrated in Table 1, and the CO-labilization effects noted for related species in solution.[66] First of all, each of the reactive ions producing $(CO)_3FeX^-$ products does so rapidly. Therefore, bis-decarbonylation in these systems must be exothermic and the energy level for $(CO)_3FeX^- + 2CO$ must lie below that for the reactants. Moreover, since both F^- and $CF_3CH_2O^-$ produce tricarbonyliron product ions in our system, and F^- did so in the earlier ICR study,[30] then ΔH_4 for these ions must be less than 43 kcal/mol and 39 kcal/mol, respectively (Table 2, column six). An average value for ΔH_4 of $\leq$38 kcal/mol appears consistent with the results and is in accord with the expectation that the Fe—CO bond strengths in $(CO)_4FeX^-$ complexes should be less than that in $Fe(CO)_5$(41.5 kcal/mol)[60] due to the well-known CO labilizing influence of "anionic ligands" in metal carbonyls.[66]

With reasonable estimates of the energetics for the separate steps of our mechanism in hand, we can now construct semiquantitative energy profiles for the negative ion-$Fe(CO)_5$ condensation reactions. Energy level diagrams for the reactions of $Fe(CO)_5$ with $C_6H_5^-$, OH^-, CH_2CN^-, and Cl^- have been assembled from the data in Table 2, and these are illustrated in Figure 5. With the exception of the Cl^- profile, each diagram is characterized by a deep (covalent) energy well associated with initial negative ion attack on $Fe(CO)_5$ that produces an acylferrate intermediate **IV** with sufficient excess internal energy to overcome the approximately 34-kcal/mol barrier for loss of the first CO. In both the $C_6H_5^-$ (Figure 5A) and OH^- (Figure 5B) profiles, the CO-dissociation barriers are far below the total energy of the reactants, so this mode of fragmentation can proceed rapidly from the excited acylferrate ion intermediates **IV**. That is, the unimolecular dissociation lifetimes in these systems are relatively short since the excitation energy exceeds the reaction barrier. In contrast, the passage over the top of the barrier is fairly "narrow" in the CH_2CN^- profile (Figure 5C), so loss of the first CO is slower. As a result, these activated ions persist somewhat longer, and collisional stabilization to form an observable adduct becomes relatively favorable. The Cl^- profile represents the limiting case in which the barrier for decarbonylation exceeds the initial activation energy in intermediate **IV**. Therefore, only back-dissociation to reactants or collisional stabilization to form $(CO)_4FeC(O)Cl^-$ may occur, *despite* the fact that formation of $(CO)_4FeCl^-$ would be exothermic.

The migratory deinsertion step which follows (or accompanies) decarbonylation releases an average of about 45 kcal/mol (maximum) extra internal energy into the chemically activated intermediate, resulting in energy levels for $(CO)_4FeX^- + CO$ that are lower than those of the corresponding acylferrate intermediates by an amount of energy given by $\Delta H(27)$. In the absence of collisional stabilization by the bath gas, all such $[(CO)_4FeX^-]^*$ intermediates formed in this way will decarbonylate since their internal energy will generally exceed the Fe—CO bond energy, as argued earlier. This was evidently the case in the earlier ICR studies in which only $(CO)_3FeX^-$ products were observed.[30]

3.1.4. Reactivity Trends

It is instructuve to consider the trends in the nucleophilic addition product distributions for $Fe(CO)_5$ in simple terms which can be applied to other substrates. The physically meaningful factors that determine the outcome for

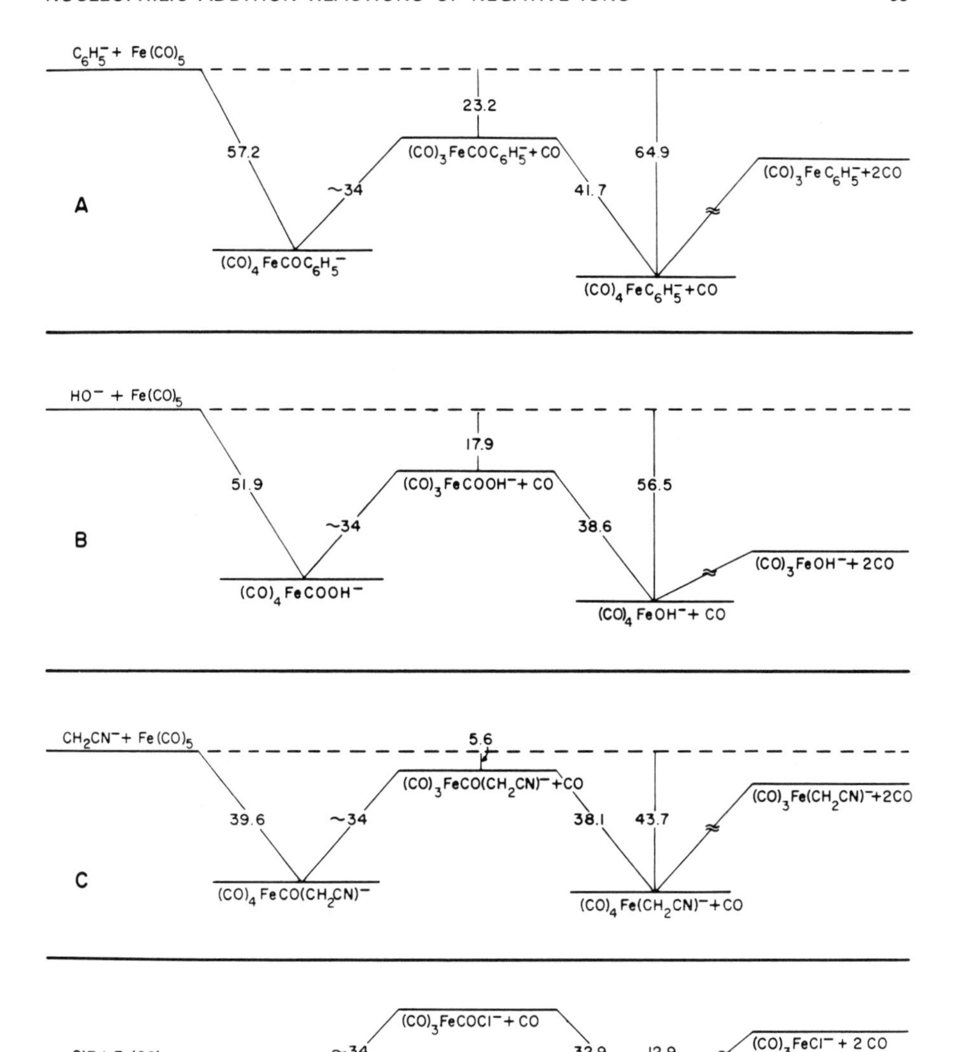

Figure 5. Semiquantative energy profiles for gas phase condensation reactions between negative ions and $Fe(CO)_5$.

a particular anion are the initial binding energy $D[(CO)_4FeCO—X^-]$, the relative lifetimes of the intermediates **IV–VI** towards fragmentation and collisional stabilization, and the overall exothermicity for formation of $(CO)_3FeX^-$.

An increased lifetime for **IV–VI** favors collisional stabilization, while fragmentation is facilitated by an increase in the overall reaction exothermicity.

We can identify certain properties of the reactant anions that can influence the reaction thermochemistry and the lifetimes and internal energies of the intermediates: anion basicity, the size of the anion, and the type of heteroatom at the nucleophilic site. It was shown earlier that the binding energy of an anion to $Fe(CO)_5$ is proportional to its basicity and, therefore, to the total excess internal energy in intermediate **IV**.[55] As the size of a reactant anion increases, the lifetime of adduct **IV** will generally increase due to the availability of a greater number of internal modes for excess energy dispersal. The lifetime toward fragmentation is prolonged since CO loss requires energy localization into specific dissociative modes. As a result, collisional stabilization becomes more likely. This size effect is clearly seen in the observed product distributions for a series of alkoxide ions with varying size, as described in the original paper.[43]

The type of heteroatom at the nucleophilic site in the anion can also influence the reaction energetics for both the rearrangement step (**V** → **VI**) and the ensuing CO loss from **VI**. Lone-pair-bearing heteroatoms such as O, N, or F will exert a (*cis*) CO-labilizing influence in **VI** that will be absent when X^- is a carbanion or hydride. This is due to both $d\pi$–$p\pi$ repulsive interactions between the lone-pair substituents and the *saturated* metal center in **VI**, as well as stabilization of the *unsaturated* metal center by these types of substituents once CO is expelled. The strong CO-labilizing effects of oxygen bases, halides, and other π-donors is well known in the condensed phase substitution chemistry of metal carbonyls.[66] An effect of this type satisfactorily accounts for the fact that all the localized oxyanions produce *only* $(CO)_3FeOR^-$ products, while H^- and carbanions of even greater basicities yield observable $(CO)_4FeR^-$ products. The low yield of $(CO)_4FeF^-$ produced by F^- may also be rationalized on this basis.

3.2. Group 5 and Group 6 Hexacarbonyls, $M(CO)_6$ (M = V, Cr, Mo, W)

Reactions between negative ions and a few other transition metal carbonyls have been examined, and a brief account of these unpublished studies is provided here.[67] The reactions observed with the group 6 hexacarbonyls are basically the same as those for $Fe(CO)_5$, but generally show a greater extent of CO expulsion from the product metal ions (Equation 29). The results for $Cr(CO)_6$ are typical. The strongly basic OH^- and NH_2^- ions displace both two and three CO ligands, resulting in 16- and 14-electron metal ion complexes, respectively (Equation 30).

$$X^- + M(CO)_6 \longrightarrow \begin{cases} (CO)_5MC(O)X^- \\ (CO)_nMX^- + (6-n)CO \quad n = 3\text{–}5 \end{cases} \quad M = \text{Cr, Mo, W} \tag{29}$$

$$HO^- + Cr(CO)_6 \longrightarrow \begin{cases} (CO)_4CrOH^- + 2CO \\ (CO)_3CrOH^- + 3CO \end{cases} \tag{30}$$

Delocalized ions with lower basicities such as $CH_2{=}CHCH_2^-$, $c\text{-}C_6H_7^-$, and $CH_3\bar{C}HCN$ produce mainly adduct, with lesser amounts of a monodecarbonylation product, $(CO)_5CrR^-$, and a bis-decarbonylation product $(CO)_4CrR^-$ in the case of the nitrile anion. Dissociative electron transfer also occurs with both hydrocarbon ions (Equation 31), along with H^- transfer from $c\text{-}C_6H_7^-$.

$$CH_2CHCH_2^- + Cr(CO)_6 \rightarrow Cr(CO)_5^- + C_3H_5 + CO \tag{31}$$

All three group 6 hexacarbonyls react with the bare H^- ion to yield exclusively the corresponding tetracarbonyl hydride, $(CO)_4MH^-$.

Experiments with $V(CO)_6$ are problematic because of the extreme reactivity of this 17-electron complex towards oxygen and other trace contaminants. However, in preliminary studies we have established that both OH^- and NH_2^- react with $V(CO)_6$ to yield exclusively the tricarbonyl products by displacement of three CO ligands:

$$X^- + V(CO)_6 \rightarrow (CO)_3VX^- + 3CO \qquad X^- = OH^-, NH_2^- \tag{32}$$

Both dissociative and nondissociative electron transfer processes are also evident with many types of reactant anions (Equation 33), although reliable characterization of these reactions is rendered difficult by the unavoidable presence of free electrons in the flow reactor.

$$\begin{aligned} X^- + V(CO)_6 &\rightarrow V(CO)_6^- + X \\ &\rightarrow V(CO)_5^- + CO + X \end{aligned} \tag{33}$$

As with $Fe(CO)_5$, a stepwise reaction mechanism involving initial attack at a carbonyl ligand followed by CO loss and migratory deinsertion of the nucleophile is considered likely for the group 6 complexes. The greater extent of CO loss compared to that observed for $Fe(CO)_5$ probably reflects the relatively weaker metal–carbonyl bonds in early transition metal carbonyl complexes,[60,,68,69] along with the greater M—X bond strengths expected for the group 5 and group 6 metals when X is an oxygen or nitrogen π-donor type ligand.[70] The observed product distributions from the reactions with OH^- shown in Table 4 support this hypothesis. While each of the group 6 hexacarbonyls produces both $(CO)_3MOH^-$ and $(CO)_4MOH^-$ products, the yields differ significantly, ranging from 95/5 for $Cr(CO)_6$ to 60/40 for $W(CO)_6$.

Table 4. Reactions of Hydroxide Ion with Metal Carbonyls

$M(CO)_n$	Primary products	Product distribution[a] (%)
$Fe(CO)_5$	$(CO)_3FeOH^-$	100
$Cr(CO)_6$	$(CO)_4CrOH^-$	5
	$(CO)_3CrOH^-$	95
$Mo(CO)_6$	$(CO)_4MoOH^-$	20
	$(CO)_3MoOH^-$	80
$W(CO)_6$	$(CO)_4WOH^-$	40
	$(CO)_3WOH^-$	60
$V(CO)_6$	$(CO)_3VOH^-$	100

[a] Determined at 0.4 torr, 298 ± 2 K.

The first (and average) metal–carbonyl bond energies in the neutral complexes are known to increase in the order Cr < Mo < W,[60,69] and, accordingly, relatively less bis-decarbonylation occurs for $W(CO)_6$ than for $Mo(CO)_6$ and $Cr(CO)_6$.

The displacement of three CO ligands by OH^- from $V(CO)_6$ is consistent with the weak metal-carbonyl bonds expected for this system.[68] However, the addition/decarbonylation mechanism may differ from that proposed for the other metal carbonyls since direct attack of the nucleophilic anion on the metal and/or single-electron transfer-mediated reactions are possible.[71]

3.3. Reactions with Partially Solvated Nucleophiles

The specific influence of solvation and counterions on anionic reactions of metal carbonyls in solution is not well defined at present, although some progress has been made recently.[20,72] In order to begin an assessment of the mechanistic and energetic consequences of solvation in these reactions, we have characterized the gas phase reactions of partially solvated negative ions with $Fe(CO)_5$.[73] Rates and product distributions have been determined for reactions involving over two dozen different homoconjugate clusters of the general formula $X(XH)_n^-$, where XH is water, ammonia, alcohols, thiols, and carboxylic acids. With few exceptions, the solvated ions react rapidly and completely with $Fe(CO)_5$ to yield exclusively the corresponding tetracarbonylacylferrate complexes:

$$X(XH)_n^- + Fe(CO)_5 \rightarrow (CO)_4FeC(O)X^- + nXH \tag{34}$$

Table 5 provides a partial listing of the data for a few representative systems. Discussion of the $HO(H_2O)_n^-$ cluster ion reactions and the $(CO)_4FeCO_2H^-$ product is deferred to the later section on catalysis intermediates (Section 5.1).

Several recurrent features can be noted in the data in Table 5 which provide clues about the origin of the differences between bare and solvated ion reactions. First of all, retention of solvent molecules by the product metal

Table 5. Reactions of Partially Solvated Anions with $Fe(CO)_5$ at 0.4 Torr[a]

Cluster ion	n	k_{obsd} (10^{-9} cm^3/s)	Efficiency[b]	Primary product ions
$HO(H_2O)_n^-$	0	2.1	0.69	$(CO)_3FeOH^-$
	1	2.3	1.0	$(CO)_4FeCO_2H^-$
	2	2.1	1.1	
	3	1.8	1.1	
	4	0.97	0.64	
$CH_3O(CH_3OH)_n^-$	0	1.8	0.76	$(CO)_3FeOCH_3^-$
	1	1.7	0.96	$(CO)_4FeCO_2CH_3^-$
	2	1.3	0.84	
	3	1.1	0.79	
$CF_3CH_2O(CF_3CH_2OH)_n^-$	0	1.3	0.86	$(CO)_3FeOCH_2CF_3^-$
	1	0.83	0.67	$(CO)_4FeCO_2CH_2CF_3^-$
	2	0.17	0.15	
$HS(H_2S)_n^-$	0	0.78	0.34	$(CO)_4FeC(O)SH^-$
	1	1.0	0.57	
$CH_3CO_2(CH_3CO_2H)_n^-$	0	0.14	0.08	$(CO)_4FeC(O)O_2CCH_3^-$
	1	<0.002		
	2	<0.002		
$NH_2(NH_3)_n^-$	0	2.3	0.72	$(CO)_3FeNH_2^-$
	1	2.1	0.90	$(CO)_3FeNH_2^-$ $(CO)_4FeNH_2^-$ $(CO)_4FeC(O)NH_2^-$
$OH(NH_3)_n^-$	0	2.1	0.69	$(CO)_3FeOH^-$
	1	1.8	0.80	$(CO)_3FeOH^-$ $(CO)_4FeCO_2H^-$ $(CO)_4FeH^-$ $(CO)_4FeOH^-$ $(CO)_4FeNH_2^-$ $(CO)_4FeC(O)NH_2^-$

[a] Taken from Reference 73.
[b] Efficiency = $k_{obsd}/k_{Langevin}$; Reference 54.

ion complexes is never observed. Formation of the acylferrate product is either accompanied by complete solvent expulsion or reaction does not occur at all [e.g., $CH_3CO_2(CH_3CO_2H)^-$]. Moreover, reactant ion solvation appears to completely suppress CO loss from the product in all cases except $NH_2(NH_3)^-$ and $HO(NH_3)^-$. The kinetic data show that addition of the first solvent molecule to an ion generally results in an *increase* in the efficiency of its reaction with $Fe(CO)_5$. For example, an approximately 30% increase occurs for $HO(H_2O)^-$ relative to OH^-, and ca. 20% increases are apparent for $CH_3O(CH_3OH)^-$, $HS(H_2S)^-$, and $NH_2(NH_3)^-$.

The behavior illustrated in Table 5 can be understood in terms of the generalized reaction mechanism shown in Scheme 5. Capture collision between

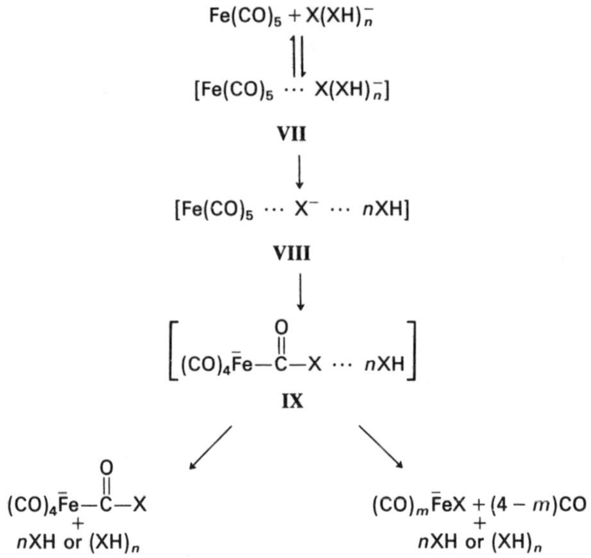

$$Fe(CO)_5 + X(XH)_n^-$$

$$\rightleftharpoons$$

$$[Fe(CO)_5 \cdots X(XH)_n^-]$$

VII

$$\downarrow$$

$$[Fe(CO)_5 \cdots X^- \cdots nXH]$$

VIII

$$\downarrow$$

$$\left[(CO)_4\bar{F}e{-}\overset{\overset{\displaystyle O}{\|}}{C}{-}X \cdots nXH\right]$$

IX

$$\swarrow \qquad \searrow$$

$$(CO)_4\bar{F}e{-}\overset{\overset{\displaystyle O}{\|}}{C}{-}X + nXH \text{ or } (XH)_n \qquad (CO)_m\bar{F}eX + (4-m)CO + nXH \text{ or } (XH)_n$$

Scheme 5

a cluster ion and $Fe(CO)_5$ initially produces an energy-rich ion-molecule complex (**VII**). The actual magnitude of the electrostatic interaction energy in the complex may be quite large since $Fe(CO)_5$ possesses a substantial bulk polarizability ($\alpha = 28\ \text{Å}^3$).[74] Energies ranging from 10 to 30 kcal/mol for ion-molecule distances of a few angstroms are not unlikely. As the excess energy distributes among the available rotational and vibrational modes of the complex, scission of the ion-solvent electrostatic bonds (H-bonds) may occur. As the anion emerges from its solvent cage (**VIII**), bonding interactions with a carbonyl ligand of $Fe(CO)_5$ develop that ultimately give rise to the acylferrate complex **IX**. If the binding energy of the negative ion to $Fe(CO)_5$ exceeds the total solvation energy in the cluster, then the exothermicity for ion transfer from the cluster to $Fe(CO)_5$ results in "boil-off" of solvent molecules.[75]

The complex absence of CO loss in most every case implies that the energy barrier for CO dissociation from the metal is greater than the barrier for solvent expulsion from the complex (*vide infra*). This is not surprising, since once the acylferrate anion is formed, electrostatic interactions involving the solvent molecules must be significantly attenuated due to the diffuse nature of the charge in the metal ion complex. Computed charge distributions for anionic metal acyls indicate substantial delocalization of negative charge from the metal to the acyl oxygen atom and, to a lesser extent, to the carbonyl oxygens.[50,76] Charge dispersal should lead to an overall weakening of metal-ion-solvent interactions compared to the hydrogen bonding interactions within $X(XH)_n^-$. In support of this, we find that preformed acylferrate complexes such as $(CO)_4Fe(O)X^-$, $X = OH$, OCH_3, F, do *not* undergo observable termolecular association reactions with added neutral solvents such as water or alcohols, even at the highest pressures (0.9 torr) and clustering neutral concentrations (ca. 5×10^{13} cm^{-3}) available to our instrument. Therefore, the metal-ion-solvent binding energies are probably quite small. In contrast, the first Fe—CO bond energy in the acylferrate anions is estimated to be nearly 34 kcal/mol, as determined from the results described in Section 3.1.3. Thus, any excess energy in complex **IX** is preferentially removed by cleavage of the weaker metal-ion-solvent bonds.

The observed increase in reaction efficiency for many of the monosolvated ions relative to the corresponding bare ion also supports this mechanism. Loss of two CO ligands is observed in the nude reactions involving OH^- and MeO^-. In contrast, the first cluster of each undergoes a *more efficient* reaction with $Fe(CO)_5$ in which only the solvent molecule is displaced. Thus, the barrier for solvent dissociation must be significantly less than the barrier for CO loss, since the latter is the rate-determining barrier in the nude reactions. The greater rate and efficiency observed with $HS(H_2S)^-$ is expected, since the nude reaction is a *termolecular* addition that is effectively converted to a *bimolecular* reaction by the presence of a "disposable" solvent molecule in the cluster. The transfer of the $CF_3CH_2O^-$ ion to $Fe(CO)_5$ from its first and second clusters must be approaching thermoneutrality, since a nearly sixfold drop in reaction efficiency occurs between the the bare ion and $CF_3CH_2O(CF_3CH_2OH)_2^-$.

The anion binding energies of $Fe(CO)_5$ established in Section 3.1.3. permit us to evaluate the energetic consequences of partial solvation of the reactant ion in Equation 34. The main thermochemical influence of reactant ion solvation in the present context is that it will act to decrease the energy of anion attachment to $Fe(CO)_5$ by an amount equal to the total solvation energy in the cluster. If the excess energy remaining from ion transfer between the cluster and $Fe(CO)_5$ exceeds the 34-kcal/mol estimate for the Fe—CO bond strength, then CO loss may accompany solvent boil-off from the product complex. If the overall exothermicity is less than 34 kcal/mol, then only $(CO)_4FeC(O)X^-$ will be produced. Moreover, if the reactant ion solvation energy exceeds the ion binding energy to $Fe(CO)_5$, then no reaction will occur, since solvation of the acylferrate product would provide little energy benefit to offset the endothermicity of ion transfer from the cluster. Therefore, with a knowledge of ion solvation energies, anion affinities of $Fe(CO)_5$ (Equation 25), and the nominal 34-kcal/mol decarbonylation barrier for an acylferrate ion, we may rationalize the observed cluster ion reactivity patterns shown in Table 5.

Table 6 provides a summary of the calculated thermochemistry for four representative cluster ion reactions for which the relevant solvation data are known, or estimable.[77] Calculated enthalpies, entropies, and free energies for Equation 34 are tabulated for each cluster in the last three columns. Discussion of the mechanistic implications of the thermochemical data for the higher methanol and water clusters ($n = 3, 4$) is beyond the scope of this chapter, and the reader is referred to the original paper[73] and related works.[78] However, it is apparent from the data in Table 6 that the presence of the first solvent molecule in $HO(H_2O)_n^-$, and $CH_3O(CH_3OH)_n^-$ drops the overall exothermicity [$\Delta G(34)$ and $\Delta H(34)$] for OH^- or CH_3O^- attachment to

Table 6. Thermochemical Data for Homoconjugate Cluster Ion Reactions with $Fe(CO)_5$ at 298 K[a]

Cluster ion $X(XH)_n^-$	n	$-\Delta H_{0,n}$ [b]	$-\Delta S_{0,n}$ [b]	$D[(CO)_4FeCO{-}X^-]$ [c]	$\Delta H(34)$	$\Delta S(34)$ [d]	$\Delta G(34)$
$NH_2(NH_3)_n^-$	0	—	—	60.4	−60.4	−31.2	−51.1
	1	11.9[e]	20.0[f]		−48.5	−11.2	−45.1
$OH(H_2O)_n^-$	0	—	—	$53.1 \le x \le 60.3$	−53.1	−32.8	−43.3
	1	25.0	20.8		−28.3	−12.0	−24.7
	2	42.9	42.0		−10.4	9.2	−13.1
	3	58.0	66.8		4.7	34.0	−5.4
	4	72.2	96.3		18.9	63.5	≤0.0
$CH_3O(CH_3OH)_n^-$	0	—	—	45.7	−45.7	−32.2	−36.1
	1	21.8	21.8		−23.9	−10.4	−20.8
	2	(37.4)	(44.0)		−8.3	11.8	−11.8
	3	(50.6)	(70.0)		4.9	37.8	−6.4
$CH_3CO_2(CH_3CO_2H)_n^-$	0	—	—	24.0	−24.0	−34.6	−13.7
	1	27.0	26.2		3.0	−8.4	5.5

[a] Energies in units of kcal/mol; entropies in eu.
[b] Solvation enthalpies and entropies taken from Reference 77. Estimated values in parentheses; see Reference 73.
[c] Estimated with Equation 5; see Reference 55. The values for OH^- are the experimentally determined limits.
[d] Entropy changes for anion attachment to $Fe(CO)_5$ estimated from $\Delta S_{trans} + \Delta S_{rot}$: NH_2^- (−31.2 eu), OH^- (−32.8 eu), CH_3O^- (−32.2 eu), $CH_3CO_2^-$ (−34.6 eu). $\Delta S(34)$ is given by the difference between these values and the corresponding $\Delta S_{0,n}$ in the fourth column.
[e] *Ab initio* HF/4-31 + G//4-31G level, Reference 73.
[f] Estimate based on known values for other homoconjugate clusters.

$Fe(CO)_5$ to a value below the estimated 34-kcal/mol decarbonylation barrier. As a result, no decarbonylation products are observed. In contrast, for the $NH_2(NH_3)^-$ cluster $\Delta G(34)$ and $\Delta H(34)$ *exceed* the decarbonylation barrier by more than 10 kcal/mol. This is consistent with the appearance of $(CO)_3FeNH_2^-$ and $(CO)_4FeNH_2^-$ in addition to $(CO)_4FeC(O)NH_2^-$ as primary products from this reaction.

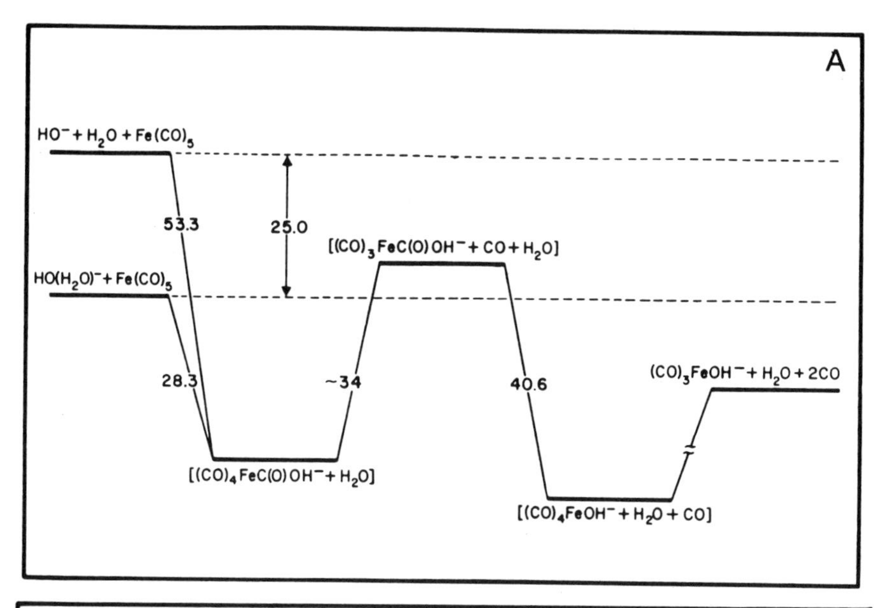

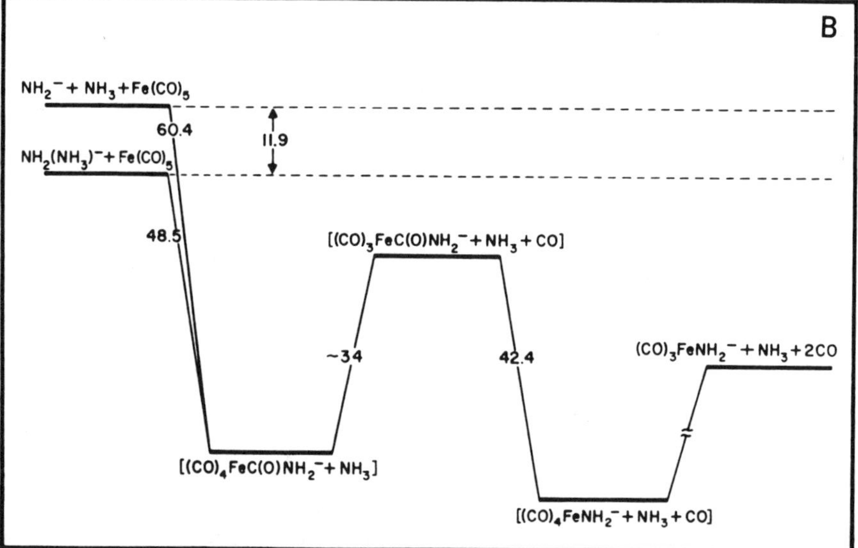

Figure 6. Semiquantitative energy profiles for reactions of $Fe(CO)_5$ with bare and monosolvated negative ions. (A) The $HO(H_2O)^-$ ion yields only the OH^- transfer product. (B) Decarbonylation accompanies NH_2^- transfer from $NH_2(NH_3)^-$.

We can best illustrate the consequences of reactant ion monosolvation on the product distributions with semiquantitative energy profiles similar to those presented earlier for the bare ion reactions. Figure 6 shows two such profiles for the $HO(H_2O)^-$ and $NH_2(NH_3)^-$ reactions. Here, the energy profiles are formulated in terms of enthalpy since the nominal 34-kcal/mol decarbonylation barrier for acylferrate ions is formally an enthalpy term and, furthermore, $\Delta G(34)$ and $\Delta H(34)$ for $HO(H_2O)^-$ and other *monosolvated* ions are quite similar (Table 6). In constructing these profiles, the interaction energies between the neutral solvent molecules (H_2O or NH_3) and each of the intermediate metal ion complexes are presumed to be negligiblly small. The bare OH^- ion energy profile (Figure 6A) is identical to the one in Figure 5 except for inclusion of a free H_2O molecule for internal consistency. It is evident in the figure that association of OH^- with H_2O drops the total energy of the reactants below the top of the barrier for CO loss from the initially formed hydroxycarbonyl adduct by approximately 6 kcal/mol. Therefore, CO loss does not occur. In Figure 6B, the lesser solvation energy in $NH_2(NH_3)^-$ and the stronger binding energy between NH_2^- and $Fe(CO)_5$ combine to produce a total energy for the reactants which is roughly 15 kcal/mol *greater* than the barrier for decarbonylation of the $(CO)_4FeC(O)NH_2^-$ intermediate. Accordingly, $(CO)_4FeNH_2^-$ and $(CO)_3FeNH_2^-$ are observed as products. The appearance of $(CO)_4FeNH_2^-$ is especially interesting since it did not appear at all in the bare NH_2^- ion reaction. Evidently, its formation in the $NH_2(NH_3)^-$ ion reaction with approximately 12 kcal/mol *less* excess internal energy than in the bare ion reaction increases its dissociation lifetime sufficiently to permit collisional stabilization and, thus, its observation as a product.

In concluding this section, we note that the presence of even a single solvent molecule can transform otherwise "exotic-looking" gas phase reactions of negative ions with $Fe(CO)_5$ into familiar processes which appear to be more closely related to the corresponding ionic reactions in solution. Just as the medium acts to rapidly dissipate heat in an exothermic condensed phase reaction, so the "disposable" solvent molecules in these gas phase cluster-ion reactions may also remove excess energy from the products and thereby inhibit fragmentation.

4. TRANSITION METAL ARENE, CYCLOPENTADIENYL, AND DIENE COMPLEXES

4.1. $(\eta^6\text{-}C_6H_6)Cr(CO)_3$

4.1.1. General Reactions

We have examined the reactions of a variety of simple negative ions with benzene chromium tricarbonyl as a prototype coordinatively saturated transition metal arene complex.[67,79] Kinetic measurements for these reactions are precluded because of the low volatility of $(C_6H_6)Cr(CO)_3$. However, qualitative observations of the products showed several interesting features. The particular reaction with H^- proved to be especially interesting, and we have carried out more detailed mechanistic studies for this system as outlined in Section 4.1.2.

The reactions of $(\eta^6\text{-}C_6H_6)Cr(CO)_3$ with negative ions differ from those of the metal carbonyls somewhat in that only proton transfer and/or adduct formation are observed, with no CO-displacement products appearing at all in the product ion mass spectra[67]:

$$X^- + (\eta^6\text{-}C_6H_6)Cr(CO)_3 \begin{cases} \longrightarrow (\eta\text{-}C_6H_5)Cr(CO)_3^- + HX \\ \longrightarrow [(\eta\text{-}C_6H_6)Cr(CO)_3 \cdot X^-] \end{cases} \tag{35}$$

Proton abstraction is observed for NH_2^-, H^-, OH^-, $CH_2{=}CHCH_2^-$, MeO^-, $^iPrO^-$, and $c\text{-}C_6H_7^-$, but not with $CH_3C(O)CH_2^-$ or anions with lesser basicities. An apparent gas phase acidity for $(\eta^6\text{-}C_6H_6)Cr(CO)_3$ of 370 ± 5 kcal/mol may be assigned from these results, corresponding to an approximately 25- to 30-kcal/mol acidity enhancement compared to free, uncomplexed benzene $[\Delta H_{acid}(C_6H_6) = 399\ \text{kcal/mol}]$.[44] An enhancement of this magnitude is consistent with the effects of metal coordination on the acidities of toluenes noted by Kahn *et al.* (Equation 7).[32]

Two likely candidates exist for the structure of the $(M{-}H)^-$ ion: an η^6-phenide type structure in which the ligand remains fully π-bonded (**X**), or a rearranged structure involving a σ-phenyl ligand (**XI**):

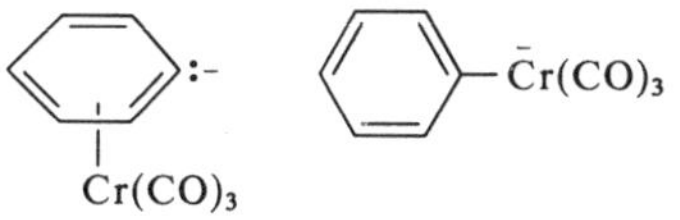

:- $Cr(CO)_3$ — **X**

$\bar{C}r(CO)_3$ — **XI**

It is noteworthy in this regard that the $(M{-}H)^-$ ion undergoes complete H/D exchange with added D_2O[80] and forms an adduct with CO_2, but not CO. This behavior is more consistent with the π-arene structure **X** since the formally 14-electron σ-phenyl complex would be expected to add both CO and CO_2, but not undergo H/D exchange.[81] In an attempt to generate an authentic σ-phenyl chromium carbonyl complex, we allowed free phenide ion (formed from reaction between NH_2^- and benzene) to react with $Cr(CO)_6$:

$$C_6H_5^- + Cr(CO)_6 \begin{cases} \longrightarrow (C_6H_5)Cr(CO)_4^- + 2CO \\ \longrightarrow (C_6H_5)Cr(CO)_5^- + CO \end{cases} \tag{36}$$

Roughly equal amounts of mono- and bis-decarbonylation products are produced, neither of which exhibit any H/D exchange in the presence of D_2O. We therefore conclude that the $(M-H)^-$ ion exists as a π-type complex, i.e., $(\eta^6\text{-}C_6H_5)Cr(CO)_3^-$.

Adduct ions are formed from benzene chromium tricarbonyl with $CH_2{=}CHCH_2^-$, $c\text{-}C_6H_7^-$, $HC{\equiv}C^-$, and $CH_3C(O)CH_2^-$. Addition of the nucleophile to the benzene ring, to a carbonyl ligand, or directly to the metal are all conceivable. However, based on our experiences with the H^- reactions (*vide infra*) and the complete absence of any CO displacement, we presently favor structures wherein the negative ions are attached to the *exo* face of the benzene ring (**XII**). Ring addition is frequently noted as the preferred mode of attack in nucleophilic reactions of $(\eta^6\text{-}C_6H_6)Cr(CO)_3$ in solution.[17,82,83]

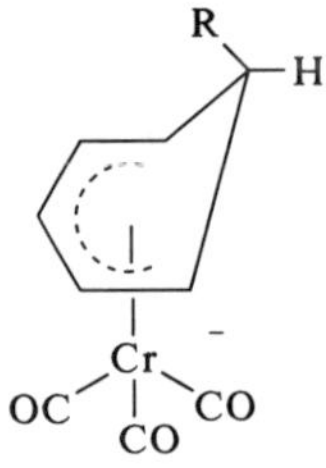

R
H
Cr
OC CO
CO
–

XII

4.1.2. Reaction with Hydride Ion

The strongly basic hydride ion $[PA(H^-) = 400\ \text{kcal/mol}]$[44] reacts in an unusual way with $(\eta^6\text{-}C_6H_6)Cr(CO)_3$ in that benzene displacement occurs in addition to proton transfer[79]:

$$H^- + (C_6H_6)Cr(CO)_3 \begin{cases} \longrightarrow (C_6H_5)Cr(CO)_3^- + H_2 \\ \longrightarrow HCr(CO)_3^- + C_6H_6 \end{cases} \quad (37)$$

The reaction proceeds rapidly and yields the two product ions in roughly equal amounts. Benzene displacement is completely unique to the H^- ion, and from the occurrence of this reaction we can infer that $D[(CO)_3Cr-H^-] > D[(CO)_3Cr-(C_6H_6)] \simeq 31\ \text{kcal/mol}$.[69,84]

Several distinct mechanistic possibilities exist for this reaction.[85-88] Direct H^- addition to the metal requires slippage of the η^6-benzene ligand to η^4 (**XIII**); subsequent or simultaneous expulsion of the hydrocarbon from the energy-rich adduct may then produce $HCr(CO)_3^-$. Alternatively, initial H^- addition to the benzene ring (**XIV**) or a carbonyl ligand (**XV**) may occur followed by intramolecular migration of hydrogen to the metal with accompanying benzene loss.

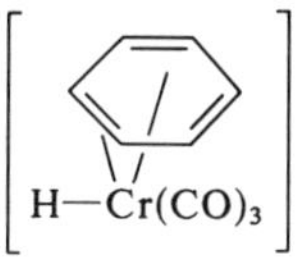

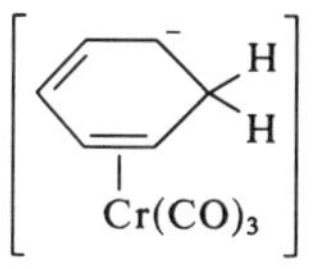

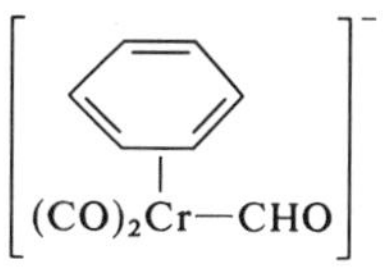

$[H{-}Cr(CO)_3(C_6H_6)]^-$ **XIII** $[(C_6H_6)Cr(CO)_3]^-$ (H, H) **XIV** $[(C_6H_6)(CO)_2Cr{-}CHO]^-$ **XV**

In order to distinguish among these possibilities, we determined the product distribution that results when D^- is the attacking nucleophile. The rationale here is that if either intermediate **XIII** or **XV** is formed exclusively, then D^- will produce only $DCr(CO)_3^-$. However, if D^- first adds to the ring, then hydrogen and deuterium may, in principle, become scrambled in the product. Moreover, if ring addition were to be followed by exclusive migration of an *endo* hydrogen to the metal (such as is the case in solution),[85-88] then *exo*-addition of D^- would yield only $HCr(CO)_3^-$. The experiment yields both $HCr(CO)_3^-$ and $DCr(CO)_3^-$ in a 7:1 ratio:

$$D^- + (C_6H_6)Cr(CO)_3 \longrightarrow \begin{cases} (C_6H_5)Cr(CO)_3^- + HD \\ HCr(CO)_3^- + C_6H_5D \\ DCr(CO)_3^- + C_6H_6 \end{cases} \tag{38}$$

No free H^- is produced as a product at low conversions of D^-, nor does $HCr(CO)_3^-$ undergo H/D exchange with the ND_3 which is present in the flow reactor as a precusor for D^-.

The appearance of $HCr(CO)_3^-$ from the D^- reaction requires that ring attachment makes a significant contribution to the mechanism. One way to interpret the fact that the observed 7:1 product ratio is close to the 6:1 statistical ratio for the system is that complete H/D scrambling occurs within the ion-molecule collision complex prior to expulsion of benzene. A mechanism that is consistent with this view is shown in Scheme 6. Initial ring addition by D^- produces a vibrationally and rotationally excited cyclohexadienyl anion complex that can undergo rapid hydrogen shifts in either a thermally allowed sigmatropic manner[89] or with mediation by the $Cr(CO)_3$ fragment. This latter possibility seems unlikely, however, since we have found that $HCr(CO)_3^-$ does *not* undergo H/D exchange in the presence of C_6D_6, although an adduct readily forms by termolecular association:

$$HCr(CO)_3^- + C_6D_6 \longrightarrow \begin{cases} HCr(CO)_3(C_6D_6)^- \\ DCr(CO)_3^- + C_6H_5D \end{cases} \tag{39}$$

This implies that a barrier exists for hydride migration from chromium to the arene ligand and, by inference, that once the metal-hydrogen bond is formed,

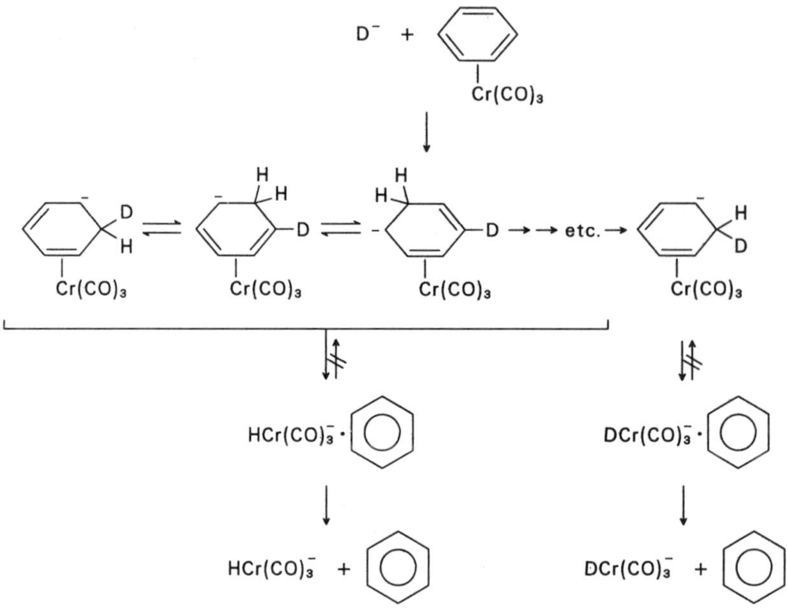

Scheme 6

expulsion of benzene follows. Thus, H/D scrambling may proceed entirely within the benzene moiety prior to a slower chromium–hydride bond formation step which is accompanied by loss of neutral benzene.

An alternative possibility not mentioned in our preliminary report[79] is for direct attack of D^- at the metal (or CO) to occur one out of every eight encounters, yielding only $DCr(CO)_3^-$, while most of the time *exo*-addition of D^- to the ring occurs followed by exclusive migration of an *endo* hydrogen to the metal to form $HCr(CO)_3^-$. Appropriate experiments to distinguish between these possibilities are now in progress. A significant finding in this regard is that an H^- adduct of $(\eta^6\text{-}C_6H_6)Cr(CO)_3$ can be generated by hydride transfer from $c\text{-}C_6H_7^-$ or $Et_3SiH_2^-$.[90] Determination of the disposition of hydrogen and deuterium in the analogous D^- adduct should be most illuminating. In any event, it is clear from the above results that nucleophilic reactions of $(C_6H_6)Cr(CO)_3$ with H^- in the gas phase can proceed by attack at the benzene ring, just as they do in solution.[82,83]

4.2. $(\eta^5\text{-}C_5H_5)Mn(CO)_3$

The gas phase reactions of cyclopentadienyl manganese tricarbonyl [$CpMn(CO)_3$] with negative ions are generally quite similar to those of $(C_6H_6)Cr(CO)_3$, with the exception of the reaction with H^-.[67] Adducts of undetermined structure are produced with allylic and dienylic carbanions, alkoxide ions, and enolates. Proton abstraction from $CpMn(CO)_3$ occurs with NH_2^-, H^-, OH^-, $CH_3CH{=}CHCH_2^-$, MeO^-, and $^i PrO^-$, but not with $c\text{-}C_6H_7^-$

or $CH_3C(O)CH_2^-$. An acidity estimate of 372 ± 5 kcal/mol is derived from these results, which is essentially the same as that for $(C_6H_6)Cr(CO)_3$. Since the gas phase acidity of the cyclopentadienyl radical has been estimated by McDonald and co-workers to be 377 ± 3 kcal/mol,[91] then we may conclude that a significantly smaller acidity enhancement accompanies metal coordination of the organic ligand in $CpMn(CO)_3$ compared to $(C_6H_6)Cr(CO)_3$.

The reaction between H^- and $CpMn(CO)_3$ provides an interesting contrast to $(C_6H_6)Cr(CO)_3$. The formation of both an adduct and a decarbonylation product is observed in addition to proton transfer:

$$H^- + (C_5H_5)Mn(CO)_3 \longrightarrow \begin{cases} (C_5H_4)Mn(CO)_3^- + H_2 \\ (C_5H_5)Mn(CO)_3(H)^- \\ (C_5H_5)Mn(CO)_2(H)^- + CO \end{cases} \tag{40}$$

The fact that CO is displaced instead of the hydrocarbon ligand, whereas the opposite is true in the chromium system, probably reflects the lesser relative stability of C_5H_5 compared to C_6H_6 as a free ligand and, hence, its greater binding energy to the metal. The location(s) of the "extra" hydrogen in the adduct and decarbonylation products are unknown at present.

4.3. *Isomeric $(C_4H_6)Fe(CO)_3$ Complexes*

Comparison of the reactions of butadiene iron tricarbonyl [(BUD)Fe(CO)$_3$] and trimethylenemethane iron tricarbonyl [(TMM)Fe(CO)$_3$] with negative ions provides additional support for the suggestion made above that a closed-shell, even-electron organic ligand is more easily displaced in a nucleophilic addition reaction than an open-shell, radical-type ligand. In

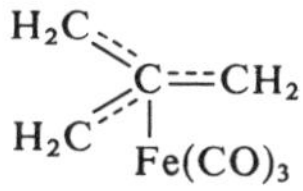

H_2C, $C{=}CH_2$, H_2C, $Fe(CO)_3$ — (TMM)Fe(CO)$_3$

$CH{=}CH_2$, HC, CH_2, $Fe(CO)_3$ 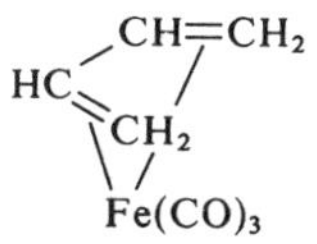 — (BUD)Fe(CO)$_3$

general, negative ions react with the two complexes by proton abstraction, electron transfer, and/or nucleophilic addition/fragmentation.[92] Proton abstraction occurs from both isomers with NH_2^-, H^-, and OH^-. Electron transfer and dissociative electron transfer is also observed for both complexes with H^-, NH_2^-, and $CH_2CHCH_2^-$. Hydride, amide, and hydroxide each react with $(BUD)Fe(CO)_3$ to displace both CO and the butadiene ligand:

$$X^- + (C_4H_6)Fe(CO)_3 \longrightarrow \begin{cases} \xrightarrow{(a)} XFe(CO)_3^- + C_4H_6 \\ \xrightarrow{(b)} XFe(CO)_2(C_4H_6)^- + CO \end{cases} \tag{41}$$

For the $(TMM)Fe(CO)_3$ isomer, path b in Equation 41 occurs only with hydride. Allyl anion also reacts with $(BUD)Fe(CO)_3$ by displacement of the hydrocarbon ligand; in this case producing an $(\eta^3\text{-}C_3H_5)Fe(CO)_3^-$ complex:

$$CH_2CHCH_2^- + (BUD)Fe(CO)_3 \longrightarrow \begin{cases} (BUD)Fe(CO)_{2,3}^- + C_3H_5 \\ H{-}C(CH_2)_2{-}Fe(CO)_3^- + C_4H_6 \end{cases} \tag{42}$$

Proton abstraction from $(BUD)Fe(CO)_3$ was also observed with $CH_3C(O)CH_2^-$, CH_2CN^-, $CH_2NO_2^-$, and PhO^-, but not with $CH_2CHCH_2^-$, N_3^-, or Cl^-. From the occurrence of proton abstraction by phenoxide we can infer that $\Delta H_{acid}[(BUD)Fe(CO)_3] < 350$ kcal/mol. Interestingly, in attempts to refine this limit with protonation studies involving reference acids, we found that only CF_3CO_2H ($\Delta H_{acid} = 322.7$ kcal/mol)[44] would readily effect neutralization of the $(C_4H_5)Fe(CO)_3^-$ ion. This suggests either an exceedingly low kinetic basicity for the complex or, possibly, rearrangement of the organic ligand from η^4-bonding to a σ-vinyl species with a reduced thermodynamic basicity.

5. CATALYSIS INTERMEDIATES

An important aspect of these studies is that gas phase ion–molecule reactions can occasionally be used to generate certain metal ion complexes that are proposed as intermediates in homogeneous catalysis cycles, but which are difficult or impossible to identify or characterize in solution because of their extreme reactivity or low thermodynamic stabilities. We have seen that gas phase nucleophilic addition reactions with metal carbonyls can provide access to a wide variety of interesting negative metal ion complexes. Many of these ions represent gas phase analogues of important intermediates in CO and CO_2 activation chemistry such as metal formyls,[93] hydroxycarbonyls (metallocarboxylic acids),[7,9] metal formates,[8] carbonates,[94] and metal hydrides.[95,96] In this section we describe our investigations of the formation, properties, and reactivity of various metal ion complexes of this type that were carried out with the aim of developing new thermochemical and mechanistic information relating to their behavior in catalytic cycles.

5.1. Hydroxycarbonyl Complexes and the Homogeneously Catalyzed Water–Gas Shift Reaction

Homogeneous catalysis of the water–gas shift reaction (WGSR) by transition metal carbonyls has received considerable attention since the first examples were reported some 50 years ago.[9–11,97] Nevertheless, certain mechanistic

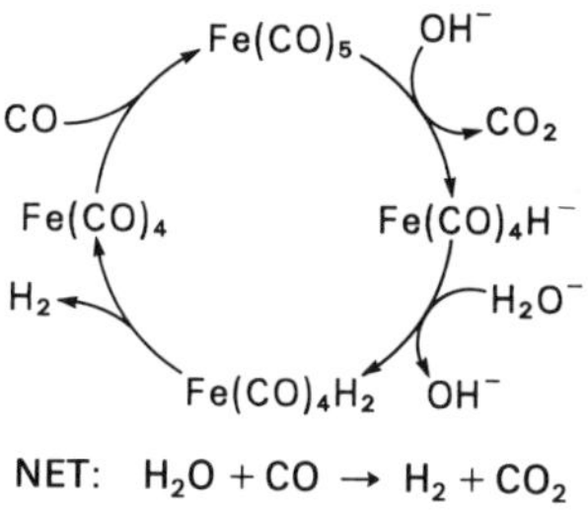

$Fe(CO)_5$ OH^-

CO CO_2

$Fe(CO)_4$ $Fe(CO)_4H^-$

H_2 H_2O^-

$Fe(CO)_4H_2^-$ OH^-

NET: $H_2O + CO \rightarrow H_2 + CO_2$

Scheme 7

details have remained unclarified, largely due to difficulties in identifying and characterizing the transient intermediate species that are postulated for many of the key steps. For example, a recent investigation of the kinetics of the WGSR catalyzed by $Fe(CO)_5$ in alkaline solution[98] (Scheme 7) questioned earlier proposals[11,99,100] regarding the mode of decarboxylation of the hydroxycarbonyl complex (**XVI**) that is commonly believed to be the initial intermediate in the catalytic cycle:

$$Fe(CO)_5 + OH^- \rightarrow \underset{\textbf{XVI}}{[(CO)_4FeCO_2H^-]} \rightarrow (CO)_4FeH^- + CO_2 \tag{43}$$

It has been suggested by Pearson and Mauermann that CO_2 loss proceeds via dianion **XVII** (Equation 44) rather than directly from the hydroxycarbonyl **XVI**.[98]

$$\underset{\textbf{XVI}}{[(CO)_4FeCO_2H^-]} \xrightarrow{\text{base}} [(CO)_4FeCO_2^{=}] \xrightarrow{-CO_2} \underset{\textbf{XVII}}{[(CO)_4Fe^{=}]} \xrightarrow{H^+} (CO)_4FeH^- \tag{44}$$

While similar conclusions have been reached for related anionic and neutral metallocarboxylic acids,[101,102] no clear trends in the general behavior of such complexes with respect to decarboxylation have been established.[9,103]

5.1.1. Formation and Reactions of $(CO)_4FeCO_2H^-$

In Section 3.3 we described the reactions of $Fe(CO)_5$ with $HO(H_2O)_n^-$ cluster ions which lead to an $Fe(CO)_5/OH^-$ adduct ion believed to be the iron hydroxycarbonyl complex:

$$HO(H_2O)_n^- + Fe(CO)_5 \begin{cases} \xrightarrow{n=0} (CO)_3FeOH^- + 2CO \\ \xrightarrow{n=1\text{-}4} (CO)_4FeCO_2H^- + nH_2O \end{cases} \tag{45}$$

The appearance of **XVI** as a stable species in our experiments has important implications concerning its proposed role in WGSR catalysis. Our assignment of the iron hydroxycarbonyl structure **XVI** to the product of reaction 45 is based on several lines of chemical evidence.[104] We rule out an electrostatically bound ion–molecule complex such as **XVIII** in favor of a structure involving stronger covalent bonding to OH^-:

$$(CO)_4\bar{Fe}-C(=O)OH \qquad [Fe(CO)_5 \cdots OH^-] \qquad (CO)_4\bar{Fe}-O-C(=O)-H$$

XVI **XVIII** **XIX**

This is a nontrivial point since ion–induced dipole attraction energies between OH^- and a large molecule like $Fe(CO)_5$ may become substantial for separation distances of a few angstroms. However, since the reaction between bare OH^- and $Fe(CO)_5$ results in expulsion of two CO ligands,[43] and the hydrated ions all react at or near the collision rate with no solvent retention by the product (Table 5), we believe that strong covalent bonding interactions must be operative. Furthermore, neutral switching type reactions such as might be expected for an electrostatically bonded cluster ion[75] are *not* observed when the adduct is allowed to react with neutral substrates possessing large permanent dipole moments, such as CH_3NO_2 ($\mu_D = 3.46$ debye).[105] Structure **XIX**, which represents the isomeric formate anion, is also ruled out by our observation that the $Fe(CO)_5/OH^-$ adduct undergoes H/D exchange in the presence of simple carboxylic acids and H_2S, as does its protonated analogue in the presence of CH_3COOD:

$$(CO)_4FeCO_2D^- + HCO_2H \rightarrow (CO)_4FeCO_2H^- + HCO_2D \tag{46}$$

$$(CO)_4FeCO_2D^- + H_2S \rightarrow (CO)_4FeCO_2H^- + HSD \tag{47}$$

$$(CO)_4FeCO_2H^- + CH_3CO_2D \rightarrow (CO)_4FeCO_2D^- + CH_3CO_2H \tag{48}$$

It is unlikely that the formyl proton in structure **XIX** would be subject to exchange under these conditions. A mechanism for the H/D exchange with CH_3COOD is proposed in Scheme 8 wherein deuteration at the acyl oxygen of the hydroxycarbonyl anion produces an intermediate dihydroxy metal carbene that subsequently transfers the original proton back to the acetate anion.

5.1.2. Decarboxylation Mechanism Studies

Given a correct structural assignment for **XVI** as the iron hydroxycarbonyl complex, the foregoing results permit several conclusions to be made regarding its role in WGSR catalysis.[104] On a millisecond time scale, isolated $Fe(CO)_4CO_2H^-$ is stable with respect to decarboxylation, despite the moderately favorable energetics, viz., a barrier exists. From the occurrence of reaction

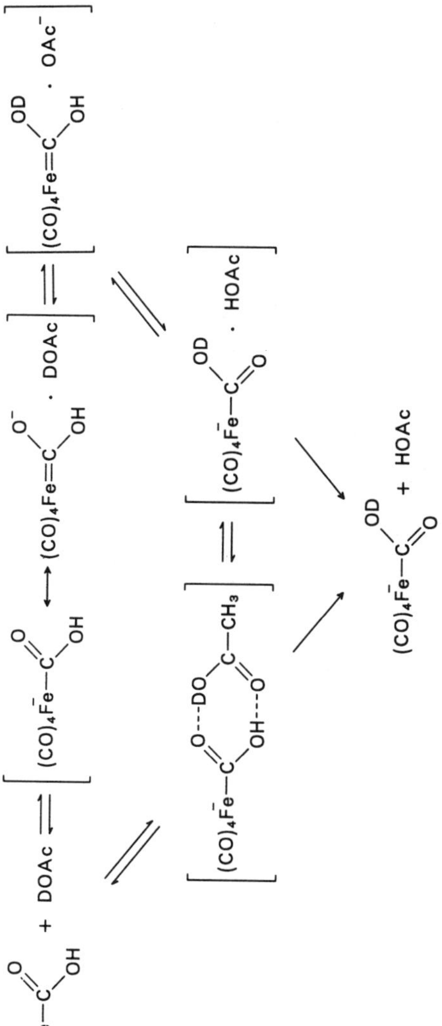

$$(CO)_4\overset{-}{Fe}{-}C(=O)OH + DOAc \rightleftharpoons \left[(CO)_4\overset{-}{Fe}{-}C(=O)OH \longleftrightarrow (CO)_4Fe{=}C(O^-)(OH) \cdot DOAc\right] \rightleftharpoons \left[(CO)_4Fe{=}C(OD)(OH) \cdot OAc^-\right]$$

$$\left[(CO)_4\overset{-}{Fe}{-}C(=O\text{---}DO)(OH\text{---}O{=})C{-}CH_3\right] \rightleftharpoons \left[(CO)_4\overset{-}{Fe}{-}C(OD){=}O \cdot HOAc\right]$$

$$\longrightarrow (CO)_4\overset{-}{Fe}{-}C(OD){=}O + HOAc$$

Scheme 8

45 for $n = 4$ we can conclude that $\Delta H_f[(CO)_4FeCO_2H^-, g] \leq -260$ kcal/mol. Combining this with an estimate for $\Delta H_f[(CO)_4FeH^-, g] = -178 \pm 6$ kcal/mol[81] leads to a lower limit to the enthalpy of decarboxylation for **XVI** of approximately -17 kcal/mol. Moreover, in the absence of a solvating medium, decarboxylation is not catalyzed by a single water molecule nor by stronger neutral bases such as amines since **XVI** sustains many hundreds of collisions with water in the course of its formation and detection in our instrument and it appears to be totally unreactive in the presence of added $(CH_3)_2NH$.

The collision-induced dissociation (CID) spectrum of ion **XVI** obtained with a flowing afterglow-triple-quadrupole apparatus[106] is also supportive of obligatory base catalysis for decarboxylation. $(CO)_4FeCO_2H^-$ undergoes fragmentation by *exclusive loss of CO* when subjected to low-energy CID. If decarboxylation of **XVI** were a thermal process, then one might expect that deposition of excess internal energy by collisional activation in the gas phase would induce CO_2 loss. That no such fragmentation is observed suggests that the barrier to *decarboxylation* from an internally excited $(CO)_4FeCO_2H^-$ ion exceeds that for *decarbonylation.*

We therefore conclude that $(CO)_4FeCO_2H^-$ is relatively stable as an isolated species in the gas phase and exhibits no intrinsic propensity to lose CO_2 as depicted in Equation 43. However, when a more strongly basic solvent molecule is present in the cluster ion from which $(CO)_4FeCO_2H^-$ is formed, decarboxylation can be observed.[73] Among the many metal ion products formed from the reaction between $HO(NH_3)^-$ and $Fe(CO)_5$ described in Section 3.3 is $HFe(CO)_4^-$:

$$HO(NH_3)^- + Fe(CO)_5 \rightarrow HFe(CO)_4 + CO_2 + NH_3 \quad (49)$$

This is especially interesting in light of the fact that NH_3 does not effect decarboxylation in a reaction with preformed $(CO)_4FeCO_2H^-$:

$$(CO)_4FeCO_2H^- + NH_3 \rightarrow [(CO)_4FeCO_2H^- \cdots NH_3] \nrightarrow HFe(CO)_4^- + CO_2 + NH_3 \quad (50)$$

The key difference in the $HO(NH_3)^-$ reaction is that the intermediate ammonia/iron hydroxcarbonyl complex is formed with a significantly greater amount of excess internal energy than the one depicted in Equation 50. Scheme 9 summarizes our view of the mechanism for the NH_3-catalyzed decarboxylation mechanism. In a collision between $OH(NH_3)^-$ and $Fe(CO)_5$, transfer of OH^- from the cluster to a carbonyl ligand occurs to produce the iron hydroxycarbonyl complex according to the mechanism outlined in Scheme 5. This endows the resulting intermediate **XX** with ca. 44 kcal/mol excess internal energy, as given by the difference between the calculated $OH(NH_3)^-$ solvation energy (13 kcal/mol)[73] and the $OH^-/Fe(CO)_5$ binding energy $\{D[(CO)_4FeC(O)—OH^-] \simeq 57 \pm 3$ kcal/mol (Table 3)$\}$. Loss of CO ligands may occur since the 34-kcal/mol decarbonylation barrier is exceeded. Moreover, direct expulsion of NH_3 followed by collisional deactivation may

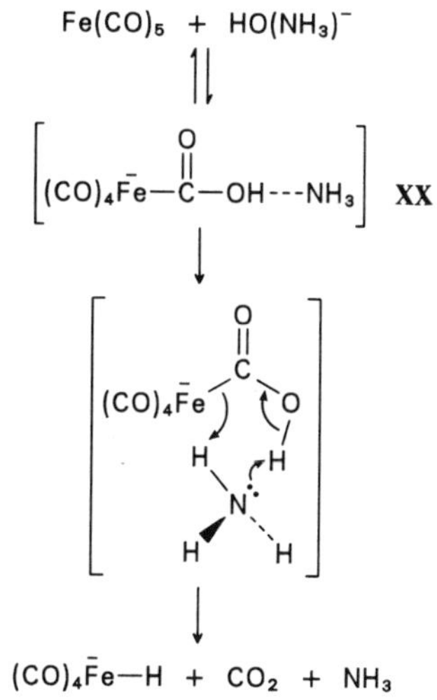

$$Fe(CO)_5 + HO(NH_3)^- \rightleftharpoons \left[(CO)_4\bar{Fe}-\overset{O}{\overset{\|}{C}}-OH\cdots NH_3\right] \quad \mathbf{XX}$$

$$\downarrow$$

$$\left[\text{cyclic transition structure: } (CO)_4\bar{Fe}-C(=O)-O-H\cdots NH_3\right]$$

$$\downarrow$$

$$(CO)_4\bar{Fe}-H + CO_2 + NH_3$$

Scheme 9

give rise to the observed $(CO)_4FeCO_2H^-$ product. Competing with these reactions is a NH_3-mediated decarboxylation via the proposed cyclic proton transfer shown in Scheme 9. Note that the mechanism requires that the hydrogen in $HFe(CO)_4^-$ is ultimately derived from the NH_3 solvent molecule. However, this would be inherently difficult to test by experiments with $DO(NH_3)^-$ or $HO(ND_3)^-$, since H/D exchange studies have suggested that all protons become scrambled in $HO(NH_3)^-$ ions.[107]

5.2. Hydride Transfer Reactions: Thermochemistry for Transition Metal Formyl Ions

Another important type of organometallic acyl complex that plays a role in CO-activation chemistry is the transition metal formyl[93]:

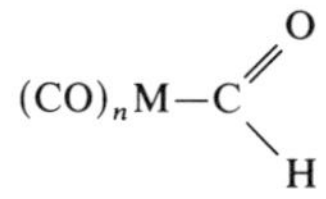

$$(CO)_nM-C\overset{\displaystyle O}{\underset{\displaystyle H}{\lessgtr}}$$

Model formyl complexes have been generated in solutions of the corresponding metal carbonyl and borohydride reagents[108-110] and have been reported to function as hydride reducing agents in the presence of organic carbonyl compounds and other metal carbonyls.[93,111] While the synthesis and reactivity of metal formyls in solution have been investigated for several years, relatively little thermochemical data have been determined for these species which might guide in the development of more effective formyl-based catalysts.[113] In the course of our examination of the gas phase reactions between negative ions and simple metal carbonyls, we found that a variety of formyl complexes could be readily generated through binary H^- transfer reactions.[114, 115] In this section we describe our measurements of the thermochemistry for a series of metal formyl complexes and discuss the relationship of these data to the relative reactivity of these species in solution.

As outlined in general terms in Section 3.1.3, the hydride affinity of a neutral compound, HA(X), is defined as the enthalpy of dissociation of the corresponding anion HX^- to free H^- and neutral X[116]:

$$HX^- \xrightarrow{\Delta H} H^- + X$$

$$HA(X) = \Delta H = \Delta H_f(H^-) + \Delta H_f(X) - \Delta H_f(HX^-) \quad (51)$$

These data may be calculated from known heats of formation for the corresponding neutral compounds and anions, the latter values being derived from gas phase acidity measurements.[44] A representative listing of hydride affinities for several familiar types of organic compound is presented in Table 7. From these known hydride affinities, values for other compounds such as the metal carbonyls may be determined by bracketing techniques and equilibrium H^- transfer reactions.

Limits on the hydride affinities of $Fe(CO)_5$, $Cr(CO)_6$, $Mo(CO)_6$, and $W(CO)_6$ have been established from a series of bracketing experiments. Table 8 presents a summary of representative results for $Fe(CO)_5$.[114] Hydride transfer yielding the formyl complex is a minor reaction channel for the 1-phenylethyl anion, formate, and isobutyronitrile anion, but not for propion-

Table 7. Hydride Affinities[a]

Compound (X)	Anion (HX^-)	HA(X) (kcal/mol)
$(CF_3)_2CO$	$(CF_3)_2CHO^-$	85.1
BH_3	BH_4^-	74.0
CS_2	HCS_2^-	67.3
$CH_2{=}C{=}O$	CH_2CHO^-	62.2
$CH_2{=}CHCHO$	CH_3CHCHO^-	61.5
CF_3COCH_3	$CF_3(CH_3)CHO^-$	59.0
O_2	HO_2^-	58.0
$CH_2{=}CHCN$	CH_3CHCN^-	57.6
$CH_2{=}C(CH_3)CN$	$(CH_3)_2CCN^-$	56.7
CO_2	HCO_2^-	51.6
$PhCH{=}CH_2$	$PhCHCH_3^-$	50.4
PhCHO	$PhCH_2O^-$	45.9
n-C_4H_9CHO	n-$C_5H_{11}O^-$	43.5
$CH_2{=}O$	CH_3O^-	41.2
CH_3CHO	$CH_3CH_2O^-$	40.7
$(CH_3)_2CO$	$(CH_3)_2CHO^-$	39.6
$CH_3N{=}CH_2$	$(CH_3)_2N^-$	26.2
c-C_6H_6	c-$C_6H_7^-$	22.8
CO	HCO^-	8.0

[a] Taken from Reference 116.

Table 8. Hydride Transfer Reactions with $Fe(CO)_5$ [a]

Anion (HX^-)	Neutral (X)	Hydride affinity, HA(X) (kcal/mol)	k_{obsd} [b]	Percent H^- transfer[c]	Other products (%)[d]
$C_6H_5\bar{C}HCH_3$	$C_6H_5CH{=}CH_2$	50.4 ± 2.5	1.08 (0.72)	3.0 (0.3)	$(C_6H_5CHCH_3)Fe(CO)_5^-$(77) $(C_6H_5CHCH_3)Fe(CO)_4^-$(10) $HFe(CO)_4^-$(10)
HCO_2^-	CO_2	51.6 ± 2.5	0.16 (0.08)	13.0 (0.1)	$(HCO_2)Fe(CO)_5^-$(75) $HFe(CO)_4^-$(12)
$(CH_3)_2\bar{C}CN$	$CH_2{=}C(CH_3)CN$	56.7 ± 2.5	1.16 (0.67)	5.0 (0.03)	$[(CH_3)_2CCN]Fe(CO)_5^-$(69) $[(CH_3)_2CCN]Fe(CO)_4^-$(23) $HFe(CO)_4^-$(3)
$CH_3\bar{C}HCN$	$CH_2{=}CHCN$	57.6 ± 2.5	1.16 (0.61)	0.5 (0.003)	$(CH_3CHCN)Fe(CO)_5^-$(78) $(CH_3CHCN)Fe(CO)_4^-$(20) $(CH_3CHCN)Fe(CO)_3^-$(2)
$CH_3\bar{C}HCOCH_3$	$CH_2{=}CHCOCH_3$	59.3 ± 2.5	1.05 (0.61)	0.1 (0.0006)	$(CH_3CHCOCH_3)Fe(CO)_5^-$(100)

[a] Taken from Reference 114.
[b] Observed bimolecular rate coefficient for disappearance of reactant anion in units of 10^{-9} cm^3/s. Overall reaction efficiency shown in parentheses.
[c] Efficiency for hydride transfer channel shown in parentheses.
[d] Product distribution shown in parentheses.

itrile anion and 2-butanone enolate. The trend suggested by the quantitative data shown in Table 8 is further supported by qualitative observations involving hydride transfer reagents spanning a 40-kcal/mol range in hydride affinity.

The reverse reaction wherein $(CO)_4FeCHO^-$ donates H^- to reference neutral compounds is not observed under our conditions, even when the reaction is substantially exothermic. Low efficiencies are commonly noted for hydride transfer reactions involving both positive and negative ions,[117] and the complete absence of the hydride transfer reaction with the $(CO)_4FeCHO^-$ ion probably results from unfavorable charge localization requirements in the large metal ion complex. That is, the formyl C—H bond is a relatively strong one, and the low electronegativity of hydrogen may inhibit the buildup of hydride character in the CHO moiety as would be required for H^- transfer to another molecule.

From the results listed in Table 8 we have assigned the hydride affinity of $Fe(CO)_5$ to be 56 ± 4 kcal/mol. This estimate may be combined with other thermochemical data to derive the heats of formation and metal-ligand bond strengths summarized in Table 9.† The corresponding data for the group 6 hexacarbonyls are also shown in the table.[115] Chromium, molybdenum, and tungsten carbonyls show identical behavior in bracketing experiments with reference H^- donors: slow hydride transfer with CH_3O^- (HA = 41.2 kcal/mol) and other anions with lower H^--binding energies, and no hydride transfer with $PhCH_2O^-$ (HA = 45.9 kcal/mol) or anions with greater H^- binding energies. A hydride affinity for each of 44 ± 4 kcal/mol is assigned from these results.

Several important conclusions can be derived from these data. First of all, they confirm earlier ideas from condensed phase studies concerning the relative stability of the iron formyl complex and the corresponding hydride.[93] That is, CO insertion into the metal-hydrogen bond in $(CO)_4FeH^-$ is thermodynamically unfavorable:

$$(CO)_4FeH^- + CO \rightarrow (CO)_4FeCHO^- \qquad \Delta H = 10 \pm 7 \text{ kcal/mol} \tag{52}$$

Moreover, the borohydride-based synthesis of the metal formyls listed in Table 9, as well as their reported action as hydride reducing agents with organic carbonyl compounds, must be strongly dependent on solvation or counterion effects. This is so because our thermochemical data show that H^- transfer *from* BH_4^- [$HA(BH_3)$ = 74 kcal/mol][119] *to* each metal carbonyl and *from* all of the formyl anions *to* simple aldehydes and ketones (HA = 38–46 kcal/mol) are intrinsically endothermic reactions. This implies that, in addition to changing ligands about the metal, variation of solvents and/or Lewis-acid catalysts would be a potentially useful approach to optimizing metal formyl-based reductions.

Finally, the equality of the group 6 hexacarbonyl hydride affinities and the fact that the value for $Fe(CO)_5$ is 12 kcal/mol larger deserves comment. Dedieu and Nakamura have emphasized the importance of the vacant d_{z^2}

† The $D[(CO)_{n-1}M{-}CHO^-]$ values have been recalculated with use of the most recent value for $\Delta H_f[HCO^-, g]$; see Reference 118.

Table 9. Thermochemical Data for Transition Metal Formyl Anions Derived from Gas Phase Hydride Transfer Reactions[a]

$M(CO)_n$	HA[b]	$\Delta H_f[(CO)_{n-1}MCHO^-, g]$[c]	$D[(CO)_{n-1}M—CO]$[d]	$D[(CO)_{n-1}M—CHO^-]$
$Fe(CO)_5$	56 ± 4	-194 ± 4	41.5	88 ± 5
$Cr(CO)_6$	44 ± 4	-226 ± 4	36.8	71 ± 5
$MO(CO)_6$	44 ± 4	-228 ± 4	40.5	75 ± 5
$W(CO)_6$	44 ± 4	-221 ± 4	46.0	81 ± 5

[a] Data in units of kcal/mol.
[b] Bracketed hydride affinity as defined in Equation 51.
[c] $\Delta H_f[FeCO)_5, g] = -173.0 \pm 1.5$ kcal/mol; $\Delta H_f[Cr(CO)_6, g] = -217.1 \pm 0.5$ kcal/mol; $\Delta H_f[Mo(CO)_6, g] = -219.0 \pm 0.4$ kcal/mol; $\Delta H_f[W(CO)_6, g] = -211.4 \pm 0.6$ kcal/mol (Reference 69).
[d] Taken from Reference 60.

orbital in the bonding of an axial formyl ligand in $(CO)_4FeCHO^-$.[50,120] The energy of this orbital was shown to correlate with the ordering of the computed H^- binding energies for $Fe(CO)_5$ and $Ru(CO)_5$, viz., the lower energy of the d_{z^2} orbital in $Ru(CO)_5$ was held responsible for its 7 kcal/mol greater hydride affinity compared to $Fe(CO)_5$.[50] The relative magnitudes of the hydride affinities for the iron and chromium carbonyls are satisfactorily accounted for on this basis, since the empty d_{z^2} orbital in $Cr(CO)_6$ is expected to be somewhat higher in energy than in $Fe(CO)_5$. However, larger hydride affinities would be predicted for $Mo(CO)_6$ and $W(CO)_6$ since the d_{z^2} orbital is substantially lower in energy in these complexes.[121,122]

It is also interesting to note that our measured hydride affinity ordering is in good accord with the relative reactivities of these complexes toward nucleophilic addition reactions in solution. Electrophilic reactivity of metal carbonyls has been correlated by Darensbourg and Darensbourg[123,124] with the C—O stretching force constants (k_{CO}), which may in turn reflect the partial positive charge at the carbonyl carbon atom(s). High k_{CO} values correlate with greater positive character and enhanced carbon electrophilicity, while low values of k_{CO} correlate with reduced reactivity towards nucleophiles.[125] The group 6 hexacarbonyls have essentially identical force constants (16.49, 16.52, and 16.41 mdyn/Å for $Cr(CO)_6$, $Mo(CO)_6$, and $W(CO)_6$, respectively), whereas k_{CO} for $Fe(CO)_5$ is somewhat larger (16.98 mdyn/Å).[124] The hydride affinities listed in Table 9 follow this same trend. On the strength of this apparent correlation, we may venture the prediction that $Ni(CO)_4$ (k_{CO} = 17.14 mdyn/Å[124] should have a greater hydride affinity than $Fe(CO)_5$. Experiments to test this prediction as well as to determine the hydride affinities of other organometallic compounds are currently in progress.

6. CONCLUDING REMARKS

We have attempted to summarize here the current status of the nucleophilic addition chemistry of gaseous negative ions and transition metal organometallic compounds. Gas phase ion-molecule techniques in general, and the flowing afterglow method in particular, can provide powerful tools for probing the intrinsic behavior of nucleophilic addition reactions involving isolated reagents. Both familiar and surprising new behavior for these reactions have been observed wherein attack by an anionic nucleophile is frequently accompanied by rapid decomposition (fragmentation) of the products. In this sense, the gas phase behavior resembles the "hot-molecule" type of chemistry which often characterizes photoactivated, pyrolytic, or sonochemical reactions.[126-128] On the other hand, it was shown that the presence of ancillary solvent molecules can significantly alter the course of gas phase anionic nucleophilic additions in such a way that they more closely resemble the analogous reactions in solution. Nucleophilic addition to the ligand in saturated, neutral metal complexes appears to be facile in gas phase reactions, just as it is in solution, while direct negative ion addition to the metal still awaits definitive experimental proof.

Finally, the potential for useful applications of gas phase negative ion–molecule reactions in the study of reactive intermediates in key catalytic systems appears excellent. The ultimate utility of thermochemical data derived for isolated gas phase ions related to catalysis will depend, in part, upon a quantitative understanding of solvation effects.[129] Nevertheless, useful new insights concerning the intrinsic reactivities of these species can develop from gas phase experiments, and considering the vast number of catalytically important anionic metal complexes beyond the simple iron acyls described here, future studies along these lines should be most fruitful.

ACKNOWLEDGMENTS

We are grateful to the National Science Foundation, the Donors of the Petroleum Research Fund, administered by the American Chemical Society, and Research Corporation for their support of this research. K.R.L. thanks procter and Gamble and R.R.S. thanks the Alfred P. Sloan Foundation for fellowships.

REFERENCES

1. Lindsell, W.E. In *Comprehensive Organometallic Chemistry*; Wilkinson, G., Ed.; Pergamon: New York, 1982; Vol. 1, Chapter 4.
2. Evans, W.J. *Adv. Organomet. Chem.* 1985, *24*, 131.
3. Marks, T. J. *Science* 1982, *217*, 989.
4. Hartley, F.R.; Patai, S., Eds. *The Chemistry of the Metal–Carbon Bond*; Wiley: New York, 1982; Vol. 1, 1985; Vol. 2.
5. Collman, J. P.; Hegedus, L.S.; Norton, J.R.; Finke, R.G. *Principles and Applications of Organotransition Metal Chemistry*; University Science Books: Mill Valley, CA, 1987.
6. Alper, H. *Transition Metals in Organic Synthesis*; Academic: New York, 1976; Vol. 2.
7. Ford, P.C., Ed. *Catalytic Activation of Carbon Monoxide*, American Chemical Society: Washington, DC, 1981; ACS Symp. Ser., Vol. 152.
8. Darensbourg, D.J.; Kudaroski, R.A. *Adv. Organomet. Chem.* 1983, *22*, 129.

9. Ford, P.C.; Rocicki, A. *Adv. Organomet. Chem.* 1988, *28*, 139.
10. Darensbourg, D.J.; Rocicki, A. *Organometallics* 1982, *1*, 1685.
11. Frazier, C.C.; Hanes, R.R.; King, A.D.; King, R.B. *Adv. Chem. Ser.* 1979, *173*, 94.
12. Parshall, G.W. *Homogeneous Catalysis*; Wiley: New York, 1981.
13. Masters, C. *Adv. Organomet. Chem.* 1979, *17*, 61.
14. Keim, W., Ed. *Catalysis in C_1 Chemistry*; D. Reidel: Dordrecht, 1983.
15. Hegedus, L.S. In *The Chemistry of the Metal-Carbon Bond*; Hartley, F.R.; Patai, S., Eds.; Wiley: New York, 1985; Vol. 2.
16. Pauson, P.L. *J. Organomet. Chem.* 1980, *200*, 207.
17. Kane-Maguire, L.A.P.; Honig, E.D.; Sweigart, D.A. *Chem. Rev.* 1984, *84*, 784.
18. Semmelhack, M.F. *Ann. N.Y. Acad. Sci.* 1977, *295*, 36.
19. Watts, D.W. *Pure Appl. Chem.* 1979, *51*, 1713.
20. Darensbourg, M. Y. *Prog. Inorg. Chem.* 1985, *33*, 221.
21. Backvall, J.E.; Nordberg, R.E.; Wilhelm, D. *J. Am. Chem. Soc.* 1985, *107*, 6892.
22. Casey, C.P.; Bunnell, C.A. *J. Am. Chem. Soc.* 1976, *98*, 436.
23. For current general reviews, see: Bowers, M.T., Ed. *Gas Phase Ion Chemistry*, Academic: New York, 1979; Vol. 1, 1979; Vol. 2, 1984; Vol. 3.
24. Olmstead, W.N.; Brauman, J.I. *J. Am. Chem. Soc.* 1977, *99*, 4219.
25. Asubiojo, O.I.; Brauman, J.I. *J. Am. Chem. Soc.* 1979, *101*, 3715.
26. Bohme, D.K.; Mackay, G.I.; Payzant, J.D. *J. Am. Chem. Soc.* 1974, *96*, 4027.
27. Caldwell, G.; Magnera, T.F.; Kebarle, P. *J. Am. Chem. Soc.* 1984, *106*, 959.
28. Foster, M.S.; Beauchamp, J.L. *J. Am. Chem. Soc.* 1975, *97*, 4808.
29. Wronka, J.; Ridge, D.P. *J. Am. Chem. Soc.* 1984, *106*, 67.
30. Foster, M.S; Beauchamp, J.L. *J. Am. Chem. Soc.* 1971, *93*, 4924.
31. Corderman, R.R.; Beauchamp, J.L. *Inorg. Chem.* 1977, *16*, 3135.
32. Kahn, S.D.; Hehre, W.J.; Bartmess, J.E.; Caldwell, G. *Organometallics* 1984, *3*, 1740.
33. Sidorov, L.N.; Nikitin, M.I.; Skokan, E.V.; Sorokin, I.D., *Int. J. Mass Spectrom. Ion. Phys.* 1980, *35*, 203.
34. Sidorov, L.N.; Zhuravleva, L.V.; Sorokin, I.D. *Mass. Spec. Rev.* 1986, *5*, 73.
35. Chilingarov, N.S.; Korobov, M.V.; Sidorov, L.N.; Mit'kin, V.N.; Shipachev, V.A.; Zemskov, S.V. *J. Chem. Thermodyn.* 1984, *16*, 965.
36. Sorokin, I.D.; Sidorov, L.N.; Nikitin, M.I.; Skokan, E.V. *Int. J. Mass Spectrom. Ion. Phys.* 1981, *41*, 45.
37. Sidorov, L.N.; Sorokin, I.D.; Nikitin, M.I.; Skokan, E.V. *Int. J. Mass Spectrom. Ion. Phys.* 1981, *39*, 311.
38. Dillow, G.W.; Gregor, I.K. *Inorg. Chim. Acta* 1984, *84, 86*, 267.
39. Gregor, I.K.; Guilhaus, M. *Mass Spec. Rev.* 1984 *3*, 39.
40. For reviews of this technique, see: (a) Ferguson, E.E.; Fehsenfeld, F.C.; Schmeltekopf, A.L. *Adv. At. Mol. Phys.* 1969, *5*, 1; (b) Graul, S.T.; Squires, R.R. *Mass Spec. Rev.* 1988, *7*, 263.
41. DePuy, C.H.; Bierbaum, V.M. *Acc. Chem. Res.* 1981, *14*, 146.
42. Bohme, D.K.; Dunkin, D. B.; Fehsenfeld, F.C.; Ferguson, E.E. *J. Chem. Phys.* 1969, *51*, 863.
43. Lane, K.R.; Sallans, L.; Squires, R.R. *J. Am. Chem. Soc.* 1986, *108*, 4368.
44. Bartmess, J.E.; McIver, R.T., Jr. In *Gas Phase Ion Chemistry*; Bowers, M.T., Ed., Academic: New York, 1979; Vol. 2, Chapter 11; Lisa, S.G.; Bartmess, J.E.; Liebman, J.F.; Holmes, J.L.; Levin, R.D.; Mallard, W.G. *J. Phys. Chem. Ref. Data* 1988, *17*, Suppl. 1.
45. Mackay, G.I.; Lien, M.H.; Hopkinson, A.C.; Bohme, D.K. *Can. J. Chem.* 1978, *56*, 131.
46. Fischer, E.O.; Kiener, V. *J. Organomet. Chem.* 1970, *23*, 215.
47. Fischer, E.O.; Maasbol, A. *Angew. Chem. Int. Ed. Engl.* 1964, *3*, 580.
48. Werner, H.; Beck, W.; Engelman, H. *Inorg. Chim. Acta* 1969, *3*, 331.
49. Tuan, F.; Hoffmann, R. *Inorg. Chem.* 1985, *24*, 871.
50. Dedieu, A.; Nakamura, S. *Nouv. J. Chim.* 1984, *8*, 317.
51. Tatsumi, K.; Nakamura, A.; Hofmann, P.; Stauffert, P.; Hoffmann, R. *J. Am. Chem. Soc.* 1985, *107*, 4440.
52. Curtis, D.M.; Shiu, K.; Butler, W. M. *J. Am. Chem. Soc.* 1986, *108*, 1550.
53. Wojcicki, A. *Adv. Organomet. Chem.* 1973, *11*, 87.

54. Su, T.; Bowers, M.T. In *Gas Phase Ion Chemistry*; Bowers, M.T. Ed.; Academic: New York, 1979; Vol. 1, Chapter 3.
55. Lane, K.R.; Sallans, L.; Squires, R.R. *J. Am. Chem. Soc.* 1985, *107*, 5369.
56. Larson, J.W.; McMahon, T.B. *J. Am. Chem. Soc.* 1985, *107*, 766, and references therein.
57. Yamdagni, R.; Kebarle, P. *J. Am. Chem. Soc.* 1972, *94*, 2940.
58. Murphy, M.K.; Beauchamp, J.L. *J. Am. Chem. Soc.* 1977, *99*, 4992.
59. Sullivan, S.A.; DePuy, C.H. *J. Am. Chem. Soc.* 1981, *103*, 480.
60. Lewis, K.E.; Golden, D.M.; Smith, G.P. *J. Am. Chem. Soc.* 1984, *106*, 3905.
61. Brown, T.L.; Bellus, P.A. *Inorg. Chem.* 1978, *17*, 3726.
62. Doxsee, K.M.; Grubbs, R.H. *J. Am. Chem. Soc.* 1981, *103*, 7696.
63. Anstock, M.; Taube, D.; Gross, D.C.; Ford, P.C. *J. Am. Chem. Soc.* 1984, *106*, 3696.
64. Darensbourg, D.J.; Gray, R.L.; Pala, M. *Organometallics* 1984, *3*, 1982.
65. McDonald, R.N.; Schell, P.L.; McGhee, W.D. *Organometallics* 1984, *3*, 182.
66. Lichtenberger, D.L.; Brown, T.L. *J. Am. Chem. Soc.* 1978, *100*, 366, and references therein.
67. Lane, K.R. Ph.D. Dissertation, Purdue University, 1986.
68. Pearson, R.G. *Inorg. Chem.* 1984, *23*, 4675.
69. Connor, J.A. *Top. Curr. Chem.* 1977, *71*, 71.
70. Cotton, F.A.; Wilkinson, G., Eds., *Advanced Inorganic Chemistry*; *A Comprehensive Text*, Wiley: New York, 1980.
71. Shi, Q.; Richmond, T.G.; Trogler, W.C.; Basolo, F. *J. Am. Chem. Soc.* 1984, *106*, 71.
72. Poweell, J.; Gregg, M.; Kuksis, A.; Meindl, P. *J. Am. Chem. Soc.* 1983, *105*, 1064, and references therein.
73. Lane, K.R.; Squires, R.R. *J. Am. Chem. Soc.* 1986, *108*, 7187.
74. Dorfman, Y.G. *Z. Fiz. Khim.* 1963, *37*, 2496.
75. Bohme, D.K. In *Ionic Processes in the Gas Phase*; Ferriera, M.A.A.., Ed.; D. Reidel, Dordrecht, 1984.
76. Blyholder, G.; Zhao, K. M.; Lawless, M. *Organometallics* 1985, *4*, 1371.
77. Keesee, R.G.; Castleman, A.W., Jr. *J. Phys. Chem. Ref. Data* 1986, *15*, 1011.
78. Henchman, M. In *Structure/Reactivity and Thermochemistry of Ions*; Ausloos, P.; Lias, S.G., Eds; D. Reidel, Dordrecht, 1987; p. 381.
79. Lane, K.R.; Squires, R.R. *J. Am. Chem. Soc.* 1985, *107*, 6403.
80. Stewart, J.H.; Shapiro, R.H.; DePuy, C.H.; Bierbaum, V.M. *J. Am. Chem. Soc.* 1977, *99*, 7650.
81. Squires, R.R. *Chem. Rev.* 1987, *87*, 623.
82. Semmelhack, M.F.; Hall, H.T., Jr.; Farina, R.; Yoshifuji, M.; Clark, G.; Bargar, T.; Hirotsu, K.; Clardy, J. *J. Am. Chem. Soc.* 1979, *101*, 3535.
83. Semmelhack, M.F.; Clark, G.R.; Garcia, J.L.; Harrison, J.J.; Thebtaranonth, Y.; Wulff, W.; Yamashita, A. *Tetrahedron* 1981, *37*, 3957.
84. Calhorda, M.G.; Frazao, C.F.; Simoes, J.A.M. *J. Organomet. Chem.* 1984, *262*, 305.
85. Faller, J.W. *Inorg. Chem.* 1980, *19*, 2859.
86. Brown, D.A.; Glass, W.K.; Ubeid, M.T. *Inorg. Chim. Acta* 1984, *98*, L47.
87. Reger, D.L.; Klaeren, S.A.; Lebioda, L. *Organometallics* 1986, *5*, 1072.
88. Fenske, R.F.; Milletti, M.C. *Organometallics* 1986, *5*, 1243.
89. Bates, R.B.; Brenner, S.; Deines, W. H.; McCombs, D.A.; Potter, D. E. *J. Am. Chem. Soc.* 1970, *92*, 6345, and references therein.
90. Hajdasz, D.J.; Squires, R.R. *J. Am. Chem. Soc.* 1986, *108*, 3139.
91. McDonald, R.N.; Chowdhury,A.K.; Setzer, D.W. *J. Am. Chem. Soc.* 1980, *102*, 6491.
92. Wang, D.: Squires, R.R. *Organometallics* 1987, *6*, 905.
93. Gladysz, J.A. *Adv. Organomet. Chem.* 1982, *20*, 1.
94. Darensbourg, D.J.; Sanchez, K.M.; Rheingold, A.L. *J. Am. Chem. Soc.* 1987, *109*, 290, and references therein.
95. Slocum, D.W.; Moser, W.R., Eds. *Catalytic Transition Metal Hydrides*; *Ann. N.Y. Acad. Sci.* 1983, 415.
96. Kaesz, H.D.; Saillant, R.B. *Chem. Rev.* 1972, *72*, 231.
97. Hiever, W.; Leutert, F. *Z. Anorg. Allg. Chem.* 1932, *104*, 145.
98. Pearson, R.G.; Mauermann, H. *J. Am. Chem. Soc.* 1982, *104*, 500.

99. Kang, H.; Mauldin, C.H.; Cole, T.; Slegeir, W.; Cann. K.; Pettit, R. *J. Am. Chem. Soc.* 1978, *100*, 2925.
100. King, A.D.; King, R.B.; Yang, D.B. *J. Am. Chem. Soc.* 1980, *102*, 1028.
101. Darensbourg, D.J.; Froelich, J.A. *Inorg. Chem.* 1978, *17*, 3300.
102. Bercaw, J.E.; Goh, L.; Halpern, J. *J. Am. Chem. Soc.* 1972, *94*, 6534.
103. Gibson, D. H.; Owens, K.; Ong, T.S. *J. Am. Chem. Soc.* 1984, *106*, 1125, and references therein.
104. Lane, K.R.; Lee, R.E.; Sallans, L.; Squires, R.R. *J. Am. Chem. Soc.* 1984, *106*, 5767.
105. McClellan, A.L. *Table of Experimental Dipole Moment*; Rahara Enterprises: El Cerrito, CA, 1979; Vol. 2.
106. Squires, R.R.; Lane, K.R.; Lee, R.E.; Wright, L.G.; Wood, K.V.; Cooks, R.G. *Int. J. Mass Spectrom. Ion. Phys.* 1985, *64*, 185.
107. Grabowski, J.J.; DePuy, C.H.; Bierbaum, V.M. *J. Am. Chem. Soc.* 1983, *105*, 2565.
108. Casey, C.P.; Neumann, S.M. *J. Am. Chem. Soc.* 1976, *98*, 5395.
109. Gladysz, J.A.; Selover, J.C. *Tetrahedron Lett.* 1978, 319.
110. Winter, S.R.; Connett, G.W.; Thompson, E.A. *J. Organomet. Chem.* 1977, *133*, 339.
111. Casey, C.P.; Neumann, S.M. *J. Am. Chem. Soc.* 1978, *100*, 25544.
112. Gladysz, J.A.; Williams, G.M.; Tam, W.; Johnson, D.L. *J. Organomet. Chem.* 1977, *140*, C1.
113. Farnos, M.D.; Woods, B.A.; Wayland, B.B. *J. Am. Chem. Soc.* 1986, *108*, 3659.
114. Lane, K.R.; Sallans, L.; Squires, R.R. *Organometallics* 1985, *3*, 408.
115. Lane, K.R.; Squires, R.R. *Polyhedron* (in press).
116. Squires, R.R. In *Structure/Reactivity and Thermochemistry of Ions*; Ausloos, P.; Lias, S.G., Eds.; D. Reidel: Dordrecht, 1987; p. 373.
117. Sharma, R.B.; Sensharma, D.K.; Hiraoka, K.; Kebarle, P. *J. Am. Chem. Soc.* 1985, *107*, 3747.
118. Murray, K.K.; Miller, T.M.; Leopold, D.G.; Lineberger, W.C., Jr. *J. Chem. Phys.* 1986, *84*, 2520.
119. Workman, D.B.; Squires, R.R. *Inorg. Chem.* 1988, *27*, 1846.
120. Nakamura, S.; Dedieu, A. *Theor. Chim. Acta* 1982, *61*, 587.
121. Schreiner, A.F.; Brown, T.L. *J. Am. Chem. Soc.* 1968, *90*, 5713.
122. Beach, N.A.; Gray, H.B. *J. Am. Chem. Soc.* 1968, *90*, 5713.
123. Darensbourg, D.J.; Darensbourg, M.Y. *Inorg. Chem.* 1970, *9*, 1691.
124. Darensbourg, M.Y.; Concor, H.L.; Darensbourg, D.J.; Hasday, C. *J. Am. Chem. Soc.* 1974, *95*, 5919.
125. For a related report, see: Bush, R.C.; Angelici, R.J. *J. Am. Chem. Soc.* 1986, *108*, 2735.
126. Weiller, B.H.; Grant, E.R. *J. Am. Chem. Soc.* 1987, *109*, 1051, and references therein.
127. Fletcher, T.R.; Rosenfeld, R.N. *J. Am. Chem. Soc.* 1985, *107*, 2203.
128. Suslick, K.S., Ed. *High-Energy Processes in Organometallic Chemistry*; American Chemical Society: Washington, DC, 1987; ACS Symp. Ser., Vol. 333.
129. Nolan, S. P.; de la Vega, R.L.; Hoff, C.D. *Organometallics* 1986, *5*, 2529.

3

Reactions in Ionized Metal Carbonyls: Clustering and Oxidative Addition

Douglas P. Ridge and Wilma K. Meckstroth

1. MASS SPECTROMETRY OF METAL CARBONYLS

1.1. Binary Metal Carbonyls

Volatile metal carbonyls were among the first transition metal compounds studied by mass spectrometry.[1-4] On electron impact the metal carbonyl molecular ions lose carbonyl ligands successively. A series of ions with increasing degrees of coordinative unsaturation result. The bare metal atomic ion (or cluster ion in the case of polynuclear carbonyls) in fact frequently dominates the mass spectrum. These coordinatively unsaturated ions represented a unique opportunity as techniques for studying gas phase ion–molecule reactions developed. Mass spectrometric studies of the metal carbonyls provide the basis on which we can exploit this opportunity.

Early investigators characterized the electron impact-produced fragmentation of a number of binary metal carbonyls. In general, the electron impact-produced mass spectrum of metal carbonyls contains a complete series of peaks corresponding to ions formed by successive loss of carbonyl ligands from the parent ion.[1-4] In polynuclear carbonyls, cleavage of metal–metal bonds competes to some extent with cleavage of metal–ligand bonds.[4,5] Similar fragment ions are observed in photoionization mass spectrometry. Single-photon ionization has been used to obtain appearance potentials for the various fragments.[6] Multiphoton processes have been shown to strip all the ligands efficiently to produce bare metal ions and bare metal cluster ions.[7,8]

Douglas P. Ridge • Department of Chemistry, University of Delaware, Newark, Delaware 19716. *Wilma K. Meckstroth* • Department of Chemistry, Ohio State University, Newark, Ohio 43055.

1.2. Other Metal Carbonyls

Mass spectra have been reported for many metal carbonyls with other ligands in addition to the carbonyls.[9] These include halides, hydrides, alkyls, nitrosyls, cyclopentadienyls, and so on. Loss of CO on electron impact is a common feature of all of these species. Another general feature of these species as well as of the binary metal carbonyls is that when the parent ion fragments, the metal-containing fragment retains the charge. These species are, therefore, also a source of interesting coordinatively unsaturated metal ions.

1.3. Electronically Excited Fragment Ions

Electron impact is an energetic process, and it produces fragments with substantial amounts of internal energy. Polyatomic fragments probably have a distribution of vibrational energies and, in some instances, electronic energies. Atomic metal ions are known to be formed with a distribution of electronic energies in a number of cases.[10-16] Information on cross sections and lifetimes of those states is limited. Recent studies suggest, however, that spin- and parity-forbidden intersystem crossings occur readily in collision complexes containing a transition metal atomic ion and a polyatomic organic molecule.[11,17] Thus, the chemistry of metal-containing ions may be relatively insensitive to the initial distribution of electronic states formed in the electron impact event.

1.4. Negative Ion Mass Spectra

Negative ions are formed by a mechanism quite different from that by which positive ions are formed. Metal carbonyls generally have positive electron affinities. They can capture thermal electrons and eject one or two carbonyl ligands.[18] At higher electron energies more extensive fragmentation occurs.[19-22] As with positive ion fragmentations, the metal-containing fragment retains the charge. At high pressures or in an ion trap, most negative ions formed from metal carbonyls result from capture of thermalized secondary electrons. The internal energy of ions formed in this way can generally be expected to be less than that of cations formed by electron impact.

1.5. Thermochemistry of Fragment Ions

Thermochemical information on fragment ions formed from metal carbonyls can be derived in a number of ways. The heats of formation of electron impact-produced fragment ions can be deduced from their appearance potentials if the heat of formation of the parent neutral is known (see Reference 23 for examples). Similar information can usually be deduced with greater precision from photoionization mass spectrometry.[6] As discussed in detail elsewhere in this volume (Chapters by Dunbar and Freiser), gas phase ions may photodissociate when exposed to sufficiently energetic photons, and bond strengths in the ions may be deduced from variation of the photodissociation cross section with wavelength. Electron affinities may be reliably deduced from photoelectron spectra of fragment ions.[24,25] The data deduced in these various ways are still quite limited. One generalization that can be made is that the M—CO bonds in the metal carbonyl ions are not generically weaker than those in neutral metal carbonyls. The total energy required to remove all the carbonyls from the corresponding molecular ion is greater than or equal to the energy required to remove all the carbonyls from neutral $Cr(CO)_6$, $Fe(CO)_5$, $Ni(CO)_4$, $Mo(CO)_6$, and $W(CO)_6$.†

2. CLUSTERING REACTIONS OF METAL CARBONYL IONS WITH METAL CARBONYLS

2.1. Early Results

Early investigators found that fragment ions of metal carbonyls react with the neutral metal carbonyls to form polynuclear ions. A study using conventional mass spectrometric techniques identified such clustering reactions in group 6 metal carbonyls.[26] $Cr_2(CO)_n^+$ ($n = 2$–11) and $Cr_3(CO)_n^+$ ($n = 6$–14) were observed in $Cr(CO)_6$. Similar ions were observed in $Mo(CO)_6$ and $W(CO)_6$. In the mass spectrum of a mixture of $Cr(CO)_6$ and $Mo(CO)_6$, ions such as $CrMo(CO)_n^+$ were observed. The pressure dependence of the abundance of these ions showed them to be ion–molecule reaction products. The first ion cyclotron resonance study of clustering reactions of this kind was done on $Fe(CO)_5$.[27] Ion cyclotron resonance techniques provide a more detailed characterization of the chemistry and allow observation of more extensive chemistry. The cationic fragments of $Fe(CO)_5$ were observed to react sequentially to form clusters with as many as four metal atoms.

2.2. Structure–Reactivity Relations in Clustering Reactions: Multiple Bonds in Iron Carbonyl Clusters

An ion cyclotron resonance study of clustering reactions of anionic fragments of $Fe(CO)_5$ turned up an apparent relationship between structure and reactivity.[28] Early reports noted the reaction of $Fe(CO)_3^-$ with $Fe(CO)_5$ to form $Fe_2(CO)_6^-$, but $Fe(CO)_4^-$ was not observed to react in the earliest studies.[27,29] The recent study[28] used trapped ion techniques and revealed, however, that $Fe(CO)_4^-$ does react slowly with $Fe(CO)_5$ to form $Fe_2(CO)_8^-$. The complete series of reactions revealed in this study is outlined in Scheme 1. Some of the ions involved in this scheme react rapidly while others react much more slowly. The relative rate constants for reaction of these ionic species with $Fe(CO)_5$ vary over three orders of magnitude. The most reactive species react at a rate approaching the collision rate.

† Based on data in Reference 23.

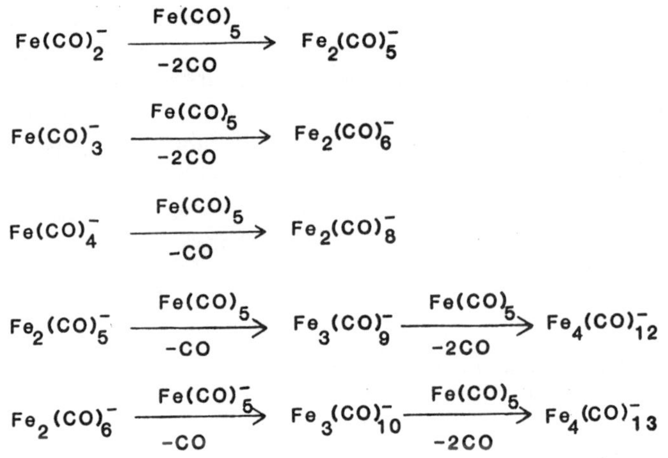

$$Fe(CO)_2^- \xrightarrow[-2CO]{Fe(CO)_5} Fe_2(CO)_5^-$$

$$Fe(CO)_3^- \xrightarrow[-2CO]{Fe(CO)_5} Fe_2(CO)_6^-$$

$$Fe(CO)_4^- \xrightarrow[-CO]{Fe(CO)_5} Fe_2(CO)_8^-$$

$$Fe_2(CO)_5^- \xrightarrow[-CO]{Fe(CO)_5} Fe_3(CO)_9^- \xrightarrow[-2CO]{Fe(CO)_5} Fe_4(CO)_{12}^-$$

$$Fe_2(CO)_6^- \xrightarrow[-CO]{Fe(CO)_5^-} Fe_3(CO)_{10}^- \xrightarrow[-2CO]{Fe(CO)_5} Fe_4(CO)_{13}^-$$

Scheme 1

It was observed that reactivity followed coordinative unsaturation. $Fe(CO)_3^-$, for example, reacts 100 times as fast as $Fe(CO)_4^-$. $Fe_2(CO)_5^-$ reacts 10 times as fast as $Fe_2(CO)_6^-$ which reacts at least 10 times as fast as either $Fe_2(CO)_7^-$ or $Fe_2(CO)_8^-$. The latter two ions are not observed to react at all. On the basis of this kind of observation, a criterion for coordinative unsaturation based on the 18-electron rule was suggested. An electron deficiency was defined as 18 less the average number of valence electrons per metal atom in each cluster. The average number of valence electrons per metal atom was taken as the total number of valence electrons in the cluster divided by the number of metal atoms. In order to determine the number of valence electrons in the cluster, it is necessary to determine the number of metal-to-metal bonds since each one adds two valence electrons to the cluster. This point is the key to establishing a connection between structure and reactivity. If reactivity depends on electron deficiency and electron deficiency depends on the number of metal–metal bonds, then reactivity depends on structure to the extent that a given structure will have a particular number of metal–metal bonds. The postulate was made that clusters with electron deficiencies greater than two would react efficiently, since such clusters would have on average a vacancy for a two-electron donor at each metal. Nearly any mechanism for the clustering reaction could operate efficiently in this case. As the electron deficiency drops below two, the reactivity of the cluster can be expected to drop. At the lower electron deficiencies some of the metals in the cluster will have a full complement of electrons and be unable to make new bonds, hence the drop in reactivity.

A plot of the log of the relative rate constant for reaction of anionic metal carbonyl clusters versus electron deficiency is shown in Figure 1. The data exhibit the expected behavior. In deriving the electron deficiencies, the Fe_3 and Fe_4 clusters were assumed to have triangular and tetrahedral structures with metal–metal bonds along the edges. The Fe_2 clusters were assumed to

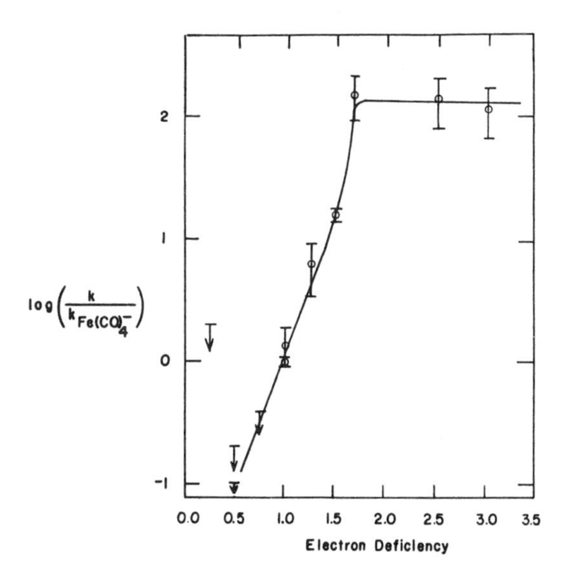

Figure 1. Relative rate constants for anionic clustering reactions in $Fe(CO)_5$ versus electron deficiencies determined assuming multiple metal–metal bonds.

have double metal–metal bonds if they had seven or fewer carbonyl ligands, and a single bond otherwise. The behavior of the data tends to support both the assumed connection between structure and reactivity and the assumed structures. It is interesting to note that if the Fe_2 species were all assumed to have single metal–metal bonds, then the plot of log of the rate constant versus electron deficiency would appear as in Figure 2. This pattern of behavior is not nearly as satisfactory as that in Figure 1, suggesting that on the basis of kinetic behavior, the Fe_2 species do have double bonds.

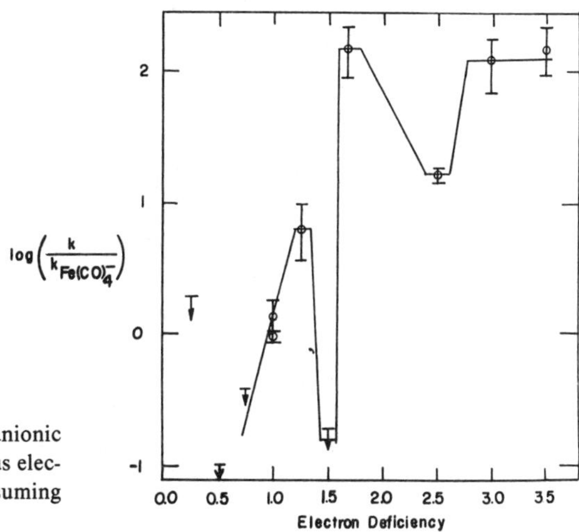

Figure 2. Relative rate constants for anionic clustering reactions in $Fe(CO)_5$ versus electron deficiencies determined assuming single metal–metal bonds.

Gas phase ions are difficult to characterize structurally and electronically so even though this structure-reactivity relationship provides only an indication of the number of metal-metal bonds, it is a welcome tool.

There are several interesting condensed phase parallels to the anionic clustering reactions in $Fe(CO)_5$. First of all, there is evidence from pulsed radiolysis studies that the solvated electron reacts efficiently with $Fe(CO)_5$ to form a species thought to be $Fe(CO)_4^-$ on the basis of its reactivity and its absorption spectrum.[30] The condensed phase and gas phase $Fe(CO)_4^-$ species both react on one encounter in 10^2 to 10^3 with $Fe(CO)_5$. The condensed phase anion has one strong absorption at ca. 300 nm, and the gas phase ion has its only nonzero photodissociation cross section in the same region. The pulsed radiolysis experiment did not reveal the identity of the products of the reaction of $Fe(CO)_4^-$ and $Fe(CO)_5$, but the reduction of $Fe(CO)_5$ in condensed phase produces species determined by ESR to include $Fe_2(CO)_8^-$, $Fe_3(CO)_{12}^-$, and $Fe_4(CO)_{13}^-$.[31] ESR studies of $^{57}Fe(CO)_5$ indicated the Fe_3 ion to have a triangular core of metal atoms and the Fe_4 ion to have a tetrahedral core of metal atoms.

While it is certainly to be expected that coordinative unsaturation is an important factor in the reactivity of these gas phase ionic metal complexes, it is not the only factor. A variety of thermodynamic, electronic, and steric constraints could limit the reactivity of these species. Thus, the simple relationship between electron deficiency and reactivity that obtains in the case of the anionic clustering reactions in $Fe(CO)_5$ may not obtain in other cases. It is important, therefore, that other systems have been examined.

2.3. Structure–Reactivity Relations in Group 7 Metal Carbonyl Clusters: Large Polyhedral Structures

Recently examined systems for which the simple relationship does seem to obtain are the group 7 carbonyls: $Mn_2(CO)_{10}$, $MnRe(CO)_{10}$, and $Re_2(CO)_{10}$. In $Re_2(CO)_{10}$, for example, sequential reactions produce ions as large as $Re_{10}(CO)_{29}^+$.[32,33] Figure 3 shows a plot of the log of the relative rate constants for the clustering reactions of the $Re_n(CO)_m^+$ ions versus electron deficiency. The plot has the same general features as Figure 1. Ions with electron deficiencies greater than two per metal atoms react efficiently. Reactivity drops off quite dramatically as the electron deficiency drops below two. It should be noted that electron counting for large clusters requires procedures somewhat more complicated than the simple 18-electron rule. A cluster containing six metal atoms arranged in an octahedron with bonds along the edges is saturated with 110 electrons rather than 108 ($=6 \times 18$), for example.[34] These requirements were accounted for in determining the electron deficiencies used in plotting Figure 3. In addition, all the reactant ions were assumed to consist of a closed polyhedral metal core with metal-metal bonds along the edges. The carbonyl ligands were assumed to be bound to the core. Since the model only involves electron count, it makes no distinction between bridging and terminal carbonyls or between mobile and fixed carbonyls. The model also

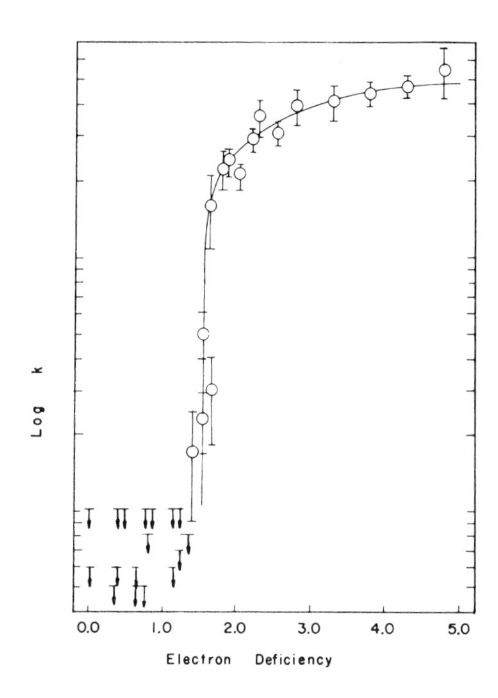

Figure 3. Relative rate constants for cationic clustering reactions in $Re_2(CO)_{10}$ versus electron deficiencies.

does not distinguish between structures with the same number of metal–metal bonds. The number of data points is large enough that the observed agreement with the expected behavior tends to support both the model and the structural assumptions. Similar behavior was observed in both the $Mn_2(CO)_{10}$ and $MnRe(CO)_{10}$ systems.[35]

There were a few ions in each of the group 7 systems that appeared to behave in contradiction to the general trend. It is possible that these ions have multiple bonds or open structures or that factors other than coordinative saturation control their reactivity. The interesting thing about the exceptions is that they tend to be the same in each system. $Re_6(CO)_{19}^+$, for example, reacts more rapidly with $Re_2(CO)_{10}$ than does $Re_6(CO)_{18}^+$ by a factor of >3. $Mn_3Re_3(CO)_{19}^+$ is similarly more reactive with $MnRe(CO)_{10}$ than is $Mn_3Re_3(CO)_{18}^+$, but in $Mn_2(CO)_{10}$ it is $Mn_6(CO)_{20}^+$ that is exceptional in that it reacts more rapidly than $Mn_6(CO)_{19}^+$.

Another interesting facet of the group 7 results is that a number of ions of very unusual stoichiometry were observed. Both $Mn_3(CO)_{15}^+$ and $Re_3(CO)_{15}^+$ were observed. These ions can have at most two metal–metal bonds, and the simplest, most symmetrical structures would have seven-coordinate metal atoms. From the point of view of isolobal analogies, it is interesting to note that one way these ions can be represented is as $[M(CO)_5]_3^+$. Since $M(CO)_5$ is isolobal with CH_3, these ions represented in this way are isolobal with $(CH_3)_3^+$. The stoichiometry of $(CH_3)_3^+$ is, of course, that of protonated propane. These $M_3(CO)_{15}^+$ ions might therefore be considered inorganic examples of coordinative supersaturation such as that found in the protonated alkanes. It is difficult to envision sensible structures for them that do not involve unusual coordination numbers for the metals.

Other interesting species observed are $Re_5(CO)_{20}^+$ and $Re_6(CO)_{22}^+$. If the former is centered around a trigonal bipyramid of metal atoms with metal-metal bonds along the edges, then it has 92 valence electrons, which exceeds 18 per metal atom. Similarly, if the latter is centered around an octahedron of metal atoms with metal–metal bonds along the edges, then it has 109 valence electrons, which also exceeds 18 per metal atom. In addition, we are unaware of any other examples of five or six metal atom clusters with as many as 20 or 22 carbonyl ligands, respectively. Various theoretical approaches to bonding in metal clusters suggest that octahedral metal clusters should have 110 valence electrons, and a number of clusters meeting that requirement have been observed.[36] The trigonal bipyramidal structures with more than 90 electrons that have been observed are axially elongated.[36] Theoretical discussions suggest that a regular trigonal bipyramid could not accommodate more than 90 electrons.[37]

A possible objection to octahedral and trigonal bipyramidal structures for $Re_6(CO)_{22}^+$ and $Re_5(CO)_{20}^+$ is that they would be sterically crowded. The analysis that has been done of crowding in metal carbonyl clusters suggests that these structures would have a very substantial amount of ligand–ligand repulsive energy,[38] although quantitative studies have not been done on these particular structures.

It is worth noting that in very recent studies even larger clusters have been observed in the $Re_2(CO)_{10}$ system.[39]

Examination of the kinetics of cationic clustering reactions in $Cr(CO)_6$,[40] $Fe(CO)_5$,[40] $Co(CO)_3(NO)$,[41] and $Ni(CO)_4$[41] reveals similar but somewhat more complicated trends than in the anions in $Fe(CO)_5$ and in the group 7 carbonyls. There is an interesting discussion of these trends in Chapter 4 of this volume. They are discussed in terms of cluster valence electrons in the manner proposed by Lauher.[34]

3. LIGAND SUBSTITUTION REACTIONS

When a second species is introduced with the metal carbonyl into the high-pressure mass spectrometer or the ion cyclotron resonance spectrometer, a completely new set of reactions occur. Of particular interest are the reactions between the second species and ions derived from the metal carbonyl. These may be divided into several categories. Ligand substitutions will be considered first.

Early examples of gas phase ligand substitution reactions of metal carbonyl ions include reactions of $C_5H_5Mn(CO)_3^+$ and $Fe(CO)_n^+$. The former was observed to react with EF_3 (E = P, As, and Sb). The trifluoride replaced two or three carbonyl ligands.[42] Such species as benzene, water, and methyl fluoride were found to displace carbonyls from $Fe(CO)_n^+$.[27] In some instances a single ligand molecule was found to displace two or more carbonyls. Ligands were categorized according to the number of carbonyls they could replace, and an order of metal–ligand bond strengths resulted. In $Co(CO)_3(NO)$, reactions with a mixture of ligands were examined.[43] Reactions of $Co(CO)_2(NO)^+$ with L_1 and L_2 produced not only $Co(CO)(NO)(L_1)^+$ and $Co(CO)(NO)(L_2)^+$ but also $Co(NO)(L_1)_2^+$, $Co(NO)(L_1)(L_2)_+$, and $Co(NO)(L_2)_2^+$. An order of ligand affinities for the metal was determined by examining the results of competitive ligand substitution. Comparison with the order of ligand affinities for Li^+ revealed inversions that showed that interactions between metal ions and ligands depend on a variety of factors. It was suggested that the "hard" and "soft" categories of condensed phase metal–ligand interaction might be useful for these gas phase ions.

Metal carbonyl anions have been observed to undergo ligand substitution.[44] Of particular interest are the reactions of 17-electron species such as $Fe(CO)_4^-$ and $Cr(CO)_5^-$ since they may exist in solution.[30,45] Maleic anhydride, for example, reacts efficiently with both of these ions to displace a CO.[46] Several $Fe(CO)_3(olefin)^-$ species have been observed in solution.[47] Other electronegative unsaturated organic compounds are found to react efficiently to displace a CO from these two ions. The best measure of reactivity for these ligands is their electron affinity.[46]

Several studies have been done to determine quantitative metal–ligand bond energies. Equilibrium constants for competitive ligand displacement reactions have been measured for many systems and scales of relative metal–ligand affinities constructed.[48] These studies do not involve metal carbonyls and are reviewed elsewhere.[49] Establishing absolute metal–ligand affinities has proved difficult. Metal–ligand bond energies can be deduced from photodissociation spectra of ionic metal–ligand complexes. Unambiguous determination of an adiabatic threshold for dissociation of the metal–ligand bond from a photodissociation spectrum requires several things. It requires knowledge of the absorption spectrum of the ion, its internal energy, the relative translation energy of the photofragments, and their internal energy. Though this information is sometimes difficult to obtain, a number of bond strengths derived from photodissociation spectra have been reported,[50] and these results are reviewed elsewhere in this volume (Chapter by Freiser). Metal–ligand bond energies can also be deduced from the threshold energy of a translationally driven, endothermic ligand transfer process. This also requires the knowledge of several things. These include the threshold law and the internal energy of the reactants and products. There has been considerable progress recently in characterizing translationally driven processes of transition metal ions, and as a result a number of reliable bond strengths have been reported.[51] These studies are also reviewed elsewhere in this volume (see Chapter by Armentrout).

4. OXIDATIVE ADDITION REACTIONS OF ATOMIC TRANSITON METAL IONS

Next we consider reactions besides ligand substitution that occur between organic molecules and ions derived from metal carbonyls. We will concentrate on the chemistry of the atomic ion, frequently a major peak in the mass spectrum of the metal carbonyl, which cannot react by ligand substitution. The remarkable propensity of a number of these atomic metal ions to form bonds to carbon is a major theme of this chemistry.

4.1. *Reactions with Alkyl Halides and Alcohols*

An informative early result was the observation of the reaction[52]:

$$Fe^+ + CH_3I \xrightarrow{(a)} FeCH_3^+ + I \qquad (1)$$
$$Fe^+ + CH_3I \xrightarrow{(b)} FeI^+ + CH_3$$

The two products were approximately equal in abundance, and the overall rate constant was nearly the collision rate. This implied that the reactions were exothermic, which implied that $D(Fe^+{-}CH_3) > D(I{-}CH_3) = 56$ kcal/mol. Subsequent ion beam results confirmed this, indicating that $D(Fe^+{-}CH_3) = 69 \pm 5$ kcal/mol.[53] This rather robust metal-to-carbon bond plays an important role in the chemistry of Fe^+. The products formed in reaction 1 suggested an oxidative addition mechanism involving a structure such as **1**:

$$I{-}Fe^+{-}CH_3$$

1

The prevalence of ligand substitution in the reactions of metal carbonyl ions provides a means of probing the mechanism of reaction 1. In a mixture of $Fe(CO)_5$, CD_3I, and CH_3I the following reactions may be unambiguously identified by ion cyclotron resonance techniques:

$$Fe(CO)^+ + CH_3I \rightarrow FeCH_3I^+ + CO \qquad (2)$$

$$FeCH_3I^+ + CD_3I \rightarrow FeCD_3I_2^+ + CH_3 \qquad (3)$$

No product other than the one indicated was observed for reaction 3. If the product of reaction 2 were an electrostatically bound complex, $(CH_3I)Fe^+$, then the methyl groups should be equivalent in reaction 3. The fact that only the one product was formed in reaction 3 suggests that the product of reaction 2 has structure **1**. This in turn suggests that reaction 1 proceeds through a structure such as **1**.

In the case of an alkyl halide with hydrogen atoms on the β-carbon, the reaction of ethyl iodide is typical[54]:

$$Fe^+ + CD_3CH_2I \xrightarrow{(a)} FeC_2H_2D_2 + DI \qquad (4)$$
$$Fe^+ + CD_3CH_2I \xrightarrow{(b)} FeC_2HD_3 + HI$$

The ratio of the two product channels is approximately the statistical ratio, 3:2 in favor of reaction 4a. This suggests that structures **2** and **3** are part of the mechanism:

$$I{-}Fe^{+}{-}CH_2CD_3 \qquad\qquad I{-}\underset{\underset{D}{|}}{Fe^{+}}\cdots{-}\begin{matrix}CH_2\\ \|\\ CD_2\end{matrix}$$

2 **3**

After an initial oxidative addition to form **2**, a D atom on the β-carbon shifts to the metal to form **3**. At this point DI may be reductively eliminated or the D atom may shift back to one of the carbon atoms. Since the D atom may shift to either carbon atom, **2** or its isotopic variant, $I{-}Fe^{+}{-}CD_2CH_2D$, may result. There are then H atoms on the β-carbon which may shift to the metal. It is then possible the HI may be reductively eliminated. If shifting H or D atoms back and forth between the metal and the β-carbon is more facile than reductive elimination, then the H and D atoms will be completely "scrambled" before elimination of HI or DI. Under these circumstances, HI and DI loss should be in a 3:2 ratio as observed.

Fe^{+}, Co^{+}, and Ni^{+} all dehydrohalogenate alkyl halides and dehydrate alcohols. The evidence suggests that the mechanisms for these processes involve oxidative addition of a C—X bond to the metal, shift of a β-hydrogen atom, and reductive elimination of HX.[54,55] The evidence includes the occurrence of displacement reactions such as given in Equations 2 and 3, correlations of reactivity with bond strengths, and isotopic labeling results such as those described in connection with reaction 4. It is worth noting that alkali metal ions also dehydrate alcohols and dehydrohalogenate alkyl halides but by a different mechanism.[55] The reactivity of the alkali metal ions follows the heterolytic bond strengths of the C—X bond of the alkyl halide or alcohol. This and other evidence suggests a mechanism involving a double-minimum potential surface where the internal barrier is associated with moving X^{-} from the carbon to the metal.

4.2. Reactions with Aryl Halides

Aryl halides also react with Fe.[56] Iodobenzene reacts to form $FeC_6H_5^{+}$ (85%) and FeI^{+}, suggesting an oxidative addition mechanism. Chlorobenzene, however, reacts to form an intriguing series of dehydrohalogenation products:

$$Fe^{+} \xrightarrow[-HCl]{C_6H_5Cl} FeC_6H_4^{+} \xrightarrow[-HCl]{C_6H_5Cl} FeC_{12}H_8^{+} \xrightarrow[-HCl]{C_6H_5Cl} FeC_{18}H_{12}^{+} \tag{5}$$

Besides indicating the generality of the tendency of Fe^{+} to insert into bonds, these reactions lead to unusual species which probably correspond to a metal-benzyne complex and metallopolyphenylenes.

4.3. Reactions with Alkanes

While oxidative addition of polar bonds to metals is a familiar phenomenon in solution, oxidative addition of the bonds in alkanes, especially C—C bonds, is unusual. The first example of such a process in the gas phase was the reaction of Fe^+ with 2-methylpropane observed in a mixture of $Fe(CO)_5$ and the alkane[57].

$$Fe^+ + i\text{-}C_4H_{10} \longrightarrow \begin{cases} FeC_3H_6^+ + CH_4 \\ FeC_4H_8^+ + H_2 \end{cases} \tag{6}$$

Methane elimination is the major channel in this reaction. The mechanism postulated for this process follows from that postulated for the alkyl halide and alcohol reactions. It is outlined in Scheme 2. The C—C bond adds oxidatively to the metal. An H atom on the β-carbon moves to the metal, and a C—H bond is reductively eliminated from the metal leaving a metal–olefin complex. An analogous mechanism can be postulated for the H_2 elimination. It was shown that Co^+[10,58] and Ni^+[10] react similarly but that Mn^+ and Cr^+ do not.[10] Cr^+ does react with 2-methylpropane to produce a $CrC_4H_8^+$ product, but that is probably the reaction of an electronically excited state of Cr^+. Evidence for the participation of such a state is discussed below.

Evidence in support of the oxidative addition mechanism for the alkane reactions was obtained using a mass spectrometric technique known as collision-induced decomposition (CID).[10,59] In this experiment an ion is prepared by electron impact on a suitable neutral or by ion–molecule reactions of such ions. After leaving the ion source of the mass spectrometer, the ion enters a mass analyzer which transmits ions of a selected mass. These ions then suffer collisions in a differentially pumped collision chamber. The resulting decomposition products are mass analyzed with a second mass analyzer. The spectrum of decomposition products reflects the structure of the original ion. A series of remarkable ligand substitution reactions make it possible to probe

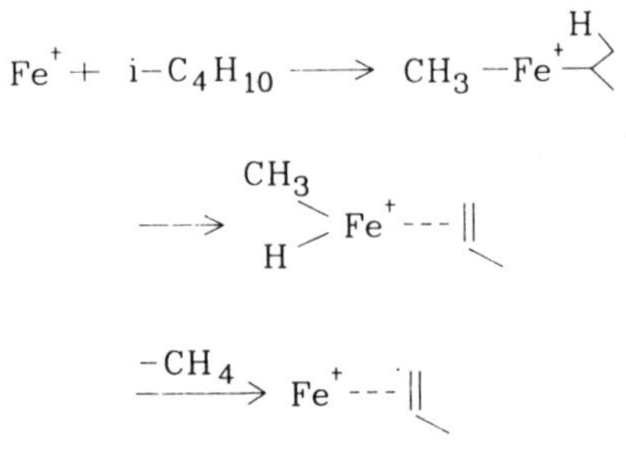

$$Fe^+ + i\text{-}C_4H_{10} \longrightarrow CH_3\text{—}Fe^+\text{—}CH(CH_3)_2$$

$$\longrightarrow (CH_3)(H)Fe^+\text{---}\|$$

$$\xrightarrow{-CH_4} Fe^+\text{---}\|$$

Scheme 2

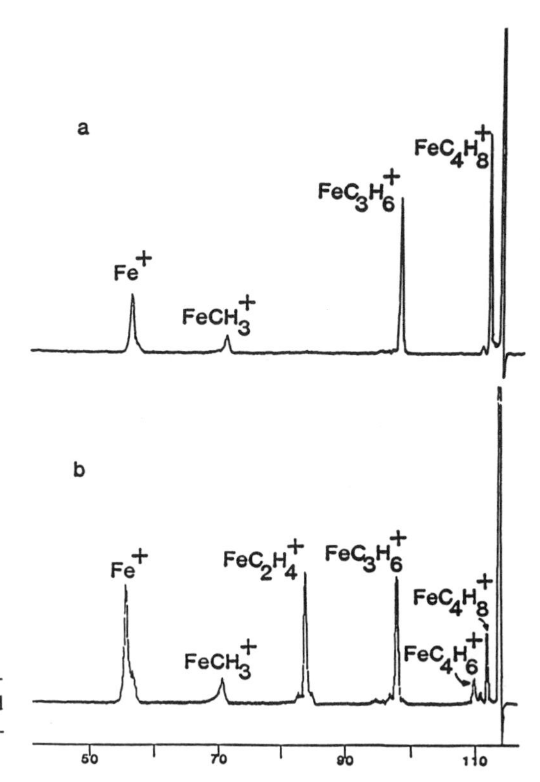

Figure 4. Collision-induced decomposition spectra of $FeC_4H_{10}^+$ formed from (a) 2-methylpropane and (b) *n*-butane.

the alkane chemistry using the CID technique:

$$M(CO)^+ + C_4H_{10} \rightarrow MC_4H_{10}^+ + CO \tag{7}$$

This reaction occurs for all $M(CO)^+$ species examined and for both C_4H_{10} isomers and for other alkanes. The reaction thus provides the simplest possible example of a metal–alkane complex. If oxidative addition of alkane bonds to the metals is facile, then the CID spectra of the products of reaction 7 should reflect that. That is, in fact, the case as illustrated in Figure 4. Shown in the figure are the CID spectra of $FeC_4H_{10}^+$ prepared from (a) 2-methylpropane and (b) *n*-butane. The fragments observed suggest clearly that the alkane decomposes when it associates with the metal ion. The observation of $FeCH_3^+$ provides direct evidence that structure **1** plays a role in the metal–butane complex. Also significant is the occurrence of the products of reaction 6 as fragments in the CID spectrum. That verifies the facility with which these products are formed. The ion prepared from *n*-butane has a different CID spectrum. Particularly notable are the $FeC_2H_4^+$ and $FeC_2H_5^+$ fragments in spectrum (b). These could readily be formed by oxidative addition to the metal of the middle C—C bond in *n*-butane. These features are absent from spectrum (a). These fragments could not readily form from 2-methylpropane by the oxidative addition mechanism. This again supports that mechanism. CID spectra of $CrC_4H_{10}^+$ show only Cr^+ as a significant fragment. This confirms the inertness of the ground state of Cr^+. It also suggests that the CID spectra reflect the potential surface on which the thermal ion–molecule reaction occurs.

Although the collisions occur at high energy (6 keV), the forward-scattered products which are sampled in a spectrum represent fragmentations resulting from the deposition of only a few eV or less in the parent ion. In addition, the short time scale of the collision tends to favor simple bond cleavages over complex rearrangements. These factors and the variations in the spectra themselves noted above support interpreting the spectra in simple structural terms as we have done.

The study of the reactions of Cr^+ has provided some insight into the role of electronically excited states in this oxidative addition reaction. Kinetic evidence suggests that electron impact on $Cr(CO)_6$ produces Cr^+ in at least two long-lived states.[10,11] The two states react at different rates with $Cr(CO)_6$. One state reacts with methane to produce $CrCH_2^+$ and the other is unreactive. Appearancc potential studies of the methane reaction suggest two states separated by 2.5 eV.[12] Careful examination of the kinetics of the methane reaction reveals that the predominant result of an encounter between excited Cr^+ and methane is to quench the reactive excited state.[11] This state has a lifetime of at least 2 s, and its decay to the ground state is probably spin and parity forbidden. Thus it is surprising that the state relaxes in a collision with methane. This suggests that in the reactions of the alkanes with atomic transition metal ions, the initial state of the ion may not be critical in determining reaction products. The ion–molecule complex may explore all energetically accessible surfaces until it finds one that leads readily to products. This is a possibility that must be considered in inferring electronic criteria for reactivity from experimental results.

4.4. The Effect of Oxidation State: Reactions of Fe^+, FeI^+, and FeI_2^+

The effect of metal ion oxidation state, metal ion excited states, and functionalization of the alkane are illustrated in the results of a single study.[60] The electron impact-produced fragments of $Fe(CO)_4I_2$ include Fe^+, FeI^+, and FeI_2^+, which are simple species containing Fe(I), Fe(II), and Fe(III), respectively. The chemistry of these species will reflect effects of the formal oxidation state of the metal on its reactivity. Examination of reactions of these species with pentane, 1-pentene, and 2-propyl alcohol shows the effects of functionalization of an alkane on these trends. Comparison of the reactivity of the Fe^+ fragment with the reactivity of Fe^+ generated in completely different ways indicates the role of excited states in the chemistry of Fe^+.

As shown in Table 1, Fe^+ reacts with pentane in much the same way that it reacts with butane. Neither FeI^+ nor FeI_2^+ reacts with pentane at all. That is to be expected if the mechanism is oxidative addition. Fe(II) and Fe(III) should be less susceptible to oxidation than Fe(I), particularly by an alkane. In the case of 1-pentene, however, there is an allylic C—C bond which might be expected to have enhanced reactivity. As shown in Table 2, both Fe^+ and

Table 1. Reactions of Pentane

Reactant ion	Products	Branching ratio: Present results	Branching ratio: Laser-desorbed Fe^+	Relative rate constant
Fe^+	$FeC_5H_{10}^+ + H_2$	0.15 ± 0.03	0.16	0.37 ± 0.08
	$FeC_4H_8^+ + CH_4$	0.24 ± 0.03	0.22	0.60 ± 0.08
	$FeC_3H_6^+ + C_2H_6$	0.20 ± 0.03	0.15	0.49 ± 0.08
	$FeC_2H_4^+ + C_3H_8$	0.41 ± 0.03	0.43	1.00 ± 0.08
FeI^+	None observed		<0.08	
FeI_2^+	None observed		<0.08	

Table 2. Reactions of 1-Pentene

Reactant ion	Products	Branching ratio	Relative rate constant
Fe^+	$FeC_5H_8^+ + H_2$	0.08 ± 0.05	0.17 ± 0.10
	$FeC_4H_6^+ + CH_4$	0.20 ± 0.05	0.40 ± 0.10
	$FeC_3H_6^+ + C_2H_4$	0.49 ± 0.05	1.00 ± 0.10
	$FeC_2H_4^+ + C_3H_6$	0.23 ± 0.05	0.47 ± 0.10
FeI^+	$FeIC_5H_8^+ + H_2$	0.34 ± 0.06	0.18 ± 0.03
	$FeIC_3H_6^+ + C_2H_4$	0.66 ± 0.06	0.35 ± 0.03
FeI_2^+	None observed		<0.05

FeI^+ react with 1-pentene with the predominant result that the activated C—C bond is cleaved. FeI_2^+ is unreactive even towards the activated bond. Again the reactivity of the three oxidation states follows the pattern to be expected for an oxidative addition mechanism.

As shown in Table 3, all three ions dehydrate 2-propanol, but probably by different mechanisms. Fe^+ probably reacts by oxidative addition of the

Table 3. Reactions of 2-Propanol

Reactant ion	Products	Branching ratio	Relative rate constant
Fe^+	$FeH_2O^+ + C_3H_6$	0.36 ± 0.06	0.55 ± 0.10
	$FeC_3H_6^+ + H_2O$	0.56 ± 0.06	0.86 ± 0.10
	$C_3H_7^+ + FeOH$	0.10 ± 0.06	0.15 ± 0.10
FeI^+	$FeIH_2O^+ + C_3H_6$	0.64 ± 0.06	1.00 ± 0.09
	$FeIC_3H_6^+ + H_2O$	0.36 ± 0.04	0.56 ± 0.05
FeI_2^+	$FeI_2H_2O^+ + C_3H_6$	0.82 ± 0.08	0.68 ± 0.06
	$FeI_2C_3H_6^+ + H_2O$	0.18 ± 0.03	0.14 ± 0.02

C—O bond followed by shift of a β-H atom and reductive elimination of water. FeI_2^+ is more likely to abstract a hydroxyl anion and then accept a proton from the incipient propyl cation. This dichotomy of behavior is analogous to that deduced from an extensive study of reactions of atomic transition metal ions and alkali metal ions with alcohols and alkyl halides.[55] This study is discussed above. An interesting trend emerges in the branching ratios observed in the 2-propanol reactions. These ratios are probably determined by competition between propylene and water to coordinate at the metal site. Water is most successful in the FeI_2^+ case and propylene in the Fe^+ case. This is consistent with the designation of the metal in FeI_2^+ as a high oxidation state 'hard" center and Fe^+ as a relatively "soft" center.

An interesting feature of the results is the good agreement of the branching ratios obtained and those obtained from a study involving Fe^+ generated by exposing the metal surface to a pulsed laser. This agreement is evident in Table 1. A possible explanation is that these two very different ionization methods produce only ground state Fe^+. Perhaps it is more likely that consistent with the conclusion derived above from the Cr^+ study, the products observed are relatively independent of the initial state of the reactant ion. Moving from one surface to another appears to occur without regard to the usual selection rules so reaction may occur on a surface different from the initial one. We conclude, therefore, that in these Fe^+ studies the branching to products occurs on the same surface regardless of how the ion is formed. The absolute rate constants in Table 4 make it clear that, at least for *n*-pentane and 1-pentene, there are no significant populations of unreactive states of Fe^+ generated by electron impact on $Fe(CO)_4I_2$.

It is interesting to note at this point an early study which illustrated the effects of metal oxidation state in a dramatic way.[61] Ti^+ and $TiCl^+$ formed by electron impact on $TiCl_4$ react with ketones to form TiO^+ and $TiClO^+$, respectively. This amounts to an increase of +2 in the formal oxidation numbers of the metal ions. $TiCl_2^+$ and $TiCl^{3+}$ react to form products of the form $TiCl_2(ketone)^+$ and $TiCl_3(ketone)^+$, where the coordinated ketone is formed from the reactant by loss of an olefin. The formal oxidation state of the metal does not change in these reactions, which probably proceed by a mechanism where the metal functions only as an acidic or electrophilic site. This is another example of the utility of the oxidation state formalism in characterizing the reactivity of these gaseous metal ion species.

Table 4. Absolute Rate Constants of Fe^+ Reactions

Reactant neutral	k (10^{-10} cm^3 s)	k_c^a (10^{-10} cm^3 s)	k/k_c
n-Pentane	13 ± 4	13.4	1.0
2-Propanol	9 ± 4	10.2	0.5
1-Pentene	14 ± 3	13.3	1.0

[a] Capture collision rate from equations in Ridge, D.P. In *Structure/Reactivity and Thermochemistry of Ions*; Ausloos, P.; Lias, S.G., Eds.; Reidel: Boston, 1987; pp. 1–14.

4.5. Reactions of Fe^+ with Cycloalkanes: Transition State Geometries

A recent examination of the reactions of Fe^+ with cycloalkanes[62] is interesting for two reasons. First, it provides a unique perspective on the oxidative insertion of a metal atom into a C—C bond. Second, it illustrates that even atomic transition metal species can react in a stereospecific way. While Fe^+ cleaves the strained bonds in cyclopropanes and cyclobutanes to form olefinic products, no evidence is found for C—C bond cleavage in cyclopentane and cyclohexane. The products observed in those cases correspond to unsaturated five- and six-membered hydrocarbon rings attached to the metal. It was postulated that activation of the C—C bonds in these rings was prevented by ring strain. That would be the case if the oxidative addition went through a C—Fe—C structure where the bond angle was 180°. Such an insertion structure could not be easily accommodated by five- and six-membered rings. The reactivity of larger rings was examined with the expectation that they would not have similar steric problems. In accord with these expectations, cyclooctane and larger cycloalkanes react with Fe^+ to give open olefins coordinated to the metal. The simplest explanation for this result is that the initial postulate was correct. This substantiates that oxidative addition of C—C bonds to atomic transition metal ions actually occurs, at least in the case of these cycloalkanes.

5. REACTIONS OF POLYNUCLEAR METAL CARBONYL IONS WITH ALKANES

5.1. Reaction of Diatomic Metal Carbonyl Ions

Next we consider several results on the reactivity of ions containing more than one metal atom. First, we consider two diatomic ions readily formed by electron impact on binuclear metal carbonyls. Electron impact on $Mn_2(CO)_{10}$ and $Co_2(CO)_{10}$ produces significant amounts of Mn_2^+ and Co_2^+, respectively. Mn_2^+ reacts with alcohols[63] and oxygen[64] but not alkanes.[10] The metal-metal bond is subject to cleavage in the reactions observed. An interesting pattern is observed in the case of Co_2^+. Although the atomic ion Co^+ reacts efficiently with alkanes,[10,58] the dimer does not.[10] Co_2CO^+, however, reacts with butane very efficiently.[65] The suggested explanation is that in order to initiate reaction, an empty orbital on Co^+, perhaps the 4*s*, must accept electron density from the σ C—M or C—C bond of the alkane. It may be that in Co_2^+ the 4*s* orbitals are involved in bonding and are effectively full or partially full. In Co_2CO^+, however, the CO may withdraw electron density from the appropriate orbital so that the metal orbital may accept electron density from the bond under attack. Much remains to be done before this or alternative explanations can be evaluated. Data on related systems are sparse, and a number of questions remain unanswered. For example, while Co_2^+ and Co^+ have quite different reactivities towards butanes, their reactivities towards molecular hydrogen are very similar. Both react endothermically to extract an H atom, and the translational excitation functions for these extraction reactions are almost superimposable.[66]

5.2. Reactions of Rhenium Carbonyl Cluster Ions with Cycloalkanes

Finally, we describe a study of the reactivity of some of the ions formed by the clustering reactions in decacarbonyl rhenium described above. In this study,[67] illustrated in Figure 5, the products of the clustering reactions are trapped in the ion cyclotron resonance cell in which they are formed. All of the product ions but one are then ejected from the cell by irradiating them at their cyclotron frequencies so that they are accelerated into the walls of the apparatus. The remaining ion reacts with cyclohexane admitted to the cell by a pulsed, electronically controlled valve. The mass spectrum of the products is then obtained by ion cyclotron resonance Fourier transform techniques. The results of a series of such experiments are summarized in Table 5. The most striking trend in the data is the decrease in reactivity with increasing coordinative saturation. In general, this parallels the behavior of the metal cluster ions in reacting with metal carbonyls. An electron deficiency per metal atom of 2.5

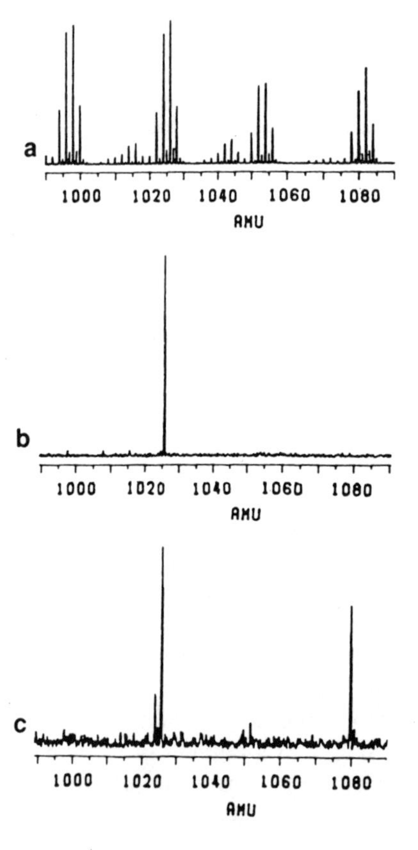

Figure 5. (a) FT-ICR spectrum of cluster ions formed by reactions of electron impact-produced fragment ions of $Re_2(CO)_{10}$ with the parent molecule. Reaction time was 2 s following a pulse of the ionizing electron beam. Pressure was 10^{-8} torr of $Re_2(CO)_{10}$. (b) As in (a) except that all ions but one are removed from the ion trap by accelerating their cyclotron motion. (c) As in (b) except that after isolating one mass ion in the trap an electronically controlled valve admits cyclohexane to a pressure of ca. 10^{-7} torr over ca. 0.5 s. The spectrum shows product ions of reaction between the isolated ion and cyclohexane.

Table 5. Product Distributions of Reactions of $Re_n(CO)_m^+$ with Cyclohexane to Give $Re_nC_6H_{2(6-x)}(CO)_{m-y}^+ + xH_2 + yCO$

n, m	x, y[a]	D[b]	H_2 yield[c]	CO yield / H_2 yield[d]
3, 7	2, 2 (0.74); 3, 2 (0.26)	4.67	2.36	0.85
3, 8	2, 1 (0.53); 2, 2 (0.29); 2, 3 (0.18)	4	2.00	0.82
3, 9	2, 1 (0.42); 2, 2 (0.38); 2, 3 (0.12); 2, 4 (0.08)	3.33	2.00	0.93
3, 10–13	No product observed	2.67–0.67		
4, 8	1, 1 (0.11); 1, 2 (0.21); 2, 1 (0.38); 2, 2 (0.30)	4.25	1.68	0.90
4, 9	1, 2 (0.70); 2, 1 (0.16); 2, 2 (0.14)	3.75	1.30	1.42
4, 10	1, 1 (0.57); 1, 2 (0.10); 1, 3(0.20); 2, 3 (0.07); 2, 3 (0.06)	3.25	1.13	1.50
4, 11	1, 2 (0.86); 2, 2 (0.14)	2.75	1.14	1.75
4, 12–16	No product observed	2.25–0.25		

[a] Branching ratio given in parentheses.
[b] The difference between 18 and the average number of valence electrons per metal atom in the reactant ion. Given by $18 - [(7n + 2m + 2b - 1)/n]$, where b, the number of metal–metal bonds, is taken to be 3 for $n = 3$ and 6 for $n = 4$.
[c] The yield of H_2 per mole of $Re_n(CO)_m^+$ reacted. In the case of $(n, m) = (4, 8)$, for example, the yield of H_2 is $0.11 + 0.21 + 2(0.38 + 0.30)$.
[d] The yield of CO is is defined analogous by to that of H_2.

or greater seems to be required for observable reaction to occur. In addition, the yield of hydrogen or the extent to which cyclohexane is dehydrogenated seems to increase with coordinative unsaturation. These results are consistent with an oxidative addition mechanism. Two metal sites may be involved. It may be that more sites are involved in multiple H_2 eliminations. It is clear from the table that quite selective chemistry can be obtained by choosing the appropriate cluster. The range of metal cluster size and extent of coordinative unsaturation available for further study is large. Significant insights into the behavior of metal clusters may be anticipated.

REFERENCES

1. Winters, R.E.; Kiser, R.W. *Inorg. Chem.* 1964, *3*, 699.
2. Winters, R.E.; Kiser, R.W. *Inorg. Chem.* 1965, *4*, 157.
3. Foffani, A.; Pignataro, S.; Cantone, B.; Grasso, F. *Z. Physik. Chem.* (*Frankfurt*) 1965, *45*, 79.
4. (a) Lewis, J.; Manning, A.R.; Miller, J.R.; Wilson, J.M. *J. Chem. Soc. A* 1966, 1663; (b) Johnson, B.J.G.; Lewis, J.; Williams, I.G.; Wilson, M.M. *J. Chem. Soc. A* 1967, 341; (c) Junk, G.A.; Svec, H.J. *J. Chem. Soc. A* 1970, 2101; (d) Meckstroth, W.K.; Ridge, D.P. *J. Am. Chem. Soc.* 1985, *107*, 2281.
5. Johnson, B.F.G. In *Mass Spectrometry of Metal Compounds*; Charalambous, J., Ed.; Butterworths: Boston, 1975; pp. 113–126.
6. Distefano, G. *J. Res. Nat. Bur. Stand., Sect. A* 1972, *74*, 233.
7. Lichtin, D.A.; Bernstein, R.B.; Vaida, V. *J. Am. Chem. Soc.* 1982, *104*, 1830.
8. Leopold, D.G.; Vaida, V. *Laser Chem.* 1983, *3*, 49.

9. For reviews, see Charalambous, J., Ed. *Mass Spectrometry of Organometallic Compounds*; Butterworths: Boston, 1975.
10. Freas, R.B.; Ridge, D.P. *J. Am. Chem. Soc.* 1980, *102*, 7129.
11. Reents, W.D., Jr.; Strobel, F.; Freas, R.B., III; Wronka, J.; Ridge, D.P. *J. Phys. Chem.* 1985, *89*, 5666.
12. Halle, L.F.; Armentrout, P.B.; Beauchamp, J.L. *J. Am. Chem. Soc.* 1981, *103*, 962.
13. Elkind, J.L.; Armentrout, P.B. *J. Phys. Chem.* 1986, *90*, 6576.
14. (a) Elkind, J.L.; Armentrout, P.B. *J. Phys. Chem.* 1986, *90*, 5736; (b) Elkind, J.L.; Armentrout, P.B. *J. Chem. Phys.* 1986, *84*, 4862.
15. Elkind, J.L.; Armentrout, P.B. *J. Am. Chem. Soc.* 1986, *108*, 2765.
16. Elkind, J.L.; Armentrout, P.B. *J. Chem. Phys.* 1987, *86*, 1868.
17. Schultz, R.H.; Elkind, J.L.; Armentrout, P.B. *J. Am. Chem. Soc.* 1988, *110*, 411–423.
18. Pignataro, S.; Foffani, A.; Grasso, F.; Cantone, B. *Z. Phys. Chem.* (*Frankfurt am Main*) 1965, *47*, 106.
19. Sullivan, R.E.; Kiser, R.W. *J. Chem. Phys.* 1968, *49*, 1978.
20. Kiser, R.W.; Sullivan, R.E.; Lupin, M.S. *Anal. Chem.* 1969, *41*, 1958.
21. Winters, R.E.; Kiser, R.W. *J. Chem. Phys.* 1966, *44*, 1964.
22. Winters, R.E.; Kiser, R.W. *J. Phys. Chem.* 1965, *69*, 1618.
23. Rosenstock, H.M.; Draxl, K.; Steiner, B.W.; Herron, J.T. *J. Phys. Chem. Ref. Data* 1977, *9*, Suppl. 1.
24. Engelking, P.C.; Lineberger, W.C. *J. Am. Chem. Soc.* 1979, *101*, 5569.
25. Stevens, A.E.; Feigerle, C.S.; Lineberger, W.C. *J. Am. Chem. Soc.* 1982, *104*, 5026.
26. Kraihanzel, C.S.; Conville, J.J.; Sturm, J.E. *J. Chem. Soc., Chem. Commun.* 1971, 159.
27. (a) Foster, M.S.; Beauchamp, J.L. *J. Am. Chem. Soc.* 1971, *93*, 4924; (b) Foster, M.S.; Beauchamp, J.L. *J. Am. Chem. Soc.* 1975, *97*, 4808.
28. Wronka, J.; Ridge, D.P. *J. Am. Chem. Soc.* 1984, *106*, 67.
29. Dunbar, R.L.; Ennever, J.F.; Fackler, J.P., Jr. *Inorg. Chem.* 1973, *12*, 2724.
30. Reed, D.T.; Meckstroth, W.K.; Ridge, D.P. *J. Phys. Chem.* 1985, *89*, 4578.
31. Krusic, P.J.; San Filippo, J., Jr.; Hutchinson, B.; Hance, R.L.; Daniels, L.M. *J. Am. Chem. Soc.* 1981, *103*, 2129.
32. Meckstroth, W.K.; Ridge, D.P. *Int. J. Mass Spectrom. Ion Proc.* 1984, *61*, 149.
33. Meckstroth, W.K.; Ridge, D.P.; Reents, W.D., Jr. *J. Phys. Chem.* 1985, *89*, 612.
34. See, for Example, Lauher, J.W. *J. Organomet. Chem.* 1981, *213*, 25, and references therein.
35. Meckstroth, W.K.; Freas, R.B.; Reents, W.D., Jr.; Ridge, D.P. *Inorg. Chem.* 1985, *24*, 3139.
36. Teo, B.K.; Longoni, G.; Chung, F.R.K. *Inorg. Chem.* 1984, *23*, 1257.
37. Teo, B.K. *Inorg. Chem.* 1984, *23*, 1251.
38. Benfield, R.E.; Johnson, B.F.G. *J. Chem. Soc., Dalton Trans.* 1980, 1743.
39. Wronka, J.; Forbes, R.; Laukien, F. *Int. J. Mass Spectrom. Ion Proc.* 1988, *83*, 23–44.
40. Fredeen, D.A.; Russell, D.H. *J. Am. Chem. Soc.* 1985, *107*, 3762.
41. Fredeen, D.A.; Russell, D.H. *J. Am. Chem. Soc.* 1986, *108*, 1860.
42. Mueller, J.; Fenderl, K. *Chem. Ber.* 1971, *104*, 2207.
43. Weddle, G.H.; Allison, J.; Ridge, D.P. *J. Am. Chem. Soc.* 1977, *99*, 105.
44. Squires, R.R. *Chem. Rev.* 1987, *87*, 623.
45. Pickett, C.J.; Pletcher, D. *J. Chem. Soc., Dalton Trans.* 1976, 749.
46. Weddle, G.; Wronka, J.; Meckstroth, W.; Pan, Ying; Ridge, D.P. Unpublished results.
47. El Murr, N.; Riveccie, M.; Dixneuf, P. *J. Chem. Soc., Chem. Commun.* 1978, 552.
48. (a) Uppal, J.S.; Staley, R.H. *J. Am. Chem. Soc.* 1982, *104*, 1235; (b) Uppal, J.S.; Staley, R.H. *J. Am. Chem. Soc.* 1982, *104*, 1238; (c) Kappes, M.M.; Staley, R.H. *J. Am. Chem. Soc.* 1982, *104*, 1813; (d) Kappes, M.M.; Staley, R.H. *J. Am. Chem. Soc.* 1982, *104*, 1819; (e) Jones, J.W.; Staley, R.H. *J. Am. Chem. Soc.* 1982, *104*, 2296; (f) Jones, R.W.; Staley, R.H. *J. Phys. Chem.* 1982, *86*, 1387.
49. Kappes, M.M.; Staley, R.H. In *Ion Cyclotron Resonance Spectrometry II, Lecture Notes in Chemistry*, Vol. 31; Hartmann, H.; Wanczek, K.-P., Eds.; Springer-Verlag: New York, 1982; pp. 119–139.
50. Hettich, R.L.; Jackson, T.C.; Stanko, E.M.; Freiser, B.S. *J. Am. Chem. Soc.* 1986, *108*, 5086.

51. Elkind, J.L.; Armentrout, P.B. *Inorg. Chem.* 1986, *25*, 1078.
52. Allison, J.; Ridge, D.P. *J. Organomet. Chem.* 1975, *99*, C11.
53. Halle, L.F.; Armentrout, P.B.; Beauchamp, J.L. *Organometallics* 1982, *1*, 963.
54. Allison, J.; Ridge, D.P. *J. Am. Chem. Soc.* 1976, *98*, 7445.
55. Allison, J.; Ridge, D.P. *J. Am. Chem. Soc.* 1979, *101*, 4998.
56. Dietz, T.G.; Chatellier, D.S.; Ridge, D.P. *J. Am. Chem. Soc.* 1978, *100*, 4905.
57. Allison, J.; Freas, R.B.; Ridge, D.P. *J. Am. Chem. Soc.* 1979, *101*, 1332.
58. Armentrout, P.B.; Beauchamp, J.L. *J. Am. Chem. Soc.* 1980, *102*, 1736.
59. Larsen, B.S.; Ridge, D.P. *J. Am. Chem. Soc.* 1984, *106*, 1912.
60. Smith, Arthur; Pan, Ying; Ridge, D.P. Submitted for publication.
61. Allison, J.; Ridge, D.P. *J. Am. Chem. Soc.* 1978, *100*, 163.
62. Ridge, D.P. In *Structure/Reactivity and Thermochemistry of Ions*; Ausloos, P.; Lias, S.G. Eds.; D. Reidel: Boston, 1987; pp. 165–176.
63. Larsen, B.S.; Freas, R.B., III; Ridge, D.P. *J. Phys. Chem.* 1984, *88*, 6014.
64. Armentrout, P.B.; Loh, S.K.; Ervin, K. *J. Am. Chem. Soc.* 1984, *106*, 1161.
65. Freas, R.B.; Ridge, D.P. *J. Am. Che. Soc.* 1984, *106*, 825.
66. Armentrout, P.B. In *Structure/Reactivity and Thermochemistry of Ions*; Ausloos, P.; Lias, S.G., Eds.; Reidel: Boston, 1987; pp. 97–164.
67. Wronka, J.; Forbes, R.; Laukien, F.; Ridge, D.P. *J. Phys. Chem.* 1987, *91*, 6450.

4

Structure–Reactivity Relationships for Ionic Transition Metal Carbonyl Cluster Fragments

David H. Russell, Donnajean Anderson Fredeen, and Ronald E. Tecklenburg

Many parallels have been made between transition metal clusters and metal surfaces. The geometric features of the metal-cluster-bound ligands are similar to those of adsorbed molecules on a metal surface, and the average metal–ligand and metal–metal binding energies of transition metal clusters are comparable to the binding energies of the metal surfaces and chemisorbed molecules.[1)] The cluster/surface analogy (in terms of chemical properties) begins to break down, however, when the coordination saturation of the transition metal cluster is considered. Typically, the coordination number for metal–ligand interactions of metal clusters is high, and the coordination number for metal–metal interactions is low. Conversely, the high coordination number for metal surfaces is due to the metal–metal interaction. Because most clusters are stable, coordinatively saturated molecules which obey the 18-electron rule,[2] experimental methods for probing coordinatively unsaturated clusters and cluster fragments, e.g., matrix isolation[3] and fast time-resolved infrared spectrscopy,[4] are being developed. The target species for many of these studies are the reactive 17-electron species and higher coordinatively unsaturated species. For the past several years the development of methods for preparing coordinatively unsatur-

David H. Russell, Donnajean Anderson Fredeen, and Ronald E. Tecklenburg • Department of Chemistry, Texas A&M University, College Station, Texas 77843.

ated ionic transition metal cluster fragments has been actively pursued in our laboratory. As a direct result of this work, ionic transition metal cluster fragments have a range of numbers of metal atoms (typicallly 2–8) and ligands (typically CO) can be prepared. In addition, the ionic cluster fragments span a large range of coordination unsaturation. Thus, the chemical and physical properties of these clusters can be studied as a function of cluster size and metal-to-ligand ratio.

Our early studies attempted to rationalize the structure and reactivity of the ionic cluster fragments $Cr_x(CO)_y^+$, $Fe_x(CO)_y^+$, $Co_x(CO)_y(NO)_z^+$, and $Ni_x(CO)_y^+$.[5,6] For example, Fe^+, formed by direct ionization of $Fe(CO)_5$, reacts with neutral $Fe(CO)_5$ to form $Fe_2(CO)_3^+$ and $Fe_2(CO)_4^+$. The latter readily reacts with $Fe(CO)_5$ to form $Fe_3(CO)_x$, $x = 5–7$, whereas $Fe_2(CO)_3^+$ is virtually inert toward $Fe(CO)_5$. In terms of metal-to-ligand ratio, $Fe_2(CO)_3^+$ is more coordinatively unsaturated than is $Fe_2(CO)_4^+$, but in terms of reactivity $Fe_2(CO)_4^+$ appears to be more coordinatively unsaturated than $Fe_2(CO)_3^+$. Such dramatic differences in reactivity can be rationalized by assuming that the reactivity of a particular ionic cluster fragment arises from differences in electron deficiency. Thus, the experimentally measured electron deficiency (based on relative reactivity) of a particular ionic cluster fragment can be used to estimate the bond order of the ionic cluster fragments. Comparison of the $Cr(CO)_6$, $Fe(CO)_5$, $Co(CO)_3(NO)$, and $Ni(CO)_4$ systems shows that the ionic cluster fragments of the type $M_x(CO)_y^+$ and $M_x(CO)_y(NO)_z^+$ can be divided into two categories: (1) ionic cluster fragments where the structure-reactivity relationship follows a simple polyhedral model, i.e., the 18-electron rule adequately describes these ionic cluster fragments, and (2) ionic cluster fragments where the structure-reactivity relationship does not obey a simple polyhedral model. That is, ionic cluster fragments of category (2) exhibit unusual binding of the ligands and/or metal centers, resulting in a high bond order. These two classes of ionic cluster fragments have been discussed in terms of the interrelationship of the reactivity/electron deficiency model[5,6] and the cluster valence molecular orbital model (CVMO) developed by Lauher.[7] The electron deficiency model and the CVMO model provide insight into the bonding of the ionic cluster fragments, but the CVMO model offers a better understanding of the sequence leading to the formation of the ionic cluster fragments.

These same ideas can be extended to the heteronuclear ionic cluster fragments of the $Co(CO)_3(NO)/Ni(CO)_4$, $Fe(CO)_5/Ni(CO)_4$, and $Fe(CO)_5/Co(CO)_3(NO)$ systems.[8] The heteronuclear ionic cluster fragments are discussed in this chapter in terms of their relative reactivity and degree of coordination unsaturation. The sequence of reactions leading to formation of the different ionic cluster fragments is discussed in terms of the CVMO model. Such studies combined with studies on metal-metal and metal-ligand bond energies, chemical reactivity (ligand exchange reaction, oxidation, etc.), and electronic spectroscopy will aid in the development of fundamental understanding of metal chemistry, e.g., gas phase metal ion/metal cluster reactivity and possibly catalysis.

1. ELECTRON DEFICIENCY MODEL

The ionic cluster fragments formed by the reaction of $M(CO)_x^+$ with $M(CO)_y$ (M = Cr, Fe, Co, and Ni) can be understood by examining the electron deficiencies of the cluster fragments. The electron deficiency (ED) of a particular ionic cluster fragment is calculated by using the 18-electron rule:

$$ED = [18n - (\text{total no. of valence electrons in the cluster})]/n \tag{1}$$

where n equals the total number of metal atoms in the cluster. Electron deficiencies calculated by using the 18-electron rule assume that the ionic cluster fragments have the structures found in the triangular polyhedra of the boron hydrides. As an example, by using this method an electron deficiency of 0.75 is obtained for the $Ni_4(CO)_9^+$ ion:

$$ED = \{18(4) - [4(13) + 9(2) - 1]\}/4 = 0.75 \tag{2}$$

Conversely, an electron deficiency of 3.5 is estimated for the $Ni_2(CO)_4^+$ ion. Since the electron deficiency is an indication of the number of open coordination sites on a metal atom, a direct correlation should exist between electron deficiency and reactivity.[5,6] Thus, $Ni_2(CO)_4^+$ (electron deficiency of 3.5) should have a higher relative reactivity than $Ni_4(CO)_9^+$ (electron deficiency of 0.75). On the basis of this simple concept, a plot of reactivity versus electron deficiency should increase in a monotonic manner. This concept is shown graphically in Figure 1, where we have plotted log relative reaction rate for the ionic cluster fragments formed by $Ni(CO)_3^+$ reacting with neutral $Ni(CO)_4$. Clearly, for this system a direct relationship between reactivity and electron deficiency is indicated. A similar plot for the $Fe(CO)^+$ ion reacting with $Fe(CO)_5$ is shown in Figure 2. Again, note the correlation between reactivity of the ionic cluster fragments and the calculated electron deficiency.

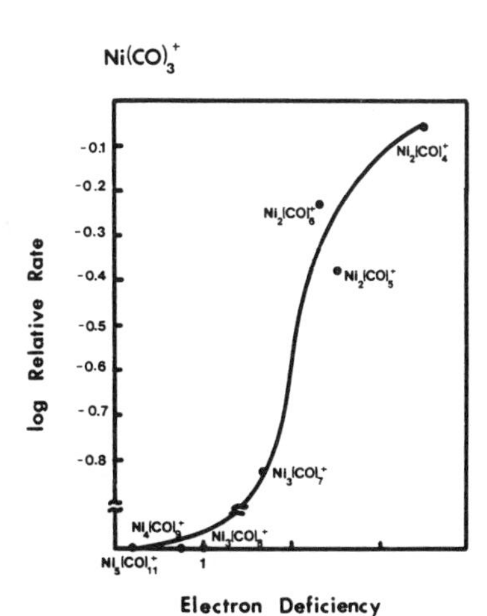

Figure 1. Reactivity/electron deficiency data for the $Ni(CO)_3^+/Ni(CO)_4$ system. The reaction rates are normalized to the reaction rate of the $Ni(CO)_3^+$ ion.

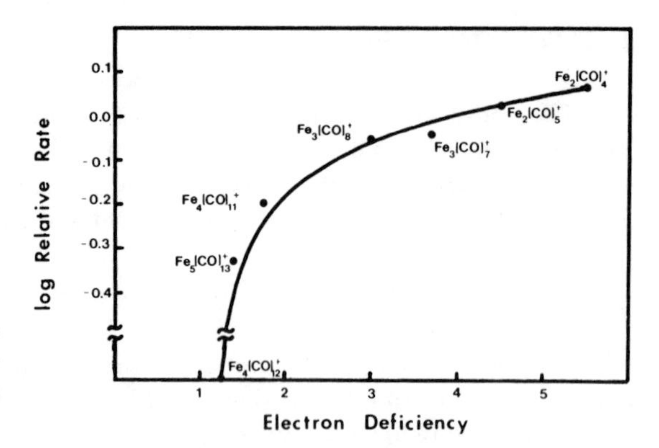

Figure 2. Reactivity/electron deficiency data for the $Fe(CO)^+/Fe(Co)_5$ system. The reaction rates are normalized to the reaction rate of the $Fe(CO)^+$ ion.

However, the correlation between reactivity and electron deficiency is not always so obvious. The reactivity versus electron deficiency data for the ionic cluster fragments formed by Ni^+ reacting with $Ni(CO)_4$ is shown in Figure 3. The simple relationship between reactivity and electron deficiency does not hold for this system. For example, $Ni_2(CO)_2^+$, $Ni_3(CO)_4^+$, and $Ni_3(CO)_{5^+}$ have electron deficiencies of 5.5, 3.7, and 3.0, respectively, based on the 18-electron rule. However, on the basis of the very low ion–molecule reactivity it is evident that these ionic cluster fragments are not highly coordinatively unsaturated. For the $Cr(CO)_6$ and $Fe(CO)_5$ [(5)] systems we suggested that the electron deficiency of a given system could be altered by the presence of multiple metal–metal bonds and/or four- or six-electron-donating CO ligands, i.e., some ionic cluster fragments appear to have higher bond order (lower degree of coordination unsaturation) than predicted by the 18-electron rule.

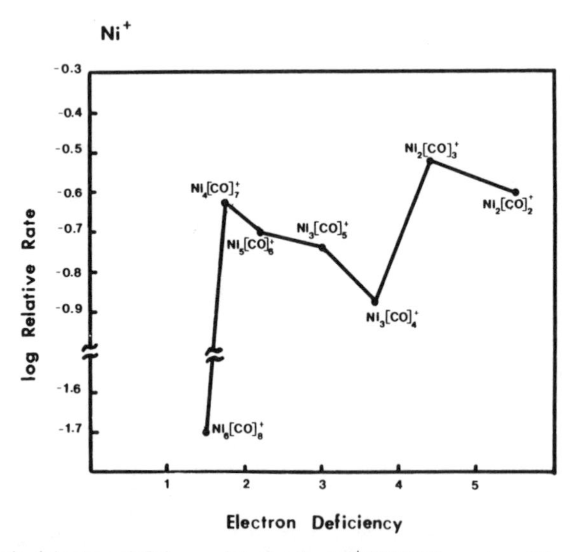

Figure 3. Reactivity/electron deficiency data for the $Ni^+/Ni(CO)_4$ system. The rates of reaction of the ionic cluster fragments are normalized to the reaction rate of Ni^+.

The ionic cluster fragments formed by Ni^+ reacting with $Ni(CO)_4$ also illustrate the interplay between coordination unsaturation and ion-molecule reactivity. For instance, the values of the electron deficiency of $Ni_2(CO)_2^+$ and $Ni_2(CO)_3^+$ calculated from Eqation 1 are 5.5 and 4.5, respectively. Such high electron deficiencies should lead to high reactivities, but this is clearly not the case for $Ni_2(CO)_2^+$. The low reactivity for the $Ni_2(CO)_2^+$ ion is consistent with the idea that this ion has a high bond order, i.e., multiple metal-metal bonds and/or CO donating four electrons. The high bond orders inferred from the reactivity/electron deficiency data can be rationalized by considering that the nickel atoms localize the metal *d* electrons rather than use them for ligand binding. The increase in the bond order for $Ni_2(CO)_2^+$ would result in a decrease in the electron deficiency. Therefore, a high bond order is assigned to the ionic cluster fragments which deviate from the reactivity versus electron deficiency curve in Figure 3. That is, the ionic cluster fragments $Ni_2(CO)_2^+$, $Ni_3(CO)_4^+$, and $Ni_3(CO)_5^+$ have low reactivities, but the calculated electron deficiencies (assigning simple polyhedral structures) of these ions are high. On the other hand, if high bond orders are assigned to these ionic cluster fragments, the reactivity versus electron deficiency curve shows a monotonic increase. Figure 4 contains the electron deficiency data for the $Ni^+/Ni(CO)_4$ system obtained by assigning high bond orders to $Ni_2(CO)_2^+$, $Ni_3(CO)_4^+$, and $Ni_3(CO)_5^+$.

It is also interesting to note the difference in reactivity for two ionic cluster fragments which differ by only a single CO ligand. For example, the pairs of ionic cluster fragments $Ni_2(CO)_2^+$ and $Ni_2(CO)_3^+$, $Ni_3(CO)_4^+$ and $Ni_3(CO)_5^+$, and $Fe_2(CO)_3^+$ and $Fe_2(CO)_4^+$ have significantly different reactivities. Considered in this manner, it is even more surprising that ionic cluster fragments

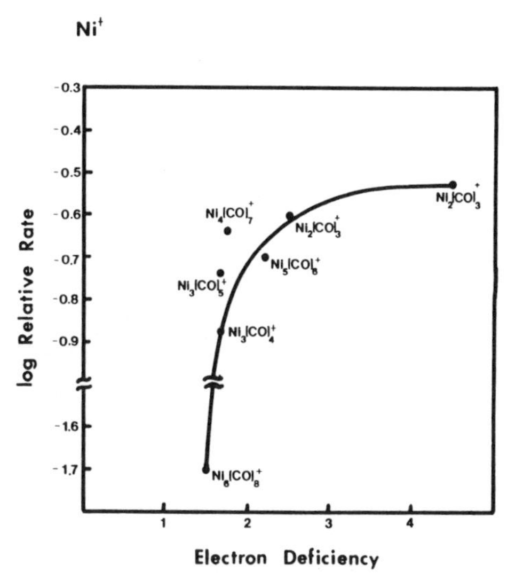

Figure 4. Reactivity/electron deficiency data for the $Ni^+/Ni(CO)_4$ system. In this graph the electron deficiencies of the ionic cluster fragments having low reactivities are adjusted to reflect high bond order for the metal-metal and/or metal-ligand interaction.

such as $Ni_2(CO)_2^+$ and $Fe_2(CO)_3^+$ are so unreactive. At least hypothetically, removal of a single CO ligand from $Ni_2(CO)_3^+$ to form $Ni_2(CO)_2^+$ should produce an ionic cluster fragment with one additional open coordination site and a corresponding increase in reactivity.

On the basis of these general concepts, we propose that ionic cluster fragments which have low reactivities have electron deficiencies which deviate from the value predicted from the 18-electron rule. Ionic cluster fragments having high bond orders have low electron deficiencies since the number of valence electrons increases (see Equation 1). That is, in cases involving multiple metal–metal bonds the number of valence electrons increases because two additional electrons are provided by each metal–metal bond.[5,6] Similarly, two additional valence electrons are provided by each four-electron-donating CO ligand. Thus, a CO molecule which donates four electrons and a double metal–metal bond decrease the electron deficiency by the same amount. (Exact determination of the structure of the ionic cluster fragments cannot be obtained by the present experimental technique.)

The electron deficiency data for the ionic cluster fragments also gives some indication of the Lewis acidity/basicity of the metal centers. In the ionic cluster fragments which have small electron deficiencies, the metals accept electrons from either the ligands and/or the surrounding metals. If multiple metal–metal bonds are formed, this suggests that π^* backbonding is unfavorable. Thus, the metal centers of ionic cluster fragments with low electron deficiencies can be classified as Lewis acids. Conversely, ionic cluster fragments having high electron deficiencies will have metal centers which act as Lewis bases, i.e., π^* backbonding is favorable.

2. CLUSTER VALENCE MOLECULAR ORBITAL MODEL

The concept of electron deficiencies and bond order of the ionic cluster fragments shares many similarities with the cluster valence molecular orbital model developed by Lauher.[7] Lauher describes the molecular orbitals of transition metal clusters to be of two types: (1) high-lying antibonding orbitals (HLAO) and (2) cluster valence molecular orbitals (CVMO). According to Lauher, the HLAO are too high in energy to accept electrons from ligands or contain nonbonding electrons. The CVMO, on the other hand, are accessible to either ligand or metal electrons. The stoichiometry of a cluster is determined by its geometry and not by the identity of the metals or the ligands.[7] Therefore, a cluster with a particular size or geometry has a certain number of bonding CVMO; a tetrahedron has 30 CVMO available for cluster bonding while a trigonal bipyramid has 36 CVMO. The number of cluster valence electrons (CVE) for a tetrahedron would then be 60 (two electrons for every CVMO). In the CVMO model, electron deficiency or unsatuation corresponds to a decrease in the number of CVE. Lauher suggests that "Unsaturated (or electron deficient) clusters can be expected to show unusual reactivity. Such a cluster would in general readily react with additional ligands to achieve the proper number of CVE."[7]

On the basis of the electron deficiency model, the reactivity of a particular ionic cluster fragment is due to the electron deficiency. This concept is consistent with the basic ideas of Lauher's CVMO model; e.g., the reactivity of a cluster or ionic cluster fragment is determined by the number of unoccupied CVMO. For example, a tetrahedron has 30 CVMO (60 CVE), but if the cluster has one unoccupied CVMO (i.e., 58 CVE) the cluster would tend to add an additional two-electron-donating ligand. Similarly, an electron-deficient ionic cluster fragment would react with additional ligands to satisfy this deficiency. The electron deficiency of the ion can also be satisfied by changing of the bonding within the cluster fragment, i.e., changes in the metal–ligand or metal–metal bond order.

The sequence of reactions leading to formation of the ionic cluster fragments can be generalized as follows. The ionic cluster fragments formed by ion–molecule reactions which follow the simple polyhedral model react with the neutral to add 14 electrons. The addition of 14 electrons is favored because this gives the proper number of ionic cluster valence electrons (ICVe). A general reaction scheme is given in Scheme 1.

The addition of 14 electrons to an ionic cluster fragment corresponds to the addition of a ligand to achieve the proper number of CVE. An excellent example of 14-electron addition can be found in the Cr^+ reaction scheme.[5] $Cr_2(Co)_4^+$ (19 ICVE) reacts with $Cr(Co)_6$ [by addition of $Cr(CO)_4$] to give $Cr_3(Co)_8^+$ (33 ICVE). The addition of $Cr(CO)_4$ to $Cr_2(CO)_4^+$ initially corresponds to addition of a ligand [$Cr(CO)_4$] to form $[Cr_2(CO)_4 \cdots L]^+$ where $L = Cr(CO)_4$. The $[Cr_2(Co)_4 \cdots L]^+$ species has the proper number of ICVE (33) needed for a dimer. However, once the proper number of ICVE is obtained, $[Cr_2(Co)_4 \cdots L]^+$ rearranges to a stable trimer, i.e., $Cr_3(CO)_8^+$. $Cr_3(CO)_8^+$ has 33 ICVE and is very unsaturated. Therefore, $Cr_3(CO)_8^+$ reacts with $Cr(CO)_6$, again adding 14 electrons [$Cr(CO)_4$]. The resulting species $[Cr_3(CO)_8 \cdots L]^+$ rearranges to a stable tetrahedron, $Cr_4(CO)_{12}^+$. According to the CVMO formalism, $Cr_2(CO)_4^+$ and $Cr_3(CO)_8^+$ are "reacting with additional ligands [e.g., $Cr(CO)_4$] to achieve the proper number of CVE."

Similar reactions corresponding to the addition of 14 electrons are observed for the Fe, Co, and Ni systems. However, in these systems the number of ICVE in the product ionic cluster fragments is higher than predicted. For example, the addition of 14 electrons to $Fe_2(CO)_4^+$ gives rise to 37 ICVE instead of 34 as expected for a dimer. This higher number of ICVE can be accommodated since the ionic cluster fragments will undergo rearrangement. That is, the Fe, Co, and Ni ionic cluster fragments add 14 electrons to obtain the proper number of ICVE. However, once the proper number of ICVE is

$$M^+ + 14e^- \rightarrow M_2^+$$

$$M_2^+ + 14e^- \rightarrow M_3^+$$

$$M_3^+ + 14e^- \rightarrow M_4^+$$

Scheme 1

obtained, the ionic cluster fragment undergoes rearrangement analogous to that found for Cr; e.g., a dimer rearranges to a trimer. This rearrangement is due to the inclusion of a metal atom in the 14 ligand electrons that are being added. $Cr_x(CO)_y^+$ ionic cluster fragments are formed by addition of $Cr(CO)_4$ [6 electrons from Cr and 8 electrons from $(CO)_4$], whereas $Fe_x(CO)_y^+$ ionic cluster fragments are formed by the addition of $Fe(CO)_3$ [8 electrons from Fe and 6 electrons from $(CO)_3$]. Once the proper number of ICVE has been achieved, the *d* electrons of $Fe(CO)_3$ interact with the *d* block CVMO of $Fe_2(CO)_4^+$, giving rise to a different metal core cluster and creating new CVMO. Similarly, $Co_x(CO)_y(NO)_z^+$ ionic cluster fragments are formed by the addition of Co(CO)(NO) (9 electrons from Co, 2 electrons from CO, and 3 electrons from No), and $NI_x(CO)_y^+$ ionic cluster fragments are formed by the addition of $Ni(CO)_2$ [10 electrons from Ni and 4 electrons from $(CO)_2$]. It should be noted that cases are observed where the NO ligand in the $Co(CO)_3NO$ system acts as a one-electron donor.

The creation of new CVMO also explains why the ionic cluster fragments continue to react once 14 electrons are added. Owing to a deficiency in the number of ICVE, $Fe_2(CO)_4^+$ reacts with $Fe(CO)_5$ to add $Fe(CO)_3$ and the resulting ionic cluster fragment, $Fe_3(CO)_7^+$, does not have the proper number of ICVE (ICVE = 37) for a trimer (CVE = 48). Thus, $Fe_3(CO)_7^+$ reacts with $Fe(CO)_5$ again to add 14 electrons and forms $Fe_4(CO)_{10}^+$. This sequence continues and terminates with the production of $Fe_6(CO)_{18}^+$. $Fe_6(CO)_{18}^+$ has 83 ICVE, which is the proper number for a cationic bicapped tetrahedron.

The reaction sequence for $Ni(CO)_3^+$ (Scheme 2) proceeds in a similar manner. Each reactant ionic cluster fragment adds 14 electrons [$Ni(CO)_2$], achieves the proper number of ICVE, creates new CVMO, and continues to react. $Ni_5(CO)_{11}^+$, the terminal ion in this reaction sequence, has 71 ICVE, which is the number of CVE found in a trigonal bipyramid. $Ni_5(CO)_{11}^+$ is electronically saturated and does not react further.

On the basis of the preceding discussion, why do some ions react to form ionic cluster fragments which have high bond order? That is, $Fe_2(CO)_4^+$ reacts with $Fe(CO)_5$ to form $Fe_3(CO)_6^+$, i.e., addition of $Fe(CO)_2$ (12 electrons). While it appears that $Fe_2(CO)_4^+$ has reacted to add 12 electrons, the actual number of added electrons may be higher if the CO molecules donate 4 electrons. That is, the addition of $Fe(CO)_2$, where one CO molecule acts as a 4-electron donor, corresponds to the addition of 14 electrons. The $Fe_3(CO)_6^+$ product ion could be formed as a result of donation of electrons from the metal center of the reactant ion. For example, the collision complex may be better represented as $\{(CO)^* \cdots [Fe_2(CO)_3]^+ \cdots Fe(CO)_3(CO)_2^*\}$; the nota-

$$Ni(CO)_3^+ + Ni(CO)_4 \rightarrow Ni_2(CO)_x^+ \qquad (x = 4\text{–}6)$$

$$Ni_2(Co)_x^+ + Ni(CO)_4 \rightarrow Ni_3(CO)_y^+ \qquad (y = 7, 8)$$

$$Ni_4(CO)_9^+ + Ni(CO)_4 \rightarrow Ni_5(CO)_{11}^+$$

Scheme 2

tion (CO)* implies this ligand is lost from the collision complex to form the $Fe_3(CO)_6^+$ product ion. Thus, formally the 14-electron $Fe(CO)_3$ moiety is added, but an additional ligand is lost from the reactant ion on formation of the product ion. Note that a similar effect is observed for the $Ni(CO)_3^+/Ni(CO)_4$ system. That is, $Ni_2(CO)_x^+$ and $Ni_3(CO)_y^+$ react to form $Ni_3(CO)_y^+$ and $Ni_4(CO)_y^+$, respectively. The product ions $Ni_3(CO)_y^+$ and $Ni_4(CO)_y^+$ appear to be formed by the addition of Ni(CO) (i.e., addition of 12 electrons) and $Ni(CO)_2$ (i.e., addition of 14 electrons).

The increase in bond order for some ionic cluster fragments, and therefore the variation in the number of electrons added, is also consistent with the CVMO theory. Lauher states that unsaturated clusters may be unusually reactive or "the metal core of the [unsaturated] clusters might isomerize to a different geometry with the proper number of CVMO or one of the existing ligands might bind in an unusual manner such that more of its electrons are formally donated to the metal."[7] Lauher uses $Os_3H_2(CO)_{10}$ and $Fe_4(CO)_{13}H^-$ as examples of this type of chemistry. The metal core of $Os_3H_2(CO)_{10}$ has two normal Os—Os bond lengths and one short Os—Os bond distance (a double bond) and deviates considerably from the D_{3h} geometry assigned to trimers. $Fe_4(CO)_{13}H^-$ appears to have 60 CVE, which is the correct number for a tetrahedron. However, one of the CO ligands of this cluster donates four electrons, resulting in 62 CVE and the C_{2v} butterfly geometry. Similar rearrangement reactions are suggested for the ionic cluster fragments with high bond orders. Thus, the variation in the number of electrons added to the ionic cluster fragment upon reaction with the neutral metal carbonyl is the result of either the CO ligands binding in an unusual manner or isomerization of the core metal cluster to a different structure. For the former case, the number of ICVE present will be greater than the number suggested by the molecular formula. In the latter case, isomerization to a different geometry changes the number of CVMO available, causing a change in the ICVE.

It is instructive to compare the differences in cluster formation patterns for Co^+ (Scheme 3) and $Co(CO)_2^+$ (Scheme 4). Co^+ reacts with $Co(CO)_3(No)$ to form $Co_2(CO)(NO)^+$, $Co_3(CO)_2(NO)_2^+$, $Co_4(CO)_3(NO)_3^+$, $Co_5(CO)_4(NO)_4^+$, and $Co_6(CO)_5(No)_5^+$, i.e., addition of Co(CO)(No) (14 electrons) to the reactant ion. These ionic cluster fragments have bond orders which are consistent with polyhedral structures. The ionic cluster fragments formed by the reaction

$$Co^+ + Co(CO)_3(NO) \rightarrow Co_2(CO)_x(NO)^+ \qquad (x = 1, 2)$$

$$Co_2(CO)_x(NO)^+ + Co(CO)_3(No) \rightarrow Co_3(CO)_2(NO)_2^+$$

$$Co_3(CO)_2(NO)_2^+ + Co(CO)_3(No) \rightarrow Co_4(CO)_{x+2}(NO)_3^+$$

$$Co_4(CO)_{x+2}(NO)_3^+ + Co(CO)_3(NO) \rightarrow Co_5(CO)_{x+3}(NO)_4^+$$

$$Co_5(CO)_{x+3}(NO)_4^+ + Co(CO)_3(NO) \rightarrow Co_6(CO)_{x+4}(NO)_5^+$$

Scheme 3

$$Co(CO)_2^+ + Co(CO)_3(NO) \rightarrow Co_2(CO)_x(NO)^+ \quad (x = 3, 4)$$

$$Co(CO)(NO)^+ + Co(CO)_3(NO) \rightarrow Co_2(CO)_y(No)_2^+ \quad (y = 2, 3)$$

$$Co_2(CO)_x(NO)^+ + Co(CO)_3(NO) \rightarrow Co_3(Co)_4(NO)_2^+$$

$$Co_3(CO)_4(No)_2^+ + Co(CO)_3(No) \rightarrow Co_4(CO)_{x+1}(NO)^+$$

$$Co_4(CO)_{x+1}(NO)^+ + Co(CO)_3(NO) \rightarrow Co_5(CO)_4(NO)^+$$

Scheme 4

of $Co(CO)_2^+$ are quite different from those formed by reaction of Co^+; e.g., $Co_3(CO)_4(No)_2^+$ reacts with $Co(CO)_3(No)$ to form $Co_4(CO)_x(NO)^+$ ($x = 4, 5$). $Co_3(CO)_4(No)_2^+$ has 36 ICVE (the NO is a one-electron donor). When $Co_3(CO)_4(No)_2^+$ reacts with $Co(CO)_3(NO)$ to form $Co_4(CO)_x(NO)^+$, the product ion isomerizes to a different geometry than expected for a tetramer. A four-metal-atom cluster can have one of four geometries: a T_d, C_{3v}, D_{2h}, or D_{4h} geometry. The T_d geometry has the smallest number of CVMO (30) while the D_{4h} geometry has the largest number of CVMO (32). The reactivity for $Co_4(CO)_4(NO)^+$ is very low, indicating that this ionic cluster fragment has a high bond order. On the basis of the reactivity/electron deficiency model, $Co_4(CO)_4(No)^+$ has a tetrahedral structure with either four double metal–metal bonds and three CO ligands which are four-electron donors or three double metal–metal bonds and four CO ligands which are four-electron donors. The presence of multiple metal–metal bonds in $Co_4(CO)_4(NO)^+$ decreases the number of CVMO and therefore the number of ICVE. As a result, $Co_3(CO)_4(NO)_2^+$ will react with $Co(CO)_3(NO)$ to add fewer than 14 electrons.

The heteronuclear ionic cluster fragments formed by ion–molecule reactions in the $Co(CO)_3NO/Ni(CO)_4$, $Fe(CO)_4$, and $Fe(CO)_5/Co(CO)_3NO$ systems follow the same general trends as observed for the homonuclear systems, i.e., each successive ionic cluster fragment is formed by the addition of an $M(CO)_a$ moiety which corresponds to adding 14 electrons to the precursor ionic cluster fragments.[8] The general reaction sequences for the heteronuclear systems have characteristic differences. For example, the ionic cluster fragments of the $Ni(CO)_4/Fe(CO)_5$ and $Co(CO)_3NO/Fe(CO)_5$ systems exhibit characteristic differences which we attribute to unusual bonding of the $Fe(CO)_x$ fragment. We are intrigued by this fact because the ionic cluster fragments which contain the $Fe(CO)_x$ moiety appear similar to stable (solution) species which contain an $Fe(CO)_4^{2-}$ moiety as a bulky analogue of an edge-bridging carbonyl ligand.[9–11] Conversely, the ionic cluster fragments of the $Co(CO)_3(NO)/Ni(CO)_4$ system follow the bonding found in the triangular polyhedra of the boron hydrides.

3. BONDING OF $Fe(CO)_x$ IN HETEROMETALLIC IONIC CLUSTER FRAGMENTS

As noted above, the heteronuclear systems containing $Fe(CO)_x$ show anomalous reactivity which we interpret as arising from unusual bonding. The relative reactivity versus cluster valence electron deficiency CVED data of the ionic cluster fragments formed in the $Co(CO)_3(NO)/Fe(CO)_5$ and $Ni(CO)_4/Fe(CO)_5$ systems suggest the presence of unusual bonding among some of the ionic cluster fragments.

It has been suggested that the $Fe(CO)_4$ group in $[Fe_3Pt_3(CO)_{15}]^{2-}$ acts as a bulky analogue of an edge-bridging carbonyl group.[9] Thus, the dianion can be viewed as $\{Pt_3(CO)_3[X\text{-}Fe(CO)_4]_3\}^{2-}$, a stabilized analogue of $\{Pt_3(CO)_3[X\text{-}Fe(CO)_4]_3\}^{2-}$. Similarly, the series of $[Na(THF_2)_2]^+\{M'[Fe(CO)_4]_2\}^{2-}$ complexes ($M' = Zn$, Cd, Hg), where the group 12-iron bond description varies from near-ionic distorted T_d d^{10} $Fe^{2-(+)}$ to near covalent d^8 Fe^0, have been studied.[10] Arndt *et al.* have suggested that $Et_4N^+Ph_3PAuFeW(CO)_9^-$ is best formulated as $Fe(CO)_4^{2-}$ interacting with two metaloelectrophiles; the 16-electron $W(CO)_3$ forms a metal-metal donor-acceptor bond with the iron nucleophile while the 12-electron Ph_3PAu^+ accepts electrons directly from the iron and indirectly from the electron-rich carbonyl carbons.[11]

The relative reactivity versus CVED data may be explained by considering the formation of dative bonds (partial ionic bonds) between $Fe(CO)_4$ and either a nickel or a cobalt cluster fragment. For example, on the basis of ion-molecule reaction rates the $NiFe(CO)_4^+$, $Ni_2Fe(CO)_4^+$, and $Ni_2Fe(CO)_6^+$ ionic cluster fragments are more coordinatively saturated than predicted by simple polyhedral models. The reactivity for $Ni_2Fe(CO)_6^+$ is relatively high (0.255), which is inconsistent with the low CVED (8) predicted for a simple polyhedral model. The high reactivity for $Ni_2Fe(CO)_6^+$ suggests that the ion is more coordinatively unsaturated than the relative reactivity versus CVED model would indicate. If a linear structure is assumed for the $Ni_2Fe(CO)_7^+$ and $Ni_2Fe(CO)_6^+$ ionic cluster fragments, a plot of log relative rate versus CVED monotonially increases.[8] A linear structure would increase the coordination unsaturation of $Ni_2Fe(CO)_7^+$ and $Ni_2Fe(CO)_4^+$ ions. Scheme 5 shows a

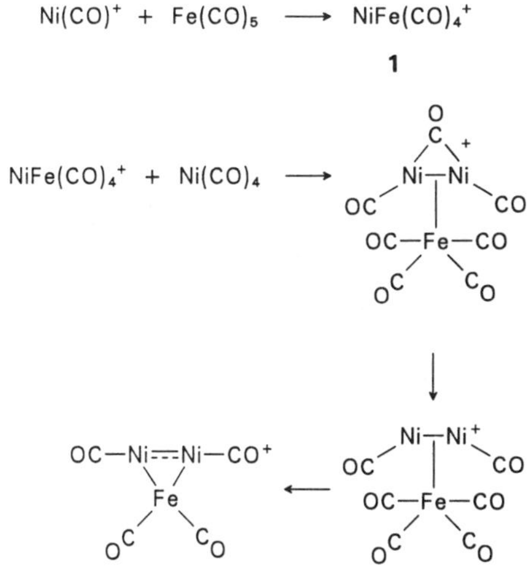

$$Ni(CO)^+ + Fe(CO)_5 \longrightarrow \underset{\mathbf{1}}{NiFe(CO)_4^+}$$

$$NiFe(CO)_4^+ + Ni(CO)_4 \longrightarrow$$

Scheme 5

possible mechanism for the formation of the triatomic ionic cluster fragments of the $Ni(CO)^+/Fe(CO)^+$ reaction sequence. The Ni^+ ion reacts with $Fe(CO)_5$ to form the $NiFe(CO)_4^+$ ionic cluster fragment. Thus, the structure of $NiFe(CO)_4^+$ is best rationalized as $[Ni \cdots Fe(CO)_4]^+$, analogous to Ph_3PAu in $Et_4N^+Ph_3PAuFeW(CO)_9^-$.[11] When $NiFe(CO)_4^+$ reacts with $Ni(CO)_4$ the linear $Ni_2Fe(CO)_7^+$ trinuclear ionic cluster fragment is produced, which can be formulated as $[Ni_2(CO)_3 \cdots Fe(CO)_4]^+$.

Our previous work showed that the $Ni_2(CO)_3^+$ ionic cluster fragment formed in the homonuclear system is quite unsaturated (Electron deficiency of 4.5).[6] $Ni_2(CO)_3^+$ contains only 26 CVE as compared to 34 CVE needed for a dimer. Owing to the coordination unsaturation of the $Ni_2(CO)_3$ moiety, no metal-metal donor bonds are formed between the two nickel atoms and the iron atom. A carbonyl ligand is lost from $Ni_2\ Fe(CO)_7^+$ to produce $Ni_2Fe(CO)_6^+$, formulated as $[Ni_2(CO)_2 \cdots Fe(CO)_4^+]$. Again, owing to the electronic unsaturation, the $Ni_2(CO)_2$ moiety accepts electron density from the $Fe(CO)_4$ group and no metal-metal donor-acceptor bonds are formed between nickel and iron. However, upon loss of two additional carbonyls, the relative reactivity versus CVED data suggest that the nickel and iron atoms do form metal-metal donor-acceptor bonds (structure **4** in Scheme 5); i.e., $Ni_2Fe(Co)_4^+$ is more coordinatively saturated. Structure **4** can result from the high coordination unsaturation of the $Ni_2(CO)_2$ moiety. Because of this high coordination unsaturation, the two carbonyl ligands would preferentially be lost from the $Fe(CO)_4$ group. The proposed loss of the carbonyl ligands from the $Fe(CO)_4$ group agrees with thermodynamic data which give $D_{Fe-CO} < D_{Ni-CO}$ ($D_{Fe-CO} = 28.1$ kcal/mol, $D_{Ni-CO} = 35.1$ kcal/mol).[12] [However, the loss of the CO ligand from the $Ni_2(Co)_3$ moiety in **2** to produce **3** can be rationalized by assuming that $Ni_2(Co)_2$ bonded to $Fe(CO)_4$ is more stable than $Ni_2(CO)_3$ bonded to $Fe(CO)_3$.] The formation of a double metal-metal bond in $Ni_2Fe(CO)_4^+$ can be a direct result of the electron density given up by the $Ni_2(CO)_2$ moiety when the donor-acceptor bonds are formed with iron. The relative reactivity data of the $Ni_2(CO)_2^+$ ionic cluster fragment formed in the homonuclear systems indicate that $Ni_2(CO)_2^+$ is coordinatively saturated and also contains a double bond.[6]

Similarly, the formation of $CoFe_2(CO)_4^+$ offers an example of $Fe(CO)_4$ acting as a ligand. If $CoFe_2(CO)_4^+$ is considered to be a trimer, the CVED of this ionic cluster fragment would be too high for its relative reaction rate, and the log relative reaction rate versus CVED plot would not show a monotonic increase. However, if $CoFe_2(CO)_4^+$ is considered to be a dimer in which a $Fe(CO)_4$ group is bound to the Co^+-Fe moiety, the log relative reaction rate versus CVED plot increases monotonically. Owing to the high unsaturation of the Co^+-Fe dimer, this ion accepts electron density directly from the iron atom and indirectly from the carbonyl ligands. As in the case of $Ni_2Fe(CO)_7^+$, the interaction of Co^+-Fe with the $Fe(CO)_4$ group is similar to the interaction of Ph_3PAu with $Fe(CO)_4$.[11] Reformulating $CoFe_2(CO)_4^+$ as $\{CoFe[\mu\text{-}Fe(CO)_4]\}^+$ also gives a structure very similar to that of $Fe_2(CO)_9$.

$CoFe_2(CO)_4^+$ is also formed by $Co(CO)^+/Fe(CO)^+$ reacting with $Co(CO)_3(NO)/Fe(CO)_5$. However, in this case the relative reactivity versus CVED data suggest that this ionic cluster fragment is indeed a trinculear cluster having three metal-metal donor-acceptor bonds. The difference in reactivity of $CoFe_2(CO)_4^+$, and therefore the difference in structure, could be a result of the added stability of the CO ligand found in the reaction sequence for $Co(CO)^+/Fe(CO)^+$.[8] Now, instead of forming a Co^+–Fe dimer a $Co(CO)^+$–Fe dimer is formed. The lower coordination unsaturation and the lower energy of the $Co(CO)^+$–Fe dimer favors formation of metal-metal donor-acceptor bonds. It should be noted that Freas and Ridge have also observed dramatic differences in the chemistry of bare metal dimer ions and ligated dimer ions, specifically Co_2^+ and $Co_2(CO)^+$.[13]

4. METAL–METAL AND METAL–LIGAND BINDING ENERGIES IN IONIC CLUSTER FRAGMENTS OF TRANSITION METAL CARBONYLS

Although changes in structure can be inferred on the basis of ion–molecule reactivity, the method is ambiguous and less than ideal for characterization of structure and bonding of ionic cluster fragments. Collision-induced dissociation (CID), another commonly used structural characterization method for gas phase ions,[14] is of very limited utility because all of the ionic cluster fragments dissociate in a similar manner, i.e., by loss of the CO ligands, and it is difficult to assign structural differences based on changes in the relative abundance of these fragment ions. A method which does have considerable promise is laser–ion beam photodissociation.

The apparatus used for laser–ion beam photodissociation is shown schematically in Figure 5. The operation of this system and details of the experiment have been described elsewhere.[15] Briefly, the ion of interest (m_1^+) is mass selected by the high-resolution mass analyzer (composed of the first electrostatic and magnetic analyzers), the laser beam intersects the mass-analyzed ion beam (perpendicular intersection), and the fragment ions (m_2^+) formed as a result of photoexcitation (a resonant photon absorption process) are energy analyzed [energy being related to mass of the photofragment ion by $m_2^+ = m_1^+(v_2/v_1)$, where v_1 is the energy of m_1^+ and v_2 is the energy of m_2^+] by the second electrostatic analyzer.

The data from this experiment are equivalent to (but not necessarily identical to) the UV–visible absorption spectrum for the ionic cluster fragment and upper limits for the metal–metal and metal–ligand binding energies. For example, compare the single-crystal absorption spectra of $Fe_3(CO)_{12}$ with the dissociation spectrum for $Fe_3(Co)_{12}^+ \rightarrow Fe_3(CO)_{10}^+$ shown in Figure 6. Note in both the neutral absorption[16] and photodissociation spectra that strong bands are observed in the 500- and 620-nm wavelength regions. Tyler *et al.* assign the 500- and 620-nm transitions of the $Fe_3(CO)_{12}$ neutral species to the $\sigma \rightarrow \sigma^*$ and $\sigma^{*\prime} \rightarrow \sigma^*$ transitions and discounts the possibility of metal–ligand

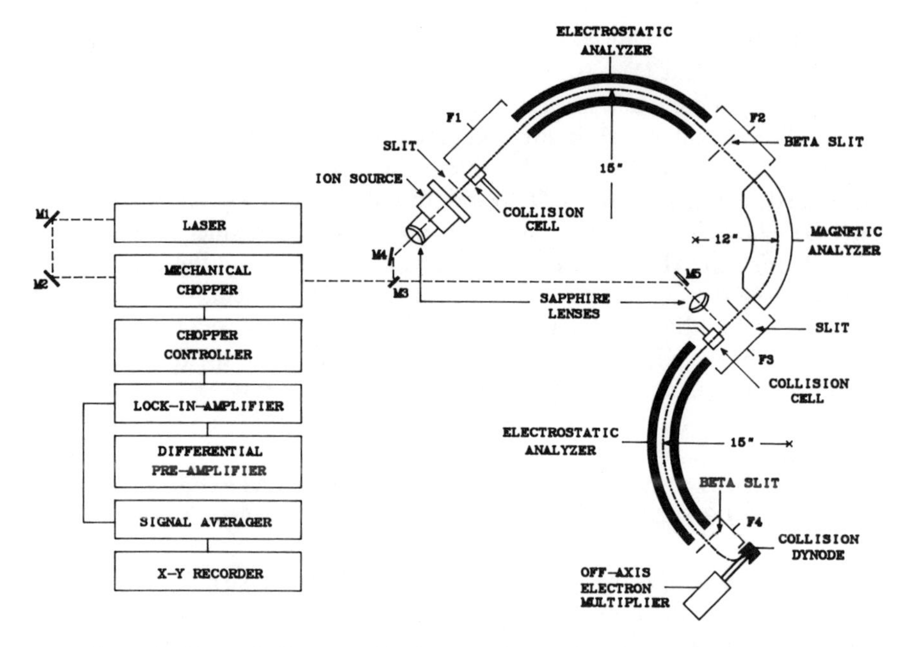

Figure 5. Schematic diagram of the laser–ion beam photodissociation apparatus.

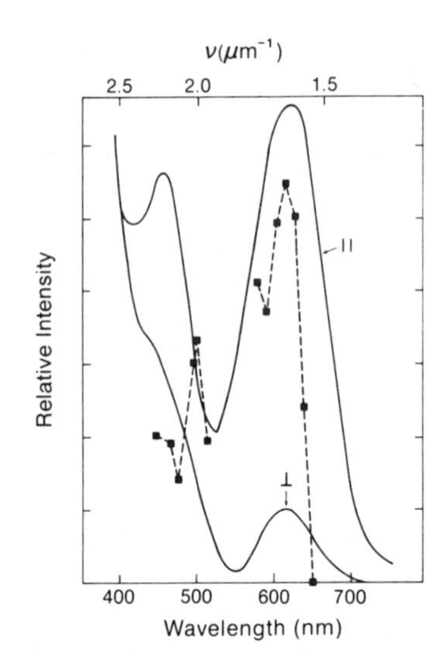

Figure 6. Comparison of the neutral $Fe_3(CO)_{12}$ single-crystal polarized absorption spectrum (solid line) and the photodissociation spectrum of the $Fe_3(CO)_{12}^{+}$ ion (dashed line).

charge-transfer bands giving rise to electronic transitions in the visible spectral range.[16] An interesting feature of the photodissociation spectrum is that the $\sigma \rightarrow \sigma^*$ (496–502 nm) band is red-shifted by approximately 60–65 nm (0.36 eV) with respect to the neutral absorption spectrum. Such a red shift could indicate that the average splitting of the xz and z^2 orbitals differs for the ion, but it could also indicate that the ions sampled by photodissociation are vibrationally hot. More detailed studies aimed at answering these questions are under way.

In general, ionic transition metal carbonyl cluster fragments readily photodissociate at visible wavelengths. The reasons for this are two fold: (1) the transition metal carbonyl ions have reasonably large absorption cross sections in the visible region, and (2) metal–metal and metal–carbonyl bond dissociation is accessible at energies in the visible spectral region. A general problem in making estimates of bond energies with use of photodissociation thresholds is that the occurrence of photodissociation only provides an upper limit to the bond energy. For example, if the excited state of $M_x(CO)_y^+$ formed by photon absorption lies above the threshold energy for dissociation, the photodissociation threshold is spectroscopically limited. Thus, accurate measurements of bond dissociation energies can only be achieved when an excited electronic state of the ion fortuitously lies close to the dissociation threshold.

It is not possible in our experiment to distinguish a thermodynamic versus a spectroscopic threshold, and no attempt has been made in this work to correct the bond dissociation energies for this anomalous behavior. Thus, we have assumed that at energies above the bond dissociation threshold, the photodissociation spectra for the $M_x(CO)_y^+$ systems are the same as those of their neutral counterparts (or the molecular ion) and that any deviations from this behavior can be attributed to spectroscopic factors. That is, we have assumed that the chromophore of the $M_x(CO)_y^+$ species is the M_x fragment and that the ligands (or number of ligands) attached to the M_x fragment do not significantly influence the absorption spectrum. Clearly, the validity of this assumption will be limited to absorption transitions that are not associated with the ligand or metal- to ligand charge-transfer bands. On the basis of Tyler *et al.*'s interpretation of the spectroscopic data for $Fe_3(CO)_{12}$ our assumption should be valid. Of course, the generality of this assumption will break down at some point. In metal cluster systems changes in the metal–ligand bonding could result in dramatic changes in the metal center, which could greatly affect the absorption spectrum.[17] It is difficult to systematically study the effects of metal/ligand ratio on the absorption spectra. However, such effects can be studied by laser–ion beam photodissociation, and these studies are currently under way.

A specific example of a photodissociation reaction threshold which can be observed is shown in Figure 6. Figure 6 compares the neutral $Fe_3(CO)_{12}$ single-crystal polarized absorption spectrum and the photodissociation spectrum of $Fe_3(CO)_{12}^+$. Note that the neutral $Fe_3(CO)_{12}$ absorbs at wavelengths greater than 700 nm, but the photodissociation cross section for $Fe_3(CO)_{12}^+$ approaches zero at 651 nm. We attribute the decrease in photofragment ion yield at 651 nm to the energy required for bond cleavage being greater than the energy of the photon; i.e., photodissociation of $Fe_3(CO)_{12}^+$ to $Fe_3(Co)_{10}^+$ is thermodynamically limited.

The potential utility of laser–ion beam photodissociation for studies of structure and bonding of ionic transition metal cluster fragments is demonstrated by the studies of $Fe_3(CO)_5^+$ and $Fe_3(CO)_7^+$ cluster fragments. The photodissociation thresholds for formation of $Fe_3(CO)_4^+$ from $Fe_3(CO)_7^+$ and $Fe_3(CO)_5^+$ are the same to within 1 kcal/mol. One possible explanation is that the measured thresholds correspond to thermodynamic thresholds, which implies that the metal–ligand bonding in $Fe_3(CO)_5^+$ differs from that in $Fe_3(CO)_7^+$. Such an interpretation of the data is consistent with the results from ion–molecule reaction chemistry.[5] On the basis of the reactivities of the ionic cluster fragments, the electron deficiency of the $Fe_3(CO)_7^+$ ion was found to be 3.67 and for $Fe_3(CO)_5^+$ a value of 1.67 was predicted. Such a low value for the electron deficiency of $Fe_3(CO)_5^+$ suggests that the carbonyl ligands are four-electron donors to the Fe_3^+ metal center or that the metal–metal bonds in the two ions are different. Thus, the photodissociation and ion–molecule reactivity data are consistent.

It was noted above that the photodissociation and UV-visible absorption spectra are similar but not necessarily identical. That is, in the photodissociation experiment the signal for a particular ion is monitored as a function of the laser wavelength, and an increase in the signal for ion m_2^+ (the photofragment ion) indicates that the ion m_1^+ has absorbed a photon and undergone a dissociation reaction. Similarly, a decrease in the signal intensity for m_1^+ indicates that photon absorption has occurred. The question that must now be addressed in more detail is the mechanism by which photodissociation occurs. The photophysics of transition metal species is not well understood. For instance, photodissociation of polyatomic species generally occurs by a vibrational predissociation mechanism; i.e., resonant photon absorption produces a bound excited state ion which undergoes internal conversion to the ground state, and thus dissociation occurs from a vibrationally excited ground electronic state ion. It is not clear that the picture for transition metal species is so simple. In a recent review, Poliakoff and Weitz discuss the photophysics of the relatively simple $Fe(CO)_4$ molecule.[18] It is clear from the work cited in this review that vibrationally "hot" fragments are formed by gas phase photolysis, and this fact cannot be overlooked when using photodissociation for determining thermodynamic quantities, e.g., metal–metal and metal–ligand bond energies. With the understanding that such estimates of bond energies are subject to errors in cases where (1) the ion contains some internal excitation prior to photodissociation and (2) the photofragments may be formed with some internal excitation, we began a study to evaluate the utility of laser–ion beam photodissociation for characterizing ionic cluster fragments.

Although the single-photon photodissociation of manganese, iron, and cobalt carbonyls has not been studied previously, the gas phase multiphoton dissociation/ionization (MPD/MPI) of $Mn_2(CO)_{10}$,[19–21] $Fe_3(CO)_{12}$,[22,23] and $Co_4(CO)_{12}$[24] has been reported. In general, the MPD/MPI studies of these metal carbonyl clusters yield bare metal cluster ions. For example, the ions Mn_2^+ and Mn^+, Fe_3^+, Fe_2^+, and Fe^+, and Co_4^+, Co_3^+, and Co_2^+ are the major products observed in the MPD/MPI studies of $Mn_2(CO)_{10}$,[20] $Fe_3(Co)_{12}$,[23] and $Co_4(CO)_{12}$,[24] respectively. One exception to the exclusive production of bare metal ions by MPD/MPI was reported by Lichtin *et al.*[25] These workers reported the production of carbonyl-containing species such as $Mn_2(CO)_y^+$ ($y = 5, 9, 10$) and $Mn_3(CO)_y^+$ ($y = 8, 10$) upon MPD/MPI of $Mn_2(CO)_{10}$. However, these ions were attributed to ion–molecule reactions of

Mn^+ with various neutrals of the type $Mn_2(CO)_y$ (formed by MPD) and not from the direct dissociation/ionization of the neutral. The removal of all carbonyl ligands from $Mn_2(CO)_{10}$ as well as ionization of the bare metal neutrals is thought to require between 7 and 10 photons in the 350- to 450-nm range.

The results presented here also find the same bare metal cluster ions produced as a result of single-photon photodissociation. The difference is, however, that we are photodissociating mass-selected ions; e.g., Mn_2^+ arises from the photodissociation of $Mn_2(CO)_y^+$ ($y = 1$–5) and not directly from $Mn_2(CO)_{10}^+$. The fact that only Mn_2^+ and Mn^+ ions are observed in the MPI mass spectra of $Mn_2(CO)_{10}$ suggests that the Mn—Mn bond strength must be greater than the Mn—CO bond strength. Of course, another critical difference between the MPD/MPI and single-photon photodissociation studies is the photon wavelength. The wavelength used for photoactivation could be important, especially if metal- ligand charge-transfer bands are populated in the ultraviolet range. The single-photon photodissociation results differ in that cleavage of the Mn—Mn bond is possible and $Mn(CO)_y^+$ product ions are observed. For example, in Figure 7, the photodissociation cross section for $Mn_2(CO)_{10}^+ \rightarrow Mn(CO)_5^+$ is quite large even at low energy (581–651 nm), a result that suggests that the Mn—Mn bond in $Mn_2(CO)_{10}^+$ is weak. Other differences between the MPD/MPI and the single-photon photodissociation studies could be attributed to the fact that the dissociation energy of the Mn—Mn bond in $Mn_2(CO)_y^+$ type ions differs from that of the neutral molecule. For example, ionization of $Mn_2(Co)_{10}$ is thought to occur from a molecular orbital with large amounts of metal character.[26,27]

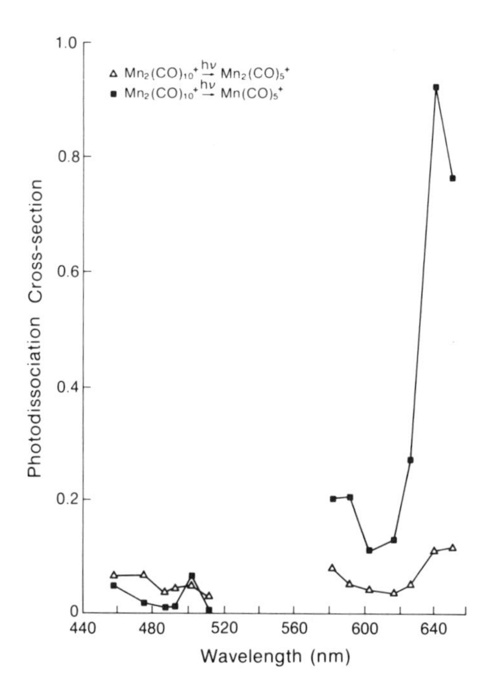

Figure 7. Photofragment ion yield as a function of laser wavelength for $Mn_2(CO)_{10}^+$. The photofragment ions formed are $Mn_2(CO)_5^+$ (△) and $Mn(CO)_t^+$ (■).

$Mn_2(CO)_{10}$ is by far the most intensely studied metal carbonyl system. Numerous reports on the Mn—Mn bond energy in $Mn_2(CO)_{10}$ derived from gas and solution phase studies have been reported. The values for the Mn—Mn bond dissociation energy vary from 22[28] to 41 ± 9[29] kcal/mol. There are fewer reports on the average Mn—CO bond dissociation energy, and the value of 24 kcal/mol reported by Connor *et al.* is generally accepted.[30] The metal-metal bond energy in $Mn_2(CO)_{10}^+$ has been reported to be 22.2 kcal/mol,[28] while that reported for Mn_2^+ varies from 19.6 ± 5[31] to $\geqslant$32.0 kcal/mol.[32] Because most $Mn_x(CO)_y^+$ ions photodissociate at wavelengths greater than 651 nm, accurate energy thresholds cannot be obtained from the data. On the other hand, however, upper limits for the Mn—CO and Mn—Mn bond dissociation energies can be made. Bond dissociation energies derived from single-photon photodissociation thresholds for homolytic Mn—Mn bond cleavage of $Mn_2(CO)_{10}^+$ precursor ions were both found to be <43.9 kcal/mol. One of the surprising results from the photodissociation data is that $Mn_2(CO)_5^+$ is formed by photodissociation of $Mn_2(CO)_{10}^+$. The $Mn_2(CO)_5^+$ photofragment ion is the dominant product at blue-green (454-514.5 nm) wavelengths, whereas when lower-energy photons are used the $Mn(CO)_5^+$ ion is formed. The formation of the $Mn_2(CO)_5^+$ photofragment ion is especially surprising because it indicates an extremely low value for the average Mn—CO bond dissociation energy, viz., <8.8 kcal/mol. Appearance potential measurements for the loss of five carbonyl ligands from $Mn_2(Co)_{10}^+$ yields an average Mn—CO bond energy of 15.4 kcal/mol,[33] also a surprisingly low value for the metal-ligand bond energy. At first, it was thought that removal of five carbonyl ligands from the molecular ion must occur by a multiphoton process. However, photodissociation experiments performed with a perpendicular laser-ion beam configuration in the third field-free region of the mass spectrometer, where the interaction time between ion and laser beam is very short, i.e., <1 ns, are inconsistent with the occurrence of a multiphoton process.

In Table 1 average metal-metal and metal-carbonyl bond dissociation energies reported in this paper are compared with available literature values. Most of the measured bond energies are substantially lower than the average bond dissociation energies taken from the literature. One possible source of error in these values is that the ions possess some excess rovibronic energy that reduces the photodissociation threshold, thus lowering the measured bond dissociation energy. For example, the bond energy of Mn_2^+ determined by collision-induced dissociation is 0.85 ($\pm$0.2) eV,[31] whereas the photodissociation data give a value of $\geq$1.39 eV.[32] Jarrold *et al.* suggested that the differences in the experimentally determined dissociation energies for Mn_2^+ arise from excess internal energy of the ion.[32]

It is not surprising that ionic metal species are formed with excess energy. Photolysis experiments on $Mn_2(CO)_{10}^+$ ($\lambda = 351$ nm) suggest that $Mn(CO)_5$ is produced in excited vibrational states, which undergo radiative decay to the ground electronic state by emission of a red photon.[34] The single-photon[35] and multiphoton[36] ionization of $Fe(CO)_5$ have produced iron atoms in excited electronic states. Yardley *et al.* recently reported that the laser photodissoci-

Table 1. Bond Dissociation Energies Measured from Photodissociation Thresholds for $Mn_x(CO)_y^+$ Ions

Transition	Threshold (nm)	Threshold (kcal/mol)	Threshold energy/bonds broken (kcal/mol)
$Mn_2(CO)_{10}^+ \rightarrow Mn_2(CO)_5^+$	>651	<43.9	<8.8
$Mn_2(Co)_{10}^+ \rightarrow Mn(CO)_5^+$	>651	<43.9	<43.9
$Mn_2(CO)_9^+ \rightarrow Mn_2(CO)_5^+$	640	44.7	11.2
$Mn_2(Co)_6^+ \rightarrow Mn_2(CO)_5^+$	>651	<43.9	<43.9
$Mn_2(CO)_5^+ \rightarrow Mn_2(CO)_3^+$	>651	<43.9	<21.9
$Mn_2(CO)_5^+ \rightarrow Mn_2(CO)_2^+$	>651	<43.9	<14.6
$Mn_2(CO)_5^+ \rightarrow Mn_2(CO)^+$	>651	<43.9	<11.0
$Mn_2(CO)_4^+ \rightarrow Mn_2(CO)_2^+$	>651	<43.9	<21.9
$Mn_2(CO)_4^+ \rightarrow Mn_2(CO\text{-}^+$	>651	<43.9	<14.6
$Mn_2(CO)_4^+ \rightarrow Mn_2^+$	>651	<43.9	<11.0
$Mn_2(CO)_4^+ \rightarrow Mn(Co)_3^+$	>651	<43.9	<21.9
$Mn_2(CO)_3^+ \rightarrow Mn_2(CO)^+$	>651	<43.9	<21.9
$Mn_2(Co)_3^+ \rightarrow Mn_2^+$	>651	<43.9	<14.6
$Mn_2(CO)_2^+ \rightarrow Mn_2(CO)^+$	>651	<43.9	<43.9
$Mn_2(CO)_2^+ \rightarrow Mn_2^+$	>651	<43.9	<21.9
$Mn_2(Co)^+ \rightarrow Mn_2^+$	>651	<43.9	<43.9
$Mn_2^+ \rightarrow Mn^+$	>651	<43.9	<43.9
$Mn(CO)_5^+ \rightarrow Mn(CO)_3^+$	>651	<43.9	<21.9

ation of $Fe(CO)_5$ yielded $Fe(CO)_{5-n}$ $(1 \leq n \leq 4)$ neutrals with retention of internal energy.[37] It is, however, surprising that the authors found that even the small neutrals [e.g., Fe(CO) and $Fe(CO)_2$] contained excess internal energy because loss of CO neutrals is thought to be effective in removing excess energy.[38] Likewise, laser photolysis/transient infrared studies of $Fe(CO)_5$[39,40] and $Cr(Co)_6$[41] demonstrate that the photoproducts have infrared stretching bands that blue-shift with time and reach a fixed maximum stretching frequency within a time period of 10 s. A long-lived excited state of Cr^+ was also invoked to explain the different kinetic data for reactions of Cr^+ formed by direct electron impact ionization of $Cr(CO)_6$.[42] Reents *et al.* proposed that 25% of the Cr^+ ions had a reaction rate 4.4 times larger than that of the remaining Cr^+ ions.[42]

One aspect of the photodissociation data that cannot be attributed to excess energy of the ion is the loss of multiple CO ligands at high energies and the exclusive loss of two CO ligands at low photon energy. Although such trends could be attributed to state-specific reactions, we prefer to rationalize the data in more general terms. Figure 8 illustrates schematically how specific reactions may occur at low photon energy while other reaction channels requiring less energy may not. At high energies the higher-lying excited electronic state (S_j) of $M_x(CO)_y^+$ can be accessed, and photodissociation yields excited state (S_i) product ions $M_x(CO)_{y-1}^+$ and $M_x(CO)_{y-2}^+$. However, at low energies, the energy of the excited state of $M_x(CO)y - 2$ that is accessed is greater than the ground state energy of $M_x(CO)_{y-1}^+$, and thus photodissociation

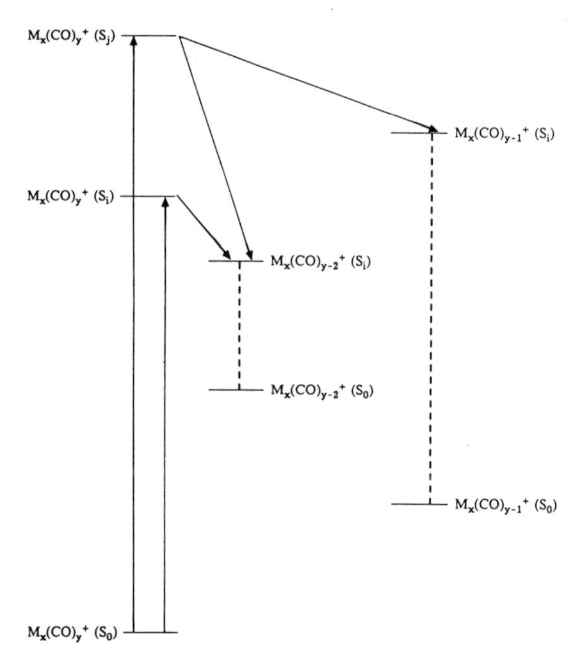

Figure 8. Energy level diagram depicting the proposed mechanism for the different photodissociation channels at high (454–514.5 nm) and low (590–620 nm) photon energy.

occurs by specific electronic transitions which lead exclusively to the loss of two carbonyl ligands.

Although state-specific reactions of large polyatomic systems are rare, such behavior is feasible and even quite probable for transition metal systems. The vibrational modes of the metal centers of cluster fragments are not effectively coupled with other modes of the molecule. That is, the low vibrational frequencies of the metal centers could reduce the rates for energy randomization into other degrees of freedom. In order to probe these ideas further, studies are presently under way to ascertain the effects of different excited states on the photodissociation process. In particular, further work will concentrate on the efficiency of photodissociation via metal- to ligand charge-transfer bands, and these reactions will be compared with photodissociation via metal–metal bonding electronic states.

ACKNOWLEDGMENTS

This work was supported by grants from the U.S. Department of Energy, Office of Basic Energy Sciences (DE-AS05-82ER13023), and the National Science Foundation (CHE-8418457).

REFERENCES

1. (a) Muetterites, E.L. *Catal. Rev.-Sci. Eng.* 1981, *23*, 69; (b) Muetterites, E.L.; Rhodin, T.N.; Band, E.; Brucker, C.F.; Pretzer, W.R. *Chem. Rev.* 1979, *79*, 91; (c) Sung, S.-S.; Hoffman, R. *J. Am. Chem. Soc.* 1985, *107*, 578.

2. (a) Wade, K. In *Transition Metal Clusters*; Johnson, B.F.G., Ed.; J. Wiley and Sons: New York, 1980; Chapter 3, p. 193; (b) Mingos, D.M.P. *Acc. Chem. Res.* 1984, *17*, 311.
3. (a) Burdett, J.K. *Coord. Chem. Rev.* 1978, *27*, 1; (b) Hitam, R.B.; Mahmoud, K.A.; Rest, A.J. *Coord. Chem. Rev.* 1984, *55*, 1.
4. Poliakoff, M. *Adv. Organomet. Chem.* 1986, *25*, 277.
5. Fredeen, D.A.; Russell, D.H. *J. Am. Chem. Soc.* 1985, *107*, 3762.
6. Fredeen, D.A.; Russell, D.H. *J. Am. Chem. Soc.* 1986, *108*, 1860.
7. Lauher, J.W. *J. Am. Chem. Soc.* 1978, *100*, 5305.
8. Fredeen, D. A.; Russell, D. H. *J. Am. Chem. Soc.* 1987, *109*, 3903.
9. Longoni, G.; Manassero, M.; Sansoni, M. *J. Am. Chem. Soc.* 1980, *102*, 7973.
10. Sosinsky, B.A.; Shong, R.G.; Fitzgerald, N.N.; O'Rourke, C. *Inorg. Chem.* 1983, *22*, 3124.
11. Arndt, L.W.; Darensbourg, M.Y.; Fackler, J.P., Jr.; Lusk, R.J.; Marler, D.O.; Youngdahl, K.A. *J. Am. Chem. Soc.* 1985, *107*, 7218.
12. Chini, P. J. *Organomet. Chem.* 1980, *200*, 37.
13. Freas, R.B.; Ridge, D.P. *J. Am. Chem. Soc.* 1984, *106*, 825.
14. MacMillan, D.K.; Gross, M.L. Chapter 12 of this text.
15. Tecklenburg, R.E.; Russell, D.H. *J. Am. Chem. Soc.* 1987, *109*, 7654.
16. Tyler, D.R.; Levenson, R.A.; Gray, H.B. *J. Am. Chem. Soc.* 1978, *100*, 7888.
17. (a) Hettich, R.L.; Freiser, B.S. *J. Am. Chem. Soc.* 1985, *107*, 62222; (b) Cassady, C.J.; Freiser, B.S. *J. Am. Chem. Soc.* 1984, *106*, 6176; (c) Leopold, D.G.; Lineberger, W.C. *J. Chem. Phys.* 1986, *85*, 51.
18. Poliakoff, M.; Weitz, E. *Acc. Chem. Res.* 1987, *20*, 408.
19. Rothberg, L. J.; Gerrity, D.P.; Vaida, V. *J. Chem. Phys.* 1981, *74*, 2218.
20. Leopold, D.G.; Vaida, V. *J. Am. Chem. Soc.* 1984, *106*, 3720.
21. Kobayashi, T.; Ohtani, H.; Noda, H.; Teratani, S.; Yamazaki, H.; Yasufuku, K. *Organometallics* 1986, *5*, 110.
22. Welch, J.A.; Vaida, V.; Geoffroy, G.L. *J. Phys. Chem.* 1983, *87*, 3635.
23. Leutwyler, S.; Even, U. *Chem. Phys. Lett.* 1981, *84*, 188.
24. Hollingsworth, W.E.; Vaida, V. *J. Phys. Chem.* 1986, *90*, 1235.
25. Lichtin, D.A.; Bernstein, R.B.; Vaida, V. *J. Am. Chem. Soc.* 1982, *104*, 1830.
26. Levenson, R.A.; Gray, H.B. *J. Am. Chem. Soc.* 1975, *97*, 6042.
27. Green, J.C.; Mingos, D.M.P.; Seddon, E.A. *Inorg. Chem.* 1981, *20*, 2595.
28. Connor, J.A.; Zafarani-Moattar, M.T.; Bickerton, J.; El-Saied, N.I.; Suradi, S.; Carson, R.; Al Takhin, G.; Skinner, H.A. *Organometallics* 1982, *1*, 1166.
29. Martinho Simoes, J.A.; Schultz, J.C.; Beauchamp, J.L. *Organometallics* 1985, *4*, 1238.
30. Connor, J.A.; Skinner, H.A.; Virmani, Y. *Faraday Symp. Chem. Soc.* 1984, *8*, 18.
31. Ervin, K.; Loh, S.K.; Aristov, N.; Armentrout, P.B. *J. Phys. Chem.* 1983, *87*, 3593.
32. Jarrold, M.F.; Illies, A.J.; Bowers, M.T. *J. Am. Chem. Soc.* 1985, *107*, 7339.
33. Svec, H.J.; Junk, G.A. *J. Am. Chem. Soc.* 1967, *89*, 2836.
34. Bray, R.G.; Seidler, P.F.; Gethner, J.S.; Woodin, R.L. *J. Am. Chem. Soc.* 1986, *108*, 1312.
35. Horak, D.V.; Winn, J.S. *J. Phys. Chem.* 1983, *87*, 265.
36. Nagano, Y.; Achiba, Y.; Kimura, K. *J. Chem. Phys.* 1986, *84*, 1063.
37. Yardley, J.T.; Gitlin, B.; Nathanson, G.; Rosan, A.M. *J. Chem. Phys.* 1981, *74*, 370.
38. Jarrold, M.F.; Misev, L.; Bowers, M.T. *J. Phys. Chem.* 1984, *88*, 3928.
39. Seder, T.A.; Ouderkirk, A.J.; Weitz, E. *J. Chem. Phys.* 1986, *85*, 1977.
40. Ouderkirk, A.J.; Seder, T.A.; Weitz, E. *Applications of Lasers to Industrial Chemistry*; SPIE–The International Society for Optical Engineering: Bellingham, Washington, 1984; Vol. 458, p. 148.
41. Seder, T.A.; Church, S.P.; Weitz, E. *J. Am. Chem. Soc.* 1986, *108*, 4721.
42. Reents, W.D., Jr.; Strobel, F.; Freas, R.B., III; Wronka, J.; Ridge, D.P. *J. Phys. Chem.* 1985, *89*, 5666.

5

Metal and Semiconductor Cluster Ions

Martin F. Jarrold

1. INTRODUCTION

By repetitively dividing an object into smaller and smaller parts, one eventually comes to a dimer. A dimer obviously has different properties than the object we started with. It is not reasonable to think of a dimer as a semiconductor or of a dimer melting! Clearly, there is a minimum particle size below which we have to change our solid state language to a molecular language. But at what cluster size must we make this change, does this critical size vary with the property being considered, and how do the chemical properties fit into this picture? Recent advances in experimental techniques have made it possible to generate very small particles or clusters (containing only a few atoms) of virtually any element in the periodic table.[1,2] Unlike the cluster compounds studied using more conventional inorganic chemistry approaches, these are isolated gas phase species that generally do not have ligands. Furthermore, they are free of substrate interactions that hinder matrix isolation studies.

The recent interest in clusters arises from several sources. As mentioned above, there is fundamental interest in following the development of bulk physical and chemical properties as the cluster size increases. Second, clusters are a unique state of matter likely to show some properties different from those of both the atomic and bulk states, and so are interesting in their own right. Clusters may be useful as models for surface phenomena and in understanding catalytic processes. Finally, there are potential commercial applications in using clusters to generate new materials with tailored properties.

Progress in studying metal and semiconductor clusters has been rapid since the development of the pulsed laser vaporization technique for generating these species.[1,2] There are, however, several other approaches available for making clusters, which are discussed in more detail below. A variety of

Martin F. Jarrold • AT&T Bell Laboratories, Murray Hill, New Jersey 07974.

techniques have been used to probe the clusters, including ionization potential measurements,[3-6] photoelectron spectroscopy,[7,8] studies of magnetic properties,[9] photodissociation,[10,11] collision-induced dissociation,[12-14] and studies of chemical reactivity.[15-17] Detailed structural information on the clusters would be fascinating. Unfortunately, attempts to apply spectroscopic techniques such as laser-induced fluorescence (LIF) and resonant two-photon ionization (R2PI) to metal clusters have not been very successful, presumably because the lifetime of the intermediate excited state is so short (which is a characteristic of metallic behavior). The largest cluster for which it has yet been possible to obtain a spectrum is a trimer.[18,19]

Cluster sources produce a range of different cluster sizes. Since one of the objectives of studying clusters is to investigate size-dependent properties, the presence of a range of cluster sizes can be a problem. Studying cluster ions provides a way of avoiding the uncertainties (such as unknown ionization cross sections and fragmentation) associated with ionization of neutral clusters for subsequent mass analysis. Furthermore, with cluster ions it is relatively easy to select a particular cluster size before probing its properties. For these reasons, the study of the chemistry of size-selected metal cluster ions has developed into a vigorous research area over the last two years. The presence of the charge might be expected to cause differences in the reactivity between ionic and neutral clusters which vanish as the cluster size increases and the charge becomes more delocalized. In the limited number of cases where comparisons can currently be made, it appears that the reactions of ionic and neutral clusters are often quite similar. The specific examples are discussed in detail below.

In this chapter the recent results of studies of size-selected cluster ions are described. The main focus is on the chemistry of these species and the recent developments in photodissociation and photoelectron spectroscopy. We consider only bare clusters with more than three metal atoms (the extensive studies on smaller species and on clusters covered with ligands are described in other chapters in this book). We focus on carbon, silicon, and aluminum clusters and some recent work on transition metal clusters. Carbon, silicon, and aluminum clusters have been quite extensively studied recently. The majority of the work mentioned is less than two years old, which gives a good indication of how rapidly this field is progressing. Several reviews have recently appeared covering different aspects of cluster research.[20-22] Some recently published conference proceedings also provide a useful broad overview of current work in this area.[23-25]

2. METHODS FOR GENERATING CLUSTER IONS

Historically, supersonic expansions have been the natural choice for making clusters.† However, if the material of interest is not a gas, we clearly

† For a historical review, see Reference 26.

need a modified approach. A natural extension would be to employ an oven to heat the material. If the resulting vapors are entrained in a carrier gas and then cooled using a supersonic expansion, or a cooled quench cell (or both), cluster growth occurs in the supersaturated vapor. This was the approach employed in many of the earlier studies of metal clusters.[23] However, while this method works well for low-boiling-point elements such as sodium, severe engineering problems can be encountered for more refractory materials. Ovens provide the most intense cluster sources. They are used in ionized cluster beam deposition, which is a technique for making thin films.[27] However, there is currently a controversy over whether the large (>2000 atom) clusters thought to be responsible for the favorable film properties can be generated in pure vapor expansions under the conditions employed.[28-30]

Fast ion beam or fast neutral beam sputtering is a commonly used technique for making cluster ions.[31-36] Kilovolt-energy particles cause sputtering by a momentum transfer mechanism, where the resulting collision cascade eventually causes material to be ejected from the surface.[37] With metals, most of the material is removed as atoms and a roughly exponential falloff in cluster intensity is observed with increasing cluster size. Sputtered cluster ions have a range of kinetic energies and are internally excited.[38,39] Anderson and co-workers have described an approach to cooling the clusters using a radio frequency "stack of plates" ion trap containing a buffer gas.[14] This method cools the clusters' kinetic energy and probably the internal energy as well.

Direct pulsed laser sputtering[40-43] is a simple approach that works well for nonmetallic materials. Figure 1 shows a mass spectrum of silicon cluster ions generated by direct laser sputtering of polycrystalline silicon.[44] Positive and negative silicon cluster ions with up to seven atoms can be produced in usable quantities. The intensities of the cluster ions decrease continuously with increasing cluster size, and there are sharp drops in the intensities of the positive ions at $n = 5$ and 7. Carbon cluster ions with over 100 atoms can be made by direct laser sputtering, although the intensities of the large clusters are quite low.[45] This approach can also be used to make mixed clusters. For example, indium phosphide clusters can be made by direct laser sputtering of the bulk III–V semiconductor.[46] However, the dominant clusters do not have the 1:1 stoichiometry of the bulk. For reasons which are not entirely clear, direct laser sputtering is less successful with metals. Virtually all the material removed appears to come off the surface as atoms, with in some cases a small amount of dimer and trimer (larger clusters have been detected but their intensities are extremely low).[41,43,47,48] The mechanism by which clusters are formed by direct laser sputtering is not well understood. Are the clusters generated directly off the surface or are they formed by collisions in the laser-generated plasma? Laser-sputtered clusters appear to be formed from those materials whose equilbrium vapor contains molecular species.[43] However, this may simply reflect cluster stability. Recently, an enhancement of silver cluster ion production has been reported for direct laser sputtering of an AgO/ZnO mixture.[49] Clusters up to Ag_{11}^{+} were observed, with the odd-numbered clusters being much more intense than the ones with an even

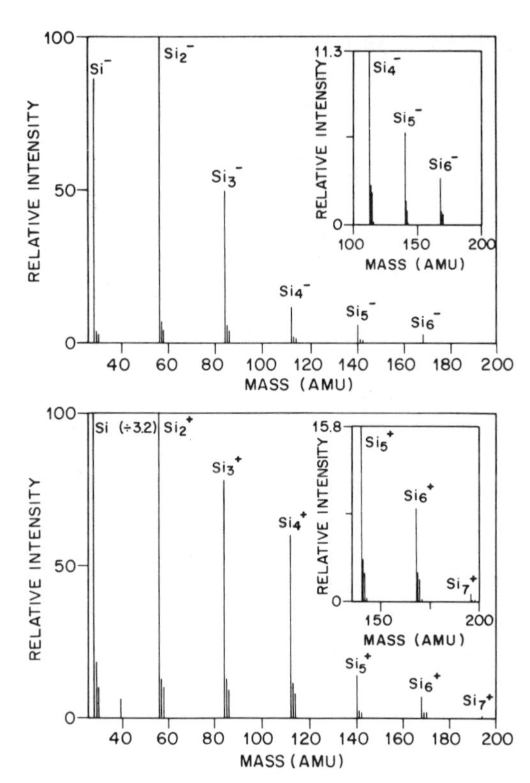

Figure 1. Mass spectra of positive and negative silicon cluster ions generated by direct laser sputtering of polycrystalline silicon. The mass spectra were recorded using Fourier transform ion cyclotron resonance (FT-ICR). (Reproduced with permission from Reference 9.)

number of atoms. The mechanism of this enhancement has not yet been explained, and it is too early to say at this time whether this will be a generally useful approach.

To generate larger metal clusters using laser evaporation, it is necessary to employ a flowing buffer gas (usually helium).[1,2] The buffer gas constrains and cools the material evaporated by the laser pulse and so promotes cluster growth in the gas phase. A rapid flow of buffer gas is required to remove the evaporated material from the vicinity of the surface and prevent redeposition. It is also necessary to slowly move the sample to prevent the laser from boring a hole into it. The choice of laser and pressure conditions does not appear to be critical. Cluster sources have been operated with pressures from tens of torr[13,50] up to several atmospheres[10,16,17,51] and with Nd-YAG,[10,16,17,51] excimer,[13,50] and copper vapor[52] lasers. With high-repetition-rate excimer lasers (over 500 Hz) and copper vapor lasers a continuous beam of clusters can be generated. Cluster ions arising from ionization in the laser plasma can be produced directly from these sources.[53] Figures 2a and 2b show mass spectra of positive and negative carbon cluster ions formed in this way. A mass spectrum of clusters photoionized after the expansion is also shown in Figure 2c for comparison. Notice that the peaks due to $n = 7$, 11, 15, 19, and 23 are slightly more intense than their neighbors in both positive ion mass spectra. This periodicity of four may reflect the aromatic stabilization expected

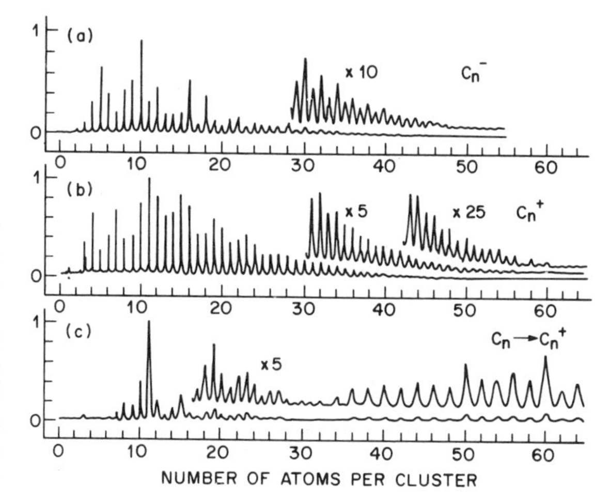

Figure 2. Mass spectra of carbon cluster ions from laser evaporation into a flowing buffer gas: (a) negative ions from the laser plasma; (b) positive ions from the laser plasma; and (c) positive ions from photoionization of neutral clusters after the expansion. (Reproduced with permission from Reference 53.)

for $4n + 2\pi$ electrons.(54–56) Surprisingly, the odd-numbered clusters with $n > 30$ are absent in the mass spectrum of photoionized neutral clusters.(57) Presumably, neutral carbon clusters with an odd number of atoms and $n > 30$ are not very stable.

The number of cluster ions produced directly from the laser plasma is limited by electron–ion recombination or space charge effects as the clusters travel through the source. In some cases (particularly metals) it is useful to enhance the degree of ionization downstream from the point of laser vaporization either by laser photoionization or by a high-energy electron beam.(13,58) If this ionization occurs in the buffer gas, then excess internal energy can be removed from the clusters by collisions. However, it is not yet possible to measure directly the internal temperatures of the clusters except in the special case of dimers and trimers where optical spectra can be obtained.(18,19,39) So the degree of cooling achieved and the amount of internal excitation remaining in the larger clusters can only be estimated. Figure 3 shows a mass spectrum of aluminum cluster ions produced by pulsed laser vaporization followed by ionization in the buffer gas by a high-energy electron beam.(13) Notice that the peaks due to Al_7^+ and Al_{14}^+ are more intense than their neighbors—they are "magic numbers" (the term "magic number" is commonly used to denote particularly intense peaks in a cluster ion mass spectrum). Similar results were obtained for aluminum cluster ions generated in different ways.(9,12,31) As will be discussed in more detail below, Al_7^+ and Al_{14}^+ are particularly stable clusters.

Another approach to making clusters is the liquid metal ion source.(59,60) With this technique, the liquid metal wets an emitter tip which is subjected to a strong electric field. Ionized atoms and clusters field evaporate from the tip. The intensities of the clusters are typically 0.001 to 0.0001 of the intensity of the atomic ion but large (>30 atom) clusters can be generated. Since large ion currents (several microamps) can be drawn from these sources, the cluster ion currents are also quite large (up to nanoamps). In order for this source to operate, the liquid metal must wet the tip. However, these sources have been

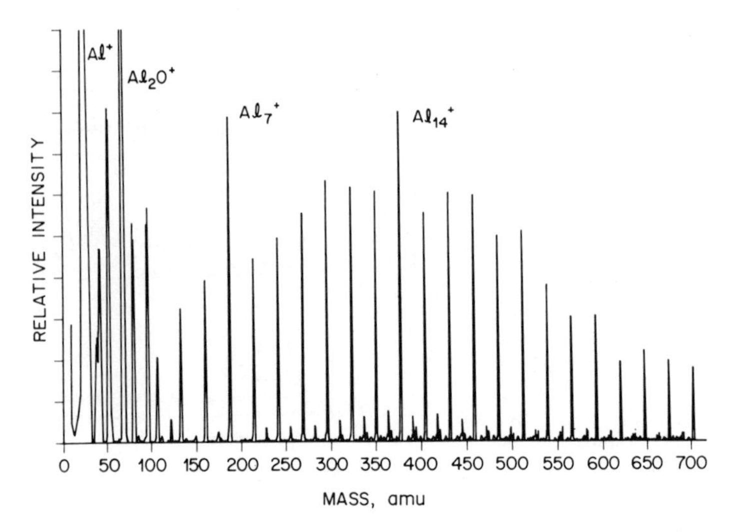

Figure 3. Mass spectrum of aluminum cluster ions produced by laser evaporation into a flowing buffer gas followed by ionization by a high-energy electron beam. (Reproduced with permission from Reference 13.)

operated with a wide range of materials. The principal drawback of the liquid metal ion source is that the clusters have a wide range of kinetic energies (several hundred electron volts), and they are probably internally excited as well. Given the large cluster ion currents that can be drawn from these sources, it is feasible to select out a narrow range of kinetic energies or to cool the clusters using a buffer gas. Some experiments along these lines are in progress.[61]

3. METHODS FOR STUDYING CLUSTER IONS

Some of the methods used to investigate the properties of cluster ions will be briefly reviewed here to give the reader an appreciation of the technology involved. Two approaches are currently used to investigate the chemical reactions of size-selected clusters: Fourier transform ion cyclotron resonance (FT-ICR)[44,62–64] and low-energy ion beam techniques.[32,34,35,65] Both techniques have been used for many years to investigate ion–molecule reactions.[66] FT-ICR and the low-energy ion beam technique are in many ways complementary. FT-ICR provides a measure of reaction rate constants under thermal (or close to thermal) conditions. Beam techniques give a measure of reaction cross sections at kinetic energies above thermal energies, but measurements can be made as a function of kinetic energy.

The principle behind the FT-ICR approach is illustrated in Figure 4.[67] Cluster ions generated by direct laser sputtering are trapped and undergo cyclotron motion in the crossed electric and magnetic fields of the ICR cell. In the specific example shown in Figure 4, a dual ICR cell is used to give

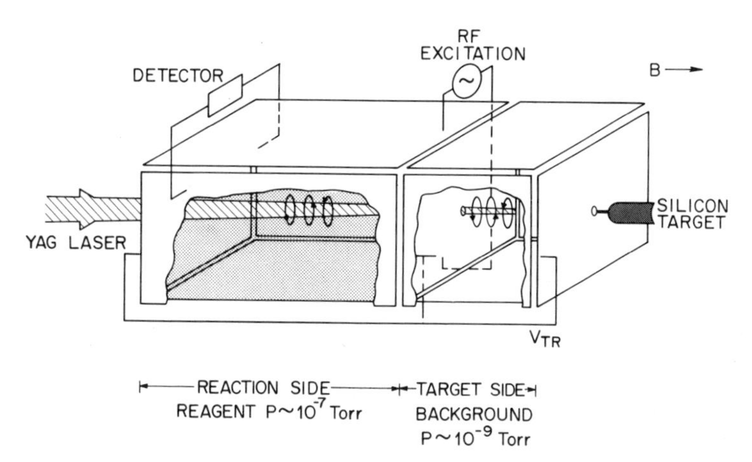

Figure 4. Schematic diagram of an FT-ICR experiment employing a dual cell and direct laser sputtering to generate the clusters. (Reproduced with permission from Reference 67.)

separate cluster generation and reaction regions. This configuration was found to reduce complications due to contamination of the laser-sputtered surface by the reagent gas. Unwanted clusters can be ejected from the cell by irradiation at their cyclotron frequencies, which increases the cyclotron orbits until the ion collides with the cell walls. Size-selected clusters then undergo reactions with gas admitted to the cell, and after a variable time the products are mass analyzed by measuring their characteristic cyclotron frequency. An attractive feature of the FT-ICR approach is the very high mass resolution that can be achieved. The resolution would be sufficient to resolve positive and negative ions of the same molecule (though it is not possible to trap positive and negative ions in the cell at the same time). This high mass resolution is more difficult to attain with high-mass species. In practice, the mass resolution that can be obtained in the ejection step is much lower because of the excitation of the kinetic energy of ions with similar masses. A potential problem with these studies is that the cluster ions generated by direct laser sputtering may have excess internal and kinetic energies. By using low trapping voltages and allowing collisions to occur before starting to investigate the reactions, it is possible to reduce the errors that could arise from these sources. Recently, Smalley and co-workers have injected cluster ions into an ICR cell from an external source.(68) This considerably extends the size and type of clusters that can be studied using the FT–ICR technique. Injecting ions into an ICR cell is a technological achievement because the magnetic field acts as a mirror.(69) Currently, it is necessary to employ multiple injections to obtain a sufficient cluster ion density in the ICR cell for reaction studies to be performed.

A schematic diagram of one of several(32,34,35,65) low-energy ion beam instruments built to investigate cluster ions is shown in Figure 5. These instruments consist of a source to generate the cluster ions (using either fast atom sputtering or laser vaporization in a buffer gas) followed by a mass spectrometer to select out a particular cluster size. Reactions occur in either

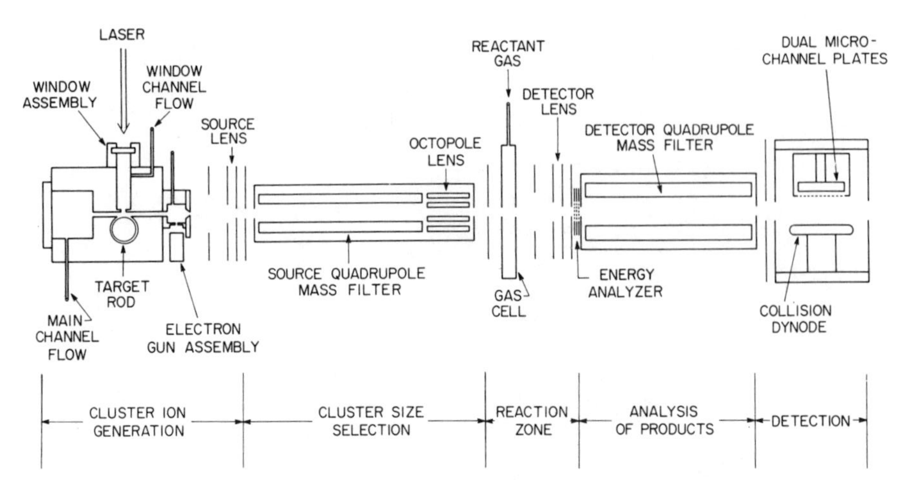

Figure 5. Schematic diagram of a low-energy ion beam instrument used to investigate the chemical reactions of size-selected cluster ions. (Reproduced with permission from Reference 13.)

a gas cell or a radio frequency ion guide. The products are analyzed by a second mass spectrometer and then detected using an electron multiplier. The products of the reactions of large cluster ions can vary widely in mass (if cluster fragmentation occurs) so care must be taken to avoid mass discrimination, both in the analyzing mass spectrometer and at the detector. The probability of mass discrimination increases with cluster size, and this problem is compounded by the difficulty of directly determining the extent of discrimination. High-voltage (up to 30 kV) collision dynodes along with pulse counting techniques are usually employed to avoid mass discrimination at the detector.

Photodissociation, photodetachment, and photoelectron spectroscopy studies have recently yielded some fascinating results for metal and semiconductor cluster ions.[7,8,10,11,70–73] Figure 6 shows a schematic diagram of an instrument constructed by Bloomfield *et al.*[10] to investigate the photodissoci-

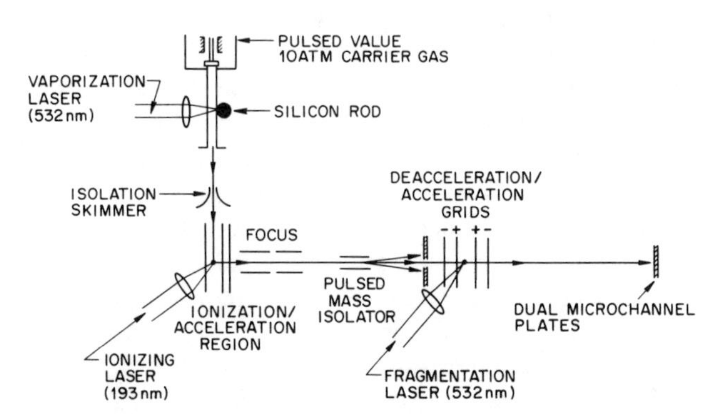

Figure 6. Schematic diagram of an experimental apparatus constructed to investigate the photodissociation of size-selected cluster ions. (Reproduced with permission from Reference 10.)

ation of size-selected cluster ions. A similar approach has been used by other workers.[11] Clusters are produced by laser vaporization in a pulsed flow of helium buffer gas. The clusters are ionized by an excimer laser and then accelerated to kilovolt energies. A particular cluster size is selected using time-of-flight mass analysis and a pulsed electrostatic mass gate which is pulsed open at the time the chosen cluster arrives. The size-selected cluster packet is then decelerated and irradiated by a third laser. The photofragments are accelerated, analyzed by time of flight, and detected using dual microchannel plates. With negative ions both photodissociation and photodetachment (removal of an electron) are observed.[70,71] Energy analysis of these electrons is photoelectron spectroscopy, which provides information on the electronic structure of the neutral clusters. These experiments are currently being performed in a number of laboratories.[7,8,72,73] The photoelectrons are often confined by magnetic fields in order to increase sensitivity.

4. CARBON CLUSTER IONS

Carbon clusters are probably important in soot formation and in interstellar space, particularly around carbon-rich stars. They have recently received considerable attention. Carbon clusters are comparatively easy to generate (large carbon clusters can be made by direct laser sputtering[45] and carbon is light so the clusters are quite easy to work with experimentally. Since carbon only has a few electrons, the clusters are attractive objects for theoretical studies as well. Recently, there has been a controversy over the suggestion by Smalley and co-workers that C_{60} has a truncated icosahedral (soccerball) structure.[74] Evidence that there is something special about C_{60} comes from observations that this is a particularly abundant species in the mass spectra of carbon clusters produced in a number of ways.[45,53,57,74,75] However, the interpretation of these data has been challenged.[76,77] We will return to the "soccerball" question towards the end of this section.

The photodissociation of size-selected carbon cluster ions with up to 20 atoms has been investigated by Geusic *et al.*[78,79] They measured product distributions and photodissociation cross sections and investigated the laser fluence dependence of photodissociation with 248-nm and 351-nm light. The product distributions recorded at 248 nm are shown in Figure 7. These results were obtained with a laser fluence of $\approx 2\ \mathrm{mJ/cm^2}$, which is sufficiently low that multiphoton effects are not important. Similar results were obtained using 351 nm light. As can be seen from Figure 7, the surprising result of these studies is that all the carbon cluster ions studied (with the single exception of C_5^+) fragment mainly by the loss of a C_3 unit. As will be discussed in more detail below, loss of C_3 is observed for carbon cluster ions excited in a number of different ways. So the photodissociation of carbon cluster ions in this size range probably involves excitation to a bound excited state followed by internal conversion to the ground electronic state and statistical dissociation on the ground electronic state potential surface to give preferentially the most stable

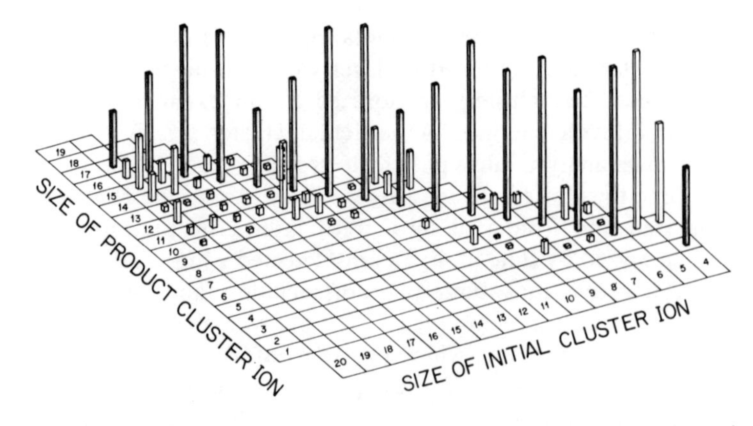

Figure 7. Histogram of the products from the photodissociation of carbon cluster ions with 248-nm light. The main product arises from the loss of a C_3 unit. The products arising from this process are shaded. Fluence $\approx 2\ \mathrm{mJ/cm^{-2}}$. (Reproduced with permission from Reference 79.)

products. An implication of this conclusion is that the C_3 channel dominates, not because C_3 is the building block of the clusters, but because the C_3 species is very stable.[80] These results illustrate an important point: *one must be extremely cautious about deducing structural information from cluster fragmentation patterns.* The photodissociation of larger carbon cluster ions has been studied by Smalley and co-workers.[81] The results of these studies are discussed below.

The chemical reactions of carbon cluster ions have been investigated by several groups. The group at the Naval Research Laboratory (NRL) has studied the reactions of carbon clusters with up to 19 atoms with oxygen and deuterium.[62,82] These experiments were performed using FT-ICR and direct laser sputtering to generate the clusters. With deuterium, only clusters up to $n = 9$ were found to react:

$$C_n^+ + D_2 \longrightarrow C_nD^+ + D \tag{1}$$

$$ \longrightarrow C_{n-3}D_2^+ + C_3 \tag{2}$$

The main product was C_nD^+, and a minor amount of $C_{n-3}D_2^+$ (loss of C_3) was observed for $n = 6$ and 8. Measured rate constants for these reactions are shown in Figure 8. The reactions of the even-numbered clusters appear to be slower than those of the clusters with an odd number of atoms. The dramatic change in reactivity at $n = 10$ correlates with an abrupt change in the photodissociation cross sections at this cluster size.[78,79] These results suggest that carbon cluster ions with up to 9 atoms are fundamentally different from those with 10 and more atoms. Some theoretical calculations predict that a change in geometric structure occurs from linear chains for $n < 10$ to monocyclic rings for $n > 9$.[54,55,83] The reactivity of the smaller clusters could then be accounted for by the carbene(=C:) character of the terminal carbon atom. The carbene

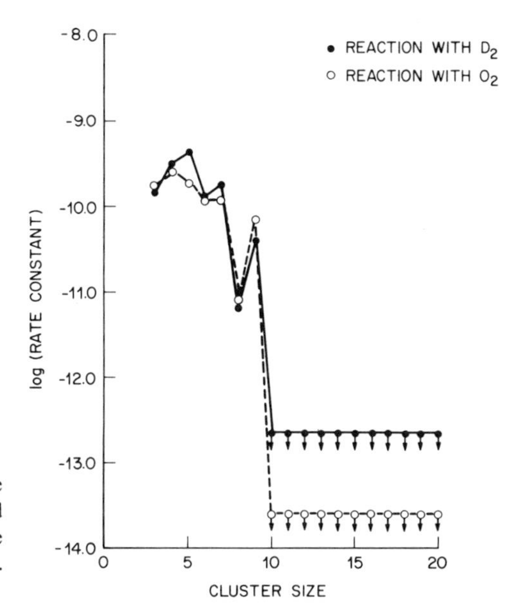

Figure 8. Plot of the rate constants for the reactions between carbon cluster ions and deuterium (●) and oxygen (○). The measurements were made using FT-ICR. (Data taken from Reference 82.)

cation is known to insert into σ bonds.[84,85] Insertion into the deuterium bond would result in a $C—C_{n-2}—CD_2^+$ complex which could subsequently fragment by loss of D or C_3 to give the observed products.

With oxygen the following reactions were observed:

$$C_n^+ + O_2 \longrightarrow C_nO^+ + O \tag{3}$$

$$\longrightarrow C_{n-1}^+ + CO_2 \tag{4}$$

$$\longrightarrow C_{n-2}^+ + 2CO \tag{5}$$

where the neutral products are inferred from thermodynamic arguments. The main product was C_nO^+. Rate constants for these reactions are plotted in Figure 8, where they can be compared to the rate constants for the reactions with deuterium. As for the reactions with deuterium, there is a general decline in the rates with increasing cluster size, and a reaction is not observed for clusters with $n > 9$.

While performing the reaction studies, it was found that there are apparently two components of C_7^+ which react with both deuterium and oxygen at different rates.[62,82] This was not observed for any of the other clusters studied. Figure 9 is a plot of the fraction of C_7^+ against reaction time with deuterium. Around 33% of the C_7^+ reacts with D_2. The remaining fraction does not appear to react over times as long as 3 s. It appears possible to rule out internal or kinetic energy as the cause of this effect, and it is unlikely that a metastable electronically excited state could be responsible because this would require a lifetime in excess of 5 s. So the most likely explanation is that

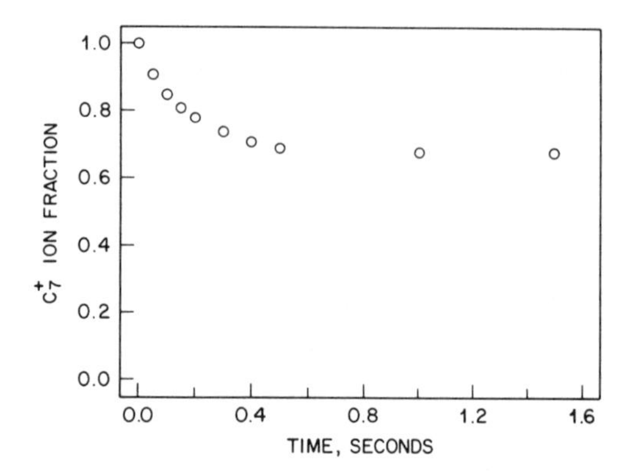

Figure 9. Plot of the fraction of C_7^+ against reaction time with deuterium. Approximately one-third of the C_7^+ reacts. (Reproduced with permission from Reference 85.)

two geometric structures are responsible (possibly linear and cyclic). Surprisingly, approximately the same amount of the unreactive isomer appears to be produced by direct laser sputtering of graphite and diamond and from the fragmentation of C_{10}^+.

Marshall and co-workers[3] have surveyed the reactions between carbon cluster ions and C_6H_{12}. They found that clusters with up to 10 atoms rapidly reacted and those with more than 10 atoms did not react. This reactivity pattern appears to be slightly different from that observed with D_2 and O_2, suggesting that it may have a different origin. The main product was $C_6H_{11}^+$.

The collision-induced dissociation of carbon cluster ions has been studied by the NRL group.[82] Low-energy collisions (eV range) result in vibrational excitation. High-energy collisions (keV range) result in electronic excitation, which for polyatomic species is generally assumed to be followed by rapid internal conversion to give the vibrationally excited ground state. In both cases dissociation generally occurs statistically on the ground state potential surface to give predominantly the most stable products. Direct (knock-out) processes are not generally considered to be important. The collision-induced dissociation of carbon cluster ions by xenon was studied using FT-ICR (low-energy collisions). The main products observed for carbon cluster ions with $n = 5$–15 was C_{n-3}^+ (loss of C_3). Attempts to dissociate C_{50}^+ and C_{60}^+ were not successful,[86] probably because (as is discussed in more detail below) large amounts of energy are required to dissociate the larger cluster ions.

The metastable dissociation of size-selected carbon cluster ions has been investigated by Bowers and co-workers[56] using mass-analyzed ion kinetic energy spectrometry (MIKES). In this experiment carbon cluster ions generated by direct laser sputtering are accelerated to 2–8 keV, mass selected by a magnet, and then allowed to spontaneously fragment. The resulting products have a nominal energy of

$$E_p = \frac{m_p}{m_r} \cdot E_r \qquad (6)$$

where E_r and E_p are the energies and m_r and m_p are the masses of the reactants and products, respectively. The products are identified by energy analysis using an electrostatic analyzer. This experiment monitors fragmentation in a time window typically 15 μs long, starting around 15 μs after formation of the clusters. Product distributions for clusters with $n = 5$–60 are shown in Table 1. For clusters larger than C_{30}^+, only the even-numbered clusters (except C_{41}^+) were investigated because the intensity of the large odd-numbered clusters was low. The product distributions for the smaller clusters are quite similar to those observed in earlier photodissociation and collision-induced dissociation studies, although there are significant differences in the region around $n = 11$. A striking change in the fragmentation pattern occurs around $n = 30$. For even clusters with $n > 30$, the main product arises from loss of C_2 rather than C_3. This sharp transition occurs at the point where the odd-numbered clusters in some mass spectra are absent or have reduced intensity (see Figure 2). So the propensity for C_2 loss probably reflects the enhanced stability of the even-numbered cluster ions over the odd ones with $n > 30$. The main product from the fragmentation of C_{41}^+ is C_{40}^+. One of the most striking features of the results shown in Table 1 is the C_{10} and C_{14} loss processes observed for $n = 24$–29. Particularly stable neutral rings have been predicted at $n = 10$, 14, 18, and 22 by some theoretical calculations. Therefore, the existence of these processes could be accounted for by the stability of the C_{10} and C_{14} neutral species.

The photodissociation of larger cluster ions (with up to 80 atoms) has recently been studied by Smalley and co-workers.[81] Measurements of the fluence dependence of photodissociation for C_{60}^+ indicated that for fragmentation to occur on the experimental time scale, at least three photons from an excimer laser operated with ArF (6.4-eV photons) are required. As the authors pointed out, this result does not mean that C_{60}^+ is bound by 19.2 eV. The RRKM theory of unimolecular reactions leads us to expect that large clusters with many vibrational degrees of freedom will require a substantial amount of energy above the dissociation threshold in order for dissociation to occur on the experimental time scale (around 10^{-5} s). In these photodissociation studies, clusters with up to 31 atoms were found to fragment by sequential loss of C_3. The odd-numbered clusters with more than 34 atoms lose a carbon atom and the even-numbered clusters fragment by sequential loss of C_2. This is basically the same fragmentation pattern as observed in the metastable studies.[56] However, there are some significant differences in the 30-atom size range. Photodissociation of C_{32}^+ generated by sequential fragmentation of larger clusters and direct photodissociation of C_{32}^+ resulted in a "catastrophic disintegration" to small cluster ions in the 10–19-atom size range. The main products were C_{18}^+ and C_{15}^+. These products could be accounted for by loss of C_{14} (as seen in the metastable studies for smaller clusters) followed by loss of C_3 to give C_{15}^+. The differences between the product distributions observed in the metastable and the photodissociation studies may be due to the different internal energy distributions of the fragmenting ions. However, different structural isomers could also be responsible.[81] Figure 10 shows the photofragment ions produced by photodissociation of C_{60}^+ with 353-nm light with a range of fluences from 8 mJ/cm^2 to 55 mJ/cm^2. At low fluence the main products are

Table 1. Branching Ratios for the Neutral Lost in the Metastable Reaction $(C_n^+)^* \rightarrow C_{n-x}^+ + C_x$ [a]

Size of initial cluster ion (n)	Size of neutral lost (x)[b]													
	1	2	3	4	5	6	7	8	9	10	11	12	13	14
5		100												
6			100											
7			100											
8	20		80											
9		29	71											
10	50		50											
11	84		16											
12	88	6	6											
13	3	4	93											
14			100											
15			84		16									
16		11	42		47									
17	12		88											
18	4		96											
19	3		69		28									
20			50		50									
21			100											
22			100											
23	4		60		36									
24	18		27		36					20				
25			50		4					32				14
26	10		78		12									
27			54		25		7			14				
28			20		15					36				29
29			17							44				39
30		70	30											
32	12	80	8											
34		100												
36		100												
38		100												
40		100												
41	70	30												
42	5	95												
44		100												
46	5	95												
48		100												
50		100												
52		100												
54		95		5										
56		70		30										
58		94		6										
60		88		12										

[a] From Reference 56.
[b] The percentage of the observed metastable fragments is given. These percentages are accurate to at least ±30%.

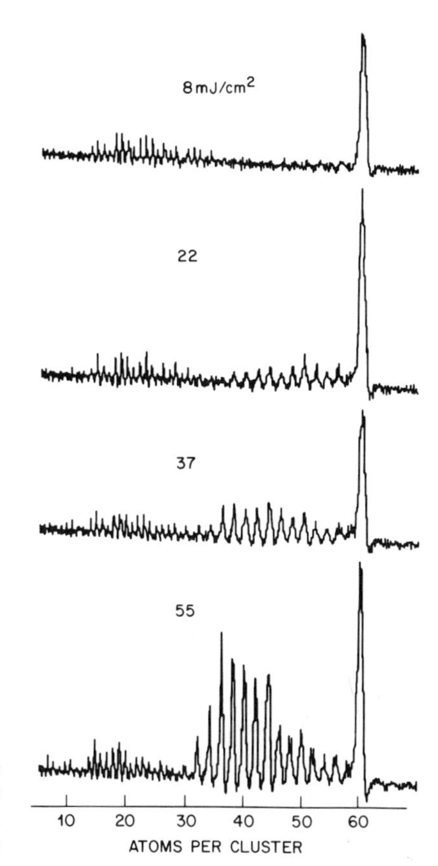

Figure 10. Mass spectra of the products from the photodissociation of C_{60}^+ using 353-nm light with a range of fluences from 8 mJ/cm² to 55 mJ/cm². (Reproduced with permission from Reference 81.)

in the 10 to 30-atom size range; at high fluence the intensity of these products has not substantially increased but new photoproducts have appeared in the 30 to 60-atom size range. It was suggested that these data indicate that there are two isomers of C_{60}^+: a poorly formed cluster which easily fragments at low fluence to give small fragment ions and a strongly bound species which dissociates only at high fluence by sequential loss of a C_2 unit. The C_{60}^+ ions for these studies were not generated using optimum source conditions, and there may be other possible explanations of these data. Identifying and characterizing the different structural isomers that might exist for these larger clusters is clearly going to be a challenge.

Ultraviolet photoelectron spectra of size-selected negative carbon cluster ions ($n = 48$–84) have been measured using an ArF excimer laser.[8] An overview of the results is shown in Figure 11. C_{60}^- and to a lesser extent C_{50}^- and C_{70}^- show nearly resolved, low-intensity, low-electron-binding-energy peaks. This type of photoelectron spectrum for a negative ion is characteristic of a closed-shell neutral species where the extra electron in the negative ion lies alone in the LUMO significantly above the HOMO in energy. These spectra spectra suggest that C_{50}^-, C_{60}^-, and C_{70}^- are closed-shell species, with a HOMO-

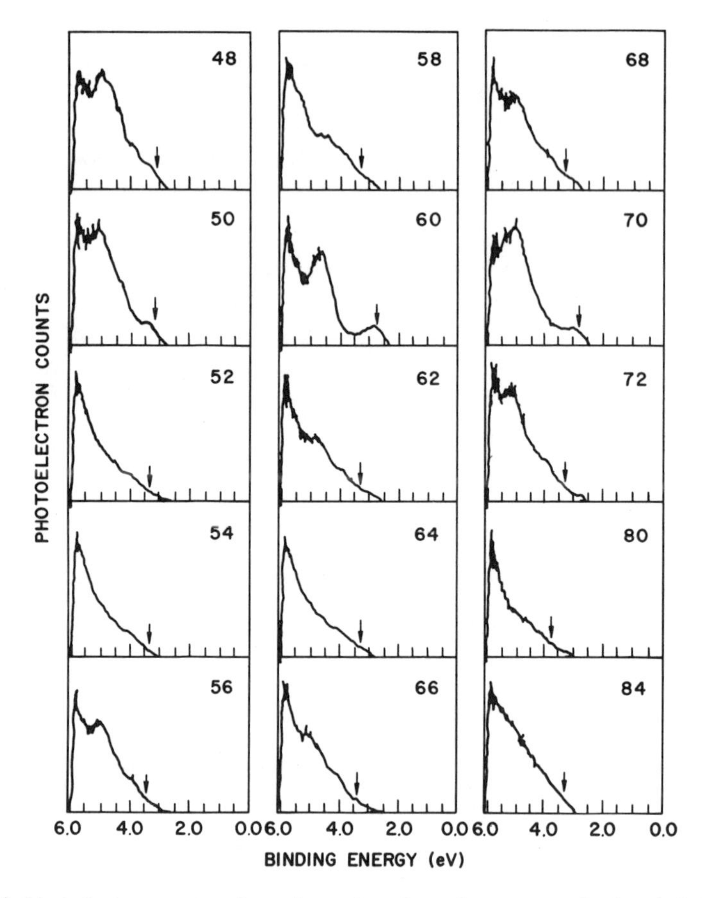

Figure 11. Photoelectron spectra of negative carbon cluster ions measured using an ArF excimer laser. Notice the partially resolved low-electron-binding-energy peaks for C_{50}^-, C_{60}^-, and C_{70}^-. (Reproduced with permission from Reference 8.)

LUMO gap of 1.5–2.0 eV for C_{60}. Given the stability of C_{60}, a closed-shell structure might be expected. A closed-electronic-shell structure is predicted for C_{60}, assuming a truncated icosahedral structure.[88–96]

Along with the experimental studies of carbon clusters, there has been a corresponding increase in theoretical activity. In the last two years a flurry of papers has appeared,[88–99] but the majority of these have been restricted to C_{60}. Most of the theoretical work has been on neutral clusters, which is unfortunate because the bulk of the experimental data is on cluster ions. Early studies using semiempirical methods considered both linear chains and monocyclic rings[83,87] and predicted a transition to monocyclic rings at n = 10.[83] Particularly stable monocyclic rings were predicted to exist for $n = 10$, 14, 18, and 22 due to aromatic stabilization. These results are in general agreement with more recent calculations using semiempirical methods.[54,55,82] Bernholc and Phillips[54,55] used these methods to determine the kinetic parameters of a model for aggregation of carbon clusters. Calculated distributions

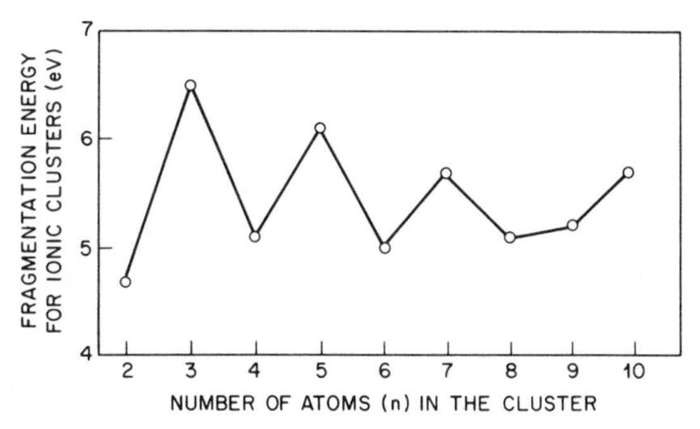

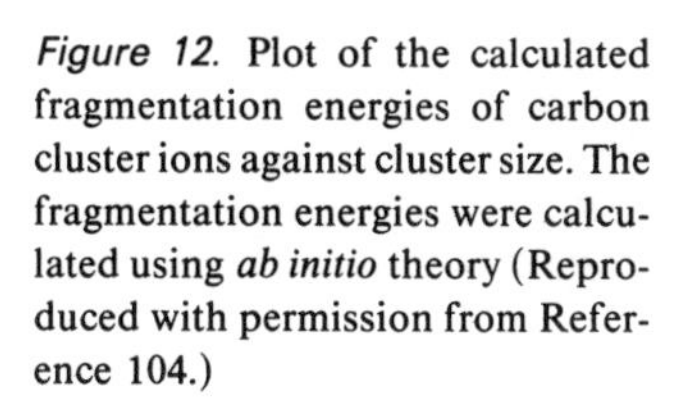
Figure 12. Plot of the calculated fragmentation energies of carbon cluster ions against cluster size. The fragmentation energies were calculated using *ab initio* theory (Reproduced with permission from Reference 104.)

were in good agreement with the observed positive ion, negative ion, and photoionized neutral cluster mass spectra.

Higher-level *ab initio* calculations suggest that the monocyclic rings are slightly more stable than the linear chains for the even-numbered clusters with $n < 10$.[100-104] As might be expected, cyclic C_6 shows more stabilization relative to the linear structure than C_4 or C_8 as it has aromatic stabilization. Detailed calculations by Raghavachari and Binkley[104] for $n < 10$, however, suggest that the monocyclic rings might not be the most stable structures for the positive ions due to the higher ionization potentials of the rings relative to the linear molecules. Their calculations also show that the loss of C_3 observed in the fragmentation of these clusters is indeed the lowest-energy fragmentation pathway available to the clusters. Calculated dissociation energies are shown in Figure 12 for clusters up to C_{10}^+. There is a striking odd–even alternation in the cluster dissociation energies. The odd-numbered clusters are significantly more strongly bound than their even-numbered neighbors.

As mentioned above, there have been several recent calculations on C_{60} and other possible spheroidal shell molecules.[88-99] These calculations indicate that a truncated icosahedral structure for C_{60} would be stable, and so this is perhaps the first example of a spherical aromatic molecule. However, it has also been pointed out that a truncated icosahedron is not the only possible spheroidal structure that might exist for C_{60}.[95,97] There are not yet any unambiguous experimental data (i.e., spectroscopic information) on the structures of the larger carbon clusters. Thus, the proposed "soccerball" structure for C_{60} is really only speculation, although the concept of spherical shell molecules is very appealing and there is clearly something special about C_{60}.

5. SILICON CLUSTER IONS

Silicon is below carbon in the periodic table. It is an intrinsic semiconductor, and it is an important material technologically. In the bulk, silicon has a diamondlike lattice with terahedrally coordinated atoms. As for carbon, it is possible to generate silicon clusters by direct laser sputtering.[43] However, it is only possible to generate fairly small clusters (with up to 9 atoms) in this way. To make larger clusters it is necessary to employ laser vaporization with

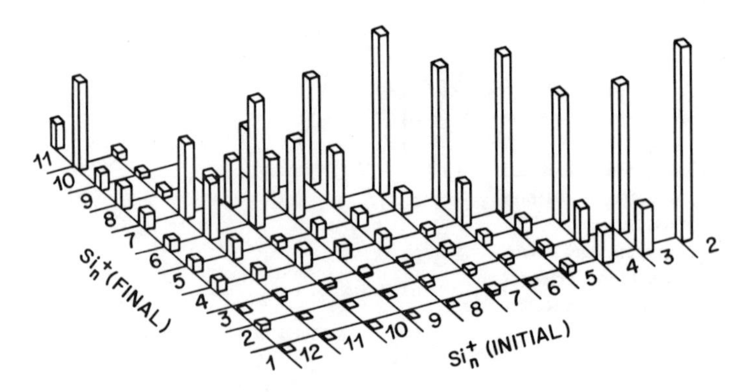

Figure 13. Histogram showing the photofragments from the photodissociation of size-selected silicon clusters. (Reproduced with permission from Reference 103.)

a buffer gas. The resulting cluster distribution shows "magic numbers" (particularly intense cluster peaks) at $n = 6$, 10, and 16,[10,105] suggesting that these clusters are particularly stable. Similar conclusions were reached from a study of silicon cluster ions generated in a radio frequency discharge.[106]

The first study of the properties of silicon cluster ions was performed by Bloomfield *et al.*[10] They investigated the photodissociation of size-selected silicon clusters. A histogram showing the photofragments for clusters with up to 12 atoms is given in Figure 13. For the small clusters the main ionic product is Si_{n-1}^+ (loss of an atom). Starting at Si_7^+ and continuing up to Si_{11}^+, formation of Si_6^+ and, to a lesser extent, of Si_7^+ become the main fragmentation pathways. For Si_{12}^+ the main product is Si_{10}^+. Similar product distributions were observed with 267-, 355-, and 532-nm light. Thus, it appears that fragmentation of silicon cluster ions is dominated by the formation of Si_6^+ and Si_{10}^+ "magic fragments." This is more evidence that these are particularly stable clusters. As mentioned above, $n = 6$ and 10 are "magic numbers" in mass spectra of silicon cluster ions. In the case of germanium, Ge_6^+ and Ge_{10}^+ were found to be the dominant photoproducts.[107] The behavior of silicon and germanium clusters is thus strikingly different from that found for carbon cluster ions, where the main product observed for the fragmentation of the smaller clusters arises from the loss of a C_3 unit.

Smalley and co-workers[108] have recently investigated the photodissociation of larger silicon clusters with 10 to 80 atoms. The observed products determined under low-fluence conditions are shown in Table 2. The observed product distribution was found to be quite insensitive to laser wavelength. In the region of overlap between this study and that of Bloomfield *et al.*[10] (10–12 atoms) there appears to be reasonable agreement except for Si_{12}^+, for which different products were observed. For clusters with 12 to 26 atoms photodissociation appears to involve fission into roughly equal-sized fragments, often by the loss of a 6-, 7-, or 10-atom neutral fragment. For clusters with $n > 30$ the fragments were always in the 6 to 11-atom size range with a subsidiary channel corresponding to loss of a single atom. The fragmentation of the larger clusters

Table 2. Fragmentation Channels of Si_n^+ and Ge_n^+ [a,b]

n	Products Si_x^+ of Si_n^+	Products Ge_x^+ of Ge_n^+
10	6, 4, 7, 5	6, 7, 4, 8, 9, 5
11	7, 6, 5, 4, 10	7, 6, 5, 10, 4
12	6	6, 7, trace 8, 10, 11
13	7, 6, 12	6, 7, 3, weaker 11, 12
14	7, 8, 10, 6	7, 8, 6, 9
15	8, 9	8, 9, 10
16	10, 6, 4	10, 6
17	10, 11, 7	10, 7, 6, 11
18	11, 15, 17, 8	11, trace 6–10
19	9, 10, 6, 7, 12, 13, 16	9, 10, 11, 12, 18, 7
20	10, trace 6–11	10, trace 6–11
21	11, trace 6–10	11, trace 6–10
22	12, 15, 10, 6	6, 8, 12, 15,10
	Weaker 11, 8, 9, 16	11, 7, 9, 5, 4
23	10, 13, 11, 16, 6, 7	6, 7, 10, 13, 16 Trace 8, 9, 11, 12
24	14, 11, 7, 10	14, 7, 10, 6, 8, 11
25	15, trace 23, 6–11	15, 8, weaker 6–11
26	10, 16, 19, 6–11	10, 16, 6–11, 15
27	10, 11, 15, 14, 17, 24, 25	10, 20, 6–11, 15, 17, 16
28	11, 10, 12, 8, 6, 18, 22, 26	11, 6–10, 18, 21, 26, 15, 16
29	10, 11, 9, 7, 6, 8, 16, 19, 27, 23	9, 6–11, 19, 22, 12, 15, 16
30	10, 11, 6–9, 14, 15, 16, 20	20, 10, trace 6–11

[a] From Reference 108.
[b] The products are listed in order of decreasing intensity. Photodissociation at 266 nm.

into fragments in the 6 to 11-atom size range is remarkable because it implies a complete disintegration of the cluster (the ionization potentials fall with increasing cluster size so the formation of a large neutral fragment is unlikely). It was suggested that this disintegration is essentially an explosion of the cluster driven by the energy gained from restructuring the cluster during the dissociation process. The photodissociation of germanium clusters was also studied. The observed fragmentation pattern for the clusters with up to 30 atoms is quite similar to that observed for silicon clusters (see Table 2). Larger germanium clusters, at low laser fluences, sequentially lose Ge_{10}, unlike the larger silicon clusters which give fragments in the 6- to 11-atom size range.

The reactions of small silicon cluster ions have been investigated by groups at Bell Labs[44,67,109,110] and at the NRL[111] using direct laser sputtering and FT-ICR. The Bell Labs group has investigated the reactions of positively charged (and in some cases negatively charged) silicon clusters with CH_3SiH_3,[44] WF_6,[109] NO_2,[67] and XeF_2.[110] With CH_3SiH_3, only positively charged clusters with up to five atoms were found to react. For Si_n^+, $n = 1$-5, the reaction probabilities (reaction rate constant divided by collision rate constant) are shown in Figure 14. The reaction probabilities show a general decrease with increasing cluster size. The decrease in the reaction probabilities was ascribed to increasing delocalization of the positive charge over the clusters.

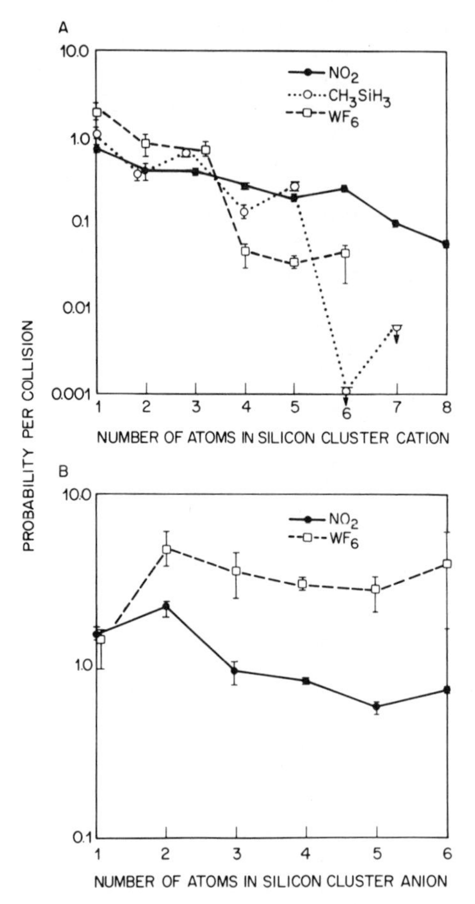

Figure 14. Plot of the reaction probabilities (reaction rate constant divided by collision rate constant) as a function of cluster size for the reactions of (A) positively and (B) negatively charged silicon clusters with CH_3SiH_3, WF_6, and NO_2. (Reproduced with permission from Reference 67.)

Table 3 shows the observed products. Most of the products can be interpreted as addition of CH_3SiH_3 to the silicon cluster followed by loss of a hydrogen atom or hydrogen molecules. Thus, the silicon clusters have gained a silicon atom; hence this is a prototypical deposition reaction. A mechanism was

Table 3. Product Distributions for the Reactions of Si_n^+ with CH_3SiH_3 [a]

	Ionic and neutral products formed						
Reactant ion, Si_n^+	$Si_{n-1}CH_5^+$ $+H$	$Si_{n+1}CH_4^+$ $+H_2$	$Si_{n+1}CH_2^+$ $+2H_2$	$Si_{n+1}H_3^+$ $+CH_3$	$Si_nCH_3^+$ $+SiH_3$	$Si_nCH_2^+$ $+SiH_4$	Si_nCH^+ $+SiH_3+H_2$
Si^+		0.53			0.47		
Si_2^+		0.28	0.40		0.13	0.16	0.03
Si_3^+		0.76	0.24				
Si_4^+		0.70	0.30				
Si_5^+	0.81			0.19			

[a] From Reference 67.

proposed to account for these observations. It was suggested that the silicon clusters contain two potential reactive sites: a trivalent silicon with a radical electron (analogous to silyl radical) and a divalent silicon with a lone pair (analogous to silylene). In the present case, the reactivity was assigned to the silylene sites, the divalent silicon inserting into an Si—H bond of CH_3SiH_3. This behavior is expected to be much less likely for larger clusters, which contain predominantly tri- and tetravalent silicon atoms.

WF_6 reacts with bulk silicon to both etch it and deposit tungsten.[112] Positive and negative silicon cluster ions react with WF_6 to give a variety of tungsten and silicon fluorides.[109] Reaction probabilities for both the anions and cations are shown in Figure 14. The anions, Si^- to Si_6^-, react at close to the collision rate. Charge transfer is the major reaction pathway (>75%). This result brackets the electron affinities of the clusters as being less than that of WF_6 (3.6 eV). The positive cluster ions Si^+ to Si_3^+ react at close to the collision rate to give a range of $Si_nF_m^+$, WF_m^+, and $WSiF_m^+$ species. At Si_4^+ the reaction rate suddenly drops (see Figure 14) and for Si_4^+ to Si_6^+ the only product observed was Si_nF^+. It was suggested that these products arise from a radical-based mechanism and that the $WSiF_m^+$ observed for the smaller clusters probably arises from silylene insertion into a tungsten-fluorine bond.

With NO_2 some fascinating reactions were observed.[67] The predominant reaction for both positive and negative silicon cluster ions is loss of a silicon atom, probably in the form of SiO:

$$Si_n^{\pm} + NO_2 \rightarrow Si_{n-1}^{\pm} + SiO + NO \tag{7}$$

Thus, the clusters are sequentially etched down to monoatomic silicon ions. For the positive ions the other major products were NO^+ (observed for all clusters except for $n = 6$ and 8) and $Si_{n-1}O^+$ (which was only observed for $n = 2$ and 3). For the negative ions the only other major product was charge transfer to give NO_2^- which was observed for $n = 1$–4. The electron affinity of NO_2 is 2.6 eV so combined with the data for WF_6 these results bracket the electron affinities of the clusters as: $EA(Si_{2\text{-}4}) < 2.6$ eV and $2.6\text{ eV} < EA(Si_{5\text{-}6}) < 3.7$ eV. Reaction probabilities are shown in Figure 14. For both the positive and negative ions the reaction probabilities gradually decrease with increasing cluster size. The reactions of the negative ions appear to occur significantly faster than those of the positive ions. It was suggested that these reactions occur by a radical-radical coupling mechanism involving the unpaired electron of NO_2 and an unpaired electron on a silicon atom at a charged or trivalent center.

The reactions between positively charged silicon clusters and XeF_2 give mainly SiF^+ and Si_nF^+ and a small amount of Si_{n-1}^+.[110] The SiF^+ and Si_{n-1}^+ products could result from fragmentation of the fluorinated cluster Si_nF^+ since fluorination is a very exothermic process. The reaction rates (see Figure 15) decrease slowly and smoothly with cluster size. The subsequent reactions of Si_nF^+ were also investigated. These reactions result in Si_{n-1}^+ and SiF^+, the relative proportions varying with cluster size. The reaction rates for Si_nF^+ (see

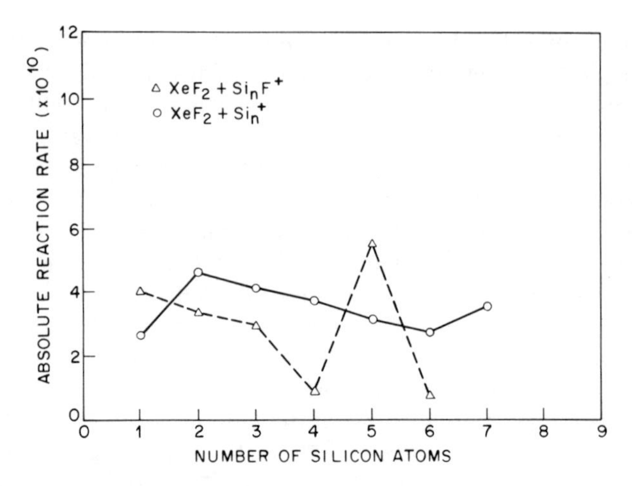

Figure 15. Plot of the rate constants (in units of $cm^3 s^{-1}$) for the reactions of Si_n^+ and Si_nF^+ with XeF_2. (Data taken from Reference 110.)

Figure 15) also decrease with increasing cluster size but show much more variation with cluster size than was observed with the bare clusters: Si_4F^+ and Si_6F^+ react slowly but Si_5F^+ reacts quite rapidly.

The NRL group has investigated the reactions between positively charged silicon cluster ions and a range of neutral gases (D_2, H_2O, CH_3OH, NH_3, O_2, and N_2O) using direct laser sputtering and FT-ICR.[111] No reaction was observed with D_2. Rate constants measured for the other species are plotted in Figure 16. There is a large drop in the reaction rates between $n = 3$ and 4. It was suggested that these results could be accounted for by reactive terminal

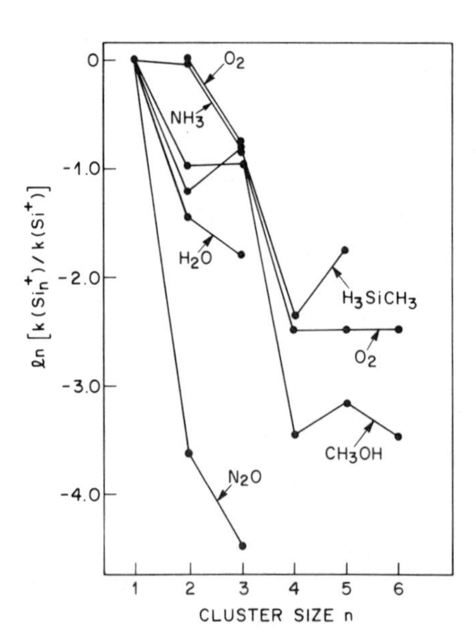

Figure 16. Plot of the relative rate constants for the reactions of positively charged silicon clusters with the neutral molecules shown. The rate constants are normalized to the rate for Si^+, except for the reaction with O_2 which is normalized to the rate for Si_2^+. (Reproduced with permission from Reference 111.)

silicon atoms in the smaller species while for $n > 3$ the cyclic structures are more stable and so the clusters are less reactive. The calculations of Raghavachari and Logovinski[(113-115)] show that the energy required for ring opening for Si_3 and Si_4 neutral clusters is around 0.4 eV and 3.4 eV, respectively. The product distributions for all the reactions studied are shown in Table 4. For the reactions with H_2O, CH_3OH, C_2H_2, and NH_3 many of the products can be accounted for by the addition of the reagent molecule to the cluster and loss of an H atom (this is also partly true for CH_3SiH_3[(44)]). It was suggested that these reactions proceed by insertion of a silicon atom into an O—H, N—H, or C—H bond of the neutral reactant followed by migration of the H atom from the inserted silicon atom to which it is initially bound. The migration may occur with or precede fragmentation of the complex into products. It was argued that rearrangement must be the rate-limiting step in these reactions.

The reactions of the clusters with O_2 and N_2O (rows 6 and 7 in Table 4) cannot be accounted for by this mechanism. With N_2O only clusters up to

Table 4. Branching Ratios for the Products Observed for the Reactions of Silicon Cluster Cations, Si_n^+, with the Neutral Molecules Shown[a]

Neutral reactant	Product ion	n = 1	2	3	4	5	6
D_2			No reaction				
H_2O	$SiOH^+$	1.0	1.0	1.0	No reaction		
CH_3OH	$SiOH^+$	0.9	0.05				
	$SiOCH_3^+$	0.1	0.8	0.9	1.0	1.0	0.1
	Si_2OH^+		0.15	0.1			
	$Si_6OCH_3^+$						0.6
	$Si_5OCH_2^+$						0.3
C_2H_2	SiC_2H^+	1.0	0.2		No reaction		
	$Si_2C_2H^+$		0.8	0.5			
	$Si_3C_2H^+$			0.25			
	$Si_3C_2^+$			0.25			
NH_3	$SiNH_2^+$	1.0	1.0	0.3	No reaction		
	Si_2N^+			0.14			
	$SiNH^+$			0.13			
	$Si_2NH_2^+$			0.13			
	Si_3NH^+			0.3			
O_2	SiO^+		0.2	0.1			
	Si_2O^+		0.3	0.1	0.4		
	Si_{n-1}^+		0.5	0.3			
	Si_{n-2}^+			0.5	0.6	1.0	1.0
N_2O	SiO^+	1.0			No reaction		
	Si_2N^+		0.7				
	Si_2O^+		0.3				
	Si_3N^+			1.0			

[a] From Reference 111.

$n = 3$ react and the main products are Si_nN^+ and Si_nO^+. With oxygen the clusters are sequentially etched, two silicon atoms at a time, to Si^+ and SiO^+. These reactions are similar to the ones found with NO_2 where etching of one silicon atom at a time was observed.[67] The only neutral products that are thermodynamically accessible in these reactions are two SiO molecules or an Si_2O_2 species which is bound by around 1.93 eV[116]:

$$Si_n^+ + O_2 \rightarrow Si_{n-2}^+ + 2SiO \text{ or } Si_2O_2 \tag{8}$$

This reaction is exothermic by 1.7-2.6 eV, even assuming that the products are two SiO molecules; yet the reactions, particularly for the larger clusters, are remarkably slow. The slow reaction rates could be accounted for by an activation barrier to Si—O bond formation. The Si—O bond is much stronger than an Si—Si bond so the energy from Si—O bond formation should be sufficient to surmount any activation barriers in the exit channel. A barrier to Si—O bond formation could arise from orbital symmetry constraints, but a more detailed theoretical analysis is required to understand the origin of this barrier. Another possibility (not necessarily inconsistent with the above explanation) is that a restrictive four-center transition state is involved in the Si—O bond formation.

The reactions of larger silicon cluster ions (with up to 65 atoms) have been investigated by Smalley and co-workers[17] using FT-ICR with the clusters injected from an external laser vaporization source. They investigated the reactions of silicon clusters with NH_3 and observed $Si_nNH_y^+$ as products. These experiments were not performed with size-selected clusters so cluster fragmentation could influence the results. Mass spectra recorded for the 36- to 45-atom size range are shown in Figure 17. Silicon has several naturally occurring isotopes so there are several peaks in the mass spectrum for each cluster size. The upper spectrum in Figure 17 is a control where no NH_3 was added, and the lower spectrum shows the effect of reactions with 1×10^{-7} torr of NH_3 for 15 s. These results suggest that silicon clusters with 39 and 45 atoms are quite unreactive and those with 42 to 44 atoms are particularly reactive. Several oscillations in reactivity were observed over the cluster size range studied. Clusters with <14, 20, 25, 33, 39, and 45 atoms were relatively unreactive, while those with 17, 18, 23, 30, 36, 42-44, and 46 atoms were particularly reactive. Above $n = 47$ the oscillations vanish and reactivity increases slowly with cluster size.

Both photodetachment and photodissociation are observed for the interaction of silicon cluster anions with light.[118] The relative proportions of fragmentation and detachment were found to depend strongly on wavelength. For example, with Si_{11}^-:

$$Si_{11}^- + h\nu \rightarrow Si_5^- + Si_6^0 \tag{9}$$

$$\qquad \rightarrow Si_{11}^0 + e^- \tag{10}$$

The e^-/Si_5^- ratio changes from 0.25 to 5.5 as the photon energy increases from

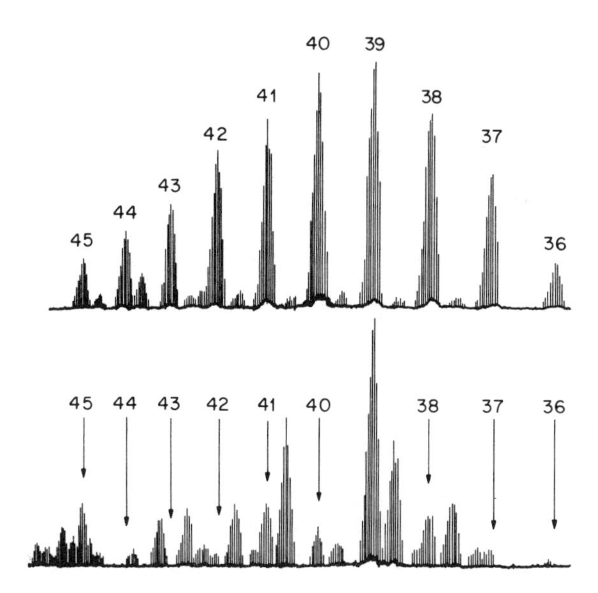

Figure 17. Mass spectra illustrating the reactivity of silicon cluster ions in the 36- to 45-atom size range with NH_3. The upper mass spectrum is a control in which no NH_3 has been added. The lower mass spectrum shows the effect of reactions with 1×10^{-7} torr of NH_3 for 15 s. (Reproduced with permission from Reference 117.)

2.82 to 3.0 eV. Generally, the binding energies of the anions (in the size range studied) are less than the electron affinities, so close to the detachment threshold dissociation dominates. As the photon energy increases, the relative amount of detachment increases, and at large photon energies detachment is probably completely dominant.

The observed photofragments are listed in order of decreasing intensity in Table 5 for both silicon and germanium cluster anions. It can be seen from these results that silicon and germanium are remarkably similar and that the 10-atom cluster anion is a favored product, which suggests that this species is particularly stable. As discussed above, the 10-atom cluster was also found to be a favored product in the photodissociation of silicon and germanium cations. Comparison of Tables 2 and 5 shows that the photofragments from silicon cluster anions and cations are quite similar.

Photoelectron spectra of silicon cluster anions in the 3- to 12-atom size range have recently been recorded using an ArF excimer laser (6.4 eV).[119] The photelectron spectra generally show several peaks. Clusters with 4, 7, and 10 atoms show small, resolved, low-electron-binding-energy peaks suggesting that these are closed-shell neutral species with HOMO–LUMO gaps of 1.0 to 1.5 eV. This HOMO–LUMO gap cannot really be compared to the band gap in the bulk semiconductor (1.09 eV) because the atoms in the cluster are not sp^3-hydridized as in the bulk. The photoelectron spectra of germanium cluster anions are remarkably similar to those for silicon, suggesting that the bonding in these closely related species is quite similar.

Table 5. Fragmentation Channels of Si_x^- and Ge_x^- [a]

Parent size, x	Daughters Si_y^- of Si_x^- [b]	Daughters Ge_y^- of Ge_x^- [b]
9	None	5
10	None	4, 6, 5
11	5, 4, 6	5, 4, 6
12	6, 5	6, 5
13	6, 7	6, 7
14	7, 10, 8	7
15	9, 5	9, 5
16	10	10
17	10	10
18	12, 6, 7, 9, 11	12, 9, 6, 11, 7, 10, 5
19	9, 10, 6	9, 10
20	10	10
21	5, 9, 10, 11, 14, 15	11, 5, 7, 10
22	10, 16, 6, 9, 5	16, 9, 10

[a] From Reference 118.
[b] The daughters are listed in order of decreasing intensity.

Electron affinities for silicon clusters derived from the photoelectron spectra are plotted in Figure 18. It is apparent that the electron affinity for Si_5 is significantly larger than those of its neighbors, and it appears from the photoelectron spectra that this is an open-shell species. As mentioned above, Mandich and co-workers[(67)] have deduced electron affinities of $2.6 \text{ eV} < EA < 3.7 \text{ eV}$ for Si_5 and Si_6 from their reaction studies. The two measurements are in reasonable agreement for Si_5, but the large discrepancy between these measurements for Si_6 is puzzling.

Several calculations of the structures of silicon clusters have been reported.[(113–115,120–124)] The calculations are generally in reasonable agreement with each other. In contrast to carbon, where the lowest-energy structures of

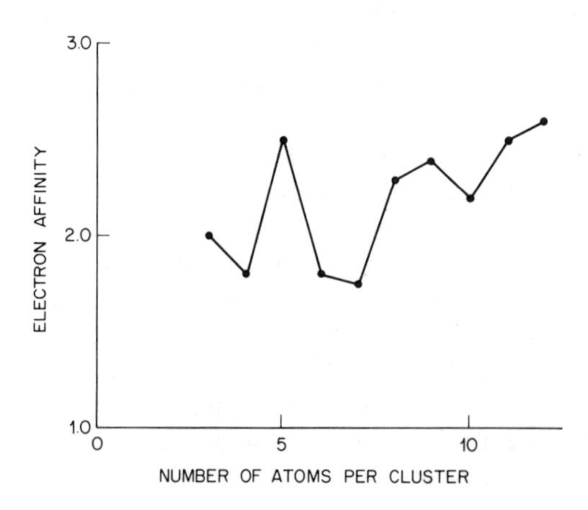

Figure 18. Electron affinities of silicon clusters with 3 to 12 atoms derived from photoelectron spectra. (Plotted from data given in Reference 119.)

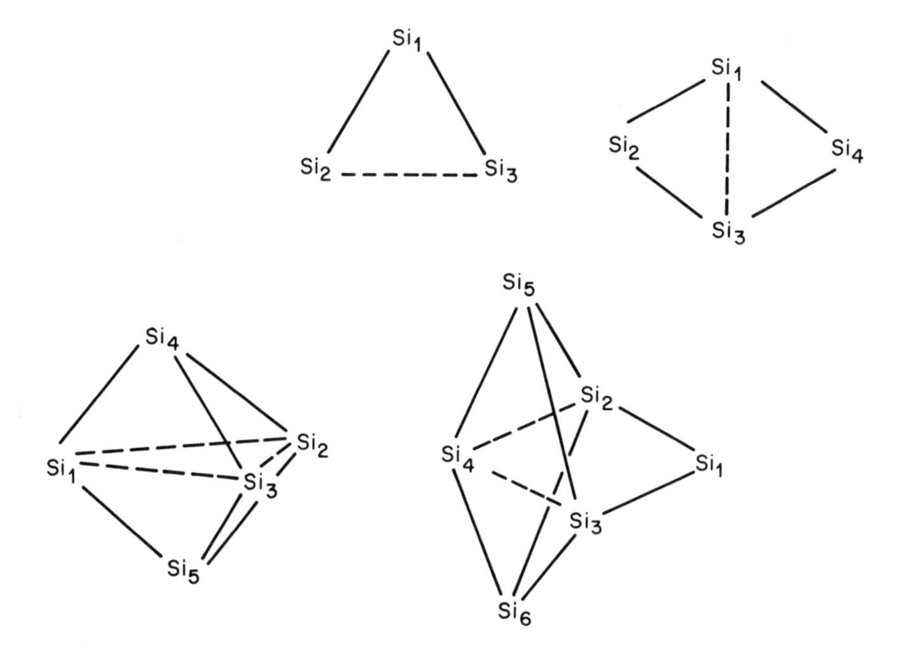

Figure 19. Calculated structures for Si_3–Si_6. Positively charged silicon clusters have very similar structures to the neutral clusters. (Reproduced with permission from Reference 122.)

the smaller clusters appear to be linear chains or monocyclic rings, three-dimensional structures appear to be important for quite small silicon clusters. Calculated structures for silicon clusters with 3 to 6 atoms are shown in Figure 19. Calculations on the positive ions[122,124] indicate that the neutral and ionic clusters have quite similar geometries because the electron is removed mainly from an orbital with significant nonbonding character. The charge on the ionic species resides mainly on the least coordinated atoms in the cluster. As can be seen from Figure 19, the calculated structure for Si_3 is cyclic (or bent since the bonding along one edge is quite weak[114]. Si_4 is a rhombus and Si_5 is a compressed trigonal bipyramid. Si_6 can be derived from Si_5 by capping one of the edges. Si_{10} probably has the structure of a tetracapped octahedron.[115] An important feature of these calculated structures is that they are extensively distorted from the microcrystalline geometries expected from consideration of the lattice of bulk silicon. The reconstruction of the microcrystalline structures is driven by the dangling bonds (unsatisfied valence) of these small silicon particles, which results in multiple bond formation and more compact structures. The final structures are determined by a balance between multiple bond formation and the strain induced in moving to more compact structures. In a very recent study, Raghavachari and Rohlfing[124] have shown from theoretical calculations that the dominant products observed in the photodissociation of silicon cluster cations with up to 20 atoms can be accounted for in terms of product stability. This is consistent with the suggestion of Smalley and co-workers that dissociation occurs statistically on the ground state potential surface.[108]

6. ALUMINUM CLUSTER IONS

Aluminum is next to silicon in the periodic table but the physical properties of the bulk materials are strikingly different. Silicon is a semiconductor and aluminum is almost an ideal free-electron metal. Aluminum is trivalent in virtually all of its known compounds. However, while the bulk element is clearly metallic, many of its compounds are on the borderline between ionic and covalent character. There have recently been several studies of aluminum clusters, and aluminum appears to be emerging as a model system for testing and developing ideas about metal clusters. Aluminum is light and only has one isotope, which makes it attractive for experimental studies. Aluminum is also widely used commercially, and there is an extensive literature on the chemical properties of aluminum surfaces. It is also possible to perform reasonably reliable calculations on fairly large aluminum clusters. This is not yet possible for transition metals, where the *d* electrons make the calculations considerably less tractable. So with aluminum clusters we are just beginning to get our first glimpses of, for example, the development of bulk activation barriers for chemisorption, and theory is playing an important role in helping to provide the physical insight required to understand the experimental data.

The dissociation of aluminum cluster ions has been studied by several groups. Meiwes-Broer and co-workers have investigated the metastable dissociation of mass-selected aluminum cluster ions generated by sputtering.[33] Clusters with up to 23 atoms were studied. Metastable fragmentation (spontaneous unimolecular dissociation) was observed for clusters with more than 7 atoms. The dominant product observed was loss of an atom to give Al_{n-1}^+. For the large clusters ($n > 14$) at short times more than 90% of the clusters fragment, and for $n > 15$ loss of two atoms is a major process. The Al_{n-2}^+ product probably arises from the sequential loss of two atoms rather than loss of a dimer. These results provide a rather striking illustration of how excited sputtered clusters are.

The collision-induced dissociation (CID) of aluminum cluster ions has been investigated on several occasions. Meiwes-Broer and co-workers[125] have investigated the CID of aluminum cluster ions with up to 7 atoms at collision energies in the kilovolt range and observed Al^+ as the main product. Jarrold *et al.* have studied the collision-induced dissociation of aluminum cluster ions with 3 to 26 atoms at a collision energy of 5.25 eV.[13] The collision gas used in these experiments was argon, and the clusters were generated by pulsed laser vaporization in a continuous flow of helium buffer gas. The observed product distribution is shown in Figure 20. Four products were observed with significant intensity: Al_{n-1}^+, Al_{n-2}^+, Al_{n-3}^+, and Al^+. The main products are Al_{n-1}^+ for the larger clusters and Al^+ for the smaller ones. The products $Al_{n-1}^+ + Al$ and $Al^+ + Al_{n-1}$ differ only by the location of the charge, so ignoring the

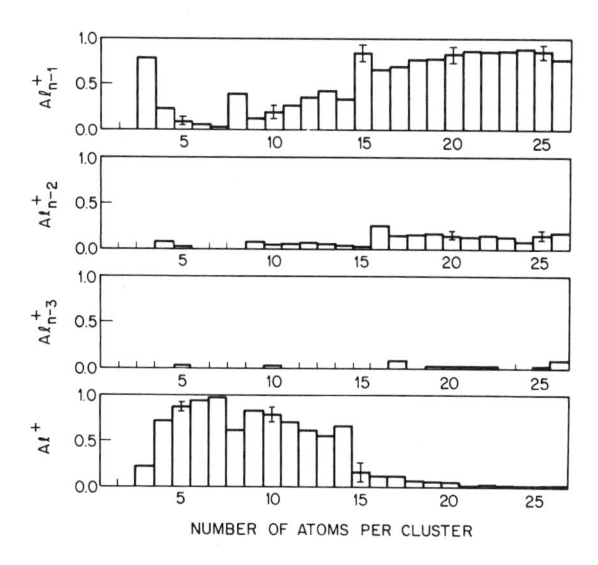

Figure 20. Histograms showing the product distributions for the collision-induced dissociation of aluminum cluster ions at a colllision energy of 5.25 eV with argon collision gas. (Reproduced with permission from Reference 13.)

charge for the moment, the main product from the collision-induced dissociation is loss of an atom from the cluster. Brucat *et al.*[11] have recently studied the photodissociation of small iron, nickel, and niobium cluster ions and concluded that the dominant dissociation pathway corresponded to loss of a single atom. Metal clusters, it appears, fragment quite differently from carbon or silicon clusters, where cluster fission and preferred "magic fragments" were observed. These results suggest that for metal clusters the binding energy per atom increases fairly smoothly with cluster size. This behavior is characteristic of three-dimensional cluster structures where the number of bonds per atom increases as the cluster size increases.

It is apparent from Figure 20 that the Al^+_{n-1} product for Al^+_8 and Al^+_{15}, the Al^+_{n-2} product for Al^+_9 and Al^+_{16}, and the Al^+_{n-3} product for Al^+_{10} and Al^+_{17} are slightly favored products. These products correspond to Al^+_7 and Al^+_{14}, which are the "magic numbers" in the mass spectra of aluminum clusters[9,12,13,31] (see Figure 3). Minima were also observed in the collision-induced dissociation cross sections at $n = 7$ and 13–14, suggesting that these clusters are particularly stable. The enhanced stability of these particular clusters can be accounted for by the electronic shell model or jellium model[5,126–129] pioneered by Ekardt and Knight and co-workers. This is an approximate model for metal clusters in which the ionic cores are replaced by a uniformly positively charged background (the jellium) and then the electronic energies are calculated self-consistently to obtain the energy levels. The energy levels show a shell structure (due to electronic angular momentum) similar to that which is well established for electronic and nuclear energy levels for atoms. Shell closings occur with 8, 18, 20, 34, 40, 58, and 68 valence

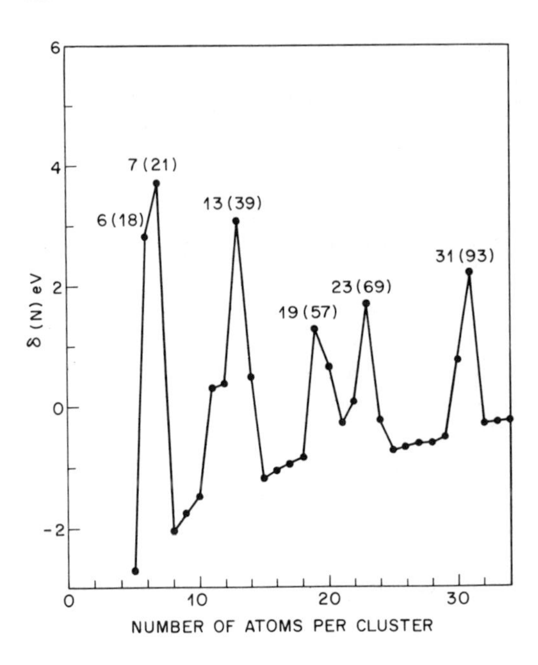

Figure 21. Jellium model calculations on aluminum clusters. The second difference in total energy (which represents the relative binding energy change for clusters with N atoms compared to clusters with N + 1 and N − 1 atoms) is plotted against the number of atoms per cluster. Numbers in parentheses are total numbers of valence electrons. Note that the calculations were performed for neutral clusters. (Adapted from reference 129.)

electrons corresponding to the 1*p*, 1*d*, 2*s*, 1*f*, 2*p*, 1*g*, and 2*d* shells. Note that the allowed values of angular momentum and the order in which the shells are filled differs from that found with atoms. The results of some jellium model calculations for aluminum clusters are plotted in Figure 21. Aluminum is trivalent and the shell closings generally do not occur with an integral multiple of three valence electrons so the effect of shell closings tends to be smeared out over several clusters. The jellium model calculations shown in Figure 21 are for neutral species and we are dealing with ions (with one less valence electron). The calculations predict that the most prominent shell closing will occur for the 2*s* and 2*p* shells with 20 and 40 valence electrons, respectively. These correspond to the "magic numbers" Al_7^+ [with 20 valence electrons: $7 \times 3 - 1$ (for the charge) = 20] and Al_{13}^+–Al_{14}^+ (38 and 41 valence electrons). Weak evidence was also found for the 2*d* (68 electrons) shell closing at Al_{23}^+.[13] No evidence was found for the 1*d* and 1*f* shell closings; these were not observed for sodium clusters either.[127,129] The electronic shell model has been a controversial topic for the last few years. It is a simple model which does a good job of accounting for many experimental observations in a semiquantitative manner. As a consequence, it is quite clear that an electronic shell structure exists for some metal clusters and no doubt plays a role in determining some of their properties. However, the shell model is obviously too simple to be able to account for all the properties of metal clusters. For example, any property that depends on structure is going to be rather poorly reproduced since the atoms (or ionic cores) are replaced by jellium. There have been attempts to modify the jellium model so that it can account for details in cluster abundance distributions. However, the model then loses its elegant simplicity and at the same time much of its credibility.

The main product from the collision-induced dissociation of aluminum clusters is Al^+ for the smaller clusters and Al^+_{n-1} for the larger ones (see Figure 20). The choice between $Al^+ + Al_{n-1}$ and $Al^+_{n-1} + Al$ products is most likely determined on energetic grounds, and thus on the relative ionization potentials (IPs) of Al and Al_{n-1}. The ionization potentials of the smaller clusters are thus larger than that of the atom and so the main product is Al^+. For the larger clusters, Al_{n-1} has a lower IP than the atom so the Al^+_{n-1} product dominates. The change from mainly Al^+ to mainly Al^+_{n-1} product occurs at Al^+_{15} so this is the point where the ionization potential of Al_{n-1} drops below the atomic value. Making some assumptions about the amount of energy transferred into the cluster during the collision and assuming that the subsequent fragmentation is statistical, it is possible to derive some semiquantitative values for the cluster ionization potentials from the CID product distributions.[13] The results of this analysis are shown in Figure 22. The ionization potentials of the clusters are above the value for the atom in the case of the smaller clusters, drop below the atom IP at Al_{14}, and then head down towards the bulk work function. Magnetic deflection experiments[9] suggest that Al_2 is a triplet so dissociation of Al^+_3 to ground state $Al_2 + Al^+$ is spin forbidden. Anderson and co-workers[14] suggest that the IP derived for Al_2 is thus probably for Al_2(singlet) and the IP for ground state Al_2 is probably around 0.38 eV higher, using the singlet–triplet energy difference calculated by Upton.[130] It is apparent from the data shown in Figure 22 that there are discontinuities in the ionization potentials at the shell closings with $n = 7$ and 14. Thus, Al_7 (with 21 valence electrons) has a low IP for essentially the same reason that a sodium atom does. Removing the electron in the highest occupied orbital leaves a closed-shell configuration. The influence of the shell closings on ionization potentials for alkali metal clusters has previously been discussed by Knight and co-workers[5,131] and Schumacher and co-workers.[3] The dashed vertical lines shown in Figure 22 are the result of laser bracketing experiments by the Exxon group.[132] Generally, their values are in reasonable agreement

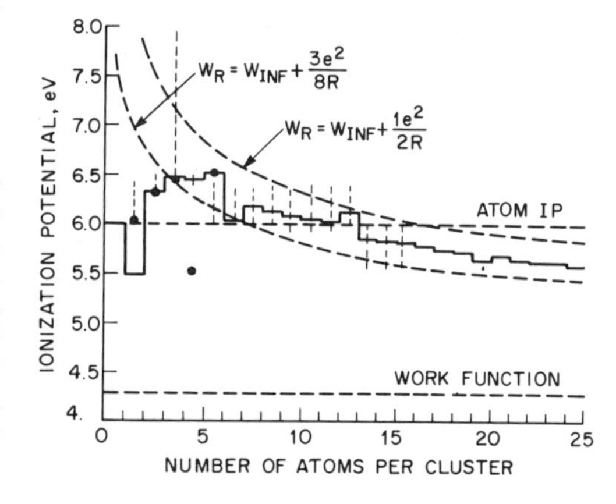

Figure 22. Ionization potentials for aluminum clusters. The values shown by the solid lines are derived from collision-induced dissociation product distributions.[13] The dashed vertical lines are from laser bracketing experiments.[132] The points are calculated values[135,136] and the thick dashed lines are from a simple classical model for the work function of a conducting sphere.[133,134]

with those derived from the CID data. Anderson and co-workers have, however, pointed out that this agreement is probably partly due to a fortuitous choice of collision energy in the CID experiments.[14] More reliable values of the ionization potentials for aluminum clusters will soon be available from Whetten's group at UCLA. The thick dashed lines in Figure 22 give the predictions of a classical model for the work function of a conducting sphere[3,133,134]:

$$W_R = W_{\mathrm{INF}} + ce^2/R \tag{11}$$

in which W_R is the work function of a sphere of radius R and W_{INF} is the work function of the bulk. There has been some confusion over the correct value for the constant c in this equation. Recently, a value of $\frac{3}{8}$ has commonly been used. However, Brus and co-workers[134] have just presented a convincing argument that the correct value for c is $\frac{1}{2}$. The predictions of this model using both values for c are shown in Figure 22, and the experimental data lie between the model predictions. Clearly, this simple classical electrostatic model does a good job of predicting the overall variation in the ionization potentials of aluminum clusters. Schumacher and co-workers have shown that the predictions of this simple model are in good agreement with measured ionization potentials of Fe_n, Hg_n, Na_n, K_n, and Pb_n clusters.[3] The ionization potentials for most metal clusters studied so far decrease fairly steadily from the atomic value to the bulk work function. Aluminum and also bismuth clusters[4] are exceptional in that the ionization potentials initially rise above the atomic value for small clusters. Upton[135,136] successfully predicted this anomaly in the ionization potentials of aluminum clusters from his theoretical calculations. His explanation for this behavior is that unlike most other metal systems studied where the highest states are of s (or s and d) character and approximately half full, the highest states for aluminum are dominated by $3p$ character and are only one-sixth full (since $6n$ orbitals arise from the p orbitals of an n-atom cluster). The Fermi levels of small clusters are thus deeper relative to the atom, leading to a higher ionization potential.

Anderson and co-workers[14] have investigated the collision-induced dissociation of aluminum cluster ions with up to 7 atoms as a function of collision energy. From the collision energy thresholds it is possible to determine thermodynamic quantities, such as the cluster dissociation energies, which are of considerable fundamental interest. The clusters for these studies were generated by sputtering and subsequently cooled with a buffer gas. The collision gas used was xenon since Aristov and Armentrout (137) had shown that in the collision-induced dissociation of VO^+, xenon gave the sharpest and lowest-energy thresholds. Collision-induced dissociation always gives an upper limit so the lowest threshold is closest to the true dissociation energy (assuming that the clusters are not internally excited). The lowest-energy fragmentation channel observed for all clusters in the 2- to 7-atom size range, except Al_3^+, was formation of Al^+. For Al_3^+, formation of Al_2^+ had a slightly lower threshold than formation of Al^+. Figure 23 shows a plot of the collision energy dependence of the CID cross section for the reaction

$$Al_5^+ + Xe \rightarrow Al^+ + Al_4 + Xe \tag{12}$$

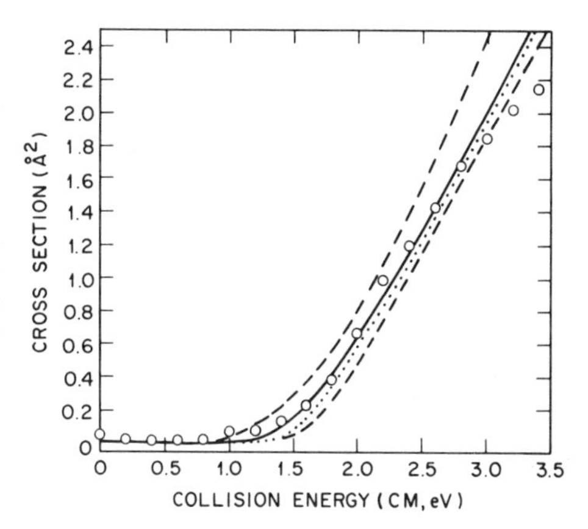

Figure 23. Plot of the cross section for the collision-induced dissociation process $Al_5^+ + Xe \rightarrow Al^+ + Al_4 + Xe$ as a function of collision energy. The points are the experimental data. The line is the result of a simulation, using an assumed cross section function, to account for threshold broadening (due to target gas motion and ion beam energy spread) and so determine the true threshold. The dotted line shows the assumed cross section function and the dashed lines show convoluted cross sections delineating the error in the simulation. (Adapted from Reference 14.)

The points are the experimental data. To determine accurate thresholds from these measurements, it is necessary to account for the broadening of the threshold region which arises from target gas motion and the ion beam energy spread. This is accomplished by simulating the experimental data with an assumed cross section function convoluted with a function to describe the threshold broadening. The solid line in Figure 23 shows the best fit to the experimental data. The dotted line gives the cross section function used, which has the form

$$\sigma(E) = A(E - E_0)^n / E^m \tag{13}$$

where A, n, and m are adjustable parameters, E is the collision energy, and E_0 is the threshold. The dashed lines in Figure 23 are convoluted cross sections delineating the error limits for the simulation.

As noted above, the thresholds from these experiments provide a measure of the dissociation energy of the clusters. Dissociation energies determined in this way are plotted in Figure 24 for clusters in the 2- to 7-atom size range. The error bars shown are conservative upper limits. The dissociation energies generally increase with cluster size, and Al_7^+ is particularly strongly bound. Note that the binding energies are really quite low. With the exception of that of Al_7^+, they are all less than half of the bulk cohesive energy (3.4 eV). This is not unexpected since these small clusters are all surface atoms and the average number of nearest neighbors is much less than in the bulk. The dashed line shown in Figure 24 gives dissociation energies derived from the theoretical calculations of Upton.[135,136] These calculations were performed before the experimental results were available and are clearly in good agreement with the experimental measurements.

Cluster dissociation energies are one of the fundamental quantities we would like to follow as a function of cluster size to determine at what point the bulk cohesive energy is approached. The data shown in Figure 24 represent the current state of the art. Clearly, we would like to extend these measurements to larger clusters, but for clusters with more than 10 atoms, determining the

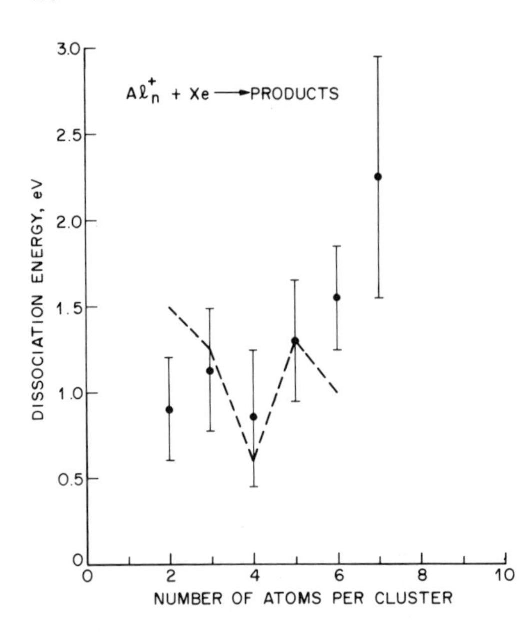

Figure 24. Plot of the dissociation energies of aluminum cluster ions as a function of cluster size. The points are from Reference 14 and are determined by collision-induced dissociation. The dashed line gives calculated dissociation energies taken from Reference 135.

dissociation thresholds becomes more difficult. At energies close to threshold the larger clusters have a significant lifetime due to the large number of internal degrees of freedom, and so they will not dissociate on the experimental time scale. In principle, it is possible to overcome this problem by waiting longer. However, at times probably not much longer than a millisecond a fundamental limit is reached—the lifetime for emision of an IR photon. The clusters will then cool without dissociating.

Hanley and Anderson[138] have investigated the collision-induced dissociation of boron cluster ions ($n = 3$–8). Boron is above aluminum in the periodic table so a comparison should be interesting. The primary fragmentation channel observed was formation of B^+ (which is similar to the finding for aluminum). However, the dissociation energies of B_n^+ are generally two to three times larger than those of the corresponding aluminum clusters. This is not unexpected since the cohesive energy of bulk boron is close to twice the cohesive energy of bulk aluminum. B_5^+ was found to be a particularly abundant species in the cluster ion mass spectrum and was an abundant fragment in the CID of the larger boron clusters.

The collision-induced dissociation of aluminum clusters with chemisorbed oxygen, $Al_nO_m^+$ ($n = 3$–26, $m = 1, 2$), has been studied by Jarrold and Bower.[139] The clusters for these studies were generated by pulsed laser vaporization in a continuous flow of helium buffer gas with a trace of oxygen added. The CID was studied at a collision energy of 5.25 eV with argon as the collision gas. Mass spectra recorded for the CID of a 19-atom aluminum cluster with one and two chemisorbed oxygen atoms are shown in Figure 25. For $Al_{19}O^+$ the main product is Al_{17}^+, which can be ascribed to the loss of an

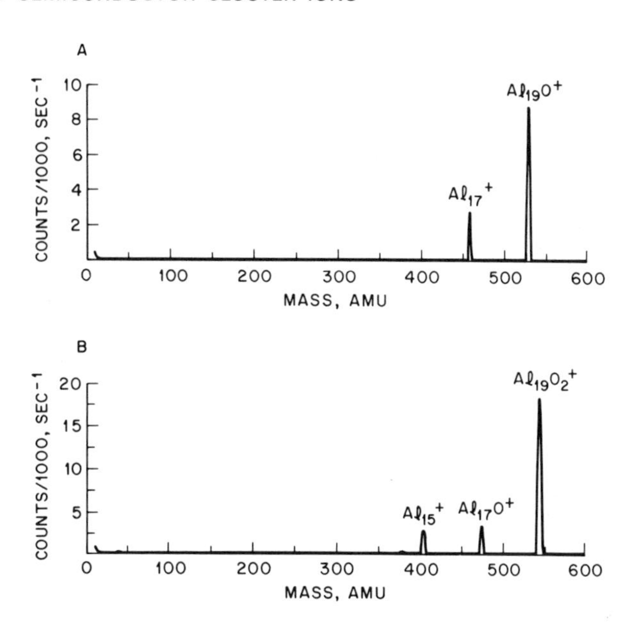

Figure 25. Mass spectra recorded for the collision-induced dissociation of (A) $Al_{19}O^+$ and (B) $Al_{19}O_2^+$. The collision energy was 5.2 eV and the collision gas was argon. (Reproduced with permission from Reference 139.)

Al_2O molecule. With $Al_{19}O_2^+$ two main products were observed—$Al_{17}O^+$ (loss of one Al_2O) and Al_{15}^+ (loss of two Al_2O molecules). The Al_2O loss channels were observed to dominate for all the clusters studied, $Al_nO_m^+$ ($n = 3$–26, $m = 1, 2$), except for Al_xO_2 ($x = 3, 4$), for which the main product was Al^+. The dominance of the Al_2O loss channel almost certainly arises because Al_2O is very strongly bound.[140] The fact that loss of Al_2O is observed rather than loss of an aluminum atom provides convincing evidence for a statistical dissociation mechanism for collision-induced dissociation rather than a direct or knockout mechanism, where loss of an aluminum atom would be expected to dominate.

The CID results were used to bracket the interaction energies between the aluminum clusters and oxygen. For Al_6^+/O this interaction energy is estimated to be 8.0 ± 1.0 eV, and for Al_{19}^+/O it is 7.5 ± 1.0 eV. These large interaction energies indicate that oxygen is multiply bonded to the cluster, probably to two or more aluminum atoms. Since the interaction energies for Al_6^+ and Al_{19}^+ are quite close, the energetics must be controlled by the few aluminum atoms at the site of chemisorption, with the neighboring atoms exerting only a minor influence. Similar conclusions were reached by Pacchioni and Fantucci[141] from their calculations.

Cross sections for the CID of $Al_nO_2^+$ are plotted in Figure 26. There are clear minima in the apparent cross sections as a function of size at $Al_9O_2^+$ and $Al_{15}O_2^+$, suggesting that these are particularly stable clusters. As noted above,

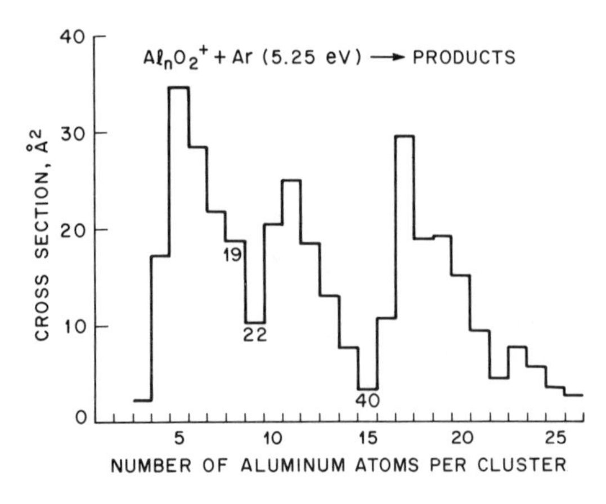

Figure 26. Cross sections for collision-induced dissociation of $Al_nO_2^+$ as a function of cluster size. The collision energy was 5.25 eV and the collision gas was argon. The numbers on the plot give the number of valence electrons near the 2*s* and 2*p* shell closings assuming that binding an oxygen atom costs the cluster two valence electrons. (Reproduced with permission from Reference 139.)

minima were observed at $n = 7$ and $n = 13$–14 for the bare clusters. In the case of $Al_{15}O_2^+$, studies of the pressure dependence of the apparent cross sections indicated that dissociation required two collisions. Clearly, there is something special about $Al_{15}O_2^+$. It was suggested that the stability of $Al_{15}O_2^+$ could be accounted for by assuming that binding an oxygen atom (as either O^{2-} or O=Al) cost the cluster two valence electrons which would normally go into cluster orbitals. Thus, $Al_{15}O_2^+$ has $15 \times 3 - 1$ (for the charge) $- 4$ (2 per oxygen atom) $= 40$ valence electrons and thus has a closed-electronic-shell configuration according to the electronic shell model. Similarly, $Al_9O_2^+$ has 22 valence electrons, which is close to the 2*s* shell closing with 20 valence electrons. Analogous studies have been performed by Schumacher and co-workers[142–144] and Brechignac and Cahuzac.[145] They studied mixed metal clusters of the type Na_nMg and K_nNa. Their findings were not always in agreement with the predictions of the electronic shell model.

The reactions between aluminum cluster ions and oxygen have been studied by Jarrold and Bower[65,146,147] and Hanley, Ruatta, and Anderson.[148,149] The oxidation of clean aluminum surfaces is a well-known phenomenon which has been extensively studied by surface science techniques.[150] Oxidation of bulk aluminum proceeds through several steps. First, dissociative chemisorption of oxygen occurs; then the oxygen migrates beneath the surface to form an underlayer, and when sufficient oxygen is present, oxide formation occurs. The reactions of aluminum cluster ions with oxygen show some similarities to the oxidation of bulk aluminum but there are some major differences.

Anderson and co-workers have measured total reaction cross sections as a function of collision energy for Al_2^+–Al_8^+ and observed small barriers to reaction.[149] The barriers are given in Table 6. No barrier was found for the dimer. The barrier height increases with cluster size from Al_3^+ to Al_7^+ and then falls for Al_8^+. Jarrold and Bower have measured total reaction cross sections and product distributions for the reactions between aluminum cluster ions, with 3 to 26 atoms, and oxygen at collision energies of 1.2 and 4.2 eV.[147] In the region of overlap there is generally good agreement between the product

Table 6. Threshold Energies Derived from Total Cross Sections for the Reaction $Al_n^+ + O_2$, $n = 2–8$ [a]

Cluster	Threshold (eV)
Al_2^+	No barrier
Al_3^+	$<0.09 \pm 0.28$
Al_4^+	$<0.07 \pm 0.28$
Al_5^+	0.11 ± 0.27
Al_6^+	0.16 ± 0.32
Al_7^+	0.25 ± 0.31
Al_8^+	0.21 ± 0.26

[a] From Reference 149.

distributions determined by these two groups. A histogram of the product distribution observed at a collision energy of 1.2 eV is shown in Figure 27. Most of the products arise from cluster fragmentation. Very little oxygen incorporation into the product ion is observed. The main oxide product is Al_2O^+, observed for clusters in the $n = 3–6$ size range. The main products found for clusters with $n > 6$ are Al_{n-5}^+ (loss of five atoms from the cluster) and Al^+. For clusters with $n > 13$ the dominant product is Al_{n-4}^+ (loss of four atoms from the cluster) except for Al_{19}^+ which loses five atoms to give the "magic number" Al_{14}^+. At higher collision energies more extensive fragmentation of the cluster ions is observed. From the results for the collision-induced dissociation of $Al_nO_m^+$ discussed above, the Al_{n-4}^+ product observed in the chemical reaction between the clusters and oxygen probably arises from the reaction

$$Al_n^+ + O_2 \rightarrow Al_{n-4}^+ + 2Al_2O \tag{14}$$

and the Al_{n-5}^+ and Al^+ products probably arise from the subsequent fragmentation of the Al_{n-4}^+ species (analogous to the reactions observed in the CID of the bare clusters). Figure 28 shows a semiquantitative potential energy diagram for the reaction. The first step in the reaction is dissociative chemisorption of oxygen onto the cluster. The measurements of Anderson and co-workers indicate that there is a small activation barrier, at least for the smaller clusters, which must be associated with cleavage of the O—O bond. As can be seen from Figure 28, dissociative chemisorption of oxygen onto the clusters is exothermic by around 10 eV. This exothermicity drives the subsequent fragmentation of the $Al_nO_2^+$ complex, resulting first in the loss of two Al_2O molecules. Overall reaction 14 is exothermic by around 7.5 eV. Assuming that this exothermicity is distributed statistically, the bulk would be deposited into the Al_{n-4}^+ fragment. As can be seen from Figure 27, clusters with $n < 13$ apparently fragment further to give Al^+ and Al_{n-5}^+ but clusters with $n > 13$ do not—the observed product is Al_{n-4}^+. Aluminum clusters in this size range are

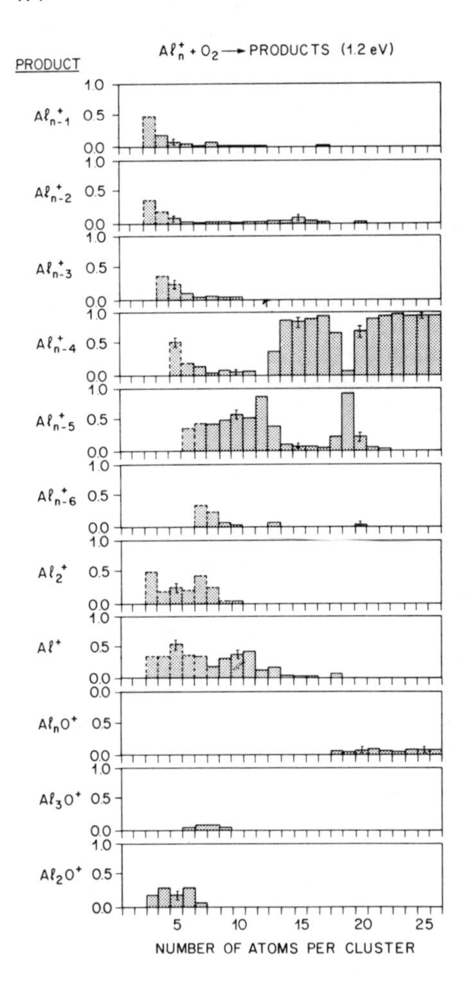

Figure 27. Histograms showing the product distributions for the reaction between aluminum cluster ions and oxygen at a collision energy of 1.2 eV. (Reproduced with permission from Reference 147.)

probably bound by 2.0–2.5 eV[151] so there appears to be enough energy available to fragment the larger clusters. We have mentioned several times that larger clusters with many internal degrees of freedom require significant excess energy in order to dissociate on the experimental time scale. In order to assess how much excess energy the Al_{n-4}^+ fragments require to dissociate further, some RRKM calculations of the dissociation rates were performed. For these calculations the cluster vibrational frequencies were obtained from a simple Debye heat capacity model. It was found that Al_{10}^+ required an excess energy of only 0.5 eV whereas Al_{25}^+ required over 3.0 eV. While for the larger clusters this excess energy is remarkably large, there still should be enough energy available to dissociate the Al_{n-4}^+ species (from the reaction with oxygen) for even the largest cluster ion studied. One possible explanation for these results is that the $Al_nO_2^+$ complex (formed by dissociative chemisorption) dissociates so rapidly that the heat of chemisorption which is initially deposited

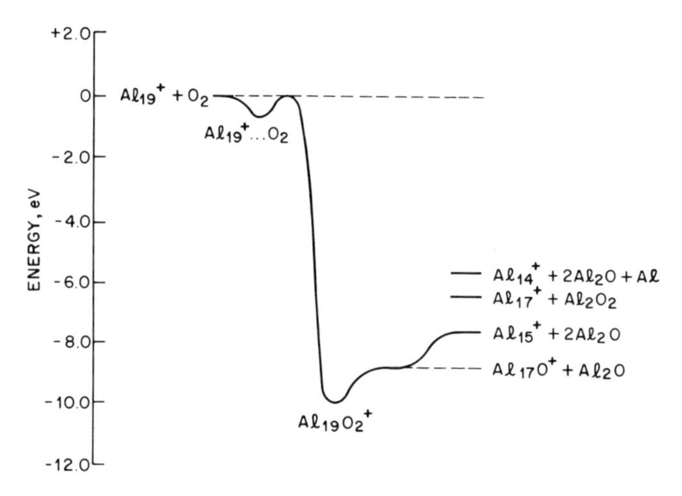

Figure 28. Semiquantitative potential energy diagram for the reaction between aluminum cluster ions and oxygen. The relative energetics are reliable to within ±2 eV. (Reproduced with permission from Reference 147.)

in the new Al—O bonds has insufficient time to flow into the cluster. Thus, a large fraction of the heat of chemisorption is carried away by the Al_2O molecules and further fragmentation of the Al^+_{n-4} product is not observed. In view of the extremely large exothermicity of these reactions and the relatively weak binding energies of the clusters, this explanation appears feasible.

Total cross sections for the reactions between aluminum cluster ions and oxygen are shown in Figure 29. The lines are the data of Jarrold and Bower[147] and the points are the results of Ruatta *et al.*[149] Both data sets show the same qualitative trends, and for the lower collision energy (1.2 eV) there is reasonable quantitative agreement. However, at the higher collision energy (4.2 eV) for the larger clusters there are significant discrepancies (approaching a factor of three) in the size of the cross sections. There appears to be no simple explanation for this discrepancy. We mentioned in Section 3 that low-energy ion beam experiments are prone to discrimination problems. However, the product distributions determined by these two groups appear to be in reasonable quantitative agreement. Other possible causes of this discrepancy are different internal energy distributions or differing cluster structures, both of which are difficult to investigate. The dashed lines in Figure 29 shows the predictions of several simple models for the cross sections. The ones labeled "Langevin" give the predictions of the Langevin, Gioumousis, and Stevenson model,[152] which assumes an ion-induced dipole potential. The line labeled "Cluster Size" gives the estimated geometric area of the clusters and the line labeled "Gas Kinetic" gives the gas kinetic cross sections.

The reactions between aluminum cluster ions (with up to 27 atoms) and deuterium have been studied using a low-energy ion beam experiment.[153] The reactions were found to have significant kinetic energy thresholds (gen-

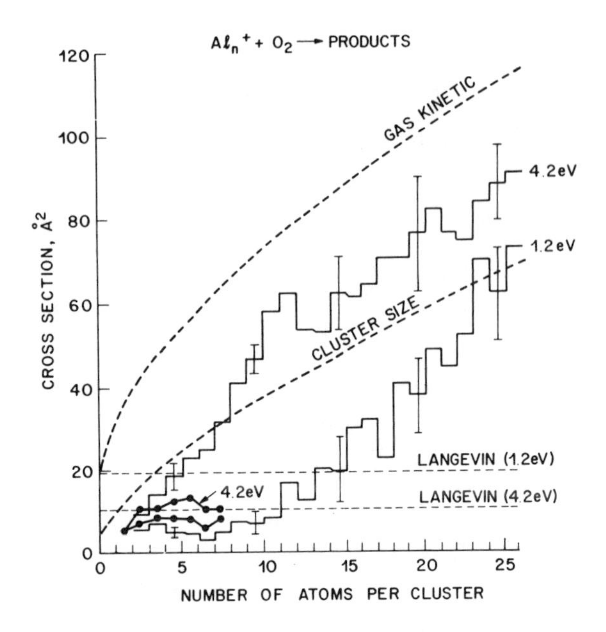

Figure 29. Total cross sections for the reactions between aluminum cluster ions and oxygen at collision energies of 1.2 and 4.2 eV. The points are data from Reference 149 and the lines are from Reference 147. The dashed lines give the predictions of various models described in the text.

erally >1 eV). However, it was not obvious whether these thresholds are due to an activation barrier or reaction endothermicity. The main products observed at a collision energy of 3.0 eV were Al_nD^+, $Al_{n-1}D^+$, Al_{n-2}^+, and, for the smaller clusters, Al^+. The neutral species associated with the $Al_{n-1}D^+$ and Al_{n-2}^+ products are probably AlD molecules.

In addition to chemical reactions, chemisorption of deuterium onto the clusters:

$$Al_n^+ + D_2 \rightarrow Al_nD_2^+ \tag{15}$$

was observed for the larger clusters ($n > 7$). Studies of the intensity of the $Al_nD_2^+$ species as a function of gas cell pressure indicated that this product arises from single collision processes (no stabilizing collisions). Thus, this adduct is metastable and only survives long enough to be detected because of the many internal degrees of freedom in the larger clusters. Plots of the cross sections for adduct formation (chemisorption) are shown in Figure 30 for Al_{25}^+ and Al_{16}^+. The points are the experimental data. For Al_{25}^+ the cross sections show a threshold and then increase linearly with increasing cluster size. For Al_{16}^+ the cross sections also show a threshold, rise to a maximum, and then fall. The decline in the cross sections observed for Al_{16}^+ at the higher collision energies is almost certainly due to the dissociation of the $Al_{16}D_2^+$ metastable adduct before it can be detected. This is observed for the smaller clusters because they have fewer internal degrees of freedom and so a shorter lifetime.

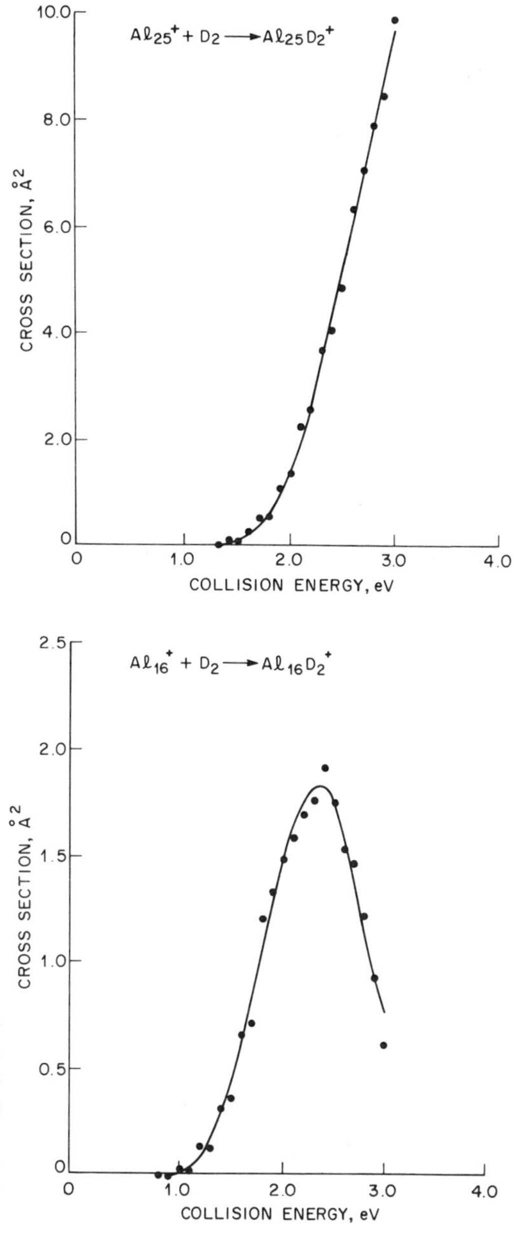

Figure 30. Plot of the cross sections for chemisorption of deuterium onto mass-selected aluminum cluster ions. The points are the experimental data and the lines are the results of a computer simulation to determine the true threshold. (Reproduced with permission from Reference 153.)

The lines in Figure 30 are the result of a computer simulation to account for threshold broadening, as described above, and determine accurate values for the thresholds. Since the $Al_nD_2^+$ adduct is directly observed in these experiments, the thresholds can be unambiguously related to the activation barriers for chemisorption. A plot of the activation barriers, determined in this way, is shown in Figure 31. The activation barriers increase from a little over 1 eV

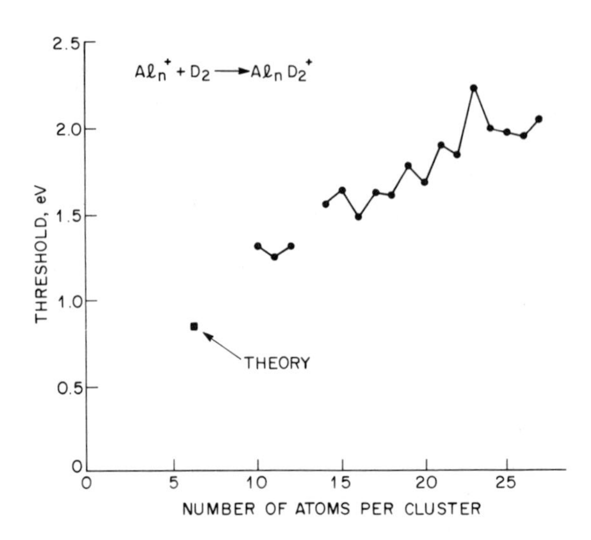

Figure 31. Plot of the activation barriers for chemisorption of deuterium onto aluminum clusters as a function of cluster size. The uncertainties in the values range from ±0.30 eV for the smaller clusters to ±0.15 eV for the larger ones. These are conservative upper limits on the absolute values of the thresholds. The point labeled theory is from the work of Upton and co-workers.[132] (Reproduced with permission from Reference 153.)

for the smaller clusters to over 2 eV. Chemisorption was not observed for Al_{13}^+, and the activation barrier is significantly higher for Al_{23}^+ than for its neighbors. There is also a clear odd–even oscillation in the height of the activation barriers. Activation barriers are not shown in Figure 31 for clusters with $n < 10$ because the observed adduct intensity for these clusters was too low to determine reliable thresholds. The low adduct intensity for the small clusters must be related to the shortening of the adduct lifetime caused by a decrease in the number of internal degrees of freedom.

Cox and co-workers[132] have investigated the chemisorption of deuterium on neutral aluminum clusters using the fast-flow reactor technique. This is a thermal energy experiment. Significant chemisorption was only observed on Al_6 and Al_7, and not on larger clusters, which is consistent with the results discussed above for the ions. Calculations by Upton and co-workers[132,135] suggest that the activation barriers for chemisorption on neutral and ionic clusters will be quite similar and that the activation barrier arises from changes in the population of the cluster molecular orbitals to minimize repulsive interactions and accommodate the new M—H bonds. Chemisorption of deuterium has not been observed on polycrystalline aluminum surfaces,[154,155] presumably because of the presence of an activation barrier as found for the larger clusters. Clearly, it would be fascinating to compare the activation barriers shown in Figure 31 to a value for the bulk. Unfortunately, the bulk value is not known. By modeling the falloff in the chemisorption cross sections for the smaller clusters at large collision energies using RRKM theory, it was possible to derive an estimate of the activation energy for desorption of deuterium from the cluster. The desorption energies deduced in this way increase from around 1.5 eV for the smaller clusters to around 2.0 eV for the larger ones. The desorption energies are thus comparable to the activation energies for chemisorption.

Meiwes-Broer and co-workers[156] have recently obtained photoelectron spectra for mass-selected negative cluster ions of aluminum. These data are not yet completely analyzed. However, preliminary conclusions are that the adiabatic electron affinities show an overall increase with cluster size and the vertical electron affinities vary irregularly with cluster size. In some cases, for example Al_{14}^-, more than one band is observed in the photoelectron spectrum. Ultimately, detailed interpretation of these data will further understanding of the electronic structure of these interesting clusters.

Several groups have performed theoretical calculations on aluminum clusters. Unfortunately, most of the work has been on neutral species. The structure of small aluminum clusters with up to five atoms has been studied by Pacchioni and co-workers.[157–159] They found the lowest-energy structures for Al_3, Al_4, and Al_5 to be linear, a rhombus, and a square pyramid, respectively. These structures can be considered as deformed sections of the aluminum fcc lattice. The ground states were found to have high spin multiplicity. Upton[130,132,135,136] has performed several theoretical studies of aluminum clusters, and the results of some of his calculations have been discussed above. The ground state structures determined by Upton for Al_3–Al_6 are shown in Figure 32. The structures differ somewhat from those found by Pacchioni and co-workers, and, furthermore, the ground states were found to be low spin, except for Al_2, in agreement with magnetic deflection experiments.[9] Upton found the molecular orbitals to be diffuse and extending over the entire cluster. The shapes and symmetries of these orbitals are comparable to solutions of a spherical well potential and thus can be compared to the predictions of the electronic shell model. The occupation of these orbitals was often found to be in disagreement with the predictions of the shell model but apparently agrees close to the shell closings. Upton accounted for these differences by introducing atomic potentials as a perturbation to the shell model.

Bauschlicher and co-workers[160–163] have recently reported several studies of aluminum clusters. They have studied Al_2—Al_6 and Al_{13}, employing both *ab initio* techniques and a simplified approach using two- and three-body

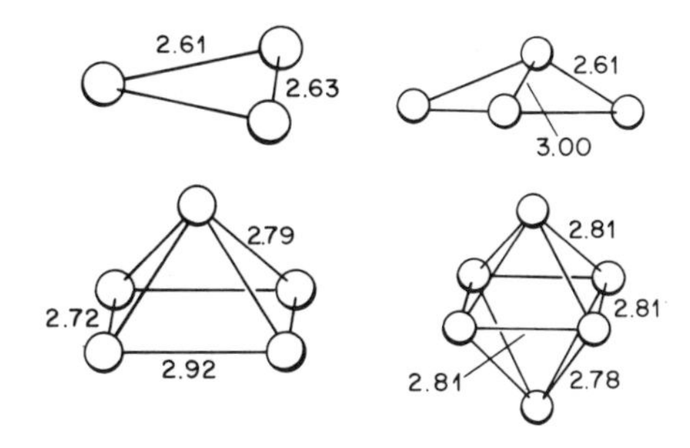

Figure 32. Calculated ground state structures of aluminum clusters. Selected nearest-neighbor distances are shown in angstroms. These can be compared to the bulk value of 2.86 Å. The point group for each structure is C_{2v}. (Reproduced with permission from Reference 135.)

interaction potentials. In the case of Al_5, they found a planar structure to be 0.2 eV lower than the pyramidal structure when extensive electron correlation was included in the calculations. For Al_{13} an icosahedron was found to have the lowest-energy structure, but this was only 0.3 eV more stable than a D_{3h} (hcp) structure and 1.2 eV more stable than an O_h (fcc) configuration. In fact, a star-shaped planar structure was only 2.5 eV above the lowest-energy structure. These calculations raise an important point: do these clusters have a well-defined structure as observed for silicon and carbon clusters or must they be considered as fluxional or molten species, rapidly isomerizing amongst several structures? Clearly, it is necessary to calculate activation barriers for isomerization between these structures. However, even if the barriers are substantial, it is almost certain that in experiments a range of different isomers will be present.

7. TRANSITION METAL CLUSTER IONS

It is well known that transition metals are important catalysts and it is believed that particle size plays a role in determining catalytic activity. The rich and varied chemistry of gas phase transition metal atomic ions is currently being explored by several groups,[164–166] and the inorganic chemistry of transition metal complexes has been studied for many years. Thus, it is easy to predict that transition metal clusters will show some fascinating chemistry. Unfortunately, the feature which provides transition metals with their varied chemistry—the *d* electrons—at the same time makes transition metal clusters excruciatingly difficult to understand. Theoretical calculations are much more difficult for transition metal clusters (except those of the noble metals) than for the clusters of the main-group elements discussed above.

Over the past three or four years there have been many studies of the reactions of neutral metal clusters using the fast-flow reactor technique. Due to the nature of this experiment, most of the reactions studied have been chemisorption reactions of the type

$$M_n + A \rightarrow M_nA^* \tag{16}$$

$$M_nA^* + He \rightarrow M_nA + He \tag{17}$$

where the heat of chemisorption is removed by collisions with the buffer gas. Since the clusters are not size selected before the reaction, cluster fragmentation reactions are obscured. Another source of uncertainty is the possibility of fragmentation when the clusters are ionized for subsequent mass analysis. However, work by groups at Argonne, Rice, and Exxon[15–17,167] has provided many spectacular examples of dramatic changes in rate constants for chemisorption as a function of cluster size. Generally, the origin of these large changes in reactivity is poorly understood. However, a particularly notable achievement has been the observation of a correlation between cluster ionization potentials and chemisorption rate constants for hydrogen on iron clusters.[168] Although this correlation does not always appear to work for other reactions, it is consistent with a mechanism for hydrogen chemisorption in which metal $\rightarrow \sigma^*$ and $\sigma \rightarrow$ metal charge donation weakens the hydrogen bond and leads to dissociative adsorption.[169]

The ionization potentials of clusters of several different transition metals have been measured using laser photoionization.[170] In some cases the ionization potentials have only been crudely bracketed using several laser lines. However, for some clusters photoionization efficiency curves have been measured using a tunable dye laser. Globally, the ionization potentials fall off in reasonable agreement with the classical model for the work function of a conducting sphere.[133,134] Superimposed on this steady decrease in ionization potential with cluster size there are quite large (up to 0.5 eV) oscillations. The origin of these oscillations is not well understood but could be related to structural or electronic changes in the clusters. The influence of chemisorption on ionization potentials has also been studied. These experiments are analogous to studies of the effect of adsorbates on work functions of surfaces. For V, Nb, and Fe clusters chemisorption of hydrogen apparently increases the ionization potential, the increase being larger for the smaller clusters.[171] This increase can be understood in terms of the removal of valence electrons from the cluster orbitals close to the Fermi level to form the M—H bonds.

The photodissociation of positive and negative silver cluster ions has been studied by Fayet and Woste[70] using a low-energy ion beam instrument and an argon-ion laser. Mass spectra recorded for the photodissociation of Ag_7^+ and Ag_7^- are shown in Figure 33. These spectra show a clear preference for the formation of product cluster ions with an odd number of atoms. This was observed for clusters with up to 16 atoms. Quite extensive fragmentation of the clusters is observed in what appear from power dependence studies to be one-photon processes. For example, Ag_{12}^+ is observed to fragment all the way down to Ag_3^+. For the negative cluster ions photodetachment competes with photodissociation, and silver seems quite unique in this regard. For other metal clusters studied (Ni_n^-, Cu_n^-, and Nb_b^-) clean photodetachment is observed.[172]

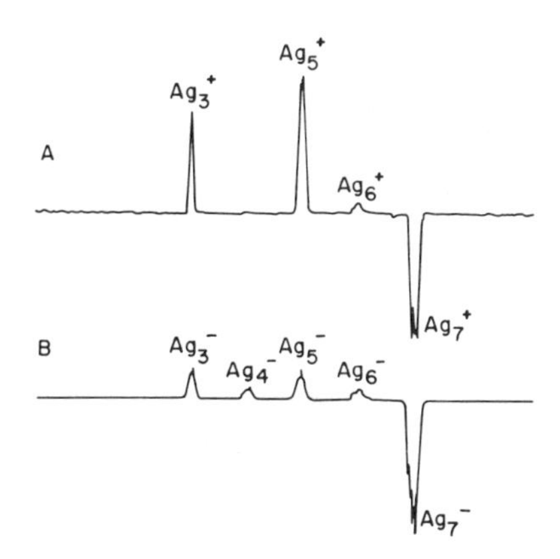

Figure 33. Mass spectra recorded for the photodissociation of (A) Ag_7^+ and (B) Ag_7^- with 488-nm light. (Reproduced with permission from Reference 70.)

Photodissociation presumably occurs for silver clusters because of their relatively high electron affinity and low binding energies.

The photodissociation of positive Fe, Ni, and Nb cluster ions with up to 10 atoms has been studied by Smalley and co-workers.[11] As mentioned above, the primary one-photon fragmentation pathway was found to be loss of a single metal atom from the cluster. The dissociation energies of Fe_2^+ and Fe_3^+ were bracketed to lie in the range 2.43–2.92 eV and 1.17–2.18 eV, respectively. Loss of an atom, as described here for Ni, Fe, and Nb clusters and as discussed above for Al clusters, is consistent with compact three-dimensional structures. Clearly, more bonds need to be broken to cleave a three-dimensional cluster than to remove a single atom. The results for silver clusters described above are clearly different from those for the other metals that have been studied so far. Extensive fragmentation is observed. A simple interpretation of these results is that silver cluster ions have one-dimensional (chain or ring) structures similar to those recently described by McAdon and Goddard[173] for lithium clusters (both lithium and silver have s^1 configurations).

Over the past few years Freiser and co-workers[174,175] have been studying the chemistry of a number of transition metal dimers and some trimers using FT-ICR techniques. The clusters for these experiments were generated using a chemical approach. For example,[75] Co_3^+ is generated by the following scheme: Co^+ from laser vaporization of a metal target reacts with $Co_2(CO)_8$ to yield predominantly $Co_3(CO)_6^+$:

$$Co^+ + Co_2(CO)_8 \longrightarrow Co_3(CO)_6^+ + 2CO \qquad (18)$$

$$\qquad\qquad\qquad\quad \longrightarrow Co_3(CO)_5^+ + 3CO \qquad (19)$$

The $Co_3(CO)_6^+$ is trapped and irradiated at its cyclotron frequency to increase its kinetic energy and then undergoes collision-induced dissociation with argon.

$$Co_3(CO)_6^+ \xrightarrow{Ar(CID)} \rightarrow\rightarrow Co_3^+ + 6CO \qquad (20)$$

Sequential loss of the CO ligands occurs to ultimately yield Co_3^+. Mixed metal clusters such as $FeCo_2^+$ can be generated in this way as well. Both $FeCo_2^+$ and Co_3^+ were found to sequentially abstract three oxygen atoms from ethylene oxide, whereas the dimers (Fe_2^+, $CoFe^+$, and Co_2^+) extract only two oxygen atoms and the atomic ions form predominantly MCO^+ and MCH_2^+. The reactions of Co_3^+ and $FeCO_2^+$ with oxygen result in cluster fragmentation yielding product ions of the type $M'MO_x^+$ ($x = 0$–2).

Freiser and co-workers have also recently investigated the reactions of Ag_3^+ and Ag_5^+ generated by pulsed laser sputtering from a ZnO/AgO mixture.[49] Ag_3^+ and Ag_5^+ were found to be quite unreactive with small alkanes, alkenes, and alcohols; only condensation products were observed. Both clusters react with *sec*-butylamine. First, condensation of two molecules of *sec*-butylamine occurs, followed by deamination and dehydrogenation of a third molecule to yield Ag_n^+—butadiene. Deamination and dehydrogenation competes with condensation of a third *sec*-butylamine molecule onto the cluster. This provides an interesting example of the influence of ligands on the chemical reactivity of metal clusters. Collision-induced dissociation of Ag_5^+ produced Ag_3^+, which is consistent with the photodissociation experiments of Fayet and Woste.[10] Collisional activation of Ag_3^+ produced no observable products. Ag_3^+ is known to be particularly stable.[176]

Campana and co-workers have studied the chemistry of copper and cobalt cluster ions.[177-179] The clusters for these studies were generated by sputtering with a beam of high-energy xenon atoms and react with gas introduced into the sputtering region. Thus, it is possible that both surface and gas phase reactions contribute to the observed products. In the case of copper clusters reacting with C_4H_{10}, it was argued that only gas phase reactions contribute. The observed products were $Cu_n(C_4H_{10})_m^+$ ($n = 1, 3, 5$; $m = 1$–3). The metastable dissociation and collision-induced dissociation of these species were studied using a double-focusing mass spectrometer. The main products arise from loss of a C_4H_{10} ligand, suggesting that the interaction between the clusters and the ligand is quite weak. For cobalt with oxygen admitted into the source three principal types of products were observed: $(CoO)_n^+$, $Co(CoO)_n^+$, and, in lower abundance, $(CoO)_nO^+$. Collision-induced dissociation of $(CoO)_n^+$ gives $(CoO)_{n-1}^+$ and $Co(CoO)_n^+$. Theoretical calculations for these clusters suggest that they have cage, ring, or ladder structures.

The reactions between mass-selected nickel clusters and CO have been studied by Woste and co-workers.[180] The nickel clusters, generated by sputtering, were mass selected and injected, at low energy, into a radio frequency quadrupole drift region. With CO introduced into the drift region the main products were $Ni_n(CO)_k^+$, with smaller amounts of $Ni_{n-1}(CO)_l^+$ and $Ni_nC(CO)_m^+$. The results are very similar to those obtained by clustering metal carbonyls using FT-ICR.[181,182] The observed pressure dependence of the product signal suggests that the $Ni_n(CO)_k^+$ species arise from a series of three-body association reactions. With increasing CO pressure the number of CO ligands increases and then saturates. For example, the maximum number of ligands binding to Ni_4^+ was 10 and the maximum number binding to Ni_{13}^+ was 22. It was found that the number of CO ligands binding to the cluster could be accounted for by the electron counting rules proposed by Wade[183] and Mingos,[184] where the contributions from both the metal skeleton and the ligands are summed to give the total number of cluster valence electrons. For example, with Ni_4^+ the number of ligands predicted for a tetrahedral Ni_4 skeleton is 10, in agreement with the experimental observation. Of course, the proposed tetrahedral arrangement of Ni_4^+ is the configuration in the complex and is not necessarily the structure of the bare cluster. For the larger clusters ($n > 9$) the electron counting rules appear to begin breaking down. In the case of Ni_{13}^+ the maximum number of CO ligands is predicted to be 20 for a nickel-centered 12-vertex polyhedron, whereas the observed maximum was 22. However, the peak for $Ni_{13}(CO)_{20}^+$ in the product ion mass spectrum was particularly intense. It is interesting that the electron counting rule used in these studies originates from inorganic chemistry and the clusters are treated as molecules with a well-defined structure. Another electron counting rule, the jellium model[5,126-129] described above for bare clusters, comes from solid state physics and ignores the structure of the cluster. This illustrates how the developing field of cluster science lies at the interface between solid state physics, surface science, chemical physics, and inorganic chemistry.

The reactions between mass-selected nickel clusters and C_4H_{10} have been studied by Michl and co-workers[35] using a triple-quadrupole mass spectrometer. For Ni_2^+-Ni_5^+ the main products were $Ni_nC_4H_6^+$ resulting from dehydrogenation, i.e., loss of two H_2 molecules:

$$Ni_5^+ + C_4H_{10} \rightarrow Ni_5C_4H_6^+ + 2H_2 \tag{21}$$

With Ni_6^+-Ni_{10}^+ more extensive dehydrogenation occurs, and the dominant products arise from the loss of three H_2 molecules to give $Ni_nC_4H_4^+$. However, for atomic nickel ion the main products result from C—C bond cleavage (or cracking) to give $NiC_2H_4^+$, although loss of an H_2 molecule to form $NiC_4H_8^+$ is also an important process. With increasing cluster size the cracking reactions are rapidly suppressed. Some products of C—C bond cleavage are seen for Ni_2^+ but not for larger clusters. Similar reactions were observed for small palladium clusters, Pd_n^+ (n = 1-4). However, the dehydrogenation activity of Pd_n^+ appears to exceed that of Ni_n^+. For example, Pd_4^+ removes three H_2 molecules where Ni_4^+ only removes two. Experiments in which Ni_2^+ and Ni_3^+ were collisionally cooled before reaction illustrate that the product distributions for these reactions were strongly influenced by the internal energy in the clusters. The results are summarized in Table 7. These data demonstrate that it is critical to work with clusters which do not contain large amounts of internal energy. Similar conclusions have been reached by Hanley and Anderson.[34]

The chemisorption of nitrogen on cobalt cluster ions has been studied by Smalley and co-workers[185] using a modification of the fast-flow reactor technique used to study neutral clusters. Thus, the clusters are not size selected before reaction. Relative rate constants for chemisorption are shown in Figure 34 for both cobalt cluster neutral and positive ions. It is apparent that there

Table 7. Reactions of Ni_n^+ with n-Butane: Cluster Ion Internal Energy Effects

	Product relative abundance [a]				
M_n^+	$M_n^+C_4H_8$ [a]	$M_n^+C_4H_6$	$M_n^+C_4H_4$	$M_{n-1}^+C_2H_4$	M_{n-1}^+
Ni_2^+ collision cooled	6.7	8.1		1.7	0.3
directly sputtered	0.8	4.1	0.02	2.3	3.8
Ni_3^+ collision cooled		3.0			
directly sputtered	0.8	1.4	0.5		2.0

[a] Normalized to an M_n^+ intensity of 100.

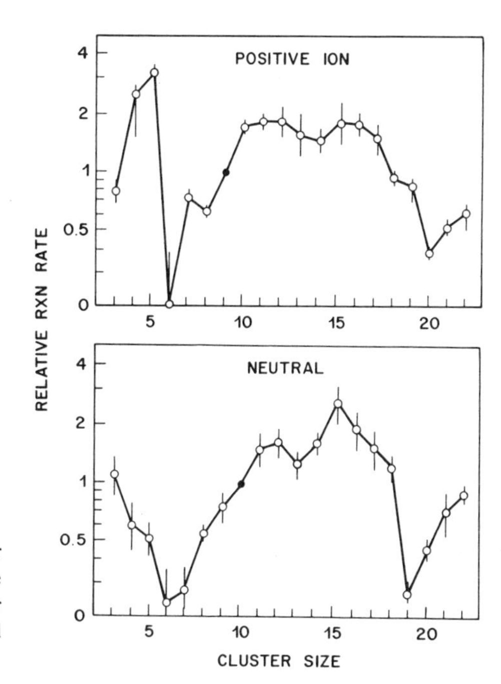

Figure 34. Relative reaction rates for chemisorption of N_2 on cobalt clusters. The upper trace is for positive ions and the lower trace is for neutral clusters. (Reproduced with permission from Reference 184.)

is a striking correspondence between the reactivity of the neutral and ionic species. In a comparable study, Smalley and co-workers[64] investigated the chemisorption of hydrogen on niobium clusters, Nb_n^+ ($n = 7$–9). In these experiments the positive cluster ions were injected into an FT-ICR cell from an external source. Nb_7^+ was found to be much more reactive than either Nb_8^+ or Nb_9^+. Similar behavior had been observed for neutral niobium clusters. It was argued that in view of the apparent similarity between the reactivity of the neutral and ionic niobium and cobalt clusters, the correlation observed between the reactivity of neutral clusters and their ionization potentials may have a more subtle origin than the simple electrostatic interpretation originally proposed.[168]

A start has also been made in investigating the photoelectron spectra of transition metal clusters. Photoelectron spectra for negatively charged copper cluster anions (with $n = 2$–10) have been measured by Lineberger and co-workers[7] using a cw argon-ion laser and a novel flowing afterglow source incorporating a cold cathode discharge. The photoelectron spectrum of Cu_2^- is shown in Figure 35. Resolved vibrational features are observed. Analysis of these features yields $r_e = 2.345$ Å, $\omega_e = 210\ \text{cm}^{-1}$, and $T = 450°\text{C}$. Vibrational features were not resolved for larger clusters, presumably because they are obscured by the increased density of vibrational states. Transitions to excited electronic states of the neutral cluster were clearly resolved for Cu_4^- and Cu_6^- and partially resolved for Cu_7^- and Cu_{10}^-. Electron affinities deduced from the photoelectron spectra were in good agreement with those measured by Smalley and co-workers[172] using laser bracketing and scans of the photodetachment

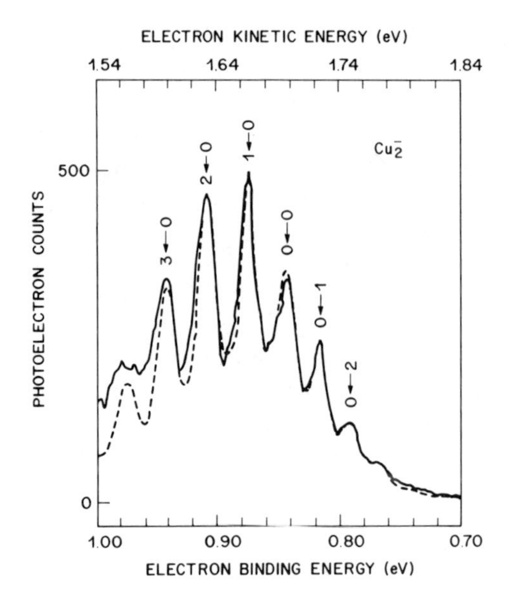

Figure 35. Photoelectron spectrum for Cu_2^- measured with 488-nm light. The solid line is the experimental data and the dashed line is the result of a Franck–Condon simulation. (Adapted from Reference 7.)

threshold region with a dye laser. The electron affinities increase with cluster size, from 0.9 eV for Cu_2^- to 2.0 eV for Cu_{10}^-. Superimposed on this global increase in electron affinity there are significant odd–even oscillations. The even-numbered clusters have higher electron affinities (except Cu_6^-). These odd–even oscillations were also observed in the ionization potentials of neutral copper clusters.[186]

8. CONCLUDING DISCUSSION

This chapter has reviewed the rapid developments that have recently occurred in the study of metal and semiconductor cluster ions. Virtually all of the studies of the chemistry of size-selected clusters described here have been performed in the last two years. A year ago there were no photoelectron spectra for these clusters and now there are at least four groups active in this area. The key technological development that has made these studies possible has been the development of sources with sufficient intensity.

It is possible to divide interest in clusters into three closely related areas: chemistry, electronic structure, and geometric structure. For the studies of chemistry, FT-ICR and low-energy ion beam techniques appear to be powerful complementary methods. Photoelectron spectroscopy of negative cluster ions has recently provided a new tool for the investigatation of the electronic structure of these species. Presumably it is only a matter of time before these techniques will be extended into the X-ray region using synchrotron sources. However, there are no techniques that can be used to directly probe the structure of the clusters. Collision-induced dissociation, photodissociation, and chemical studies do not provide direct structural information, and the prospects for measuring, let alone interpreting, an optical spectrum of a metal cluster with more than a few atoms do not appear to be good. One approach which shows some promise is scanning tunneling microscope studies of clusters on a surface. So far these experiments have been performed with a range of cluster sizes,[187] but extension to size-selected clusters seems feasible. Even this approach is far from ideal because the support interaction could influence the structure, as could the tunneling current.

A more fundamental question than what is the cluster structure is whether the cluster has a well-defined structure at all. Certainly for small silicon and carbon clusters the answer from theoretical calculations is yes. But we know these elements have highly directed covalent bonds. Metallic bonding is more diffuse, and the question of whether metal clusters rapidly isomerize between several low-energy structures has yet to be answered. It seems likely that we will have to modify our conception of structure for these species, if not for the reason discussed above then because as the cluster size increases, there comes a point when knowledge of the exact atomic coordinates of all the atoms in the cluster may not be very useful.

Another way in which clusters are unique molecules in chemical physics is a simple consequence of their physical size. The dynamics of a cluster with a large number of atoms are going to be quite different from those of normal (small) molecules. Due to their many internal degrees of freedom, the large clusters will be able to absorb substantial amounts of energy and not dissociate. We have seen glimpses of this behavior with aluminum, where deuterium molecules were observed to stick to the clusters in bimolecular collisions in much the same way that molecules chemisorb on surfaces. Many other examples of this behavior will certainly be observed. Metal clusters also have a high density of electronic states. It is less clear how this will influence the chemical properties. However, one clear manifestation is the extremely short lifetimes of excited electronic states.

It is apparent that the properties of the small clusters discussed in this chapter are not very closely related to those of the bulk materials, although, knowing what to look for, it is possible to see precursors to bulk behavior. For example, small silicon and aluminum clusters appear to have structures that can be considered as highly distorted fragments of the bulk lattice. On the other hand, the linear chains and rings of small carbon clusters do not appear to be related to either the structure of graphite or diamond. In the limited number of cases where values are available, physical properties such as ionization potentials and cluster dissociation energies are quite different from those of the bulk, but appear to be approaching the bulk values as the cluster size increases. With the chemical properties there are also parallels to bulk behavior. The reaction between oxygen and aluminum clusters is facile and very exothermic but, unlike the reaction at an aluminum surface, results in formation of Al_2O molecules. The cluster reaction may be more closely related to the explosive interaction between finely powdered aluminum and oxygen. Chemisorption of deuterium onto the clusters has a substantial activation barrier and chemisorption of deuterium has not been observed on polycrystalline aluminum. With silicon clusters, prototypical etching and deposition reactions have been observed. However, the reaction rates decrease with increasing cluster size, suggesting that another mechanism may be important for the bulk. Clearly, the chemistry of these small clusters parallels that observed on surfaces but there are significant differences.

Given the momentum that studies of clusters have achieved over the last few years, it is easy to predict that many fascinating discoveries will be made in this area in the next few years. These discoveries should lead us to extend and modify ideas and models taken from the much more extensive studies of molecules and bulk materials and will hopefully result in a better understanding of molecules, clusters, and the solid state.

ACKNOWLEDGMENTS

I am grateful to the people who supplied preprints of their work and so helped to make this chapter as current as possible. I am also grateful to my colleagues at AT&T Bell Laboratories who provided useful comments on the manuscript.

REFERENCES

1. Dietz, T.G.; Duncan, M.A.; Powers, D.E.; Smalley, R.E. *J. Chem. Phys.* 1981, *74*, 6511.
2. Bondybey, V.E.; English, J.H. *J. Chem. Phys.* 1981, *74*, 6978.
3. Kappes, M.; Schar, M.; Radi, P.; and Schumacher, E. *J. Chem. Phys.* 1986, *84*, 1863, and references therein.
4. Walstedt, R.E.; Bell, R.F. *Phys. Rev. A* 1986, *33*, 2830.
5. Knight, W.D.; de Heer, W.A.; Saunders, W.A. *Z. Physik. D*, 1986, *3*, 109.
6. Rohlfing, E.A.; Cox, D.M.; Kaldor, A.; Johnson, J.H. *J. Chem. Phys.* 1984, *81*, 3846.
7. Leopold, D.G.; Ho, J.; Lineberger, W.C. *J. Chem. Phys.* 1987, *86*, 1715.
8. Yang, S.H.; Pettiette, C.L.; Conceicao, J.; Cheshnovsky, O.; Smalley, R.E. *Chem. Phys. Lett.* 1987, *139*, 233.
9. Cox, D.M.; Trevor, D.J.; Whetten, R.L.; Rohlfing, E.A.; Kaldor, A. *J. Chem. Phys.* 1986, *84*, 4651.
10. Bloomfield, L.A.; Freeman, R.R.; Brown, W.L. *Phys. Rev. Lett.* 1985, *54*, 2246.
11. Brucat, P.J.; Zheng, L.-S.; Pettiette, C.L.; Yang, S.; Smalley, R.E. *J. Chem. Phys.* 1986, *84*, 3078.
12. Begemann, W.; Dreihofer, S.; Meiwes-Broer, K.H.; Lutz, H.O. *Z. Physik. D* 1986, *3*, 183.
13. Jarrold, M.F.; Bower, J.E.; Kraus, J.S. *J. Chem. Phys.* 1987, *86*, 3876.
14. Hanley, L; Ruatta, S.A.; Anderson, S.L. *J. Chem. Phys.* 1987, *87*, 260.
15. Riley, S.J.; Parks, E.K.; Nieman, G.C.; Pobo, L.C.; Wexler, S. *J. Chem. Phys.* 1984, *80*, 1360, Parks, E.K.; Neiman, G.C.; Pobo, L.G.; Riley, S.J. *J. Chem. Phys.* 1987, *86*, 1066.
16. Geusic, M.E.; Morse, M.D.; O'Brian, S.C.; Smalley, R.E. *Rev. Sci. Instrum.* 1985, *56*, 2123; Morse, M.D.; Geusic, M.E.; Heath, J.R.; Smalley, R.E. *J. Chem. Phys.* 1985, *83*, 2293.
17. Trevor, D.J.; Whetten, R.L.; Cox, D.M.; Kaldor, A. *J. Am. Chem. Soc.* 1985, *107*, 518, (1985); Whetten, R.L.; Cox, D.M.; Trevor, D.J.; Kaldor, A. *J. Phys. Chem.* 1985, *89*, 566.
18. Herrmann, A.; Hoffmann, M; Leutwyler, S.; Schumacher, E.; Woste, L. *Chem. Phys. Lett.* 1979, *62*, 216.
19. Morse, M.D.; Hopkins, J.B.; Langridge-Smith, P.R.R.; Smalley, R.E. *J. Chem. Phys.* 1983, *79*, 5316.

20. Morse, M.D. *Chem. Rev.* 1986, *86*, 1049.
21. Castleman, A.W.; Keesee, R.G. *Annu. Rev. Phys. Chem.* 1986, *37*, 525.
22. Phillips, J.C. *Chem. Rev.* 1986, *86*, 619.
23. Benneman, K.H.; Koutecky, J., Eds., *Surf. Sci.* 1985, *156* (Proceedings of the Third International Meeting on Small Particles and Inorganic Clusters, Berlin, West Germany, 1984).
24. Trager, F.; zu Putlitz, G., Eds. *Metal Clusters*; Springer: Berlin, 1986 (Proceedings of the International Symposium on Metal Clusters, Heidelberg, West Germany, 1986).
25. Jena, P.; Rao, B.K.; Khanna, S.N. Eds., *The Physics and Chemistry of Small Clusters*; Plenum: New York, 1987 (Proceedings of the International Symposium on the Physics and Chemistry of Small Clusters, Richmond, VA, 1986).
26. Becker, E.W. In *Metal Clusters*; Trager, F.; zu Putlitz, G., Eds.; Springer: Berlin, 1986.
27. Yamada, I.; Inokawa, H.; Takagi, T. *J. Appl. Phys.* 1984, *56*, 2746; Jin, H.-S.; Yapsir, A.S.; Lu, T.-M.; Gibson, W.M.; Yamada, I.; Takagi, T. *Appl. Phys. Lett.* 1987, *50*, 1062.
28. Kuiper, A.E.T.; Thomas, G.E.; Schouten, W.J. *J. Crystal Growth* 1981, *51*, 17.
29. Stein, G.D. In *Characterization and Behavior of Materials with Submicron Dimensions*; Waber, J.T., Ed.; (World Scientific: Singapore, 1985).
30. Yamada, I.; Usui, H.; Takagi, T. *J. Phys. Chem.* 1987, *91*, 2463.
31. Devienne, F.M.; Roustan, J.-C. *Org. Mass Spectrom.* 1982, *17*, 173.
32. Fayet, P.; Woste, L. *Surf. Sci.* 1985, *156*, 134.
33. Begemann, W.; Meiwes-Broer, K.H.; Lutz, H.O. *Phys. Rev. Lett.* 1986, *56*, 2248.
34. Hanley, L.; Anderson, S.L. *Chem. Phys. Lett.* 1985, *122*, 410.
35. Magnera, T.F.; David, D.E.; Michl, J. *J. Am. Chem. Soc.* 1987, *109*, 936.
36. Freas, R.B.; Ross, M.M.; Campana, J.E. *J. Am. Chem. Soc.* 1985, *107*, 6195.
37. Sigmund, P. *Phys. Rev.* 1969, *184*, 383; Sigmund, P. *Phys. Rev.* 1969, *187*, 768.
38. Wittmaack, K. *Phys. Lett.* 1979, *69A*, 322.
39. Fayet, P.; Wolf, J.P.; Woste, L. *Phys. Rev. B* 1986, *33*, 6792.
40. Berkowitz, J.; Chupka, W.A. *J. Chem. Phys.* 1964, *40*, 2735.
41. Furstenau, N.; Hillenkamp, F. *Int. J. Mass Spectrom. Ion Phys.* 1981, *37*, 135.
42. Tsong, T.T. *Appl. Phys. Lett.* 1984, *45*, 1149.
43. Reents, W.D.; Bondybey, V.E. *Chem. Phys. Lett.* 1986, *125*, 324.
44. Mandich, M.L.; Reents, W.D.; Bondybey, V.E. *J. Phys. Chem.* 1986, *90*, 2315.
45. O'Keefe, A.; Ross, M.M.; Baronavski, A.P. *Chem. Phys. Lett.* 1986, *130*, 17.
46. Bondybey, V.E.; Reents, W.D.; Madich, M.L. Unpublished results.
47. Weil, D.A.; Wilkins, C.L. *J. Am. Chem. Soc.* 1985, *107*, 7316.
48. Moini, M.; Eyler, J.R.; *Chem. Phys. Lett.* 1987, *137*, 311.
49. Buckner, S.W.; Gord, J.R.; Freiser, B.S. *J. Chem. Phys.* 1988, *88*, 3678.
50. Richtsmeier, S.C.; Parks, E.K.; Liu, K.; Pobo, L.G.; Riley, S.J. *J. Chem. Phys.* 1985, *82*, 3659.
51. Powers, D.E.; Hansen, S.G.; Geusic, M.E.; Michalopoulos, D.L.; Smalley, R.E. *J. Chem. Phys.* 1983, *78*, 2866.
52. Loh, S.K.; Hales, D.A.; Armentrout, P.B. *Chem. Phys. Lett.* 1986, *129*, 527.
53. Bloomfield, L.A.; Geusic, M.E.; Freeman, R.R.; Brown, W.L. *Chem. Phys. Lett.* 1985, *121*, 33.
54. Bernholc, J.; Phillips, J.C. *Phys. Rev. B* 1986, *33*, 7395.
55. Bernholc, J.; Phillips, J.C. *J. Chem. Phys.* 1986, *85*, 3258.
56. Radi, P.P.; Bunn, T.L.; Kemper, P.R.; Molchan, M.E.; Bowers, M.T. *J. Chem. Phys.* 1988, *88*, 2809.
57. Rohlfing, E.A.; Cox, D.M.; Kaldor, A. *J. Chem. Phys.* 1984, *81*, 3322.
58. Zheng, L.-S.; Brucat, P.J.; Pettiette, C.L.; Yang, S.; Smalley, R.E. *J. Chem. Phys.* 1985, *83*, 4273.
59. Joyes, P.; Sudraud, P. *Surf. Sci.* 1985, *156*, 451.
60. Barr, D.L. *J. Vac. Sci. Technol. B* 1987, *5*, 184.
61. DiCenzo, S.; Berry, S.D. and Hartford, E.; *Phys. Rev. B* (in press).
62. McElvany, S.W.; Creasy, W.R.; O'Keefe, A.; *J. Chem. Phys.* 1986, *85*, 632.
63. Knight, R.D.; Walch, R.A.; Foster, S.C.; Miller, T.A.; Mullen, S.L.; Marshall, A.G. *Chem. Phys. Lett.* 1986, *129*, 331.
64. Alford, J.M.; Weiss, F.D.; Laaksonen, R.T.; Smalley, R.E. *J. Phys. Chem.* 1986, *90*, 4480.
65. Jarrold, M.F.; Bower, J.E. *J. Chem. Phys.* 1986, *85*, 5373.

66. For examples, see Bowers, M.T., Ed. *Gas Phase Ion Chemistry*; Academic: New York, 1979; Vols. I and II.
67. Mandich, M.L.; Bondybey, V.E.; Reents, W.D. *J. Chem. Phys.* 1987, *86*, 4245.
68. Alford, J.M.; Williams, P.E.; Trevor, D. J.; Smalley, R.E. *Int. J. Mass Spectrom. Ion Proc.* 1986, *72*, 33.
69. McIver, R.T.; Hunter, R.L.; Bowers, W.D. *Int. J. Mass Spectrom. Ion Proc.* 1985, *64*, 67.
70. Fayet, P.; Woste, L. *Z. Physik. D* 1986, *3*, 177.
71. Zheng, L.-S.; Karner, C.M.; Brucat, P.J.; Yang, S.H.; Pettiette, C.L.; Craycraft, M.J.; Smalley, R.E. *J. Chem. Phys.* 1986, *85*, 1681.
72. Coe, J.V.; Snodgrass, J.T.; Friedhoff, C.D.; McHugh, K.M.; Bowen, K.H. *J. Chem. Phys.* 1985, *83*, 3169.
73. Posey, L.A.; Deluca, M.J.; Johnson, M.A. *Chem. Phys. Lett.* 1986, *131*, 170.
74. Kroto, H.W.; Heath, J.R.; O'Brian, S.C.; Curl, R.F.; Smalley, R.E. *Nature* (*London*) 1985, *318*, 162.
75. Liu, Y.; O'Brian, S.C.; Zhang, Q.; Heath, J.R.; Tittel, F.K.; Curl, R.F.; Kroto, H.W.; Smalley, R.E. *Chem. Phys. Lett.* 1986, *126*, 215.
76. Cox, D.M.; Trevor, D.J.; Reichmann, K.C.; Kaldor, A. *J. Am. Chem. Soc.* 1986, *108*, 2457.
77. Hahn, M.Y.; Honea, E.C.; Paguia, A.J.; Schriver, K.E.; Camarena, A.M.; Whetten, R.L. *Chem. Phys. Lett.* 1986, *130*, 12; also see, O'Brian, S.C.; Heath, J.R.; Kroto, H.W.; Curl, R.F.; Smalley, R.E. *Chem. Phys. Lett.* 1986, *132*, 99.
78. Geusic, M.E.; McIlrath, T.J.; Jarrold, M.F.; Bloomfield, L.A.; Freeman, R.R.; Brown, W.L. *J. Chem. Phys.* 1986, *84*, 2421.
79. Geusic, M.E.; Jarrold, M.F.; McIlrath, T.J.; Freeman, R.R.; Brown, W.L. *J. Chem. Phys.* 1987, *86*, 3862.
80. Drowart, J.; Burns, R.P.; DeMaria, G.; Inghram, M.G. *J. Chem. Phys.* 1959, *31*, 1131.
81. O'Brian, S.C.; Heath, J.R.; Curl, R.F.; Smalley, R.E. *J. Chem. Phys.* 1988, *88*, 220.
82. McElvany, S.W.; Dunlap, B.I.; O'Keefe, A. *J. Chem. Phys.* 1987, *86*, 715.
83. Hoffmann, R.; *Tetrahedron* 1966, *22*, 521.
84. Bohme, D.K.; Raksit, A.B.; Fox, A. *J. Am. Chem. Soc.* 1983, *105*, 5481.
85. Raksit, A.B.; Bohme, D.K. *Int. J. Mass Spectrom. Ion Proc.* 1983, *55*, 69.
86. McElvany, S.W.; Nelson, H.H.; Baronavski, A.P.; Watson, C.H.; Eyler, J.R. *Chem. Phys. Lett.* 1987, *134*, 214.
87. Strickler, S.J.; Pitzer, K.S. In *Molecular Orbitals in Chemistry, Physics and Biology*; Lowdin, P.-O.; Pullman, B., Eds.; Academic: New York, 1964.
88. Haymet, A.D.J. *J. Am. Chem. Soc.* 1986, *108*, 319.
89. Klein, D.J.; Schmalz, T.G.; Hite, G.E.; Seitz, W.A. *J. Am. Chem. Soc.* 1986, *108*, 1301.
90. Newton, M.D.; Stanton, R.E. *J. Am. Chem. Soc.* 1986, *108*, 2469.
91. Haddon, R.C.; Brus, L.E.; Raghavachari, K. *Chem. Phys. Lett.* 1986, *125*, 459.
92. Disch, R.L.; Schulman, J.M. *Chem. Phys. Lett.* 1986, *125*, 465.
93. Fowler, P.W.; Woolrich, J. *Chem. Phys. Lett.* 1986, *127*, 78.
94. Hale, P.D. *J. Am. Chem. Soc.* 1986, *108*, 6087.
95. Schmalz, T.G.; Seitz, W.A.; Klein, D.J.; Hite, G.E. *Chem. Phys. Lett.* 1986, *130*, 203.
96. Luthi, H.P.; Almlof, J. *Chem. Phys. Lett.* 1987, *135*, 357.
97. Stone, A.J.; Wales, D.J. *Chem. Phys. Lett.* 1986, *128*, 501.
98. Haymet, A.D.J. *Chem. Phys. Lett.* 1985, *122*, 421.
99. Fowler, P.W. *Chem. Phys. Lett.* 1986, *131*, 444.
100. Whiteside, R.A.; Raghavachari, K.; DeFrees, D.J.; Pople, J.A.; Schleyer, P.v.R. *Chem. Phys. Lett.* 1981, *78*, 538.
101. Ewing, D.E.; Pfeiffer, G.V. *Chem. Phys. Lett.* 1982, *86*, 365.
102. Raghavachari, K.; Whiteside, R.A.; Pople, J.A. *J. Chem. Phys.* 1986, *85*, 6623.
103. Brown, W.L.; Freeman, R.R.; Raghavachari, K.; Schluter, M. *Science* 1987, *235*, 860.
104. Raghavachari, K.; Binkley, J.S. *J. Chem. Phys.* 1987, *87*, 2198.
105. Heath, J.R.; Liu, Y.; O'Brian, S.C.; Zhang, Q.-L.; Curl, R.F.; Tittel, F.K.; Smalley, R.E. *J. Chem. Phys.* 1985, *83*, 5520.
106. Martin, T.P.; Schaber, H. *J. Chem. Phys.* 1985, *83*, 855.

107. Bloomfield, L.A.; Freeman, R.R.; Brown, W.L. Unpublished results.
108. Zhang, Q.-L.; Liu, Y.; Curl, R.F.; Tittel, F.K.; Smalley, R.E. *J. Chem. Phys.* 1988, *88*, 1670.
109. Reents, W.D.; Mandich, M.L.; Bondybey, V.E. *Chem. Phys. Lett.* 1986, *131*, 1.
110. Reents, W.D.; Mujsce, A.M.; Bondybey, V.E.; Mandich, M.L. *J. Chem. Phys.* 1987, *86, 5568.*
111. Creasy, W.R.; O'Keefe, A.; McDonald, J.R.; *J. Phys. Chem.* 1987, *91*, 2848.
112. Broadbent, E.K.; Ramiller, C.L. *J. Electrochem. Soc.* 1984, *131*, 1427.
113. Raghavachari, K.; Logovinsky, V. *Phys. Rev. Lett.* 1985, *55*, 2853.
114. Raghavachari, K. *J. Chem. Phys.* 1985, *83*, 3520.
115. Raghavachari, K. *J. Chem. Phys.* 1986, *84*, 5672.
116. Zmbov, K.F.; Ames, L.L.; Margrove, J.L. *High Temp. Sci.* 1973, *5*, 235; Snyder, L.C.; Raghavachari, K. *J. Chem. Phys.* 1984, *80*, 5076.
117. Elkind, J.L.; Alford, J.M.; Weiss, F.D.; Laaksonen, R.T.; Smalley, R.E. *J. Chem. Phys.* 1987, *87*, 2397.
118. Liu, Y.; Zhang, Q.-L.; Tittel, F.K.; Curl, R.F.; Smalley, R.E. *J. Chem. Phys.* 1986, *85*, 7434.
119. Cheshnovsky, O.; Yang, S.H.; Pettiette, C.L.; Craycraft, M.J.; Liu, Y.; Smalley, R.E. *Chem. Phys. Lett.* 1987, *138*, 119.
120. Tomanek, D.; Schluter, M.A.; *Phys. Rev. Lett.* 1986, *56*, 1055.
121. Pacchioni, G.; Koutecky, J. *J. Chem. Phys.* 1986, *84*, 3301.
122. Raghavachari, K. *J. Chem. Phys.* (in press).
123. Martin, T.P.; Schaber, H. *Z. Physik. B.* 1979, *35*, 61.
124. Raghavachari, K.; Rohlfing, C.E. *Chem. Phys. Lett.* 1988, *143*, 428.
125. Begemann, W.; Dreihofer, S.; Meiwes-Broer, K.H.; Lutz, H.O. In *The Physics and Chemistry of Small Clusters*; Jena, P.; Rao, B.K.; Khanna, S.N. Eds., Plenum: New York, 1987.
126. Ekardt, W. *Phys. Rev. B* 1984, *29*, 1558.
127. Knight, W.D.; Clemenger, K.; de Heer, W.A.; Saunders, W.A.; Chou M.Y.; Cohen, M.L. *Phys. Rev. Lett.* 1984, *52*, 2141.
128. Cohen, M.L.; Chou, M.Y.; Knight, W.D.; de Heer, W.A. *J. Phys. Chem.* 1987, *91*, 3141.
129. Chou, M.Y.; Cohen, M.L. *Phys. Lett. A* 1986, *113*, 420.
130. Upton, T.H. *J. Phys. Chem.* 1986, *90*, 754.
131. Knight, W.D.; de Heer, W.A.; Saunders, W.A.; Clemenger, K.; Chou, M.Y.; Cohen, M.L. *Chem. Phys. Lett.* 1987, *134*, 1.
132. Upton, T.H.; Cox, D.M.; Kaldor, A. In *The Physics and Chemistry of Small Clusters*; Jena, P.; Rao, B.K.; Khanna, S.N. Eds.; Plenum: New York, 1987.
133. Woods, D.M. *Phys. Rev. Lett.* 1981, *46*, 749.
134. Makov, G.; Nitzan, A.; Brus, L.E. *J. Chem. Phys.* 1988, *88*, 5076.
135. Upton, T.H. *Phys. Rev. Lett.* 1986, *56*, 2168.
136. Upton, T.H. *J. Chem. Phys.* 1987, *86*, 7054.
137. Aristov, N.; Armentrout, P.B.; *J. Am. Chem. Soc.* 1986, *108*, 1806.
138. Hanley, L.; Anderson, S.L.; *J. Phys. Chem.* (submitted).
139. Jarrold, M.F.; Bower, J.E. *J. Chem. Phys.* 1987, *87*, 1610.
140. Drowart, J.; DeMaria, G.; Burns, R.P.; Inghram, M.G. *J. Chem. Phys.* 1959, *32*, 1366.
141. Pacchioni, G.; Fantucci, P. In *The Physics and Chemistry of Small Clusters*; Jena, P.; Rao, B.K.; Khanna, S.N. Eds.; Plenum: New York, 1987.
142. Kappes, M.M.; Radi, P.; Schar, M.; Schumacher, E. *Chem. Phys. Lett.* 1985, *119*, 11.
143. Kappes, M.M.; Radi, P.; Schar, M.; Yeretzian, C.; Schmumacher, E. In *Metal Clusters*; Trager, F.; zu Putlitz, G., Eds.; Springer: Berlin, 1986.
144. Kappes, M.M.; Schar, M.; Schumacher, E. *J. Phys. Chem.* 1987, *91*, 658.
145. Brechignac, C.; Cahuzac, Ph. In *Metal Clusters*; Trager, F.; zu Putlitz, G., Eds.; Springer: Berlin, 1986.
146. Jarrold, M.F.; Bower, J.E. In *The Physics and Chemistry of Small Clusters*; Jena, P.; Rao, B.K.; Khanna, S.N. Eds.; Plenum: New York, 1987.
147. Jarrold, M.F.; Bower, J.E. *J. Chem. Phys.* 1987, *87*, 5728.
148. Hanley, L.; Ruatta, S.A.; Anderson, S.L. In *The Physics and Chemistry of Small Clusters*; Jena, P.; Rao, B.K.; Khanna, S.N. Eds.; Plenum: New York, 1987.
149. Ruatta, S.A.; Hanley, L.; Anderson, S.L. *Chem. Phys. Lett.* 1987, *137*, 5.

150. Batra, I.P.; Kleinman, L. *J. Electron Spectrosc. Relat. Phenom.* 1984, *33*, 175.
151. Freund, H.J.; Bauer, S.H. *J. Phys. Chem.* 1977, *81*, 994.
152. Gioumousis, G.; Stevenson, D.P. *J. Chem. Phys.* 1958, *29*, 294.
153. Jarrold, M.F.; Bower, J.E. *J. Am. Chem. Soc.* 1988, *110*, 70.
154. Flodstrom, S.A.; Petersson, L.-G.; Hagstrom, S.B.M. *J. Vac. Sci. Technol.* 1976, *13*, 280.
155. Pellerin, F.; LeGressus, C.; Massignon, D. *Surf. Sci.* 1981, *111*, L705.
156. Begemann, W.; Dreihofer, S.; Gantefor, G.; Siekmann, H.R.; Meiwes-Broer, K.H.; Lutz, H.O. In *Elemental and Molecular Clusters*; Martin, T.P., Ed.; Springer: Berlin, 1987.
157. Pacchioni, G.; Plavsic, D.; Koutecky, J. *Ber. Bunsenges. Phys. Chem.* 1983, *87*, 503.
158. Pacchioni, G.; Koutecky, J. *Ber. Bunsenges. Phys. Chem.* 1984, *88*, 242.
159. Koutecky, J.; Pacchioni, G.; Jeung, G.H.; Hass, E.C. *Surf. Sci.* 1985, *156*, 650.
160. Bauschlicher, C.W.; Pettersson, L.G.M. *J. Chem. Phys.* 1986, *84*, 2226.
161. Partridge, H.; Bauschlicher, C.W. *J. Chem. Phys.* 1986, *84*, 6507.
162. Bauschlicher, C.W.; Pettersson, L.G.M. *J. Chem. Phys.* 1987, *87*, 2198.
163. Pettersson, L.G.M.; Bauschlicher, C.W.; Halicioglu, T. *J. Chem. Phys.* 1987, *87*, 2205.
164. Armentrout, P.B.; Halle, L.F.; Beauchamp, J.L. *J. Am. Chem. Soc.* 1981, *103*, 6624; Halle, L.F.; Houriet, R.; Kappes, M.M.; Staley, R.H.; Beauchamp, J.L. *J. Am. Chem. Soc.* 1982, *104*, 6293.
165. Jacobson, D.B.; Freiser, B.S. *J. Am. Chem. Soc.* 1983, *105*, 7485; Cassady, C.J.; Freiser, B.S. *J. Am. Chem. Soc.* 1985, *107*, 1573.
166. Aristov, N.; Armentrout, P.B. *J. Am. Chem. Soc.* 1986, *108*, 1806; Elkind, J.L.; Armentrout, P.B. *J. Am. Chem. Soc.* 1986, *108*, 2765.
167. Trevor, D.J.; Kaldor, A. In *High Energy Processes in Organometallic Chemistry*; Suslick, K.S., Ed.; American Chemical Society, Washington, DC, 1987; ACS Symp. Ser. 333.
168. Whetten, R.L.; Cox, D.M.; Trevor, D.J.; Kaldor, A. *Phys. Rev. Lett.* 1985, *54*, 1494.
169. Saillard, J.-Y., Hoffmann, R. *J. Am. Chem. Soc.* 1984, *106*, 2006.
170. Kaldor, A.; Cox, D.M.; Trevor, D.J.; Zakin, M.R. In *Metal Clusters*; Trager, F.; zu Putlitz, G., Eds.; Springer: Berlin, 1986.
171. Zakin, M.R.; Cox, D.M.; Whetten, R.L.; Trevor, D.J.; Kaldor, A. *Chem. Phys. Lett.* 1987, *135*, 223.
172. Zheng, L.-S.; Karner, C.M.; Brucat, P.J.; Yang, S.H.; Pettiette, C.L.; Craycraft, M.J.; Smalley, R.E. *J. Chem. Phys.* 1986, *85*, 1681.
173. McAdon, M.H.; Goddard, W.A. *J. Phys. Chem.* 1987, *91*, 2607.
174. Jacobson, D.B.; Freiser, B.S. *J. Am. Chem. Soc.* 1985, *107*, 1581; Hettich, R.L.; Freiser, B.S. *J. Am. Chem. Soc.* 1987, *109*, 3537.
175. Jacobson, D.B.; Freiser, B.S. *J. Am. Chem. Soc.* 1986, *108*, 27.
176. Andreoni, W.; Martins, J.L. *Surface Sci.* 1985, *156*, 635.
177. Freas, R.B.; Campana, J.E. *J. Am. Chem. Soc.* 1985, *107*, 6202.
178. Freas, R.B.; Campana, J.E. *J. Am. Chem. Soc.* 1986, *108*, 4659.
179. Freas, R.B.; Dunlap, B.I.; Waite, B.A.; Campana, J.E. *J. Chem. Phys.* 1987, *86*, 1276.
180. Fayet, P.; McGlinchey, M.J.; Woste, L.H. *J. Am. Chem. Soc.* 1987, *109*, 1733.
181. Anderson Fredeen, D.; Russell, D.H. *J. Am. Chem. Soc.* 1985, *107*, 3762.
182. Meckstroth, W.K.; Ridge, D.P. *Int. J. Mass Spectrom. Ion Proc.* 1984, *61*, 149.
183. Wade, K. *Adv. Inorg. Chem. Radiochem.* 1976, *18*, 1.
184. Mingos, D.M.P. *Acc. Chem. Res.* 1984, *17*, 311.
185. Brucat, P.J.; Pettiette, C.L.; Yang, S.; Zheng, L.-S.; Craycraft, M.J.; Smalley, R.E. *J. Chem. Phys.* 1986, *85*, 4747.
186. Powers, D.E.; Hansen, S.G.; Geusic, M.E.; Michalopoulos, D.L.; Smalley, R.E. *J. Chem. Phys.* 1983, *78*, 2866.
187. Sattler, K. In *The Chemistry and Physics of Small Clusters*; Jena, P.; Rao, B.K.; Khanna, S.N. Eds.; Plenum: New York, 1987.

6

Atomic Clusters in the Gas Phase

Robert L. Whetten and Kenneth E. Schriver

1. ATOMIC CLUSTER PROPERTIES AND THEIR SIZE DEPENDENCE

1.1. Interest in Atomic Clusters

Atomic clusters are systems represented by the simple chemical symbol A_N, where A is the identity of the atom and N is the number of atoms constituting the cluster. They are related on one front (small N) to the variegated atoms and homonuclear diatomics, on another front to bulk elemental solids and liquids, and on yet another to the behavior of interfaces, since a high fraction of their atoms must be at the cluster surface.

The present decade has seen a remarkable surge of interest in the intrinsic properties of atomic clusters, which can be at least partly accounted for by the sudden emergence of experimental techniques for measuring the properties of individual clusters. Certain discoveries (Table 1) made using these techniques have already begun to be regarded as common knowledge, while at the same time changing our perspective on material systems in subtle, fundamental ways. (References 1–7 are review issues or volumes devoted to atomic and molecular clusters.)

Apart from the emergence of new techniques, there has been a confluence of interest from workers in many fields now concentrating on atomic clusters. Thus, one finds atomic and molecular physicists and chemists seeking new challenges for their mature experimental and theoretical methods. Solid state scientists recognize in atomic clusters the compound constituents of bulk matter and have found stimulating unresolved questions. The statistical mechanician hopes to find clues or evidence regarding nucleation phenomena and the size-scaling theory of critical phenomena. Materials scientists and specialists

---

Robert L. Whetten and Kenneth E. Schriver • Department of Chemistry and Biochemistry, University of California, Los Angeles, California 90024.

Table 1. Some Discoveries of Atomic Cluster Properties

Phenomenon	Period	Reference
Rare-gas clusters (A = Ne, Ar, Kr, Xe) pack in icosahedral symmetry	1981–87	10
Atomic clusters have critical sizes for stability when multiply charged	1981–84	11
Electronic properties of simple metal clusters (Groups 1, 2, and 13) are predicted by electron counting rules	1984–87	12, 13
Magnetic properties of metal clusters parallel atomic and bulk magnetism	1985–86	14, 15
Chemical reactivity of neutral transition metal clusters corresponds strongly to ionization potentials	1985–86	16, 17
Certain-size metal clusters are often found to be very inert toward chemical reactions	1984–86	18
Larger carbon clusters are able to satisfy all valence requirements	1985	19
Particular size clusters can display chemical behavior unknown for atom and bulk forms of the same metal	1986–87	20

in catalysis see the atomic clusters of certain elements as essential components for advances in their fields.

In inorganic chemistry, one also recognizes atomic clusters as systems of fundamental interest. From a chemical standpoint, these systems of identical atoms have severe difficulties with regard to valence, i.e., they typically are highly unsaturated or "open shell" in character. The rules of bonding and cohesive properties of such systems have long been a matter of speculation by inorganic chemists,[8] as is the existence or nonexistence of special collectivity in their chemical behavior.[9]

The purpose of this chapter is to bring to the attention of inorganic chemists the remarkable progress now being made in the study of atomic clusters in the gas phase or vacuum. This is an infant field of high experimental and theoretical activity which has made progress largely through the particular route of investigating the precise N-dependence of various physical properties of A_N clusters under known temperature or energy conditions (see Figure 1).

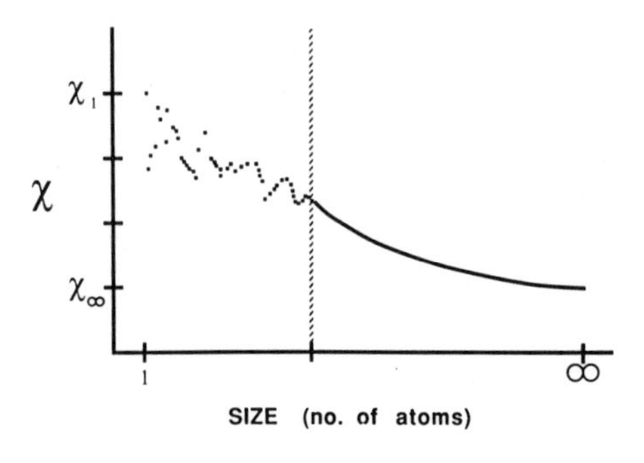

Figure 1. Hypothetical dependence of the value of a physical or chemical property on the size of a material system. The property values χ range from χ_1, the atomic value, through a discrete region ("clusters") where the number of atoms are countable, toward a continuous extrapolation to the thermodynamic limit χ_∞. The measurement and interpretation of these size dependences is a major goal of atomic cluster research.

1.2. N-Specific Properties

We will be interested in A_N clusters, where N is a known variable under the conditions of any particular measurement, and we will often refer to the N-dependence as the *size* dependence, although usually the actual physical dimensions are not known. In particular, in experiments in which the size (diameter, shape) *is* known, for example, through electron microscopy, the precise number of atoms constituting the cluster is usually not known nor is the influence of the support understood. Because it is now realized that cluster properties can have a critical dependence on N, for N as large as 200, these latter measurements are of secondary interest here. The interested reader is referred to reviews of work on the properties of small particles and droplets, in which smooth (usually monotonic) dependences of properties on size are observed.[21] (It is implicit that in these measurements a range of N is encountered; the exception is the recent use of the scanning tunneling microscope.[22])

The most important experiments to date have been those which measure the *variation* of properties with N, for a particular choice of atom A, over a wide size range. This is to be contrasted with the computational theoretical approaches most often used, which typically attempt to predict the properties for a specific, strategically chosen value of N.

For this reason, it is the (discrete) functional form of the dependence of cluster properties on N that is sought and gives perspective to what might otherwise be mundane individual measurements. Correspondingly, much significance is attached to models, which necessarily neglect many aspects of reality, that predict certain N- or size dependences.

In the search for a simplified understanding of cluster systems, one employs the traditional chemical strategy of varying the atom identity A, either within a group (Na, K, Rb) or across a series (Fe, Co, Ni). Small or systematic changes in properties (with change in A for fixed N) provide critical clues in identifying the crucial underlying physical mechanisms.

1.3. Theoretical Guidelines

There is no one successful theory of clusters, in the sense that one has a band theory of solids or molecular orbital theory of the singlet molecular ground state, or even hard-sphere theory for liquids. Nonetheless, there are important existing guidelines, and in particular cases (rare gases, alkali metals) one has the framework of a successful theory in place.

It is evident that as N becomes very large, bulk properties are approached, and much effort has been expended to determine the correct asymptotic form of the N-dependence of certain properties for specific A.[23] It must also always be asked, however, whether one is in the asymptotic regime at all, whether specific-N fluctuations in the value of properties such as work function or reaction rate constants are not comparable to the difference between bulk and cluster values. It is the dual nature of this paradigm, the search for *both* asymptotic and specific behavior, which has attracted so much theoretical attention to clusters.

Atomic clusters thus pose a severe challenge to traditional quantum chemistry, and indeed for the usual molecular viewpoint of counting vibrational modes and electronic excitations. It is often asked why a usefully complete description of a cluster, say of Ar or Na atoms, should be so much more difficult to obtain than such a description of an organic or inorganic molecule containing the same number of atoms. The resolution of this paradox lies in recognizing that strict rules of chemical valence imply a large number of *constraints* on the degrees of freedom in ordinary, closed-shell molecules. For example, a 20-atom Na cluster has no known constraints so that 57 degrees of freedom exist. In a saturated 20-atom organic or inorganic molecule one knows that each atom is nearly fixed in distance from each of its nearest neighbors, so that only *groups* of atoms possess significant freedom.

The quantum chemistry literature on clusters is vast,† but because of computational limitations nearly all studies adopt one of two approaches:

1. The electronic properties are calculated for certain clusters at single particular geometries, often chosen to be symmetrical fragments of the bulk crystal or of bulk surfaces. In this way the (discrete) size dependence of electronic properties (e.g., band structure) can be examined, but properties at actual cluster geometries remain unknown.
2. A single cluster size (specific N) is chosen and its energy minimized with respect to shape or geometrical parameters. Some insight regarding isomerism may be gained.

A common theme announced in much of the earlier work is that the asymptotic regime is achieved very early, so that even with $N \approx 10$ a crude model of bulk behavior (electronic, magnetic, chemical) is obtained.

A surprising alternative to these methods has emerged since the 1985 work of Car and Parinello,[(25)] which has allowed *ab initio* computations on large clusters with simultaneous structural energy minimization, or even finite-temperature dynamics.[(26)] Clusters of Na and Si atoms have been simulated in this way.

A drastic alternative to fully *ab initio* theory is to construct *models* in which clusters have all the properties of the bulk state, except for incorporating a small number of size-dependent features. Thus, one has as examples:

1. Alkali clusters are free-electron metals, just as in the bulk, *except* that the conduction electrons are discrete in number and quantized in a "box" the size of the cluster.[(13)]
2. Rare-gas clusters (Ar_N, Kr_N, Xe_N) are van der Waals solids, with atoms interacting pairwise through short-range potentials, *except* that for finite sizes icosahedral symmetry is possible, and in fact preferred.[(27)]
3. Magnetic metal clusters (e.g., Fe_N) are magnetic just as in the bulk, *except* that surface atoms have greatly enhanced magnetism because of their low coordination number.[(28)]

† For a review of some of the methods used, see Reference 24.

4. Network-forming clusters (e.g., C_N, Si_N) follow valence rules just as in the bulk, *except* that in clusters the drive to satisfy all valence requirements (minimize dangling bonds) forces them to adopt very low dimensionality structures (linear chains, monocyclic rings, then shells or planar structures).[29,30]

In the following we will critically summarize the experimental evidence for these models, but theoretical results will *not* be reviewed, except for a very few special correspondences. Also, results on the so-called compound clusters, such as $(NaCl)_N$, are omitted in the interest of brevity. In the following section the synthetic and preparation methods are described; these are recent in origin and are assumed to be unfamiliar to the reader. In Section 3, very brief descriptions of the methods used to measure properties are given. Section 4 reviews existing results, grouped according to chemical element, and Section 5 gives a brief prospectus of future work.

2. CONTROLLED PREPARATION (SYNTHESIS) OF ATOMIC CLUSTERS

2.1. Requirements for Property Measurement

Methods for synthesizing and equilibrating (thermalizing) atomic clusters vary in much the same way that cohesive energies of the corresponding bulk solids do.[1-7] One common feature, however, is that it is always assumed that atomic clusters are chemically very reactive. This implies that measurements should be carried out on the unsupported cluster in flight through vacuum or thermalized in a flowing inert gas. In the case of metal clusters, this means high dilutions in a rare gas, while for Ar_N clusters only cold He or vacuum is suitable. The same factors also imply a necessarily transient character in the experiment, with at best only 10^{-3} s to perform a measurement on neutral species, or up to a few seconds in ion trap experiments (see other reviews in this volume), which are beginning to see many applications in atomic cluster research.

In surveying the periodic table, one finds three classes of elemental solids:

1. Those which have a high atomic vapor pressure at ambient temperatures (rare gases and, with ovens, the noble and alkali metals);
2. The metals and semimetals, which typically are very high melting;
3. Others, including hydrogen, nitrogen, and the halogens, have dimeric (A_2) vapors and solids, while P, S, and As form polymeric vapors.

The last of these classes is not of interest here but falls instead in the growing field of *molecular clusters*, which is the study of clusters of closed-shell entities. The first class is usually synthesized by the *free-jet condensation method*,[31] and the second class by the *laser vaporization–flow condensation method*,[32] both of which are described below. Other methods specific to *ionic* cluster experiments are described elsewhere in this volume. Our intent is not to present the interesting history of these techniques, which can be found elsewhere, but simply to explain their mechanism and qualities.

2.2. The Existing Methods

The three techniques to be described each successfully generate clusters, in a large and crudely controlled size distribution, in quantities sufficient for sensitive measurement techniques. The clusters are separated by N as part of the measurement or (in the case of ions) just prior to measurement. A very important variable is the temperature (cluster internal energy content) during the measurement, which in most cases should be low (narrow in distribution). The techniques follow.

2.2.1. Free-Jet Condensation

In this method clusters form from the atomic vapor when the inert "carrier" gas into which it is mixed expands from high pressure into vacuum (Figure 2).

2.2.1a. Mechanism. The atomic vapor A (typically 5–200 torr) is diluted in inert gas (often He or Ar) at temperature T_0 (77–1000 K) and pressure P_0 (1–20 atm). The vapor mixture passes through a small orifice (or "nozzle," if shaped) into a chamber held under high vacuum (10^{-3}–10^{-7} torr) by high-speed pumps. The expansion into vacuum results in rapid isentropic cooling, during which atomic clusters nucleate and grow, or even coalesce.[33] Nascent clusters are cooled by collisions with the inert carrier gas until the expansion terminates in free flight. The distribution of A_N clusters obtained is strongly dependent on T_0, P_0, and the nozzle characteristics. The theory of its operation is complex, and most parameters are empirically determined.[34]

2.2.1b. Typical Applications. In the case of rare-gas clusters, T_0 is often less than room temperature to facilitate condensation. Sometimes, no carrier gas is used, but this results in ineffective cooling of nascent clusters. Abundance distributions characteristic of cohesive properties are measured, as is the probability of collisional breakup or fission. In the case of noble metals (e.g.,

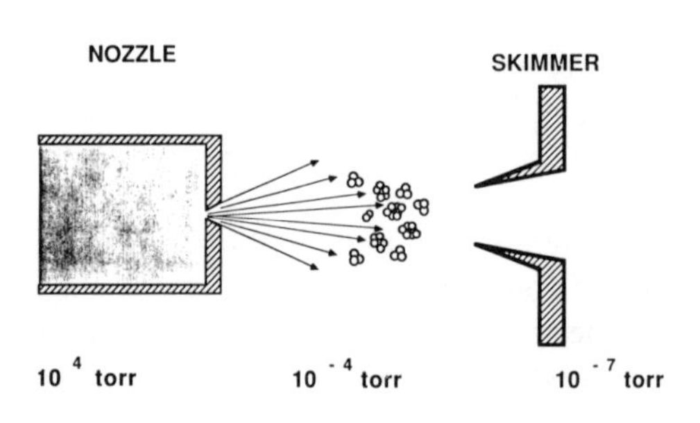

Figure 2. Illustration of the supersonic expansion method of synthesizing beams of atomic clusters. See text for description.

Hg) or alkali metals (Na or K), T_0 ranges from 300 to 1000°C, usually with Ar as a carrier. Properties measured during free flight in vacuum include ionization potentials, inner-shell ionization cross sections, and magnetic and electronic moments.

2.2.1c. Limitations. This method has had only limited success with the more refractory elements. Measurements must be performed in free flight, so T is not a useful variable. Oven-nozzle methods are technically difficult because of corrosion and clogging, and they use large quantities of starting material. Chemical reactivity is particularly difficult to study, and unless the seed ratio is very high, the clusters are not as cold as those produced by the method described in Section 2.2.2.

2.2.2. Laser Vaporization Flow Condensation

This method, applicable to all the high-melting elements, has seen widespread use since its introduction in 1981–83 by Smalley and co-workers.[32] High-density atomic vapor is produced by vaporizing a thin layer of the elemental solid with a pulsed laser. The atomic vapor is confined, cooled, and directed as desired by a high-pressure flowing inert gas, usually He, H_2, or Ne. See Figure 3 for details.

2.2.2a. Mechanism. Visible or ultraviolet radiation pulses, in which the intensity flux exceeds 10^7 W/cm^2 incident on an opaque solid surface, produce clean, highly efficient vaporization. Typical yields are 10^{14}–10^{15} atoms from a 0.01-cm^2 surface area in a 10^{-8}-s pulse. The atomic vapor density can exceed 10^{18} atoms per cm^3 (100 torr equivalent pressure) in the microseconds following the laser pulse. At the same time, He flowing over the vaporized surface traps the vapor and removes heat, inducing very rapid growth of clusters. Growth is both self-limited by depletion of the smaller clusters and externally limited by diffusion along the flow channel or to the channel walls.

2.2.2b. Typical Applications. The intense pulses of clusters, produced by this source in an inert flow stream, make this a highly versatile method. Once the metal vapor is so dilute that cluster growth terminates, chemical kinetics are studied by introducing a flow of reagents (flow reactor method).[36,37]

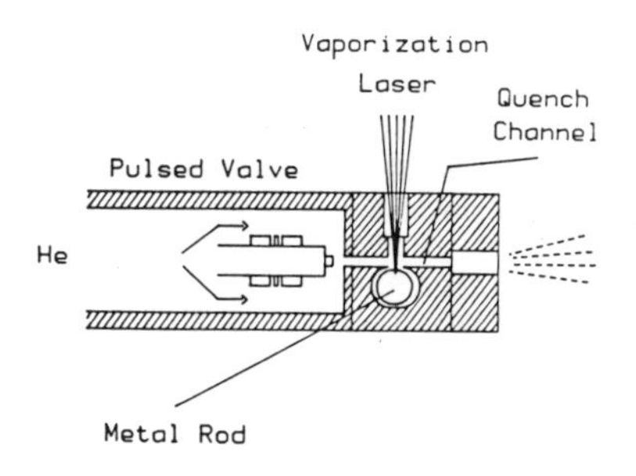

Figure 3. Illustration of the laser vaporization flow condensation method for synthesis of meal or semimetal clusters. The pulsed gas valve is placed in a high-vacuum chamber, and synchronous pulses from a focused laser beam repeatedly irradiate the rotating target. At right, the gas jet expands into vacuum. (Adapted from Reference 35.)

Alternatively, the flow may terminate in a nozzle expansion into vacuum. The intense molecular beam or ion beams (using the small charged fraction) of cold clusters have been used for numerous spectroscopic experiments and for reactivity studies.

2.2.2c. Limitations. It has not yet proved possible to extend this method to softer metals, such as groups 1, 2, or 12.

2.2.3. Surface Sputtering with Trapped-Ion Cooling

This method is described elsewhere in this volume.[38] As is well known, molecular or cluster ions sputtered directly from surfaces (by ion beam impact or focused-laser ablation) are internally very hot, to the point that nearly all are metastable. Such ions are not immediately useful in measurement of chemical or physical properties of clusters. Nonetheless, the intense, continuous ion beams produced by sputtering have initial advantages, so that several ingenious methods have been applied to cooling the ions in various ion traps, either with or without buffer gas.

2.3. Comparison among Methods

All three methods described above (see Table 2) are in active use and are likely to see continued application. The most versatile of the three, the laser vaporization method, has many advantages but is relatively expensive. Its extension to softer metals, and the entire periodic table excluding rare gases and the class 3 elements mentioned above, will probably soon be realized, as will its routine scaling to very high pulse repetition rates. For example, Armentrout and co-workers have recently reported a 8-kHz source.[39] The free-jet adiabatic expansion method will remain the favorite for rare-gas clusters and molecular clusters, while the sputtering/trap-cooling method is a favorite for coupling to ICR/FTMS instruments. Thus, the methods should

Table 2. Cluster Sources Used for Gas Phase Property Measurements

	Free-jet condensation	Laser vaporization flow condensation	Ion sputtering/ trap cooling
Elements studied	Low melting (Groups 1, 2, 12, 18)	Refractory	Refractory and group 12
Synthetic environment	Adiabatic expansion	He flow stream (10^2–10^3 torr)	Sputtered particles in vacuum
Cooling environment	Adiabatic expansion	Equilibrium with ambient, or adiabatic expansion	Ion trap or guide
Charge state	Neutral; positive or negative ion (rarely)	Neutral; positive and negative ions	Positve or negative ions

be regarded as complementary with regard to the periodic table, but unfortunately less complementary when it comes to properties to be measured.

3. THE BASIC EXPERIMENTAL MEASUREMENTS

3.1. N-Specific Detection of Measured A_N Properties

As indicated above, one of two routes is usually followed to measure the physical or chemical properties of atomic clusters in the gas phase:

1. Mass-selected (cold) cluster *ions* are investigated. The properties usually investigated then depend on the charge state of the ion. In the case of positive cluster ions, these often include either rate constants or cross sections for chemical reactions, including cluster fission, and limited physical properties, notably photofragmentation cross sections (to be connected to absorption intensities) and ultimate photochemical pathways. In the case of negative ions, the above measurements are still possible but are rarely carried out, because most attention has been directed toward determining the electron affinity (photodetachment threshold) or, more recently, the photoelectron spectrum.
2. Experiments on *neutral clusters* are always done with a mixture of sizes (distribution over N). It is implicit that during property measurement the mixture must be so well diluted, in the gas phase or in a collision-free beam, that no further cluster growth takes place. Similarly, as the clusters are detected by ionizing and mass spectrometry, fragmentation on the time scale of mass measurement must be avoided. For this reason, photoionization is the method of choice, using vacuum ultraviolet radiation just above the ionization threshold.[40]

It is important to note that mass spectra themselves are not of interest here. Instead the following strategy is used: The intensity of an A_N^+ peak in a mass spectrum is observed as a function of some experimental parameter that is varied to measure the property, with all other conditions held constant. For example, one varies the reagent (L) concentration in the flow reactor and observes the decrease in A_N^+ signal as the N-cluster reacts. Although new peaks corresponding to $A_N L_M$ reaction products may appear, the difference in photoionization probability between A_N and $A_N L_M$ makes quantitative assessments of dubious value.

Once a property such as the magnetic moment or chemical reaction rate constant or ionization threshold has been determined, it is usually plotted as a function of N. Obtaining values for N over a useful range ($N = 3$ to 30 or much more) thus involves a reduction of copious amounts of data. This process has been greatly facilitated by use of time-of-flight mass spectrometers, combined with fast waveform digitizers and averagers, so that entire mass spectra are recorded on each pulse of the cluster source. For details on these specialized techniques, the reader is directed to more technical articles.[41]

3.2. Natural Abundances and Cluster Growth Rates

Mass spectra have been obtained for clusters of over half the elements in the periodic table, including all the commonly available elements. One can ask whether any useful information on cluster properties is contained in mass spectra. In the laser vaporization/condensation methods, the early growth is often assumed to be by an atom-by-atom mechanism:

$$A_N + A \xrightarrow{k_N} A_{N+1}$$

Conversely, the final step in the adiabatic expansion method is believed to be a choice between decay of the hot, newly formed cluster or stabilization of the same by carrier gas collisions. The easiest decay process is usually:

$$A_N \xrightarrow{k'_N} A_{N-1} + A$$

When either of these two cases hold, it is profitable to look at local variations in abundance as reflecting cluster stability with respect to atom loss. In the most favorable case one can hope to relate pairwise cluster abundances, through a local equilibrium assumption, to cluster thermodynamics:

$$\frac{[A_N]}{[A_{N-1}]} = \frac{k_N}{k'_N} = \exp(-\Delta G/kT)$$

In practice, however, these quantities are coupled to all other cluster sizes and even varying $[A_1]$ is insufficient. Nonetheless, this scheme has proved useful in *reverse*, wherein model theoretical computations are used to compute an approximation to $\Delta G_{N,N-1}$ and the predicted abundances are compared to experiment.[42,43]

From an experimental standpoint, such an analysis is fraught with uncertainty, and it is not often attempted except in the case of charged cluster growth, where at least the uncertainties of ionization probability and fidelity are excluded. It is in any case preferable to obtain thermochemical quantities from other experiments, wherever possible.

For this reason, mass spectral observations are *not* included in this review unless they are helpful in explaining other observed properties or can themselves be explained by these properties. For example, the variations in ionization potentials of metal clusters such as Na, K, and Al versus N are of great help in explaining the so-called magic number anomalies in the abundance spectra of these clusters.

3.3. The Ionization Potential and Electron Affinity

Figure 4 schematically indicates the methods used for measurement of the ionization potential IP (or ϕ_N^+) and electron affinity EA (or ϕ_N^-) of an

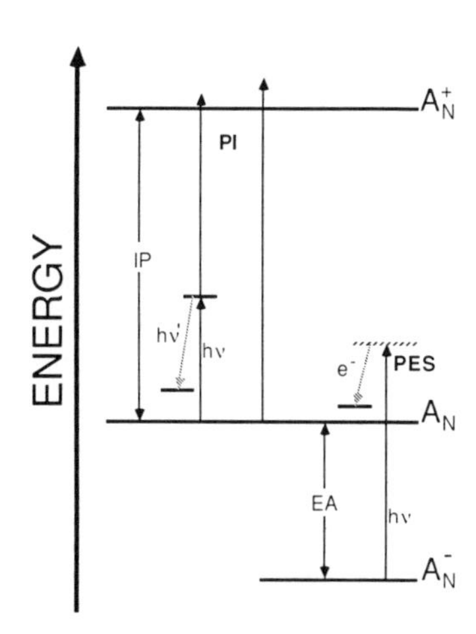

Figure 4. An energy diagram illustrates some often-measured quantities and the methods used to obtain them, for a generic N-atom cluster of element A. The three commonly encountered charge states (A_N^-, A_N, and A_N^+) are indicated at right, with the electron affinity (EA) and ionization potential (IP) defined as usual with respect to the neutral cluster. At left, threshold photoionization (PI) and resonant two-photon ionization spectroscopy are indicated as well as fluorescence emission ($h\nu'$). At lower right, photoelectron spectroscopy is indicated, where the kinetic energy of the ejected electron is given by the downward arrow.

arbitrary N-cluster. Ionization potentials are usually extracted from photoionization experiments on neutral beams, detecting positive ions mass spectrometrically.[44] Both nonlinear laser devices and synchrotron sources have been used in addition to ultraviolet lamp/monochromator systems. There is no generally accepted method for deriving an IP from the photoion yield versus wavelength curve other than that the same procedure must be used for all N.

Alternatively, electron impact ionization with a monoenergetic electron beam has been used.[45] The primary difficulty with this approach is its poor sensitivity, compared to photon absorption using available light sources, and poorer resolution.

3.4. Electric and Magnetic Moments

For molecules about which very little is known, classical properties such as the existence of a magnetic dipole moment (paramagnetism) or permanent electric dipole moment can be very valuable. Atomic clusters are certainly regarded as members of this class of molecules, and the laser vaporization/supersonic jet cooling approach to their synthesis lends itself well to classical molecular beam methods[46] of determining electric and magnetic moments.

In these experiments the neutral cluster beam is collimated into a narrow rectangular beam, typically much less than one millimeter in height. This beam passes between the pole faces of an intense electromagnet or charge electrodes. The pole faces are not flat but highly curved, to produce high field inhomogeneities for deflecting clusters from the beam axis. This deflection is measured far downstream by ionizing only a portion of the beam. Typical properties that can be measured in this way are:

1. *Magnetic dipole moments (paramagnetism)*: Unpaired electron spins or orbital angular momenta usually give rise to magnetic dipole

moments (often simply referred to as magnetic moments) of the order of one μ_B and are easily detected.

2. *Permanent electric dipole moments*: Unsymmetrical structures of high rigidity may give rise to a permanent electric dipole moment, even if all atoms are identical, as in an A_N cluster. Large deflections are obtained for even a small fraction of a debye.
3. *Electric polarizability*: In the absence of a permanent electric dipole, polarizability is detected by electric deflection going as $|E|^2$.

It may be noted that the polarizability will tend to depend linearly on the cluster size (an extensive property), so that the polarizability-to-mass ratio (and hence deflection) is size independent. For most materials, dipole moments will tend to remain small and constant, with the very notable exception of ferromagnetic clusters (see below).

3.5. Optical Spectroscopy

Optical absorption and emission spectroscopy in the standard spectral regions (microwave, infrared, and visible–ultraviolet) are the most powerful tools available for the rigorous study of gas phase molecules, and atomic clusters are not expected to be much different. The strong selection rules and almost arbitrary resolution available in modern spectrometers should provide a match to many of the fundamental questions on atomic clusters. However, thus far they have been relegated to a minor role, with the exception of A_3 spectroscopy in neutral beams[47,48] and recent photodissociation studies of van der Waals cluster ions or metal cluster ions.[49,50]

Methods used have included resonance two-photon ionization spectroscopy, laser-induced fluorescence, or stimulated emission pumping and ion beam photofragmentation. Because these have not yet seen widespread use in the study of clusters, we omit description here, except for noting that when properly carried out, they are capable of providing a faithful representation of the ordinary optical absorption or emission spectrum under very cold conditions.

3.6. Photoelectron Spectroscopy (PES)

One of the latest entries to the arsenal of experimental methods is the photoelectron spectrometer.[51,53] Reviews of PES methods can be found elsewhere; the important feature here is that the experiment is carried out on mass-selected beams of *negatively* charged cluster ions. This is illustrated in Figure 4. The difference between the (fixed) photon energy and the observed electron kinetic energies gives the energy levels (electronic and vibrational) of the neutral cluster. It may be noted that the selection rules are relatively unrestricted as far as electronic transitions are concerned and that the vibrational pattern within a band is governed by the Franck–Condon principle.

The method has been applied in both electrostatic and time-of-flight modes, to measure, respectively, the kinetic energy or velocity of the ejected electron. For further details of these remarkable instruments, the reader is referred to the technical descriptions in References 51–53.

3.7. Chemical Flow Reactors and Ion–Molecule Reactors

Chemical reactivity studies have tended to be of three types. In the case of neutrals (and sometimes ions), reactions are studied under thermal equilibrium conditions in chemical flow reactors. This technique will be described below. Reactions of ions are studied in ion beam and FTMS/ICR instruments at much lower pressure. These latter techniques are described in detail in other chapters of this volume, and for the study of larger clusters are altered primarily in source design (to inject large, cold cluster ions) as described above.

The design of a typical flow reactor experiment is given in Figure 5. Clusters are formed by pulsed laser vaporization of a solid in a dense He flow stream which may be either synchronously pulsed[35,36] or steady state[54] in operation. Both neutral clusters and, to a lesser extent, negative and positive ions are formed, and are considered to be so diluted in He that all growth has terminated before the point of reagent injection.

It is of interest to compare the strengths and weaknesses of the flow reactor experiments to those of ion beam or trap methods. This is done in Table 3. It is seen that while the flow reaction conditions and analysis are closest to that usually done in gas phase and condensed phase chemical kinetics, the ion beam or trap experiments have advantages with regard to quantitative product determination and control of reaction energetics.[38] A more subtle difference is that flow reactor experiments are more favorable for reactions of the type

$$A + B \rightarrow C \qquad \text{(association reaction)}$$

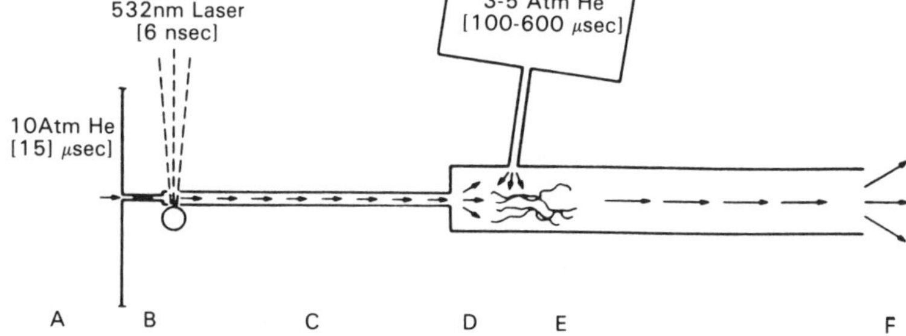

Figure 5. The flow reactor method of studying cluster chemical reactivity starts (A, B) with a laser vaporization cluster source (see Figure 3) and introduces reactant gas via a secondary channel (D). Both cluster and reactant gas are typically highly diluted in He. At the end of the reactor (F), the gas expands into vacuum and is analyzed by a mass spectrometer. (Adapted from Reference 36.)

Table 3. Methods for Studying Chemical Reactivity

Parameter	Characteristics		
	A. Flow reactor	B. Ion beam	C. Ion trap
Cluster cooling	Determined by source conditions	Determined by source conditions and expansion cooling	Determined by source conditions and expansion cooling
Reaction environment	Ambient, 20–200 torr He	Single collision, high vacuum	Multiple collisions, up to 10^{-5} torr
Kinetics analysis	High-pressure limit	Single collision, cross sections	Thermal, low-pressure limit
Cluster charge state	Neutral, positive, or negative	Usually positive	Positive or negative
Determination of reacted fraction	Photoionization mass spectrometry	Quadrupole mass spectrometry	ICR
Accurate assessment of reaction products	Sometimes	Usually	Always

while ion beam and trap experiments are often sensitive only to reactions of the type

$$A + B \rightarrow C + D \qquad \text{(exchange reaction)}$$

4. REVIEW OF RESULTS

4.1. Main-Group Metals (Groups 1, 2, and 13)

4.1.1. Overview

The main-group metals Na, K, and Al are among the earliest and most often studied sets of clusters, and both experimental and theoretical work has seen sustained progress. Clusters of Na, K, and sometimes Li, Rb, and Cs have been synthesized in seeded free-jet expansions from high-temperature ovens and studied primarily as neutrals. Recently, Coe and co-workers have attached electrons in the jet to form Na, K, Rb, and Cs cluster negative ions.[53] Aluminum clusters have been produced by laser vaporization and by sputtered/trap-cooling methods and studied in all three charge states. More recently, B_N^+ clusters have also been investigated. Group 2 metal clusters have not yet been the subject of N-specific property measurements; this is partly attributable to the expected weak (van der Waals-like) binding of dimers and trimers, which lie on the path to large cluster growth.

The measurement techniques applied to these clusters provide good coverage of those in use on all atomic clusters, and the relatively advanced state of theoretical work here enables a more in-depth discussion of the likely meaning of observed trends.

4.1.2. Thermodynamics; Abundances

The starting point of the most successful approach yet to main-group metal clusters was the 1984–85 experiments on the abundances of Na_N and K_N clusters from seeded free-jet expansions (Figure 6). These were notable because of a series of ledgelike discontinuities in observed abundances. After ruling out effects such as photoionization cross section variations, these observations were associated with an abrupt drop in stability for those clusters immediately *after* the ledges. These effects were then loosely interpreted in terms of cluster stability with respect to atom loss.

The most important aspect of these observations, however, is that the pattern of cluster sizes at the shallow maxima before each abrupt drop found a very simple explanation in terms of valence electron counting. Because this model forms the working hypotheses for many of the subsequent experiments on main-group and noble metals, we recount it briefly here. Valence electrons in simple (so-called free electron) metals are highly delocalized and fill spatially delocalized orbitals. There are zN such electrons in an A_N cluster of valence z ($z = 1$ for Na, 2 for Mg, 3 for Al). If a cluster consisting of more than a few atoms can be taken to be approximately spherical, then the delocalized orbitals are classified according to their angular momentum as $S, P, D, F, \ldots$ orbitals, just as in isolated atoms. The energy level pattern of these orbitals, appropriate to a flat-bottomed cluster potential, is given in Figure 7. Clusters whose total number of valence electrons just fills an electronic shell have unusual electronic stability compared to those just exceeding that number. Hence the stability (and abundances) vary smoothly up to each major shell closing (for Na, at $N = 8, 20, 40, 58, 92$) and then discontinuously beyond.

Naturally, it is appropriate to ask how far such a simple model can be extended: One modification includes the arrangement of atomic cores as a weak (crystal-field-like) potential to lift the spherical degeneracies.[55] This has the effect of reversing the ordering of some of the closely spaced levels so that higher angular momentum states float higher. Another path has been to allow the spherical potential to deform, while remaining continuous.[56]

It is interesting to note that specific electronic computations on alkali metal clusters using a spheroidal potential (modified jellium) model are said to faithfully reproduce abundances using local equilibrium assumptions (Section 3.2),[56] but no such abundance discontinuities are seen for neutral aluminum clusters. This has been attributed to the efficiency of the laser vaporization flow condensation method, in which small energy differences in stability probably do not influence the largely diffusion-limited growth.[57]

This electron counting model, commonly called the electronic shell model, and its physical basis have been the subject of two major articles by Cohen and co-workers.[13]

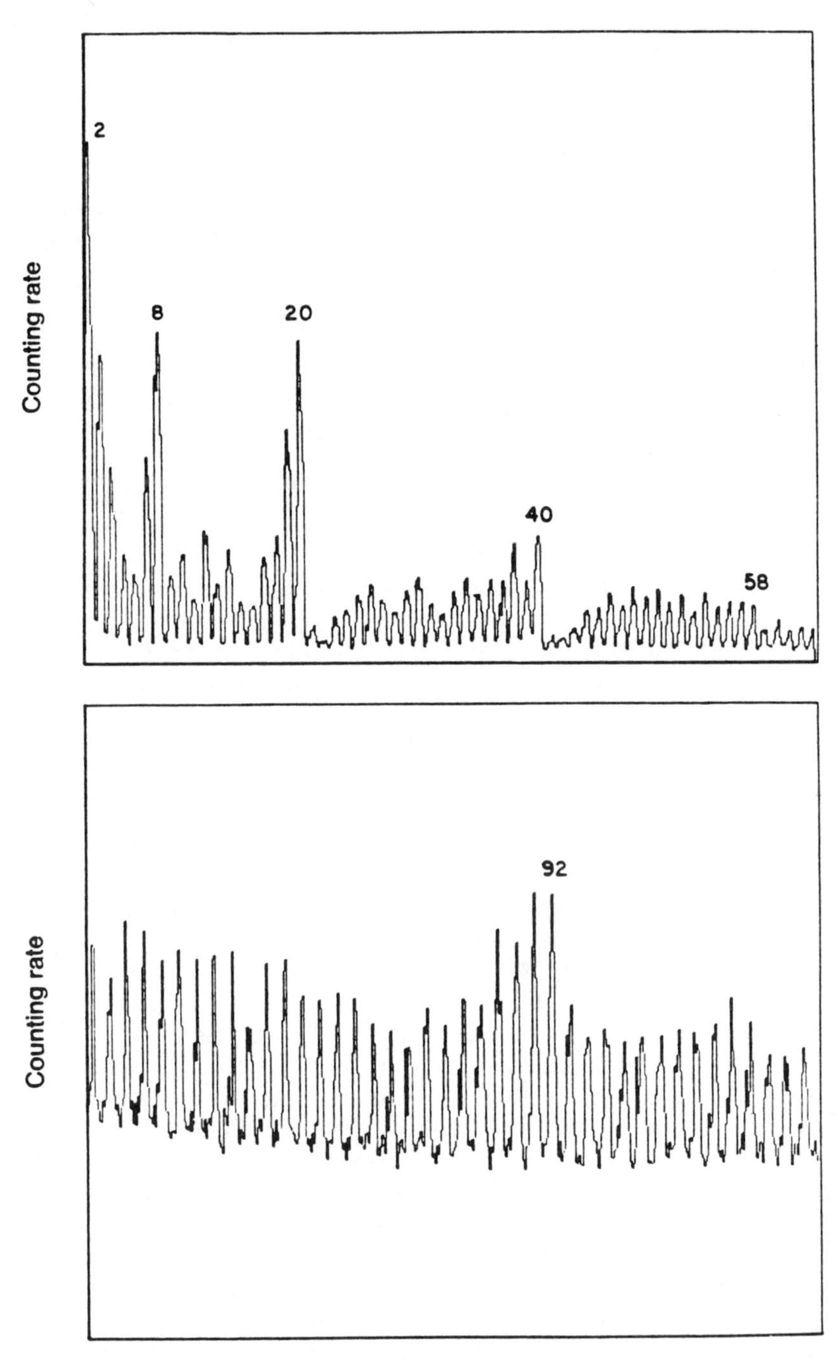

Figure 6. Abundance of Na_N clusters synthesized by free-jet expansion of Na/Ar vapor and detected by photoionization mass spectrometry. See text for discussion. (Adapted from Reference 43.)

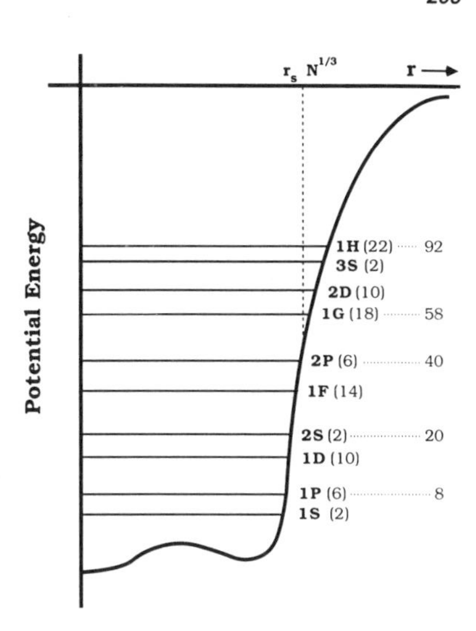

Figure 7. Illustration of the energy levels and effective potential energy felt by a valence electron in a free-electron N-cluster. The atomic cores in the cluster are assumed to provide a spherically symmetric potential so that only the radial potential is provided. The cluster radius is indicated by the dashed vertical line and is given by $r_s N^{1/3}$, where r_s is the Wigner–Seitz radius of the bulk element. At right, the various levels are labeled according to their radial $(1, 2, \ldots)$ and angular $(S, P, D, F, G, \ldots)$ quantum numbers, and the degeneracy of each level is in parentheses. The total electron filling of certain major shells is indicated at far right.

4.1.3. Ionization Potential and Electron Affinities

The ionization potentials of smaller Na_N and K_N clusters have been the subject of repeated study since 1980, and measurements were recently extended to $N = 40$ for Na and $N = 101$ for K.[44,58–60] Ionization potentials of Al_N clusters have been reported very recently by two groups.[61,62] All these measurements were made by determining the photoionization threshold using arc lamp/monochromator or tunable laser systems. Electron affinities of smaller Na_N, K_N, and Rb_N clusters (to $N \approx 8$) were recently determined by photodetachment/photoelectron spectroscopy,[63] while a preliminary report on certain Al_N cluster EAs has recently appeared.[64]

A major motivation for all the more recent IP and EA measurements has been to test the electronic shell model. In principle, the IP is the most direct test, because it measures the orbital energy of the last electron; at shell closings it should therefore drop abruptly. The results from the Berkeley group show such drops at the expected sizes for large K_N clusters (see Figure 8),[58,60] but the discontinuities are three to five times smaller than predicted by the simple electronic shell models. [The underlying smooth drop toward the bulk work function can be accounted for by an entirely classical model.)

In fact, even the results of precise *ab initio* computation on smaller Na_N clusters,[65] when compared to the precise IP determinations of Kappes *et al.*,[44] show fluctuations significantly exceeding those observed in experiment, while faithfully reproducing the overall pattern. These discrepancies have yet to be resolved.

The case of Al_N clusters is, in principle, more favorable to observing large shell closings, because the greater valence electron density of this trivalent metal translates into a larger Fermi energy and greater gaps between shells. The situation is slightly more complicated, however, because the number of valence electrons increments by three, each time a metal atom is added. The

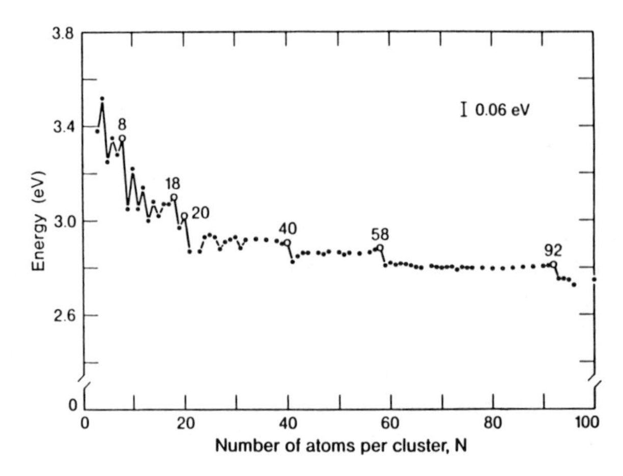

Figure 8. Ionization potential of K_N clusters as a function of N, for $N = 3$–101. Note the odd–even alternations at small N and the major drops following $N = 8, 18, 20, 40, 58$, and 92 (see Figure 7). (Adapted from Reference 58.)

IP is expected to show large drops at $N = 7$ (21 valence electrons), $N = 14$ (42 valence electrons), $N = 20$ and 23 (60 and 69), $N = 31$ (93), and $N = 47$ (149), corresponding to the electrons just filling each new orbital shell. Early measurements on clusters of up to $N = 30$ indicated that the expected gaps appear.[61] In the important case of the $N = 13$–14 IP gap, a value near 0.8 eV is indicated, larger than any other yet observed, and approximately two-thirds the predicted level splitting.

The situation regarding electron affinities is too premature for much analysis yet. Begemann *et al.* state that the Al_N affinities from $N = 2$ to 20 show the expected increase with size,[64] while Bowen *et al.* have stressed the close similarity of the (Na, K, Rb, Cs) series in the $N = 3$–8 region.[63]

4.1.4. Electric Polarizabilities and Magnetic Moments

Electric and magnetic deflection experiments have been carried out on neutral beams of Na, K, and Al clusters. The two alkalis behave very similarly in electric deflection experiments, exhibiting small deflections varying as the square of the electric field. Therefore, the polarizability is extracted, which is in turn divided by N for comparison with atomic and bulk properties. Both metals exhibit an overall slow decrease from atomic toward bulk values, with local minima at expected electronic shell closings.[66] The values at shell closings can be compared to the electronic shell model predictions, and are in poor agreement as to magnitude. The enhancement of the polarizability between shell closings could be interpreted either in terms of electrons in unfilled orbitals or by assuming the cluster deforms (Jahn–Teller deformation) and analyzing it as an ellipsoidal closed-shell species.[43,56] The latter correction is said to yield good agreement with experiment.

Electric deflection of Al_N clusters was carried out to search for permanent dipole moments in the supercooled clusters, but no large deflections were observed in preliminary experiments (in contrast to the findings for Al_NC clusters).[67] Polarizability measurements have not yet been reported.

Magnetic deflection of K_N and Na_N beams were repeatedly attempted, with disappointing results.[68] All odd clusters to $N = 7$ are found to be paramagnetic, as expected, but the deflection magnitude decreases rapidly; this effect was later interpreted in terms of spin-rotation coupling at high rotational quantum number. The higher temperature of clusters produced by free-jet condensation could be responsible for this effect.

Magnetic deflection on Al_N clusters ($N = 1, 2, 6$–20) produced by laser vaporization flow condensation and cooled by expansion yielded surprising results, as a number of the even-N clusters are also paramagnetic.[15] These include Al_2, A_6, Al_8, and probably $Al_{16,18,20}$. In contrast to the case for Fe, large clusters exhibit a decreasing moment/mass ratio, indicating no magnetic ordering. Detailed deflection profiles have not yet been obtained for these clusters.

4.1.5. Optical and Photoelectron Spectroscopy

Optical spectroscopy of main-group metal clusters has been limited largely to the A_3 trimer systems, particularly for the case Na_3, which is probably the best-studied cluster of all.[48,69] This work on trimers belongs more properly to the domain of high-resolution molecular spectroscopy, although some of the themes—isomerism, nonrigidity, and Jahn–Teller instabilities associated with open-shell electronic structure—appear to generalize to large clusters.[43,56] (See also the recent paper on Al_3 by Fu *et al.*[70])

Several papers[59,60,71,72] have emphasized the evolution of the above-IP absorption (photoionization) spectrum of K_N clusters, which exhibits a pronounced odd-N/even-N alternation in ion peak intensities, in agreement with calculations done for sodium clusters.[65]

Photoelectron spectra of A_N^- (A = Na, K, Rb, Cs; $N = 1$–8) have recently been obtained up to 2.5 eV by Bowen *et al.*[63] The beautiful results of this investigation clearly reveal the decrease of the electron affinities and gaps to excited electronic states as one proceeds down the group 1 column, for any particular N.

4.1.6. Chemical Reactivity

The reactions of neutral Al_N clusters have been extensively studied by the flow reactor method at ambient temperatures.[20] Reactivity of Al_N toward H_2 or D_2, D_2O, CO, CH_3OH, and O_2 increased in that order for all N, while CH_4 was much less reactive. The pattern of reactivity as a function of N varies strongly from one reactant to the next, and except for the case of hydrogen no explanation has been given. The reaction $Al_N + D_2 \rightarrow Al_ND_2$ was repeatedly studied because of the striking result that only for $N = 6$ and 7 was significant reactivity observed (Figure 9), while neither the atom nor bulk Al are capable of reacting with hydrogen at low temperatures. Upton[55] performed *ab initio* calculations on the reactive pathways of $Al_N + H_2$ ($N = 2$–6), and found only the reaction of Al_6 to be exothermic, which in turn was analyzed in terms of the stable, 20-electron structure achieved by the product.

The reactions of Al_N+ clusters, and more recently B_N+ clusters, have been investigated under single-collision conditions by ion beam techniques,

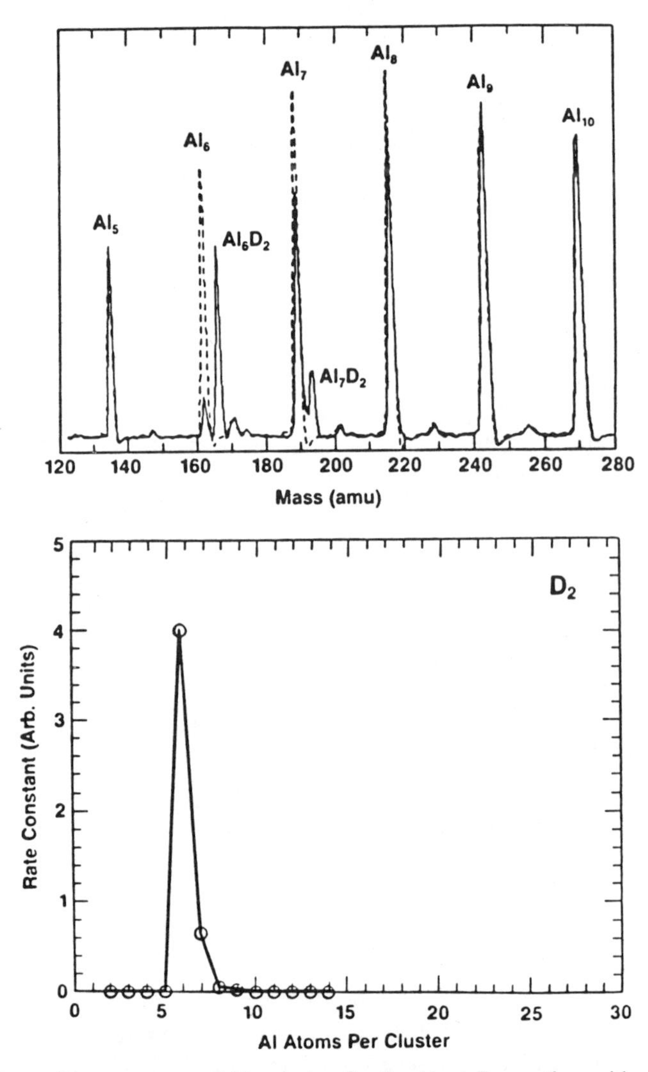

Figure 9. Above: Mass spectrum of Al_N clusters for the $Al_N + D_2$ reaction, with no D_2 in the reactor (dashed lines) and with D_2 flowing (solid line). Note the strong depletion of Al_6 and the concomitant appearance of Al_6D_2. Below: Rate constants for this bimolecular reaction, plotted against *N*. (Adapted from Reference 20.)

in a highly interesting series of experiments.[64,73–78] The unique stability (nonreactivity) of Al_7^+ (20 electrons), Al_{13}^+ and Al_{14}^+ (38 and 41 electrons), and Al_{23}^+ (68 electrons) has already been noted in connection with the shell model and ionization potential measurements, and was confirmed in ion–molecule reactivity studies. The particular advantage of the ion beam method in studying endoergic reactions (where the activation barrier is overcome by the collision energy) has been put to use to find the barriers for several reactions.[73–75]

4.2. Noble Metal Clusters (Groups 11 and 12)

4.2.1. Overview

Although not nearly as well studied as the main-group (1, 2, and 13) metals, there has been a large number of recent gas phase studies on N-specific groups 11 and 12 metal clusters, particularly Hg and Cu clusters. Hg_N beams are synthesized in seeded or neat free-jet expansions at moderate oven temperatures and ionized by electron impact, resonant lines from lamps, or synchrotron radiation, with particular regard to ionization potentials (first and second), and the photoionization spectrum. Cu_N, Ag_N, and Au_N clusters are synthesized by laser vaporization or sputtering, and the first two have been studied as positive ions, negative ions, or neutrals. The recent sputtering experiments of Katakuse and co-workers are particularly revealing in indicating that noble metal cluster ions conform to the magic number anomalies of the electronic shell model, at least in the cases of Cu, Ag, and Au and perhaps the divalent Zn and Cd as well.[79]

The major goal of experiments on Hg_N clusters has been to understand the so-called insulator-to-metal transition that occurs in dense Hg vapor near the gas–liquid critical point.[80] In chemical terms, one expects that this is brought about by a change from van der Waals to metallic bonding, which can in turn be understood in terms of hybridization of the filled s-band with the unoccupied p-band above some critical cluster size.[81] Regarding Cu_N clusters, recent results have been discussed in terms of conformity to the shell model predictions for the case of valence $z = 1$.

4.2.2. Ionization Potentials and Electron Affinities

The ionization potentials of Hg_N clusters have been repeatedly measured in the past decade, culminating in the recent experiments of Rademann *et al.* up to $N \approx 70$.[81] These latest results have been interpreted in terms of a gradual transition from van der Waals to metallic bonding over the $N = 20$–70 range, as evidenced by the drop in the IP from high, atomlike values to values near the predicted work function of a small metal droplet in that range. The IPs of Cu_N clusters have been bracketed,[82] but no other results on noble metal clusters have been reported as of mid-1987.

Electron affinities of Cu_N^- clusters were first reported in 1986 by Smalley and co-workers[83] in the $N = 7$–30 range. The values in the large size range have since been confirmed and extended by the photoelectron spectroscopy method, discussed below.

4.2.3. Optical and Photoelectron Spectroscopy

In contrast to the situation regarding chemical and physical properties, the noble metals and particularly Cu_N and Hg_N systems are among the best studied by optical methods. Bréchignac *et al.*[84,85] have studied the photoionization spectrum of Hg_N up to $N \approx 40$ using synchrotron radiation, with the goal of observing the insulator-to-metal transition in the inner-shell excitation lineshapes. Very recently, success in this endeavor has been claimed,[85] and the detailed correspondence with the IP measurements will be interesting to explore. Earlier data on clusters up to $N = 8$ had shown the $6p$-derived band to be essentially isolated and empty.[84]

The photoelectron spectra of Cu_N^- clusters have been thoroughly investigated by Leopold *et al.* (for $N = 1$–10)[86] and more recently by Pettiette *et al.* (for N = 6–41).[87] The erratic convergence of the electron affinity and the gap between ground and excited states of the neutral has been loosely interpreted in terms of the shell model, but some consideration of deformations to an ellipsoidal shape in the $N = 14$ region was required.

4.2.4. Chemical Reactivity

At present, the only reactivity measurements that have been carried out are on the $Cu_N + CO$ reaction using the flow reactor method.[88] Unsurprisingly, reactions of Cu clusters (and probably those of other noble metals) are much slower than those of the transition metal clusters (toward the same reactants), and indeed they have been used as control experiments for those systems. The interested reader is referred to the recent review of Kaldor *et al.*[89] for more details of these studies.

4.3. Clusters of Transition Metal Atoms

4.3.1. Overview

The area of transition metal clusters has experienced explosive growth starting with the realization that intense, supercooled beams of these clusters could be reliably prepared by the laser vaporization method. The overwhelming technological significance of several of these metals in various applications has generated a sustained high level of interest in experiments on ionization potentials (starting in 1983), chemical reactivity (1984), chemical saturation (1985), cluster ion photodissociation (1985), ion reactivity in traps and flow reactors (1986), and photoelectron spectroscopy (1987). An early review article[90] is now completely outdated, while a later one[91] focused more strongly on the metal dimers. The comprehensive review of Kaldor *et al.*[89] should be consulted for a detailed survey of chemical reactivity.

4.3.2. Ionization Potentials and Electron Affinities

Several experimental papers have appeared on the ionization potentials of transition metal clusters prepared using the laser vaporization flow condensation technique. In some cases the IPs or EAs were first bracketed by using fixed-frequency laser lines, but in the cases of Fe_N, Nb_N, and V_N, complete photoyield versus photon energy curves were obtained for clusters up to about 30 atoms.[17,92–94] A general feature of these curves is that they give sharply defined thresholds for the smaller clusters (to $N \approx 10$) and broader thresholds, necessitating some analysis, for larger clusters.

The pattern of IP variation as N increases has been very important in attempting to understand cluster electronic structure and perhaps even to infer geometrical structures. In general, $\phi_N +$ versus N curves are *not* monotonically decreasing, although the behavior has been crudely fit to the classical $N^{-1/3}$ functional form in order to extrapolate to the bulk work function.[44] The origin of the fluctuations from this smooth behavior has not yet succumbed to theoretical explanations. In V_N they partly take the form of even–odd oscillations (similar to the alkali clusters)[92,93] in Fe_N smooth oscillations appear[94], and in Nb_N sharp discontinuous drops occur at $N = 8, 10, 16$, and 26 (Figure 10).[17] The correlation of these patterns to the corresponding patterns of other physical and chemical properties, particularly chemisorption reactivity, is discussed below.

Some insight into electronic structure and bonding can also be gained by examining how the presence of an added atom (such as O) or molecule (H_2) affects the IP; such experiments have obvious analogies to those valuable in surface and catalysis science. Reference 95 describes the first experiments of this type.

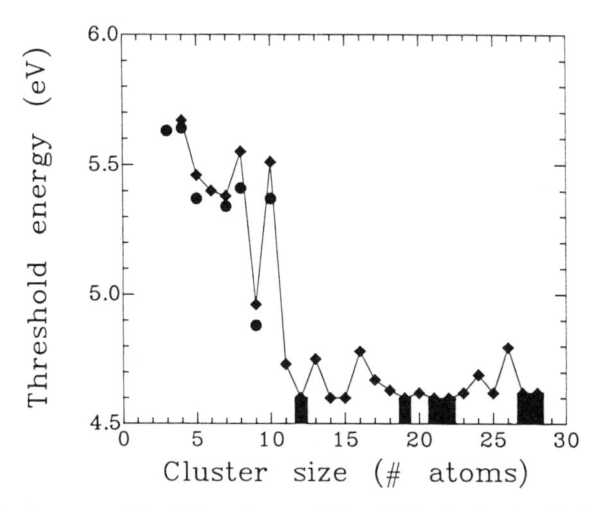

Figure 10. Ionization potential of Nb_N clusters (◆) and Nb_NO clusters (●) in the $N = 2$–28 range. Note the major local maxima at $N = 8, 10, 16$, and 26, corresponding to those clusters most unreactive with hydrogen gas. To $N = 10$ the observed Nb_NO thresholds follow the strong major trend. (Adapted from Reference 17.)

Electron affinity measurements have been carried out very recently on Pd_N and Pt_N clusters up to $N = 4$ using the photoelectron technique, and no summary is available at this time.[96] Measurements of transition metal cluster electronic structure, including IP, EA, and photoelectron spectra, will certainly be crucial in understanding the remarkable chemical and magnetic properties of these systems.

4.3.3. Magnetic Moments

The magnetic properties of several transition metal cluster systems have been subjected to preliminary investigations by the Exxon group using the beam magnetic deflection technique. Interesting results on Fe_N clusters and their oxides have been published.[14] The most important result is that the most probable moment increases approximately linearly with *N*. In the experiment this is strikingly manifested in the way that the spatial deflection (or the deplction of the on-axis beam intensity) is nearly independent of cluster size for $N > 3$. The oxides appear to be even more highly magnetic. Since the time of these experiments, theoretical predictions have multiplied concerning the enhanced magnetism of transition metal clusters of several systems,[28] and new experiments to test these ideas are eagerly awaited.

4.3.4. Chemical Reactivity

The largest number of papers on transition metal clusters concern their chemical reactivity toward various gases as a function of *N*. A well-developed debate exists regarding the nature and convergence of the reactivity toward bulk values, the correlation of reactivity to other electronic structure observables (e.g., IPs) and changes in charge state, and the ability to infer structural information from reactivity patterns and saturation levels. Because much of this work has recently been subjected to comprehensive review, we give here instead a brief outline of the main developments and a survey of the most recent experiments and interpretations.

The flow reactor experiments[18] on the $Nb_N + D_2$ reaction were the first to demonstrate qualitatively that some cluster sizes ($N = 8, 10, 16$) could be completely unreactive while all others react readily. At the same time, Whetten *et al.*[16] and Richtsmeier *et al.*[54] measured the initial rates of $Fe_N + H_2$ over several decades of reactivity. The former group claimed that the reactivity pattern was quantitatively correlated to the previously measured ionization potentials (Figure 11). When similarly quantitative data on $Nb_N + D_2$ or N_2 reactions became available, these were said to be explicable only in terms of geometrical structures.[97] However, subsequent ionization potential measurements, described above, revealed that the major features of the reactivity pattern also appeared in the IP pattern.[17] Namely, the large IP maxima at $N = 8, 10$, and 16 corresponded to unreactive Nb_N clusters, while $N = 26$ turned out to be a smaller local IP maximum and corresponded to the least reactive of any large ($N > 16$) cluster.

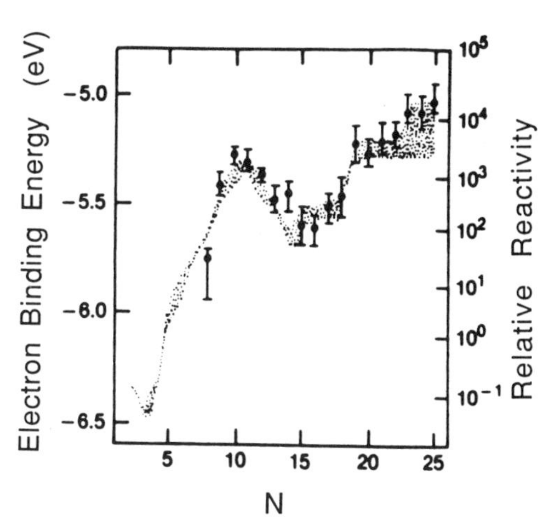

Figure 11. Initial reactivity (gray band) of iron clusters in the $Fe_N + H_2$ reaction compared with measured ionization potentials of Fe_N clusters. Note the logarithmic scale for reactivity and the negative scale for IP. This strong correlation was explained in terms of metal-to-hydrogen charge transfer in the activation of the H_2 bond. (Adapted from Reference 16.)

Because a strong correlation between IP and reactivity patterns can only *support* models based on electron counting or structure-independent wavefunction topology, rather than prove them, an experiment to determine whether structural packing or purely electronic factors dominate is important. Smalley and co-workers initiated this agenda by studying the reactions of positively charged niobium clusters Nb_N+ in flow reactor and ion trap experiments.[98] The retention of reactivity minima at $N = 8$ and 10 indicated that structural effects may dominate in this particular system, since a change in the electron number failed to alter the basic features.

However, Zakin *et al.*[99] responded with experiments on the initial reactivity of Fe_N, V_N, and Nb_N clusters in all three pertinent charge states $(-1, 0, +1)$, and found that typically the reactivity minima undergo a shift in N with change of charge state. In the case of Nb_N, this could be demonstrated only by analysis of the full pattern, and in the course of this work the first convincing evidence for isomers in a metal cluster system was found: Nb_{12} (neutral or positive) has both a reactive and a relatively nonreactive form (Figure 12).† These experiments, which have simple analogue inexperiments on the effects of charging on surface reactions, have thus begun to reach a high state of reliability and analysis.

Many other reactions of transition metal clusters have been reported, particularly for the catalytic metals, and in some cases the attention has been turned to such complex phenomena as coverage-dependent reactivity (i.e., subsequent reaction rates), isotope exchange, spectroscopic signatures of bound molecules, desorption energetics, and so on. The reader is referred to Reference 89 and the original papers for details of these interesting and important investigations, which lie beyond the scope of this survey.

† Similar behavior has also been reported very recently for Nb_{19}^+ reactions in the FTMS.[100]

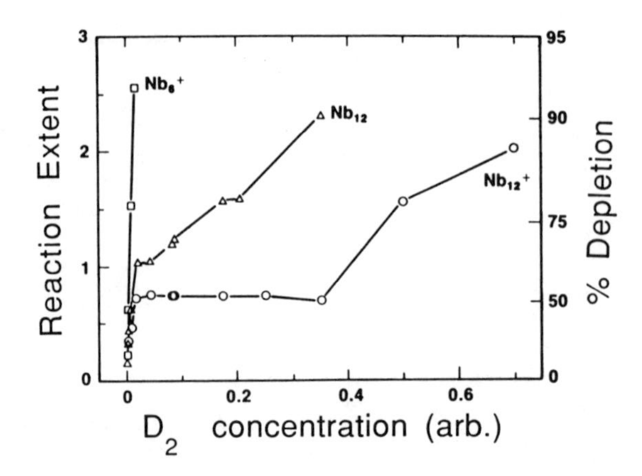

Figure 12. Evidence for two isomeric forms of Nb_{12} in the $Nb_{12} + D_2$ reaction is provided by the middle trace. Note the sharp initial slope (accounting for about half the Nb_{12}) and subsequent shallower slope. The slopes are linearly proportional to the rate constants for the reactions and thus indicate a reactive and a nonreactive form of Nb_{12}. (Adapted from Reference 99.)

4.4. Clusters of Groups 14 and 15 Elements

4.4.1. Overview

Clusters of Groups 14 and 15 elements have also been actively studied, first by ion sputtering or oven-flow methods and later using the laser vaporization methods. Clusters of the atoms C, Si, Ge, and Pb as well as Bi have been studied extensively, and the highlights of experimental and theoretical investigations on the first three have been recently reviewed in some detail.[29,30,101] A full review of this literature, particularly for the dominant case of C_N clusters, is not warranted here. Instead, we emphasize certain highlights from this work.

While interest in Si_N and Ge_N has been stimulated largely because of their role as elemental semiconductors, the extraordinary attention given to C_N derives largely from early studies and speculations on the clusters themselves.

4.4.2. Abundances and Thermochemistry

The striking abundance patterns seen in mass spectra of C_N clusters have attracted an unusual degree of attention to the bonding and structural energetics of charged and neutral carbon clusters. This field originated in early mass spectral observations of the carbon vapor at equilibrium, which led to a flurry of theoretical computations on small carbon molecules. In brief, the following had to be explained:

1. Small negative or positive ion cluster abundances demonstate odd–even alternations.

2. For positive ions in the region $N = 7\text{–}23$, the abundance maxima show a $(4N + 3)$ periodicity.
3. For $N > 32$, both the neutral (via photoionization) and positive ion distributions show an absence of odd-N clusters, while negative ion distributions show them to be of comparable intensity.
4. There are persistent abundance maxima at $N = 60$ and (to a lesser extent) $N = 50$ and $N = 70$.

To date, it has not been possible to model the abundance of large-N clusters in a systematic manner, although the hollow-shell hypothesis has received considerable numerical support. The small clusters ($N < 3$) have been successfully treated in some detail by Bernholc and Phillips, who modeled the kinetics using computed thermochemical data.[30]

4.4.3. Ionization Potentials and Electron Affinities

The higher IPs of clusters of these elements has made measurement of their ionization potentials difficult, although several low-resolution studies have appeared to Bi and Pb.[45,102] Very recently, the photoelectron technique has been used to obtain EAs of C_N, Si_N, and Ge_N clusters,[52,103] and conclusions from this work are just becoming available. Figure 13 offers an example for the smaller C_N clusters.

4.4.4. Optical and Photoelectron Spectroscopy

Optical spectroscopy has been concentrated primarily on carbon clusters, with considerable speculation centered on the proposal that linear carbon chain molecules might be responsible for the interstellar absorption bands.[104] Despite this interest, gas phase work to date has been largely unsuccessful, and the only spectrum ($N > 3$) reported is that of C_{60} weakly bound to a small molecule in the near-ultraviolet region.[105] The presence of this weak band was taken to be consistent with calculations on the proposed icosahedral-shell C_{60} structure.

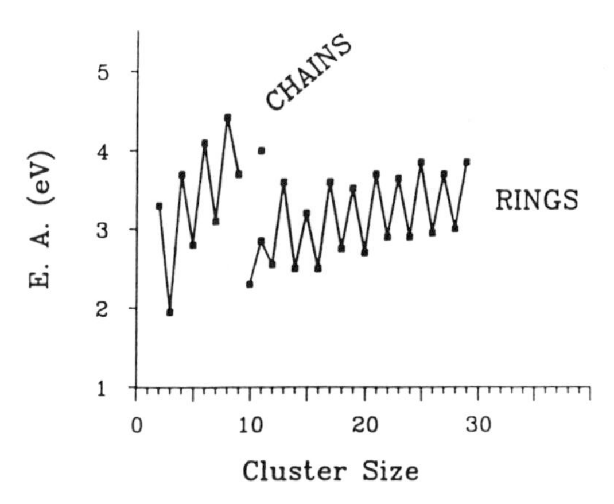

Figure 13. The electron affinity (EA) of C_N clusters as measured by the photoelectron technique applied to cold C_N^- beams. Note the odd-even alternations and the break between $N = 9$ and 10, indicative of the predicted transition from linear chain structures to monocyclic ring structures. (Adapted from Reference 103.)

Quite recently, a large number of results on photoelectron spectroscopic results have appears[51,103,106] for the C_N, Si_N, and Ge_N systems. These reveal the convergence of the electron affinity toward the bulk values and indicate large optical gaps (e.g., ~2 eV for C_{60}) for particular cluster sizes. Full reports on these studies are expected to be available soon.

4.4.5. Chemical Reactivity

The reactivity of C_N clusters has been studied by the flow reactor technique, while C_{N^+}, Si_{N^+},and Ge_{N^+} have been repeatedly investigated in ion traps.[107-10] The reaction of C_{7^+} with H_2 presents a clear example of a case of two isomers, probably a ring (unreactive) structure and a linear chain (reactive) structure.[107] The unreactivity of large clusters has been taken as consistent with the hypothesis of closed, hollow shells of carbon atoms, the most celebrated of which is the C_{60} truncated icosahedron structure.[119] More recently, unusually stable clusters such as Si_{39^+} have been found using similar techniques.[110]

4.5. Rare-Gas Clusters

4.5.1. Overview

The rare-gas clusters (particularly Ar_N, Kr_N, and Xe_N) were the first atomic clusters to be synthesized in (neat) free-jet expansions, and they have been the subject of more theoretical work than all other cluster systems combined. Despite this, almost nothing had been measured experimentally on an N-specific basis for any of the rare-gas cluster systems until the past two years. Now one has precise experimental results on the growth of charged clusters,[111] and measurements of the photodissociation cross sections of Ar_N^+ clusters at several visible laser wavelengths.[112] Also, scattering fragmentation experiments on the smallest Ar_N clusters have been reported,[113] and tunable laser spectroscopy measurements seem not far away. Gradually, it has become possible to overcome the twin difficulties of rare-gas cluster experiments: transparency in all the usual spectral regions and susceptibility to ultrafast fragmentation in the ionization/mass detection process.

The latter feature is inherent in rare-gas bonding, because in the neutral state chemical bonding is absent (only van der Waals interaction holds the atoms together), while in the ionic or excited states a relatively strong single bond can form (dimer ion energies) and ion–induced dipole forces are created. This change in bonding implies a large energization upon ionization, and the (multiple) metastable decay of Ar_N^+ and Xe_N^+ clusters has been much studied.[114-115]

4.5.2. Abundance; Thermodynamics

Theoretical work on rare-gas clusters is simplified by the absence of chemical bonding effects and the existence of good two-body and three-body potentials. Early papers concentrated on the structure/packing question and revealed the now well-established result that short-range potential energy considerations dominate the thermodynamics and stability of these clusters.

In 1981, Echt and co-workers first demonstrated the magic number pattern corresponding to icosahedral packing in Xe_N^+ clusters.[10] Despite minor differences among the Ar, Kr, and Xe series, partly attributable to different time scales of mass measurement, this picture has held up well through several generations of experiment. The pattern of anomalies in beams is firmly attributed to rapid fragmentation after ionization to yield a stable ion, confirmed by growth and photofragmentation experiments. Electron diffraction experiments, by contrast, are performed on beams of neutral clusters, yet show packing which clearly points to the polyicosahedral motif.[116]

4.5.3. Scattering, Fragmentation, and Spectroscopy

Buck and Meyer have studied the elastic scattering of Ar_N clusters by He using momentum transfer analysis and used this information to look at the probability of fragmentation following electron impact ionization.[113] These results have been recently reviewed.[117]

Lineberger and co-workers have investigated the optical absorption/photofragmentation spectrum of Ar_N^+ using cold ions extracted from a pulsed cluster jet intersected by an electron beam.[112] The ions are size selected before absorption of a visible or near-infrared photon (1.5–2.5 eV), and multiple evaporation events result. A most significant result is the large change in the red absorption cross section in the $N = 14$–20 range, indicating a change in the nature of the chromophore, whose identity is as yet unknown.

Since most theoretical studies are carried out on the neutral clusters, including the recent simulations of melting, it is highly desirable to perform spectroscopy directly on the neutrals themselves. Echt *et al.* recently reported nonresonant multiphoton ionization of Xe_N clusters at several ultraviolet and visible wavelengths.[118] Large-scale fragmentation after ionization was inferred from these results, as might be expected in 10^8- to 10^9-W/cm^2 laser fields. Hahn and Whetten were recently able to demonstrate resonant two-photon ionization of a Xe_N cluster jet at $\lambda = 157.6$ nm, at just lower energy than the first absorption in crystal or fluid Xe.[119] With tunable UV sources now becoming available, spectroscopic studies of these clusters are likely to flourish.

5. PROSPECTUS: OUTSTANDING QUESTIONS

5.1. Where Are the Isomers?

This review has considered a large number of specialized experimental results, many of which have in common little more than that the studies were *feasible* to carry out using existing clusters sources! A particularly important omission, at least from a modern chemical view, is the near absence of discussion of the geometrical structure of clusters. Because this question was at one time put in the opposite sense, i.e., it was assumed that experimental results on clusters could be meaningless because of the existence of many coexisting structures, it is more revealing to phrase it so: Where are all the isomers? As described above, experimental data have nearly always revealed just *one* sharp ionization threshold, just *one* chemical reaction rate constant, and so on. The major exceptions are C_7^+ and, more recently, Nb_{12}, Nb_{12}^+, and Nb_{19}.

One can respond to this situation in one of several ways:

1. Contrary to expectations, synthetic methods are producing clusters of just one, presumably lowest-energy, structure, for each size N; to account for the observed behavior one should determine what these specific structures are and rationalize properties in terms of them.
2. There are many structures simultaneously present, but the properties being measured are insensitive to the differences among them. This however appears to be in conflict with much of established complex chemistry and surface chemistry.
3. Under the conditions of these experiments (typically $T < 400$ K and often nearer 100 K), the clusters are structurally nonrigid entities, isomerizing continually with most atoms involved in some kind of generalized pseudorotation motions—one might call these clusters melted. However surprising this last proposition may appear, it has obtained support from experiments showing lowered melting point or spectroscopic patterns of nonrigidity at 0 K, and from the newer *ab initio*/dynamical calculations of Car *et al.* showing that large Na_N clusters effectively melt below 80 K, and Si_N below 300 K.[26]

To date, it has not been possible to disprove these hypotheses in any general way. The search for isomers and, given their existence, the structures underlying their behavior will likely continue to be a major activity for experimentalists and theorists alike.

5.2. The Energetics of Chemical Reactions

A large number of flow reactor experiments were surveyed, and for any given metal these often reveal very strong dependences of the chemical reactivity on cluster size toward at least one small-molecule reagent. However, these experiments were all carried out near ambient temperatures; at lower temperatures otherwise less-selective reagents may yet show similar dependences.

In general, relative or even absolute rate constants at a given temperature are incomplete without temperature-dependent measurements to give an enthalpy and entropy of reaction and the barrier to reaction. Experiments to determine these properties have not yet been successful, and will certainly constitute a major future avenue of research.

In the case of experiments using the ion beam/gas cell techniques, the situation is reversed, because one then obtains directly the reaction cross section as a function of collision energy. The threshold for nonzero values of the reaction cross section occurs when the translational energy is just sufficient to overcome the barrier to reaction. Only recently have threshold measurements been reported.

However, most chemical reasoning and quantum chemistry is still carried out on *neutral* species. In order to connect ion and neutral thermochemistry, it is necessary to measure ionization thresholds. Hence, one will probably see an increased correspondence of physical and chemical measurements on the same clusters, aimed toward a better general understanding of cluster reactivity.

5.3. Back to the Support

Although the study of size-selected cluster properties in the unperturbing environment of the gas phase is not yet a decade old, and has really seen widespread progress only since 1983, it already appears that much of the future interest will be directed toward size-selected clusters deposited on surfaces or in supporting materials. Increasingly often, cluster beam instruments are being redirected toward producing size-selected cluster materials. This occurrence can be interpreted both as a sign of the strength of the gas phase approach to atomic cluster science and as an admission of weakness.

First, cluster properties easily measured in the gas phase are now being studied at a breathtaking rate; the recent introduction of the photoelectron spectroscopy technique for the cluster valence levels completes a well-rounded repertoire of useful molecular beam and cell methods. Thus, the next challenge is to begin to see how well our knowledge carries over into the properties of clusters in a complex environment. Some progress has already been made in this direction, because reactivity and saturation studies already give clues regarding how various size clusters will interact with a support, for example.

Second, many properties, including precise structures, are difficult or extremely tedious to obtain from gas phase experiments and may be better analyzed using long-lived samples on rigid, quasi-inert supports.

Third, while the behavior of isolated atomic clusters remains an enticing fundamental challenge, similar in complexity and richness of pattern to the physics of the atomic nucleus, understanding the presumed properties of clusters embedded in real-world systems constitutes a longer scientific and technical challenge of great magnitude.

ACKNOWLEDGMENTS

This article was greatly aided by the many workers who sent manuscripts prior to publication. We would also like to thank our research group for assistance in gathering and summarizing articles: A.M. Camarena, M.Y. Hahn, E.C. Honea, A.J. Paguia. The work in our group has been generously supported by a Dreyfus DNF Award, a Petroleum Research Fund Grant, Office of Naval Research contract, the University of California Academic Senate, UCLA Solid State Science Center, and an NSF Presidential Young Investigator Award.

REFERENCES

1. Sugano, S.; Nishina, Y.; Ohnishi, S., *Microclusters*; Eds. Springer: New York, 1987.
2. Jena, P.; Rao, B.K.; Khana, S.N., Eds. *Physics and Chemistry of Small Clusters*, NATO-ASI Series No. 158; Plenum: New York, 1987.
3. *J. Phys. Chem.* 1987, *91*(10).
4. *Chem. Rev.* 1986, *86*(3).
5. *Z. Phys. D* 1986, *3*(2/3).
6. *Surf. Sci.* 1985, *156.*
7. *Ber. Bunsenges. Phys. Chem.* 1984, *88*(3).
8. See, for example, Moskovits, M. Ed. *Metal Clusters*; Wiley: New York, 1986; Johnson, B.F.G., Ed. *Transition Metal Clusters*; Wiley: New York, 1980.
9. Muetterties, E.L.; Burch, R.R.; Stolzenberg, A.M. *Annu. Rev. Phys. Chem.* 1982, *33*, 89.
10. Echt, O.; Sattler, K.; Recknagel, E. *Phys. Rev. Lett.* 1981, *47*, 1121.
11. Sattler, K.; Mühlbach, J.; Echt, O.; Pfau, P.; Recknagel, E. *Phys. Rev. Lett.* 1981, *47*, 160.
12. Knight, W.D.; Clemenger, K.; deHeer, W.A.; Saunders, W.A.; Chou, M.Y.; Cohen, M.L. *Phys. Rev. Lett.* 1984, *52*, 2141.
13. Cohen, M.L.; Chou, M.Y.; Knight, W.D.; deHeer, W.A. *J. Phys. Chem.* 1981, *91*, 3141; deHeer, W.A.; Knight, W.P.; Chou, M.Y.; Cohen, M.L. *Solid State Phys.* 1987, *40*, 94.
14. Cox, D.M.; Trevor, D.J.; Whetten, R.L.; Rohlfing, E.A.; Kaldor, A. *Phys. Rev. B* 1985, *32*, 7290.
15. Cox, D.M.; Trevor, D.J.; Whetten, R.L.; Rohlfing, E.A.; Kaldor, A. *J. Chem. Phys.* 1986, *84*, 4651.
16. Whetten, R.L.; Cox, D.M.; Trevor, D.J.; Kaldor, A. *Phys. Rev. Lett.* 1985, *54*, 1494.
17. Whetten, R.L.; Zakin, M.R.; Cox, D.M.; Trevor, D.J.; Kaldor, A. *J. Chem. Phys.* 1987, *85*, 1697.
18. Geusic, M.E.; Morse, M.D.; Smalley, R.E. *J. Chem. Phys.* 1985, *82*, 590.
19. Zhang, Q.L.; O'Brien, S.C.; Heath, J.R.; Liu, Y.; Curl, R.F.; Kroto, H.W.; Smalley, R.E. *J. Phys. Chem.* 1986, *90*, 525.
20. Cox, D.M.; Trevor, D.J.; Whetten, R.L.; Kaldor, A. *J. Phys. Chem.* 1988, *92*, 421.
21. Kubo, R.; Kawabata, A.; Kobayashi, S. *Annu. Rev. Mater. Sci.* 1984, *14*, 49; Halperin, W. P. *Rev. Mod. Phys.* 1986, *58*, 533.
22. Abraham, D.W.; Sattler, K.; Ganz, E.; Mamin, H.J.; Thomson, R.E.; Clarke, J. *Appl. Phys. Lett.* 1986, *49*, 853.
23. Binder, K.; Stauffer, D. In *Topics in Current Physics 7*, "Monte Carlo Methods in Statistical Physics", Binder, K. Ed.; Springer-Verlag: New York, 1979.
24. Koutecký, J.; Fantucci, P. *Chem. Rev.* 1986, *86*, 539.
25. Car, R.; Parinello, M. *Phys. Rev. Lett.* 1985, *55*, 2471.
26. Car, R.; Parinello, M.; Andreoni, W. *Microclusters*; Sugano, S.; Nishina, Y.; Ohnishi, S., Eds.; Springer-Verlag: New York, 1987; Car, R.; Parinello, M.; Andreoni, W. To be published.
27. Farges, J.; DeFeraudy, M.F.; Raoult, B.; Torchet, G. *J. Chem. Phys.* 1983, *78*, 5067; *ibid.* 1986, *84*, 3491.
28. Fu, C.L.; Freeman, A.J.; Oguchi, T. *Phys. Rev. Lett.* 1985, *54*, 2700, and references therein.
29. Brown, W.L.; Freeman, R.R.; Raghavachari, K.; Schlüter M. *Science* 1987, *235*, 860.
30. Bernholc, J.; Phillips, J.C. *J. Chem. Phys.* 1986, *85*, 3258.
31. Bauer, S.H.; Frurip, D.J. *J. Phys. Chem.* 1977, *81*, 1015, and references therein.
32. Dietz, T.G.; Duncan, M.A.; Powers, D.E.; Smalley, R.E. *J. Chem. Phys.* 1981, *74*, 6511; Smalley, R.E. *Laser Chem.* 1983, *2*, 167.
33. Soler, J.M.; Garcia, N.; Echt, O.; Sattler, K.; Recknagel, E. *Phys. Rev. Lett.* 1982, *49*, 1857.
34. Ryali, S.B.; Fenn, J.B. *Ber. Bunsenges. Phys. Chem.* 1984, *88*, 245.
35. LaiHing, K.; Wheeler, R.G.; Wilson, W.L.; Duncan, M.A. *J. Chem. Phys.* 1987, *87*, 3401.
36. Whetten, R.L.; Cox, D.M.; Trevor, D.J.; Kaldor, A. *J. Phys. Chem.* 1985, *89*, 566.
37. Geusic, M.E.; Morse, M.D.; O'Brien, S.C.; Smalley, R.E. *Rev. Sci. Instrum*, 1985, *56*, 2123.
38. See Chapter 9 in this volume.
39. Loh, S.K.; Hales, D.A.; Armentrout, P.B. *Chem. Phys. Lett.* 1986, *129*, 527.
40. Herrman, A.; Schumacher, E.; Wöste, L. *J. Chem. Phys.* 1978, *68*, 2327.
41. Bahat, D.; Cheshnovsky, O.; Even, U.; Lavie, N; Magen, Y. *J. Phys. Chem.* 1987, *91*, 2460, and references therein.

42. Doll, J.D.; Freeman, D.L. *Science* 1986, *234*, 1356.
43. Clemenger, K. Ph.D. thesis, University of California, Berkeley, 1984, (unpublished). See Reference 12 also.
44. Kappes, M.M.; Schär, M.; Radi, P.; Schumacher, E. *J. Chem. Phys.* 1986, *84*, 1863.
45. Walstedt, R.E.; Bell, R.F. *J. Chem. Phys.* 1987, *87*, 1423.
46. Ramsey, N.F. *Molecular Beams*; Oxford University Press: Oxford, 1956.
47. Morse, M.D.; Hopkins, J.B.; Langridge-Smith, P.R.R.; Smalley, R.E. *J. Chem. Phys.* 1983, *79*, 5316.
48. Delacrétaz, G.; Wöste, L. *Surf. Sci.* 1985, *156*, 770; Broyer, M.; Delacrétaz, G.; Labastie, P.; Whetten, R.L.; Wolf, J.P.; Wöste, L. *Z. Phys. D* 1986, *3*, 131.
49. Alexander, M.L.; Johnson, M.A.; Levinger, N.E.; Lineberger, W.C. *Phys. Rev. Lett.* 1986, *57*, 976.
50. deHeer, W.A.; Selby, K.; Kresin, V.; Masui, S.; Vollmer, M.; Châtelain, A.; Knight, W.D. *Phys. Rev. Lett.* 1987, *59*, 1805.
51. Leopold, D.G., Murray, K.K.; Miller, A.E.S.; Lineberger, W.C. *J. Chem. Phys.* 1985, *83*, 4849
52. Cheshnovsky, O.; Yang, S.H.; Pettiette, C.L.; Craycraft, M.J.; Liu, Y.; Smalley, R.E. *Chem. Phys. Lett.* 1987, *138*, 119; *Rev. Sci. Instrum.* 1987, *58*, 2131.
53. Coe, J.V.; Snodgrass, J.T.; Freidhoff, C.B.; McHugh, K.M.; Bowen, K.H. *J. Chem. Phys.* 1986, *84*, 618.
54. Richtsmeier, S.C.; Parks, E.K.; Liu, K.; Pobo, L.G.; Riley, S.J. *J. Chem. Phys.* 1985, *82*, 3659.
55. Upton, T.H.; *Phys. Rev. Lett.* 1986, *56*, 2168; *J. Chem. Phys.* 1987, *86*, 7054.
56. Clemenger, K. *Phys. Rev. B* 1985, *32*, 1359.
57. See, for example, Phillips, J.C. *Chem. Rev.* 1986, *86*, 619.
58. Saunders, W.A. Ph.D. thesis, University of California, Berkeley, 1986, (unpublished).
59. Knight, W.D.; deHeer, W.A.; Saunders, W.A. *Z. Phys. D* 1986, *3*, 109.
60. Saunders, W.A.; Clemenger, K.; deHeer, W.A.; Knight, W.D. *Phys. Rev. B* 1985, *32*, 1366.
61. Schriver, K.E.; Persson, J.L.; Honea, E.C.; Whetten, R.L. To be submitted.
62. Fuke, K.; Nonose, Sh.; Kikuchi, N.; Kaya, K. *Chem. Phys. Lett.* 1988, *147*, 479.
63. Bowen, K.H.; Coe, J.V.; Snodgrass, J.T.; Freidhoff, C.B.; Hugh, K.M.; Eaton, J.; Lee, G. In preparation.
64. Begemann, W.; Dreihöfer, S.; Ganteför, G.; Siekmann, H.R.; Meiwes-Broer, K.-H.; Lutz, H.O. In *Elemental and Molecular Clusters*; Martin, T.P., Ed.; Springer-Verlag: New York, 1987.
65. Martins, J.L.; Buttet, J.; Car, R. *Phys. Rev. B* 1985, *31*, 1804.
66. Knight, W.D.; Clemenger, K.; deHeer, W.A.; Saunders, W.A. *Phys. Rev. B* 1985, *31*, 2539.
67. Schriver, K.E.; Whetten, R.L. In preparation.
68. Knight, W.D.; Monot, R.; Dietz, E.R.; George, A.R. *Phys. Rev. Lett.* 1978, *40*, 1324; Knight, W.D. *Surf. Sci.* 1981, *106*, 172.
69. Broyer, M.; Delacrétaz, G.; Labastie, P.; Wolf, J.P.; Wöste, L. *Phys. Rev. Lett.* 1986, *57*, 1851; *J. Phys. Chem.* 1987, *91*, 2626.
70. Fu, Z.; Lemire, G.W.; Hamrick, Y.M.; Taylor, S.; Shui, J.-C.; Morse, M.D. *J. Chem. Phys.* 1988, *88*, 3524.
71. Bréchignac, C.; Cahouzac, Ph. *Z. Phys. D* 1986, *3*, 121; *Chem. Phys. Lett.* 1985, *117*, 365.
72. Bréchignac, C.; Cahouzac, Ph.; Roux, J. Ph. *J. Chem. Phys.* 1987, *87*, 229.
73. Jarrold, M.F.; Bower, J.E. *J. Chem. Phys.* 1986, *85*, 5373; *ibid.* 1987, *87*, 5728.
74. Jarrold, M.F.; Bower, J.E. *J. Am. Chem. Soc.* 1988, *110*, 70.
75. Hanley, L.; Anderson, S.L. *Chem. Phys. Lett.* 1986, *129*, 429; Ruatta, S.A.; Hanley, L.; Anderson, S.L. *ibid.* 1987, *137*, 5.
76. Hanley, L.; Ruatta, S.A.; Anderson, S.L. *J. Chem. Phys.* 1987, *87*, 260.
77. Jarrold, M.F.; Bower, J.E.; Kraus, J.S. *J. Chem. Phys.* 1987, *86*, 3876; Jarrold, M.F.; Bower, J.E. *J. Chem. Phys.* 1987, *87*, 1610.
78. Hanley, L.; Anderson, S.L. *J. Phys. Chem.* 1987, *91*, 5161.
79. Katakuse, I.; Ichihara, T.; Fujita, Y.; Matsuo, T.; Sakurai, T.; Matsuda, H. *Int. J. Mass Spectrom. Ion Proc.* 1985, *67*, 229; *ibid.* 1986, *74*, 33; Katakuse, I. In *Microclusters*; Sugano, S.; Nishini, Y.; Ohnishi, S., Eds.; Springer-Verlag: New York, 1987.
80. Turkevich, L.A.; Cohen, M.H. *Phys. Rev. Lett.* 1984, *53*, 2323.

81. Rademann, K.; Kaiser, B.; Even, U.; Hensel, F. *Phys. Rev. Lett.* 1987, *59*, 2319.
82. Powers, D.E.; Hansen, S.G.; Geusic, M.E.; Michalopoulos, D.L.; Smalley, R.E. *J. Chem. Phys.* 1983, *78*, 2866.
83. Zheng, L.-S.; Karner, C.M.; Brucat, P.J.; Yang, S.H.; Pettiette, C.L.; Craycraft, M.J.; Smalley, R.E. *J. Chem. Phys.* 1986, *85*, 1681.
84. Bréchignac, C.; Broyer, M.; Cahuzac, Ph.; Delacrétaz, G.; Labastie, P.; Wöste, L. *Chem. Phys. Lett.* 1985, *120*, 559.
85. Bréchignac, C.; Broyer, M.; Cahuzac, Ph.; Delacrétaz, G.; Labastie, P.; Wolf, J.P.; Wöste, L. *Phys. Rev. Lett.* 1988, *60*, 275.
86. Leopold, D.G.; Ho, J.; Lineberger, W.C. *J. Chem. Phys.* 1987, *86*, 1715.
87. Pettiette, C.L.; Yang, S.H.; Craycraft, M.J.; Conceicao, J.; Laaksonen, R.T.; Cheshnovsky, O.; Smalley, R.E. *J. Chem. Phys.* 1988, *88*, 5377.
88. Cox, D.M.; Reichmann, K.C.; Trevor, D.J.; Kaldor, A. *J. Chem. Phys.* 1988, *88*, 111.
89. Kaldor, A.; Cox, D.M.; Zakin, M.R. *Adv. Chem. Phys.* (in press).
90. Whetten, R.L.; Cox, D.M.; Trevor, D.J.; Kaldor, A. *Surf. Sci.* 1985, *156*, 8.
91. Morse, M.D. *Chem. Rev.* 1986, *86*, 1049.
92. Cox, D.M.; Whetten, R.L.; Zakin, M.R.; Trevor, D.J.; Reichmann, K.C.; Kaldor, A. In *Advances in Laser Science-I, Optical Science and Engineering Series, 6*; Stwalley, W.C.; Lapp, M., Eds.; American Institute of Physics: New York, 1986.
93. Whetten, R.L.; Cox, D.M.; Zakin, M.R.; Trevor, D.J.; Kaldor, A. To be submitted.
94. Rohlfing, E.A.; Cox, D.M.; Kaldor, A.; Johnson, K.H. *J. Chem. Phys.* 1984, *81*, 3846.
95. Zakin, M.R.; Cox, D.M.; Whetten, R.L.; Trevor, D.J.; Kaldor, A. *Chem. Phys. Lett.* 1987, *135*, 223.
96. Lineberger, W.C. Private communication.
97. Morse, M.D.; Geusic, M.E.; Heath, J.R.; Smalley, R.E. *J. Chem. Phys.* 1985, *83*, 2293.
98. Brucat, P.J.; Pettiette, C.L.; Yang, S.; Zheng, L.-S.; Craycraft, M.J.; Smalley, R.E. *J. Chem. Phys.* 1986, *85*, 4747.
99. Zakin, M.R.; Brickman, R.O.; Cox, D.M.; Kaldor, A. *J. Chem. Phys.* 1988, *88*, 3555; *ibid.* 1988, *88*, 6605.
100. Elkind, J.L.; Weiss, F.D.; Alford, J.M.; Laaksonen, R.T.; Smalley, R.E. *J. Chem. Phys.* 1988, *88*, 5215.
101. Smalley, R.E.; *Acc. Chem. Res.* (in press).
102. Saito, Y.; Yamauchi, K.; Mihama, K.; Noda, T. *Jpn. J. Appl. Phys.* 1982, *21*, *L*396.
103. Yang, S.; Taylor, K.J.; Craycraft, M.J.; Conceicao, J.; Pettiette, C.L.; Cheshnovsky, O.; Smalley, R.E. *Chem. Phys. Lett.* 1988, *144*, 431.
104. Douglas, A.E. *Nature* 1977, *269*, 130.
105. Heath, J.R.; Curl, R.F.; Smalley, R.E.; *J. Chem. Phys.* 1987, *87*, 4236.
106. Yang, S.; Pettiette, C.L.; Conceicao, J.; Cheshnovsky, O.; Smalley, R.E. *Chem. Phys. Lett.* 1987, *139*, 233.
107. McElvaney, S.W.; Creasy, W.R.; O'Keefe, A. *J. Chem. Phys.* 1986, *85*, 632; McElvaney, S.W.; Dunlap, B.I.; O'Keefe, A. *J. Chem. Phys.* 1986, *86*, 715.
108. See, for example, Reents, W.D.; Mandich, M.L.; Bondybey, V.E. *Chem. Phys. Lett.* 1986, *131*, 1.
109. Reents, W.D.; Mujsce, A.M.; Bondybey, V.E.; Mandich, M.L. *J. Chem. Phys.* 1987, *86*, 5568.
110. Elkind, J.L.; Alford, J.M.; Weiss, F.D.; Laaksonen, R.T.; Smalley, R.E. *J. Chem. Phys.* 1987, *87*, 2397.
111. Harris, T.A.; Kidwell, R.S.; Northby, J.A. *Phys. Rev. Lett.* 1984, *53*, 2390.
112. Levinger, N.E.; Ray, D.; Murray, K.K.; Mullin, A.S.; Schulz, C.P.; Lineberger, W.C. *J. Chem. Phys.* 1988, *89*, 71; *idem.* To be published.
113. Buck, U.; Meyer, H. *J. Chem. Phys.* 1986, *84*, 4854.
114. Märk, T.D.; Scheier, P.; Leiter, K.; Ritter, W.; Stephen, K.; Stamotovic, A. *Int. J. Mass Spectrom. Ion Proc.* 1986, *74*, 281.
115. Kreisle, D.; Echt, O.; Knapp, M.; Recknagel, E. *Phys. Rev. A* 1986, *33*, 768.
116. Lee, J.W.; Stein, G.D. *J. Phys. Chem.* 1987, *91*, 2450.
117. Buck, U. *J. Phys. Chem.* 1988, *92*, 1023.
118. Echt, O.; Cook, M.C.; Castleman, A.W., Jr. *Chem. Phys. Lett.* 1987, *135*, 229.
119. Hahn, M.Y.; Whetten, R.W. Unpublished.

7

Time-Resolved Kinetics of Organometallic Reactions in the Gas Phase by Transient Infrared Absorption Spectrometry

Bruce H. Weiller and Edward R. Grant

1. INTRODUCTION

Chemical change is a science of intermediates. Reactive molecules and fragments guide the chemistry of multicomponent systems and determine the stereospecific outcomes of individual reactions. Organometallic chemistry is among the richest of all for its transient species. Metal-centered intermediates variously appear and disappear via oxidative addition, insertion, metallocyclization, and their reverse reactions.[1] Progress in stoichiometric synthesis as well as catalysis depends on the accurate knowledge of the intermediate forms at work in these and other organometallic transformations.

Despite their importance, organometallic intermediates are among the most rarely observed. They are quite often highly reactive and exist only in infinitesimal concentration in steady-state and catalytic sequences. In solution, coordinatively unsaturated species interact significantly with virtually all solvents including, for example, liquefied rare gases.[2] Solution phase characterization of uncomplexed intermediates is therefore beyond the reach of even the fastest spectroscopic methods.[3]

Gas phase approaches to the problem of chemical transients do not suffer this limitation. Indeed, the detection and characterization of reactive intermedi-

---

Bruce H. Weiller and Edward R. Grant • Department of Chemistry, Purdue University, West Lafayette, Indiana 47907. *Present address for B.H.W.:* Department of Chemistry, University of California, Berkeley, California 94720.

ates has been a cornerstone of fundamental knowledge in main-group chemical kinetics for more than 50 years.[4] As exemplified by this and accompanying chapters in this volume, a small but growing body of comparable data now exists for gas phase organometallic intermediates.

Some of the earliest of this work on gaseous organometallics[5a] extended simple flash photolysis/UV-visible absorption methods that had been extensively developed for solution phase systems. By incorporating laser sources, such techniques achieve excellent sensitivity and time resolution.[5b,c] However, the UV-visible absorption spectra of organometallics are generally too broad and featureless to characterize structures or definitively resolve contributions from spectrally similar compounds.[6]

Infrared spectroscopy is free of these limitations. For carbonyl compounds in particular, which exhibit strong CO stretching transitions in the region from 2100 to 1900 cm^{-1}, the number and relative intensities of the bands observed establish the group symmetry of the CO ligands.[7] Applicability to intermediates is well illustrated by the wealth of information provided by the infrared spectroscopy of organometallic transients captured in matrix isolation.[8] This latter work in particular has spurred the refinement of infrared methods for detection and identification of organometallic fragments. It has also provided a base of information critical to the assignment of infrared spectroscopic transients observed under other conditions. Despite this broad scope and high spectroscopic resolution, however, matrix isolation techniques offer only a limited view of mechanistic pathways for organometallic transformations. Chemistry is restricted to that which occurs at low temperature, so that very often only a static picture is obtained, while the solid environment of the matrix makes it difficult to connect resolved intermediates with their kinetic function.

The most complete information, structural and kinetic, on chemically important intermediates is obtained by transient infrared spectroscopy on reactive gas phase systems. Progress along these lines, however, has been comparatively slow. Gas phase organometallic reaction systems place demands on detection sensitivity and bandwidth that are much greater than those presented by solution or matrix environments. For example, recombination reactions of many organometallic fragments proceed with cross sections approaching those of hard-sphere collisions. The same reactions in solution, limited by diffusion, are hundreds of times slower. Diffusion control of very fast competitive nucleation also allows the production of much higher densities of liquid phase, solvated, coordinatively unsaturated species than can be maintained in gas phase systems. Fortunately, recent technological advances in infrared sources and detectors have increased both sensitivity and time resolution of gas phase transient infrared experiments. As a result, fundamental data of great importance on the elementary properties of a number of key organometallic systems are now emerging.

A motivating factor in much of this work is the rich photochemistry and unique photocatalytic behavior for which organometallic complexes are well known. The photochemistry of organometallics has been reviewed by Geoffroy and Wrighton[6] and Koerner von Gustorf *et al.*,[9] while recent progress in photocatalysis is described in articles by Moggi *et al.*[10] and by Wrighton and co-workers.[11] As noted above, time-resolved gas phase IR spectroscopy is also stimulated and supported by results from matrix isolation experiments. Representative work has been recently reviewed by Hitam *et al.*[8a] and Perutz.[8b] Work through 1985 on transient infrared spectroscopy of organometallic intermediates under dynamic conditions in gas and solution phase systems has been reviewed by Poliakoff and Weitz.[12] In the present article we give an overview of methods in gas phase organometallic transient infrared spectroscopy and review work since 1985. We conclude by giving a detailed example which illustrates the information content of gas phase transient infrared absorption experiments, by drawing from recent work in our own laboratory on the gas phase CO-for-C_2H_4 dissociative substitution kinetics of $Cr(CO)_4(C_2H_4)_2$.

2. METHODS FOR THE DETECTION AND INFRARED SPECTRAL CHARACTERIZATION OF ORGANOMETALLIC INTERMEDIATES IN THE GAS PHASE

Experiments employing time-resolved infrared absorption spectrometry generally seek to record the evolving infrared spectrum of a sample as a function of time, following pulsed photogeneration of transient reactants. Studies to date have probed almost exclusively the intense and characteristic 5-μm CO stretching bands associated with organometallic carbonyl complexes. Thus, our discussion focuses on photoinitiated processes of carbonyls probed at 5 μm.

In schematic terms the necessary equipment includes: (1) a pulsed UV light source for time-resolved photogeneration of transients; (2) a tunable source of infrared light in the region of 5 μm; (3) a sensitive, high-bandwidth infrared detector; and (4) electronics for acquiring and averaging the time-resolved signal. Let us briefly consider some of the important aspects of each of these elements in turn.

2.1. Pulsed UV Photolysis Sources

The time resolution in these experiments can be no shorter than the initial temporal distribution of transient, so short UV pulses are desired. Broadband flashlamps are adequate for certain systems but are limited in pulsewidth to be no shorter than about 10^{-6} s. Unwanted UV wavelengths can be problematic, and high optical divergence limits the UV pathlength. Pulsed UV lasers are far superior because: (1) they have pulsewidths that are much shorter than those of conventional sources; (2) they are monochromatic; and (3) they have low divergence. The excimer laser is the most useful of these as a photolysis source because its output with various discharge gas mixtures spans the UV region (351, 308, 248, 222, and 193 nm). Moreover, such devices are easy to operate and maintain and can deliver within 10 ms as much as 500 mJ per pulse.

2.2. Infrared Sources

In transient infrared absorption spectroscopy, the characteristics of the source most often determine sensitivity. For absorption experiments the signal-to-noise ratio (S/N) is proportional to the wavelength-resolved intensity of the probe source. The relatively low intensity of conventional IR sources, typically 100 times weaker then their UV-visible counterparts,[13] has been the factor most limiting the use of infrared radiation for transient absorption spectroscopy. The high infrared extinction coefficients of metal carbonyls (10^3 to 10^4 M^{-1} cm^{-1})[14] make these experiments possible with conventional IR sources.

In the wavelength range of interest about 5 μm, one can choose from among several incoherent sources as well as several coherent (laser) sources. Conventional incoherent sources are thermal emitters, variously approaching ideal blackbodies. Globars and Nernst glowers have reasonably true blackbody character with color temperatures ranging from 1000 to 2000°C.[15] Low emissivity at 5 μm works against other, more powerful but less ideal, sources such as tunsten filaments,[16] carbon arcs,[17] Xe arc lamps,[18] and pulsed laser-generated plasmas.[19] When wavelength resolved with a monochromator, broadband incoherent sources provide stable, continuously tunable infrared light that is very convenient for spectroscopy. However, to achieve probe intensity sufficient for a good signal-to-noise ratio, spectral resolution and temporal bandwidth must often be compromised. For example, our apparatus, which employs a 2000°C Nernst glower to produce 5 μW of intensity, resolved to 8 cm^{-1}, gives a single shot signal-to-noise ratio of 5 with a bandwidth of 50 kHz on a 0.30% absorption. Each of these factors is interdependent so that with higher transient density and lower spectral resolution, higher temporal resolution can be obtained with the same signal-to-noise ratio.

Lasers offer distinct advantages for transient infrared absorption spectrometry. Laser sources are highly monochromatic and therefore provide high spectroscopic resolution. The CO laser, with a linewidth of 0.001 cm^{-1}, is the source in widest use, though it is only line tunable in 4-cm^{-1} steps. Diode lasers, which are continuously tunable and have even narrower spectral bandwidth, are a potentially useful alternative.

Intensity and resolution gains possible with laser sources are substantial. For example, a CO laser gives at least 200 times more usable power within a spectral bandwidth more than 1000 times narrower than a Nernst glower/monochromator combination.[20] This high intensity of usable light greatly increases the temporal bandwidth within which transient waveforms can be captured. With CO lasers, time resolution has been pushed to the limiting bandwidth of the fastest InSb detectors, about 30 MHz.

As a third advantage, IR laser sources can be configured to have a low-divergence beam. A well-collimated probe enables the use of a long-pathlength cell, which increases the signal-to-noise ratio. Cells as long as 1 m have been used with CO lasers.[21] The pathlength obtainable with blackbody sources is limited by the laws of geometric optics, which dictate a compromise between light collection efficiency and beam quality (divergence) for large sources. In practice, pathlengths are limited to a few centimeters.

Despite their apparent great utility, laser sources also have disadvantages which should not be overlooked. Acquisition and maintenance costs vary, but lasers are generally much more expensive than incoherent sources. To operate on the $v = 1 \rightarrow 0$ transition in the region of $2000\ cm^{-1}$, CO lasers must be cooled to 77 K. At this low temperature, O_3, generated by the discharge, condenses.[20,22] Associated safety hazards have prevented the commercial manufacture of such lasers and so they must be home-built. Even with cooling, the intensity varies greatly amongst CO laser lines, and for meaningful spectroscopic band shapes, the IR signal must be normalized. Diode lasers are available commercially, but they are expensive and each diode has only a very limited tuning range (typically less than $10\ cm^{-1}$).[23] Thus, the choice of an infrared source requires a careful weighing of the various advantages and disadvantages of each against the requirements of the particular experiment at hand.

2.3. *Infrared Detectors and Signal Processing Electronics*

The infrared detector is a central facilitating component of contemporary transient IR absorption experiments. Techniques once very difficult have now been made routine by recent advances in semiconductor detector technology. The detector most widely used for experiments in the carbonyl stretching region is photovoltaic InSb. It is highly sensitive at $5\ \mu m$, but is strongly wavelength dependent, falling sharply in response for wavelengths longer than $5.5\ \mu m$. HgCdTe is an alternate choice that compromises sensitivity for a wider wavelength range. Both of these detectors must be cooled to 77 K for good signal-to-noise and can have response times in the 30- to 100-ns range.[24]

In order to achieve the sensitivity required to detect small photoinduced changes in transmitted infrared intensity, the detector signal is capacitively coupled to a high-gain amplifier. This ac coupling brings the large dc component to zero, so that the informative change in the signal intensity can be selectively amplified. The signal can also be low-pass filtered before amplification for high-frequency noise reduction. The signals are usually captured by a transient digitizer which is normally interfaced to a lab computer for averaging and storage.

The sections to follow review recent results on IR detection of transient gaseous organometallics including results from our laboratory on the dissociative substitution of $Cr(CO)_4(C_2H_4)_2$. The basic apparatus is similar for all laboratories. Thus, we illustrate further details by making specific reference to our experiment.

2.4. Experimental Setup for Gas Phase Transient Infrared Absorption Using an Incoherent Source

The experimental apparatus in operation in our laboratory is diagramed in Figure 1. A pulsed UV excimer laser beam crosses an incoherent infrared beam in a static gas cell. A cylindrical lens ensures that the UV beam fills the 5-cm-diameter cell. The infrared source is a 2000°C Nernst glower driven at 1 kHz to prevent thermal fluctuations in infrared intensity. The IR beam is collimated by an off-axis parabolic mirror and focused by a CaF_2 lens through the cell and onto the slits of a 0.3-m monochromator. The wavelength-resolved IR beam is focused onto a 77 K InSb detector. A 4.5-μm long-wave pass filter removes high-order grating reflections. The detector, its matched preamplifier, and a low-pass filter are all enclosed in a radio frequency shield. The photovoltaic signal is passed through a high-gain ac-coupled amplifier and then to an 8-bit digital oscilloscope for digitizing and averaging. High-resolution, single-beam IR spectra of the sample before and after irradiation are obtained by use of a chopper, lock-in amplifier, and chart recorder.

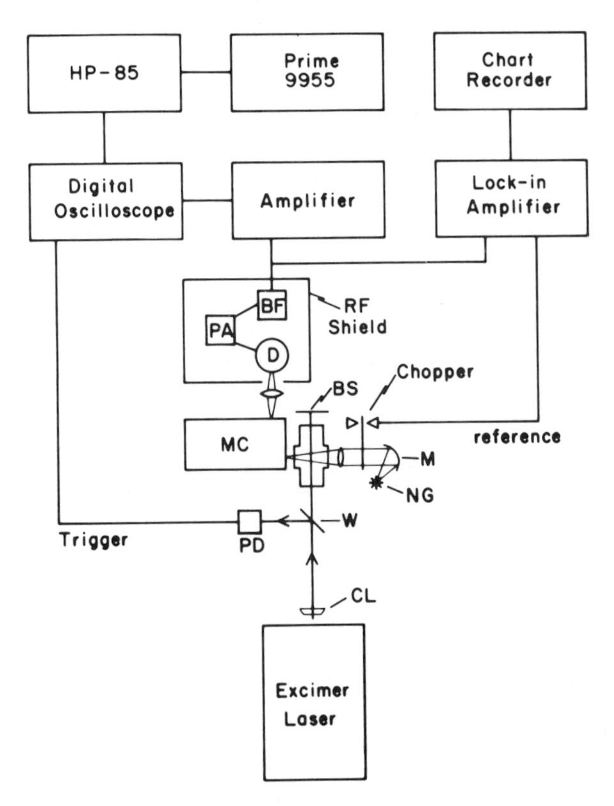

Figure 1. Experimental apparatus with an incoherent source. CL = cylindrical lens, W = Pyrex flat, NG = Nernst glower, PD = photodiode, MC = monochromator, BS = beam splitter, M = off-axis parabolic mirror, D = InSb detector, PA = preamplifier, BF = low-pass filter. See text for details.

The bandwidth and wavelength resolution limits of the apparatus are 1 MHz and 0.5 cm^{-1}, though better signal-to-noise is achieved at 30 kHz and 8-cm^{-1} resolution. The active infrared pathlength is 3.3 cm. The laser wavelength, pulse width, and pulse energy are 351 nm, $\simeq$10 ns, and 170 mJ, respectively. Thermal lens effects often observed in gas phase, time-resolved absorption spectrometry are not found in this configuration because the infrared beam is much larger than the monochromator entrance slit.

Regardless of the details of the experimental arrangement, the basic objective of gas phase organometallic transient infrared absorption experiments has been the same—to record spectra and characterize relaxation kinetics of photogenerated organometallic intermediates. Attention has focused on relatively few systems. Progress has been rapid, however, yielding abundant new information on elementary kinetics and dynamics for these complex and important systems. The section to follow reviews work since 1985. In each case an excimer laser has been used to produce short-lived transition metal complexes and a CO laser or Nernst glower has been employed to probe their fast kinetics. Results are organized by parent compound. We then conclude with a more detailed account of work from our laboratory, employing the methods just described to characterize the elementary kinetics of the reactions of $Cr(CO)_4(C_2H_4)_2$.

3. REVIEW OF RECENT PROGRESS IN GAS PHASE ORGANOMETALLIC TRANSIENT INFRARED ABSORPTION SPECTROSCOPY

3.1. $Fe(CO)_5$

The gas phase photochemistry of $Fe(CO)_5$ was one of the first organometallics to be investigated by time-resolved infrared absorption spectroscopy.[20,25,26] Early experiments made the first direct spectral observation of unsolvated $Fe(CO)_4$ and measured the rate constant for

$$Fe(CO)_4 + CO \rightarrow Fe(CO)_5$$

More recent work[27] has focused on $Fe(CO)_3$ and $Fe(CO)_2$, which are the photoproducts that dominate the gas phase photochemistry of $Fe(CO)_5$ at shorter wavelengths (e.g., $\leq$249 nm). Multiply unsaturated fragments such as these are unique to gas phase photolysis. Rapid collisional quenching and cage effects preclude such species in solution.

Rate constants for the recombination reactions of $Fe(CO)_3$ and $Fe(CO)_2$ with CO are found to be about one-tenth the gas kinetic collision frequency, while that of $Fe(CO)_4$ is much slower. This difference in reactivity is rationalized by requiring a spin multiplicity change in the final recombination step: $Fe(CO)_2$, $Fe(CO)_3$, and $Fe(CO)_4$ are thought to be triplets, while $Fe(CO)_5$ is a singlet.

Gas phase structures, deduced for $Fe(CO)_3$ and $Fe(CO)_2$ from transient IR spectra, agree well with matrix isolation results, indicating that matrix perturbations in these cases are small. Photofragment spectra are observed which show rapid shifts to higher frequency. These are assigned to vibrationally excited photofragments that collisionally relax before reacting, presaging the application of transient spectroscopy to organometallic dynamics.

Gas phase irradiation of $Fe(CO)_5$ in the presence of ethylene produces the π-bound complexes $Fe(CO)_4(C_2H_4)$ and less stable $Fe(CO)_3(C_2H_4)_2$.[28-32] Each has been used as a parent compound in transient infrared studies of photoinitiated organometallic kinetics.[31,32]

3.2. $Fe(CO)_4(C_2H_4)$

Irradiation of $Fe(CO)_4(C_2H_4)$, either synthetically prepared[33] or photogenerated from $Fe(CO)_5$, produces $Fe(CO)_3(C_2H_4)_2$ in the presence of ethylene. In CO/C_2H_4 mixtures, this *bis*-ethylene complex decays to $Fe(CO)_4(C_2H_4)$. The kinetics of this process have been studied by Weiller and co-workers.[31] Conversion is slow enough to be followed by FTIR. Observed rates [decay of $Fe(CO)_3(C_2H_4)_2$ and formation of $Fe(CO)_4(C_2H_4)$] depend inversely on C_2H_4 pressure and exhibit saturation in the partial pressure of CO. These facts are consistent with a dissociative substitution mechanism. Quantitative measurements under systematically varying conditions of CO and C_2H_4 concentrations confirm this, and return rate constants for dissociation of photoproduced parent $Fe(CO)_3(C_2H_4)_2$ and recombination of unsaturated $Fe(CO)_3(C_2H_4)$ with CO relative to C_2H_4. The room temperature decay constant for $Fe(CO)_3(C_2H_4)_2$ is found to be $(2.9 \pm 0.3) \times 10^{-3}\ s^{-1}$, while the unsaturated complex exhibits a preference for CO relative to C_2H_4 by a ratio of 35:1. These parameters predict a 1-h half-life for $Fe(CO)_3(C_2H_4)_2$ in a typical gas phase mixture consisting of complex together with 700 torr of C_2H_4 and 1.5 torr of CO. The decay rate of $Fe(CO)_3(C_2H_4)_2$ and the branching ratio for $Fe(CO)_3(C_2H_4)$ in its reactions with CO versus ethylene are observed to be temperature dependent. As a preliminary result,[34] the composition dependence of decay rates at a single elevated temperature (43°C) suggests a slight activation energy difference for CO versus C_2H_4 recombination with $Fe(CO)_3(C_2H_4)$ favoring that with CO as the lower-barrier (or barrierless) process by 8 ± 6 kcal/mol. The same single higher-temperature experiment places an approximate barrier for unimolecular dissociation to cleave the Fe–ethylene bond in isolated $Fe(CO)_3(C_2H_4)_2$ at 27 ± 6 kcal/mol, with an A factor of $10^{17.1}\ s^{-1}$ ($\Delta S^{\ddagger} = 20$ eu). The *bis*-ethylene complex has been shown to be a thermal catalyst precursor in the gas-phase photoinitiated hydrogenation of ethylene by $Fe(CO)_5$ and $Fe(CO)_4(C_2H_4)$.[34]

3.3. $Fe(CO)_3(C_2H_4)_2$

In the presence of ethylene-to-CO ratios as low as 400:, gas phase samples of $Fe(CO)_4(C_2H_4)$ can be photodriven almost completely to

$Fe(CO)_3(C_2H_4)_2$.[31] Metastable *bis*-ethylene iron tricarbonyl produced in this way has been used as a starting material for photopreparation and decay kinetics studies of $Fe(CO)_2(C_2H_4)_3$.[32] The *tris*-ethylene product, formed by irradiation of the *bis*-ethylene parent in the presence of ethylene, decays in typical gas phase mixtures of C_2H_4 and CO on a millisecond time scale. The kinetics of this decay have been followed by transient infrared absorption using a Nernst glower source in a manner comparable to that discussed in Section 2.4.[32] Evidence for two isomers of $Fe(CO)_2(C_2H_4)_3$ is found. As with $Fe(CO)_3(C_2H_4)_2$, observed decay rates depend inversely on C_2H_4 pressure and saturate in CO. The kinetics suggest distinct decay paths for the two isomers with unimolecular rate constants $(3.6 \pm 0.9) \times 10^3\ s^{-1}$ for *cis*-$Fe(CO)_2(C_2H_4)_3$ and $(1.2 \pm 0.4) \times 10^3\ s^{-1}$ for the slightly more stable *trans* isomer. Branching ratios for CO versus C_2H_4 recombination with the respective unsaturated intermediates are found to strongly favor CO, with deduced values of 370 ± 130 and 480 ± 220.

3.4. $Mn_2(CO)_{10}$

The photochemistry of $Mn_2(CO)_{10}$ has been extensively studied in solution, where an unresolved issue is the relative importance of competing Mn—Mn scission and CO loss, viz:

$$Mn_2(CO)_{10} \xrightarrow{h\nu} 2Mn(CO)_5$$

$$Mn_2(CO)_{10} \xrightarrow{h\nu} Mn_2(CO)_9 + CO$$

This question has motivated new work in the gas phase, where the branching ratio is unbiased by caged $Mn(CO)_5$ recombination reactions, which are possible in solution. Seder *et al.* have found that photolysis of $Mn_2(CO)_{10}$ at 351 nm in the gas phase leads predominantly to the formation of $Mn(CO)_5$.[35] The observed second-order decay of $Mn(CO)_5$,

$$2Mn(CO)_5 \rightarrow Mn_2(CO)_{10}$$

exhibits a reaction cross section near the hard-sphere limit. All previous measurements in solution have been diffusion controlled.

Further work[36] has shown that CO loss can be increasingly favored by decreasing the photolysis wavelength from 351 to 248 to 193 nm. The CO loss product, $Mn_2(CO)_9$, is identified by a bridging CO ligand carbonyl stretching band at $1745\ cm^{-1}$. The rate constant for the CO recombination reaction of $Mn_2(CO)_9$,

$$Mn_2(CO)_9 + CO \rightarrow Mn_2(CO)_{10}$$

is found to be orders of magnitude smaller than those for CO recombination of other photolytically generated organometallic fragments [due presumably to the stability of $Mn_2(CO)_9$ in its bridged form].

As with $Fe(CO)_5$ photofragments, shifting infrared bands are taken as evidence for vibrationally excited photofragments such as $Mn(CO)_5$. Further evidence for these excited states comes from observations by Bray and coworkers of infrared fluorescence by photofragments of $Mn_2(CO)_{10}$.[37] This latter group has made similar measurements of photolytically induced infrared fluorescence as well as infrared chemiluminescence in gaseous (η^5-C_5H_5)$Mn(CO)_3$.[38]

3.5. $Co(CO)_3NO$

Nitrosyls are useful alternative chromophores to carbonyls because they absorb at a lower infrared frequency, which is within reach of conventional CO lasers. One study of transients produced in the 248-nm photolysis of $Co(CO)_3NO$ has been conducted.[39] The kinetics in this case are complicated by competitive loss and recombination processes involving both CO and NO and reaction of fragments with parent $Co(CO)_3NO$. The predominant photofragment is CoCO, and rate constants have been determined for its reaction with CO and $Co(CO)_3NO$. Not uncharacteristically, the CO recombination cross section is near hard-sphere.

3.6. $Cr(CO)_6$

Early work[21,40,41] established that photolysis of $Cr(CO)_6$ at 249 nm yields $Cr(CO)_5$ and $Cr(CO)_4$ together with vibrationally excited CO. Both fragments were found to react with CO at near gas kinetic rates. In addition, $Cr(CO)_4$ shown to react with parent $Cr(CO)_6$ to form $Cr_2(CO)_{10}$, an unstable complex with a lifetime greater than 1 ms.

More recent experiments[42] have extended the photolysis wavelengths to include 193 and 351 nm. As with $Fe(CO)_5$, higher-energy photolysis is observed to create smaller fragments [e.g., $Cr(CO)_2$], which are internally excited. Infrared bands observed for gas phase $Cr(CO)_x$ species are in accord with structures deduced from matrix isolation data. The bimolecular rate constants for the reactions of $Cr(CO)_5$, $Cr(CO)_4$, and $Cr(CO)_3$ with CO are found to be very large, though disagreement exists on the precise value of that for $Cr(CO)_4$. It is worth noting that, apart from $Fe(CO)_4$ and $Mn_2(CO)_9$, CO recombination rate constants for all unsaturated metal carbonyls studied thus far are near corresponding gas kinetic values.

Other work[43] has focused on the recombination reactions of doubly unsaturated $Cr(CO)_4$ with a series of ligands ranging from $(CH_3)_2NH$ to H_2. The disappearance of $Cr(CO)_4$ in the presence of ligand L, spectrally monitored by transient IR absorption, yields the bimolecular rate constant for

$$Cr(CO)_4 + L \rightarrow Cr(CO)_4L$$

A correlation of these rates with ligand ionization potential is suggested and is rationalized by relating this parameter to the σ-donating power of the ligand. However, it should be noted that the ionization potential cannot be the sole determining factor. For example, measured rate constants for CO and C_2H_4 are almost identical, while their ionization potentials differ by 3.5 eV.

An alternative perspective on reactions related to these latter two in particular is offered by our own work. In recent experiments, we have determined the relative magnitudes of CO versus C_2H_4 recombination with $Cr(CO)_4(C_2H_4)$, in a study of the gas phase dissociative substitution kinetics of $Cr(CO)_4(C_2H_4)_2$. The following section presents a somewhat more detailed summary of this work.

4. TIME-RESOLVED INFRARED ABSORPTION SPECTROSCOPY AS A PROBE OF DISSOCIATIVE CO-FOR-C_2H_4 SUBSTITUTION IN $Cr(CO)_4(C_2H_4)_2$

4.1. Background

An intriguing difference exists in the strength with which olefins are coordinates to the 14-electron carbonyl fragments of the iron and chromium groups. Group 8 metal tricarbonyls preferentially bind conjugated dienes in an η^4 fashion over nonconjugated dienes or unconnected olefin pairs, whereas group 6 metal tetracarbonyls apparently prefer nonconjugated or unconnected olefins over conjugated ones. This observed difference has been rationalized in terms of changes in the metal-cent⁻r orbital occupancies for group 8 compared with group 6.[44]

The unimolecular decay of $Cr(CO)_4(\eta^4$-butadiene) has been measured at room temperature in solution.[45] In accord with the qualitative observations above, the reported decay constant, $(2.1 \pm 0.2) \times 10^{-3}\ s^{-1}$, is apparently larger than those of nonconjugated $Cr(CO)_4$ (norbornadiene) and $Cr(CO)_4(\eta^2$-*trans*-cyclooctene$)_2$, which are both stable at room temperature.[46,47] Stability in the latter case, however, is undoubtedly affected by the unique character of the *trans*-cyclooctene ligand. Relief of ring strain in the uncomplexed *trans*-cyclooctene molecule favors olefin coordination.

Motivated by this situation, we have investigated the gas phase reactivity of the simplest *bis*-olefin chromium carbonyl complex, $Cr(CO)_4(C_2H_4)_2$. This section presents the results of our observations on the dissociative CO substitution kinetics and spectroscopy of this complex, as determined by transient infrared absorption. In contrast to the solution phase results just cited, we find that $Cr(CO)_4(C_2H_4)_2$ is unstable at room temperature, decaying with a time constant that is orders of magnitude faster than that reported for the conjugated diene. This result suggests that approximating nonconjugated dienes as two ethylene units for the purpose of orbital analysis may be inadequate, and thus has important implications for the understanding of bonding in olefin and diene complexes of $Cr(CO)_4$. The method used in these experiments is that described in Section 2.4. Presented below are data on the lifetimes of pulsed laser-generated $Cr(CO)_4(C_2H_4)_2$ transients as a function of CO and C_2H_4 concentrations that establish the elementary gas phase unimolecular decay rate of parent $Cr(CO)_4(C_2H_4)_2$ as well as the relative elementary bimolecular rate constants for recombination of unsaturated $Cr(CO)_4(C_2H_4)$ with CO versus C_2H_4.

4.2. Results and Discussion

To observe $Cr(CO)_4(C_2H_4)_2$, we first convert $Cr(CO)_6$ into metastable $Cr(CO)_5(C_2H_4)$ by photolysis with excess C_2H_4. When monitored by FTIR spectroscopy, photolysis of a mixture of $Cr(CO)_6$ and C_2H_4 produces a new complex with bands at 2085 (0.03), 1980 (0.82), and 1975 (1.00) cm^{-1}. These bands are assigned to $Cr(CO)_5(C_2H_4)$, agreeing well with recent observations in liquid Xe solution.[48]

After $Cr(CO)_6$ is converted to $Cr(CO)_5(C_2H_4)$, photoinduced transient absorptions, which we assign to $Cr(CO)_4(C_2H_4)_2$, are clearly observed. Transient waveforms presented below are displayed as the change in absorbance with time referenced to zero at the laser pulse. Signals are proportional to the change in transmitted infrared intensity. However, the photoinduced changes are extremely small so that the negative of the change in transmission approximates very well the change in absorbance. Negative changes in absorbance correspond to photolytic destruction of the initial species followed by reformation while positive changes correspond to formation of photoproducts followed by their decay.

We observe five transient absorptions as displayed in Figure 2. Figures 2a and 2b show negative transients at 1975 and 2084 cm^{-1}, respectively, that correlated with absorptions of $Cr(CO)_5(C_2H_4)$. Figures 2c–e show positive transients due to the new species formed by photolysis of $Cr(CO)_5(C_2H_4)$ with C_2H_4. These occur at 2045 cm^{-1} (Figure 2e), 1961 cm^{-1} (Figure 2c), and 1931 cm^{-1} (Figure 2d). The new species is assigned as *cis*-$Cr(CO)_4(C_2H_4)_2$ based on spectroscopic and kinetic data,[48] again in good agreement with recent similar photopreparative studies carried out in liquid Xe.[49]

By stepping the wavelength of the monochromator and collecting waveforms as in Figure 2, we obtain a time-resolved infrared spectrum. For this, the resolution is narrowed to 5 cm^{-1} and traces are taken every 4 cm^{-1} from 2088 to 1898 cm^{-1}. Point spectra are constructed by taking instantaneous amplitudes from each waveform at fixed time intervals of 13.7 ms from 30 μs to 78.5 ms after the laser pulse. A spline fit is used to smooth the data in the frequency dimension.[50] The data points are shown along with the spline fits in Figure 3, again as change in absorbance. The five transients of Figure 2 are clearly evident.

Substitution reactions of chromium carbonyl complexes are generally observed to occur by dissociative ligand loss.[51] We therefore expect $Cr(CO)_4(C_2H_4)_2$ to relax to $Cr(CO)_5(C_2H_4)$ by the process:

$$Cr(CO)_4(C_2H_4)_2 \underset{k_2}{\overset{k_1}{\rightleftarrows}} Cr(CO)_4(C_2H_4) + C_2H_4 \tag{1}$$

$$Cr(CO)_4(C_2H_4) + CO \overset{k_3}{\rightarrow} Cr(CO)_5(C_2H_4) \tag{2}$$

Application of the steady-state assumption to the intermediate $Cr(CO)_4(C_2H_4)$ leads to a prediction of first-order kinetics for the decay of $Cr(CO)_4(C_2H_4)_2$:

$$[Cr(CO)_4(C_2H_4)_2] = [Cr(CO)_4(C_2H_4)_2]_0 \exp(-k_{obs}t) \tag{3}$$

and for the recovery of $Cr(CO)_5(C_2H_4)$:

$$[Cr(CO)_5(C_2H_4)] = [Cr(CO)_5(C_2H_4)]_0 + [Cr(CO)_4(C_2H_4)_2]_0\{1 - \exp(-k_{obs}t)\} \tag{4}$$

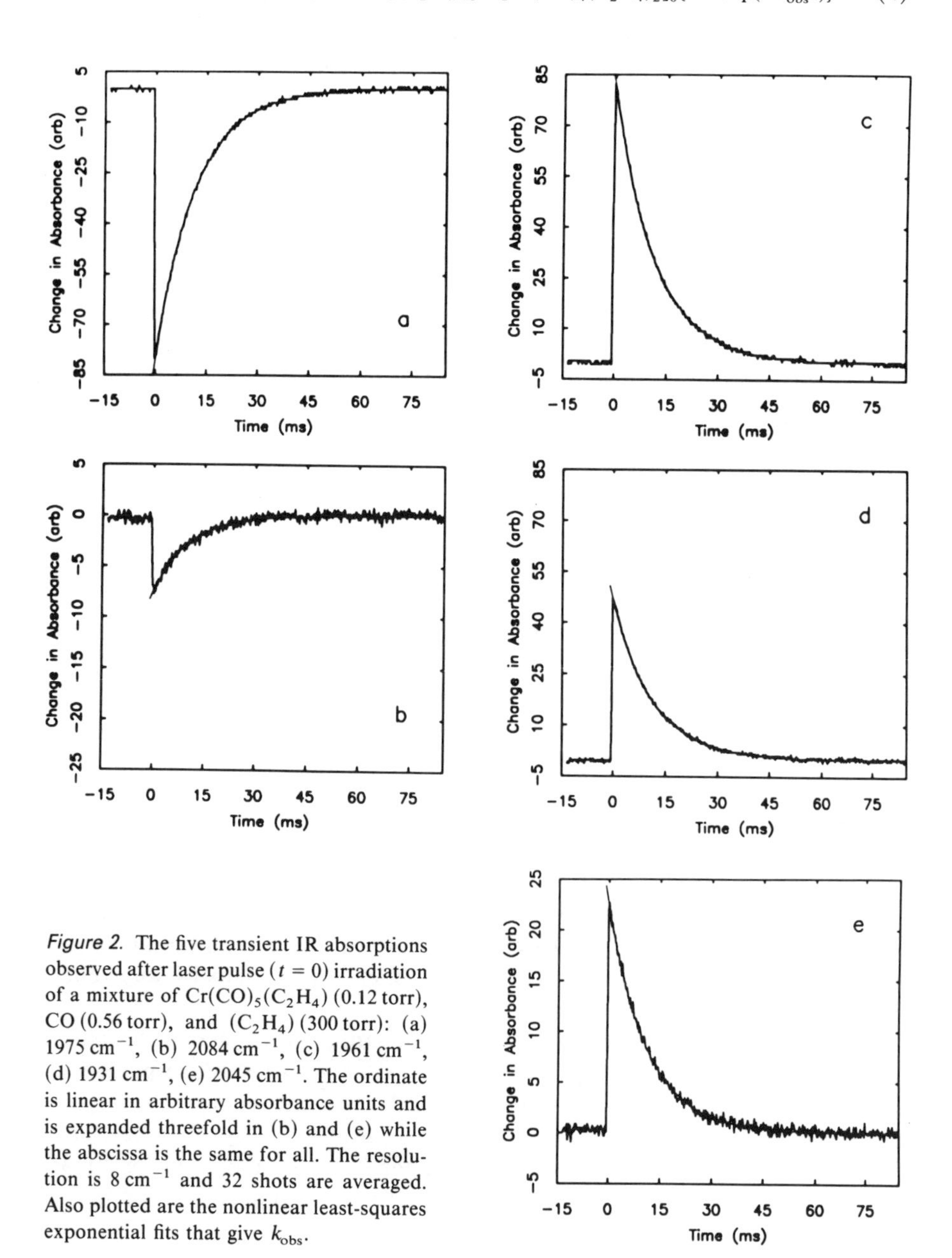

Figure 2. The five transient IR absorptions observed after laser pulse ($t = 0$) irradiation of a mixture of $Cr(CO)_5(C_2H_4)$ (0.12 torr), CO (0.56 torr), and (C_2H_4) (300 torr): (a) 1975 cm^{-1}, (b) 2084 cm^{-1}, (c) 1961 cm^{-1}, (d) 1931 cm^{-1}, (e) 2045 cm^{-1}. The ordinate is linear in arbitrary absorbance units and is expanded threefold in (b) and (e) while the abscissa is the same for all. The resolution is 8 cm^{-1} and 32 shots are averaged. Also plotted are the nonlinear least-squares exponential fits that give k_{obs}.

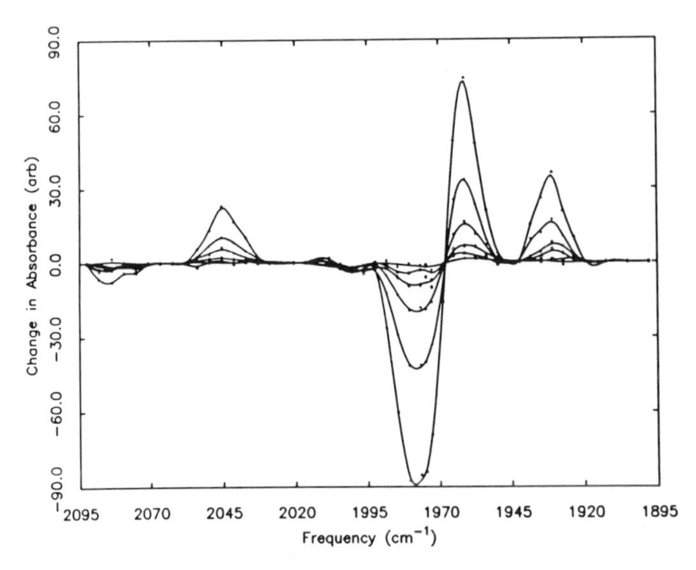

Figure 3. The transient IR absorption spectrum due to photolysis of a mixture of $Cr(CO)_5(C_2H_4)$-(0.12 torr), CO (0.56 torr), and C_2H_4 (500 torr). The initial spectrum (largest amplitude) is 30 μs after the laser pulse while subsequent ones are separated by 13.7 ms. The resolution is 5 cm^{-1} and 16 shots are averaged.

where

$$k_{obs} = \frac{k_1 k_3[CO]}{k_2[C_2H_4] + k_3[CO]} \tag{5}$$

This analysis predicts a simple exponential decay and recovery for $Cr(CO)_4(C_2H_4)_2$ and $Cr(CO)_5(C_2H_4)$, respectively. Figure 2 shows that all of the observed waveforms are well fit by single exponentials. Moreover, all observed time constants are identical within experimental error under each set of conditions.

Equation 5 predicts the observed decay constants k_{obs} to increase with CO pressure. Figure 4 shows that this is observed for both the decay of $Cr(CO)_4(C_2H_4)_2$ and the recovery of $Cr(CO)_5(C_2H_4)$. In addition, the data are well fit to the functional forms of Equation 5.

Equation 5 also predicts an inverse dependence of k_{obs} on $[C_2H_4]$. Figure 5 shows that this is precisely observed. This result is most important, because an inverse rate dependence on $[C_2H_4]$ is difficult to explain by any other mechanism.

From the data of Figures 4 and 5 in combination with Equation 5, values for k_1 and k_2/k_3 are found to be $(6 \pm 2) \times 10^4\ s^{-1}$ and 0.7 ± 0.2, respectively. The result that k_2 and k_3 are nearly equal shows that the behavior of $Cr(CO)_4(C_2H_4)$ parallels that of $Cr(CO)_4$ as found by Fletcher and Rosenfeld.[43] In that case, slight preference for C_2H_4 is also found for a pair of reactions with very large absolute rate constants, indicating essentially barrierless processes.

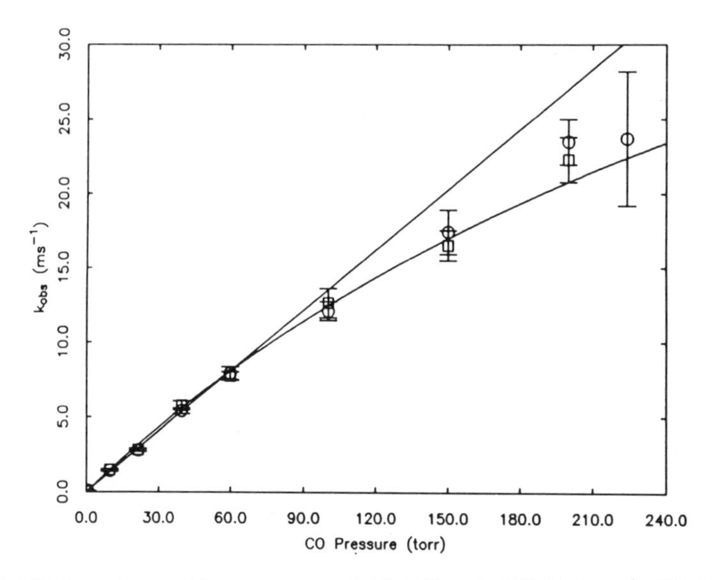

Figure 4. CO dependence of k_{obs} at constant C_2H_4 (300 torr): (□) 1975 cm^{-1}, (○) 1961 cm^{-1}. The straight line is the weighted fit to the data below 60 torr CO. The curved line is the weighted nonlinear fit to all the data except the last point, which is only for reference.

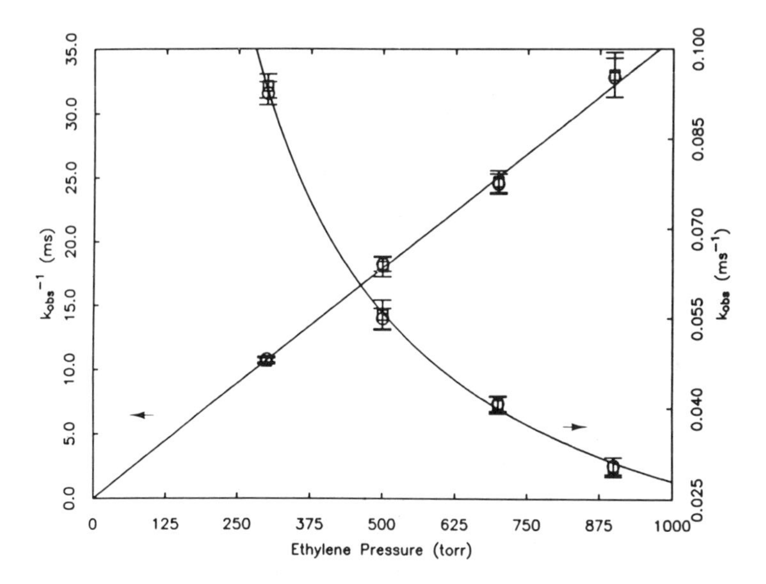

Figure 5. Ethylene dependence of k_{obs} and k_{obs}^{-1} at constant CO (0.56 torr): (□) 1975 cm^{-1}, (○) 1961 cm^{-1}. The lines are from the weighted linear least-squares fit.

The similarity between k_3 and k_2 for chromium contrasts with our previous results for the analogous group 8 complexes $Fe(CO)_3(C_2H_4)_2$[31] and $Fe(CO)_2(C_2H_4)_3$.[32] From similar dissociative substitution kinetics of these compounds, we have found ratios of bimolecular rate constants for the reactions of $Fe(CO)_3(C_2H_4)$ and $Fe(CO)_2(C_2H_4)_2$ with C_2H_4 versus CO to strongly favor CO by 35:1 for $Fe(CO)_3(C_2H_4)$ and about 400:1 for $Fe(CO)_2(C_2H_4)_2$.[32] We have noted that for barrierless reactions in which the rotations of the free ligand are frozen out in the transition state, one can expect reaction with CO to be favored by about 10:1 over reaction with C_2H_4.[31] Apparently, statistical factors that follow from the assumption of a tight transition state are not appropriate for unsaturated chromium complex recombinations.

The gas phase unimolecular decay constant for *cis*-$Cr(CO)_4(C_2H_4)_2$ [$(6 \pm 2) \times 10^4\ s^{-1}$)] quantifies the intrinsic instability of this complex. In contrast, the *trans bis*-olefin complexes of $Mo(CO)_4$ and $W(CO)_4$ are quite stable and in competitive photochemical preparation seem to be preferred over conjugated diene complexes. For example, photolysis of $Mo(CO)_6$ or $W(CO)_6$ with butadiene produces $M(CO)_4(\eta^2$-butadiene$)_2$ but not $M(CO)_4(\eta^4$-butadiene).[52] Recently, the trans *bis*-ethylene complex of $Cr(CO)_4$ has been prepared in solution by low-temperature photolysis.[53] Like the *trans*- Mo and W complexes, it is stable. Interestingly, we detect no evidence for formation of *trans*-$(C_2H_4)_2Cr(CO)_4$ in the transient infrared spectrum of our gas-phase system, despite the many ethylene dissociation-recombination cycles that accompany the decay of $(C_2H_4)_2Cr(CO)_4$. This conveys two additional important pieces of information about the energetics and dynamics of chromium carbonyl recombination and isomerization reactions: (1) The recombination of C_2H_4 with unsaturated $(C_2H_4)Cr(CO)_4$ must be highly stereoselective, yielding exclusively *cis*-$(C_2H_4)_2Cr(CO)_4$; and (2) the energy released in the isolated gas-phase *cis*- product by the formation of the Cr-ethylene bond must be insufficient to overcome the barrier for isomerization to *trans*-.

The unimolecular decay constant for $Cr(CO)_4(C_2H_4)_2$ is 3×10^7 larger than the reported value for $Cr(CO)_4(\eta^4$-butadiene) as measured in solution by the substitution of a nonconjugated diene for butadiene.[45,54]† Thus, both types of diene complex are apparently much more stable than the *cis-bis*-ethylene complex.

Thus, in summary, our results from transient infrared kinetic spectroscopy have (1) identified the gas phase structure of photoprepared $Cr(CO)_4(C_2H_4)$ as *cis*, (2) characterized its unimolecular decay as much faster than that of comparable, but more complicated, solution phase nonconjugated $Cr(\eta^4$-diolefin) complexes, as well as second- and third-row *bis*-olefin compounds, and (3) established that relative rates for recombination of unsaturated

† The data of Reference 45 can be interpreted by an alternate mechanism (see Reference 54) that does not give the elementary unimolecular rate constant for the η^4 to η^2 conversion of $Cr(CO)_4(\eta^4$-butadiene). Thus, the solution kinetic data are inconclusive. Nevertheless, $Cr(CO)_4(\eta^4$-butadiene) clearly decays faster than $Cr(CO)_4(\eta^4$-norbornadiene) but lives much longer than $Cr(CO)_4(C_2H_4)_2$.

$Cr(CO)_4(C_2H_4)$ with CO and ethylene are comparable, favoring C_2H_4 slightly. Important work remains, including studies of temperature dependence to obtain Arrhenius parameters for these reactions, as well as faster time-resolved kinetic studies to obtain absolute rates for bimolecular CO and ethylene recombinations. Work to investigate the gas phase dissociative CO substitution rate for $Cr(CO)_4(\eta^4\text{-}C_4H_6)$ is in progress.

Note added in proof: Since this review was written in 1986, several more gas-phase systems have been studied: $W(CO)_6$,[55-56] $W(CO)_5(H_2)$,[57] $Cr(CO)_6$,[58] $V(CO)_6$,[59] and $(\eta^5\text{-}C_5H_5)Co(CO)_2$[60].

ACKNOWLEDGMENT

We are grateful to the National Science Foundation for support of research described herein from our laboratory under grant No. CHE-8614702.

REFERENCES

1. Collman, J.P., Hegedus, L.S. *Principles and Applications of Organotransition Metal Chemistry*; University Science Books: Mill Valley, CA, 1980.
2. Perutz, R.N., Turner, J.L. *J. Am. Chem. Soc.* 1975, *97*, 4791.
3. Welch, J.A.; Peters, K.S.; Vaida, V. *J. Phys. Chem.* 1982, *86*, 1941.
4. (a) Gladstone, S.; Laidler, D.J.; Eyring, H. *The Theory of Rate Processes*; McGraw-Hill: New York, 1941; (b) Kondratiev, V.N. *Chemical Kinetics of Gas Reactions*; Pergamon: Oxford, 1964; (c) Baniford, C.H.; Tipper, C.F.H. Eds. *Comprehensive Chemical Kinetics*; Elsevier: Amsterdam, 1969; (d) Levine, R.D.; Bernstein, R.B. *Molecular Reaction Dynamics*; Oxford University Press: New York, 1974.
5. (a) Callear, A.B.; Oldman, R.J. *Nature* 1966, *210*, 730; (b) Breckenridge, W.H.; Sinair, N. *J. Phys. Chem.* 1981, *85*, 3557; (c) Breckenridge, W.H.; Stewart, G.M. *J. Am. Chem. Soc.* 1986, *108*, 364.
6. Geoffroy, G.L.; Wrighton, M.S. *Organometallic Photochemistry*; Academic: New York, 1979.
7. Braterman, P.S. *Metal Carbonyl Spectra*; Academic: London, 1975.
8. (a) Hitam, R.B.; Mahmoud, K.A.; Rest, A.J. *Coord. Chem. Rev.* 1984, *55*, 1; (b) Perutz, R.N. *Chem. Rev.* 1985, *85*, 77.
9. Koerner von Gustorf, E.A.; Leenders, L.H.G.; Fischler, I.; Perutz, R.N. *Adv. Inorg. Chem. Radiochem.* 1976, *19*, 65.
10. Moggi, L.; Juris, A.; Sandrini, D.; Manfrin, M.F. *Rev. Chem. Intermed.* 1981, *4*, 171.
11. (a) Wrighton, M.S.; Ginley, D.S.; Schroeder, M.A.; Morse, D.L. *Pure Appl. Chem.* 1975, *41*, 671; (b) Wrighton, M.S.; Graff, J.L.; Reichel, C.L.; Sanner, R.D. *Ann. N.Y. Acad. Sci.* 1980, *333*, 188, (c) Wrighton, M.S.; Graff, J.L.; Kazlauskas, R.J.; Mitchener, J.C.; Reichel, C.L. *Pure Appl. Chem.* 1982, *54*, 161.
12. Poliakoff, M.; Weitz, E. *Adv. Organomet. Chem.* 1986, *25*, 277.
13. Oriel Corporation Catalog, 1986.
14. Kettle, S.F.A.; Paul, I. *Adv. Organomet. Chem.* 1972, *10*, 199.
15. Artcor Corp., Costa Mesa, CA.
16. Taylor, J.H.; Ruppert, C.S.; Strong, J. *J. Opt. Soc. Am.* 1951, *41*, 626.
17. Spanbauer, R.; Farley, P.E.; Rao, K.N. *Appl. Opt.* 1963, *2*, 1340.
18. "Cermax" Xenon Illuminators, ILC Technology, Inc., Sunnyvale, CA.
19. Adamson, A.W.; Cindino, M.C. *J. Phys. Chem.* 1984, *88*, 488.
20. Ouderkirk, A.J. Ph.D. thesis, Northwestern University, 1983.
21. Fletcher, T.R.; Rosenfeld, R.N. *J. Am. Chem. Soc.* 1983, *105*, 6358.

22. Djeu, N. *Appl. Phys.* 1973, *23*, 309.
23. Laser Analytics Division, Spectra Physics, Inc.
24. Santa Barbara Research Center, Santa Barbara, CA.
25. Ouderkirk, A.J.; Weitz, E. *J. Chem. Phys.* 1983, *79*, 1089.
26. Ouderkirk, A.J.; Wermer, P.; Schultz, N.L.; Weitz, E. *J. Am. Chem. Soc.* 1983, *105*, 3354.
27. Seder, T.A.; Ouderkirk, A.J.; Weitz, E. *J. Chem. Phys.* 1986, *85*, 1977.
28. Whetten, R.L.; Fu, K.-J.; Grant, E.R. *J. Chem. Phys.* 1983, *79*, 2626.
29. Miller, M.E.; Grant, E.R. *J. Am. Chem. Soc.* 1984, *106*, 4635.
30. Miller, M.E.; Grant, E.R. *J. Am. Chem. Soc.* 1985, *107*, 3386.
31. Weiller, B.H.; Miller, M.E.; Grant, E.R. *J. Am. Chem. Soc.* 1987, *109*, 352.
32. Weiller, B.H.; Grant, E.R. *J. Am. Chem. Soc.* 1987, *109*, 1051.
33. Murdoch, H.D.; Weiss, E. *Helv. Chim. Acta* 1963, 1588.
34. Miller, M.E.; Grant, E.R. *J. Am. Chem. Soc.* 1987, *109*, 7951; Miller, M.E. Ph.D. thesis, Cornell University, 1986.
35. Seder, T.A.; Church, S.P.; Weitz, E. *J. Am. Chem. Soc.* 1986, *108*, 1084.
36. Seder, T.A.; Church, S.P.; Weitz, E. *J. Am. Chem. Soc.* 1986, *108*, 7518.
37. Bray, R.G.; Seidler, P.E.; Gethner, J.S.; Woodin, R.L. *J. Am. Chem. Soc.* 1986, *108*, 1312.
38. Bray, R.G.; Seidler, P.F. Unpublished results.
39. Rayner, D.M.; Nazran, A.S.; Drouin, M.; Hackett, P.A. *J. Phys. Chem.* 1986, *90*, 2882.
40. Seder, T.A.; Church, S.P.; Ouderkirk, A.J.; Weitz, E. *J. Am. Chem. Soc.* 1985, *107*, 1432.
41. Fletcher, T.R.; Rosenfeld, R.N. *J. Am. Chem. Soc.* 1985, *107*, 2203.
42. Seder, T.A.; Church, S.P.; Weitz, E. *J. Am. Chem. Soc.* 1986, *108*, 4721.
43. Fletcher, T.R.; Rosenfeld, R.N. *J. Am. Chem. Soc.* 1986, *108*, 1686.
44. Elian, M.; Hoffman, R. *Inorg. Chem.* 1975, *14*, 1058.
45. Dixon, P.T.; Burkingham, P.M.; Howell, J.A.S. *J. Chem. Soc., Dalton Trans.* 1980, 2237.
46. Bennet, M.A.; Pratt, L.; Wilkinsen, G. *J. Chem. Soc.* 1961, 2037.
47. Grevels, F.W.; Skibbe, V. *J. Chem. Soc., Chem. Commun.* 1984, 681.
48. Weiller, B.H.; Grant, E.R. *J. Am. Chem. Soc.* 1987, *109*, 1252; Weiller, B.H.; Grant, E.R. *J. Phys. Chem.* 1988, *92*, 1458.
49. Gregory, M.F.; Jackson, S.A.; Poliakoff, M.; Turner, J.J. *J. Chem. Soc. Chem. Commun.* 1986, 1175.
50. ICSCCU, International Mathematics and Statistics Library, Inc.
51. Darensbourg, D.J.; *Adv. Organomet. Chem.* 1982, *21*, 113; Howell, J.A.S.; Burkinshaw, P.M. *Chem. Rev.* 1983, *83*, 557; Angelici, R.J. *Organomet. Chem. Rev.* 1968, *3*, 173.
52. Stolz, J.W.; Dobson, G.R.; Sheline, R.K. *Inorg. Chem.* 1963, *2*, 1264; Grevels, Lindemann, M.; Benn, R.; Goddard, R.; Kruger, C. *Z. Natur. Forsch.* 1980, *35B*, 1298.
53. Grevels, F.-W.; Jacke, J.; Özkar, S.J. *J. Am. Chem. Soc.* 1987, *109*, 7536.
54. Zingales, F.; Conziani, F.; Basolo, F. *J. Organomet. Chem.* 1967, *1*, 461; Zingales, F.; Granziani, M.; Belluea, V. *J. Am. Chem. Soc.* 1967, *89*, 256.
55. Holland, J.P.; Rosenfeld, R.N. *Chem. Phys. Lett.* 1988, *145*, 481.
56. a) Ishikawa, Y.; Hackett, P.A.; Rayner, D.M. *J. Phys. Chem.* 1988, *92*, 3863, b) Ishikawa, Y.; Hackett, P.A.; Rayner, D.M. *Chem. Phys. Lett.* 1988, *145*, 429.
57. Isikawa, Y.; Weersink, R.A.; Hackett, P.A.; Rayner, D.M. *Chem. Phys. Lett.* 1987, *142*, 271
58. Flechter, T.R.; Rosenfeld, R.N. *J. Am. Chem. Soc.* 1988, *110*, 2097.
59. Ishikawa, Y.; Hackett, P.A.; Rayner, D.M. *J. Am. Chem. Soc.* 1987, *109*, 6644.
60. Wasserman, E.P.; Bergman, R.G.; Moore, C.B. *J. Am. Chem. Soc.* 1988, *110*, 6076.

8

Characterization of Metal Complex Positive Ions in the Gas Phase by Photoelectron Spectroscopy

Dennis L. Lichtenberger and Glen Eugene Kellogg

1. INTRODUCTION

The remarkable contributions of metals and metal-containing species to chemical and material behavior are rooted in the electronic and bonding interaction capabilities of the metal atoms themselves. Diverse theoretical and experimental investigations have been directed toward advancing the understanding of these interactions. Significant new insights have been accomplished in recent years as a consequence of major improvements in instrumentation and modeling methods. One area in which the progress is particularly noteworthy is the study of metal–metal and metal–molecule interactions in the gas phase. Gas phase studies make possible detailed and high-resolution spectroscopic characterization of metal-containing species without the complicating (and spectroscopic broadening) factors of solution or solid state investigations. In the gas phase it is possible to characterize reactive intermediates, excited states, and other short-lived species that are essential to developing the full picture of the properties of these systems. Furthermore, the species in the gas phase contain a discrete number of metal atoms and bound molecules or fragments, and therefore are amenable to detailed theoretical modeling using the most advanced methods available.

Much of the progress in gas phase metal chemistry has involved the study of metal complex ions (as shown by the contributions to this volume). A

Dennis L. Lichtenberger and Glen Eugene Kellogg • Department of Chemistry, University of Arizona, Tucson, Arizona 85721.

variety of approaches have been developed to create individual metal atom and cluster ions that can be mass-analyzed, characterized by a variety of spectroscopies, trapped, and/or reacted with other molecules. The purpose of this chapter is to explain and illustrate the unique information available from photoelectron spectroscopy for characterizing the structure, stability, and chemistry of metal complex ions in the gas phase and how this information is related to "real world" chemical behavior.

It should always be remembered that photoelectron spectroscopy is a direct experimental measure of the relative energies of the ground and all excited electronic states of the ion. There are numerous excellent texts on the subject.[1-7] In the simplest terms, the intial state of the neutral metal species (designated $[M_0]$) is ionized by a monochromatic source of radiation with energy $h\nu$ to produce the ith electronic state of the positive ion ($[M_i]^+$) and the free electron e^-. The kinetic energy of the free electron is measured with an electron energy analyzer, the energy of the excitation radiation is known, and all excitations are from the same state(s) of the neutral species. The expression for the energy of the ith electronic state of the positive ion is thus

$$E([M_i]^+) = E(h\nu) + E([M_0]) - \mathrm{KE}(e^-) \tag{1}$$

The difference in energy between two electronic states of the positive ion is simply the difference in kinetic energy between the photoelectrons. This relationship is often used to correlate the electronic spectral properties and photochemistry of a cation with the photoelectron spectrum of the corresponding neutral molecule.[8]

The more familiar quantity from photoelectron spectroscopy is the ionization energy (IE), which is defined as

$$\mathrm{IE} = E([M_i]^+) - E([M_0]) = E(h\nu) - \mathrm{KE}(e^-) \tag{2}$$

The ionization energy is a fundamental and well-defined thermodynamic quantity that often contributes to thermochemical cycles (See section 4.1). The lowest ionization energy (also called the first ionization potential) corresponds to the ground electronic state of the positive ion.

In addition to the relative energies of the ground and excited electronic states of the positive ion, the ionization band features also give a measure of the distribution of charge in the positive ion, the formal oxidation state of the metal, the electronic symmetry about the metal center, and the strengths of individual bonding interactions with the ligands. This chapter will show how this information is provided by the relative intensities, fine structures, shapes, and widths of the ionization bands and by the dependencies of these features on chemical substitutions in the complex and the frequency (and angular effects) of the excitation radiation. This information requires high-quality spectra and utilization of a variety of ionization techniques, including He I and He II UPS, XPS, and Auger spectroscopy in the gas phase. Perhaps most significant is that high-quality photoelectron data can give information on the bond distances, angles, and distortions in the molecular complex positive ion. This type of structural information is crucial to the understanding of chemical behavior, but is generally lacking for positive metal complex ions that are created and studied in the gas phase by other techniques.

The focus of this chapter on the direct information about stabilities, electron distributions, and geometries of positive ion states is somewhat different from that of traditional discussions of the value of photoelectron spectroscopy. Photoelectron spectroscopy studies of transition metal systems have been reported since 1969, and there are numerous reviews of the subject.[9-11] Most often the discussions have centered on the information related to the electronic structure and bonding in the ground state of the *neutral* molecule. The language is commonly in terms of ionizations from molecular orbital bonds, lone pairs, antibonds, or atomic cores, and it reflects the relationship of ionization energies to calculated molecular orbital eigenvalues (ε_i) as shown by Koopmans' approximation[12]:

$$\mathrm{IE} = -\varepsilon_i \tag{3}$$

This model of ionization energies does not include the additional, and purely calculational, adjustments associated with electron correlation and electron relaxation in the positive ion.[13] We have discussed experimental approaches to interpreting the electronic structure of neutral molecules by photoelectron spectroscopy in previous reviews.[14,15] Koopmans' approximation is implicated whenever an orbital picture is involved, but is not necessary when the focus is on the total electronic states of the positive ions.

The positive metal complex ions discussed in this chapter are generated by ionization of neutral transition metal complexes that are sufficiently stable for thermal vaporization. In most cases this means that the metal or cluster is coordinatively saturated with ligands, and the strength of the bonds between the metal and the molecules or fragments must be sufficient to maintain coordination at the temperature of vaporization. This is in contrast to molecular beam or mass spectrometry type techniques, which often involve metals interacting with a single molecule (leaving high degree of coordinative unsaturation) and which can include very weak metal-molecule interactions. The size of a metal complex with several coordinated ligands may at first appear to be a disadvantage for detailed experimental characterization and theoretical modeling. However, the coordinative saturation serves to reduce the number of electronic states that are close to the ground state and simplify the electronic description of the ion. A single metal atom with a single attached ligand often requires more sophisticated theoretical treatment because of the larger number of low-lying electronic states and the greater amount of configuration interaction necessary to adequately describe the system even qualitatively.

Metal complexes with a variety of ligands are also attractive for studying the trends in bonding and stabilization as a function of the electron richness or poorness at the metal center, the oxidation states of the metal, the fluidity of charge, and the electronic symmetry about the metal. The dependencies are measured by comparing ligand substitutions that change the formal *d* electron count at the metal without changing the metal, comparing related complexes that have the same *d* electron count but different metals, comparing complexes with different "hard" and "soft" ligands, and investigating the effects of other electronic perturbations of the metal complex. Examples of these relationships have been summarized in a previous account.[14]

A frequent question in the study of interactions in the gas phase concerns the extent that the information learned from discrete, isolated species transfers to large-scale synthetic or catalytic processes and macroscopic behavior. Knowledge of the ways in which fundamental metal-molecule interactions respond to electronic perturbations assists in extending the information on these interactions beyond isolated metal species. A further point is that gas phase photoelectron spectroscopy experiments start with stable transition metal complexes that are well characterized both structurally and chemically by the standard tools of the chemist. Many of these metal species are actually involved in important synthetic and catalytic processes. Because the ionization involves transition from a stable neutral molecule to a positive ion in the gas phase, photoelectron spectroscopy helps provide a bridge between study of metal species involved in large-scale solution or solid state processes and study of metal complex ions that are amenable to detailed characterization in the gas phase.

2. STRUCTURAL INFORMATION ON METAL POSITIVE IONS WITH SMALL MOLECULES IN THE GAS PHASE

It is helpful to begin with a discussion of the information provided by photoelectron spectroscopy for carbonyls bound to positive metal ions. These metal-carbonyl systems are relatively simple and well understood, and they will assist in illustrating the basic priciples of ionization spectroscopy. In addition, the observations on these systems often form the foundation for interpretation of more complex spectra.

2.1. Background

We will focus on the first ionization energy bands of $Cr(CO)_6$, $Mo(CO)_6$, and $W(CO)_6$ (see Figure 1), since these ionizations correspond to transition to the lowest electronic states of the positive ions.[16] In the language of a ground state orbital description, these ionizations are assigned to the removal of an electron from primarily metal *d* t_{2g}-symmetry orbitals. The ground electronic states of $Cr(CO)_6^+$ and $Mo(CO)_6^+$ are reasonably represented as $^2T_{2g}$ since the spin-orbit coupling in each case is small (atomic ζ_{Cr} is 0.027 eV and atomic ζ_{Mo} is 0.07 eV)[17] compared to the width of the ionization band. The $W(CO)_6^+$ ion shows spin-orbit splitting into a twofold degenerate $^2E''$ ground state centered at 8.33 eV above the neutral molecule and a fourfold degenerate $^2U'$ excited state 0.26 eV higher in energy. The description in terms of the ion

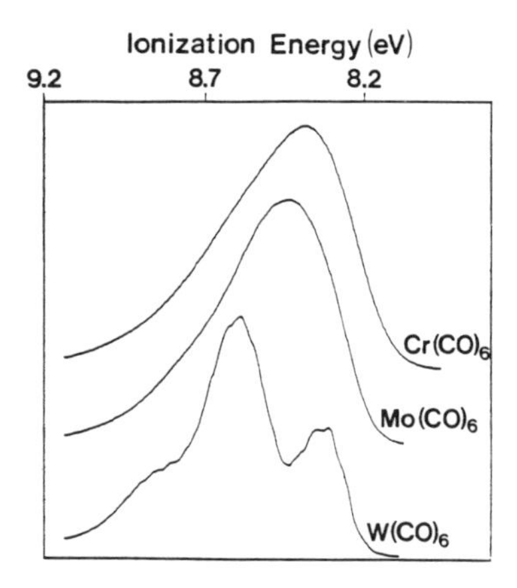

Figure 1. First ionization bands in the He I photoelectron spectra of $Cr(CO)_6$, $Mo(CO)_6$, and $W(CO)_6$.

states is certainly more awkward than the description in the language of ground state orbitals, but it is more strictly correct and, as will be shown, can lead to a more complete understanding of the electronic structure and bonding interactions of metal complex ions.

The lowest ionizations of transition metal complexes are most often (but not always; see Section 3) associated with removal of electrons from metal-based orbitals. This is understandable considering the lower electronegativity of metals compared to most of the main-group ligand atoms that coordinate to the metal center. Excited positive ion states that correlate with ionization of the carbonyls occur higher in energy (by several eV) than ionization of these metals.[18] In a previous account we listed a number of experimental approaches that can be applied to transition metal complexes to identify the primary character of the ionizations.[14] In the case of these $M(CO)_6$ complexes the first ionizations show the expected increases in ionization cross sections relative to ligand-based ionizations when the excitation source is changed from He I radiation ($h\nu = 21.2$ eV) to He II radiation ($h\nu = 40.8$ eV).[9-11] The relative cross sections of the first ionizations also increase down the series from Cr to Mo to W (the heavy atom effect).[19] In addition, stepwise substitution of the carbonyls with other ligands produces the expected reduction in electronic symmetry at the metal center and splitting of the initial ionization bands.[20,21]

The spin–orbit splitting between the ground and first excited electronic state of the $W(CO)_6^+$ ion is about 80% of the atomic spin–orbit value,[17] and in the zero differential overlap approximation this means that the electron density is about 80% localized on the metal center. The common interpretation is that the metal density is partially delocalized through backbonding to the carbonyl π^* orbitals. Considerable indirect experimental support for this backbonding has been provided by the interpretation of vibrational,[22,23] mass spectrometric,[24] and core-level XPS data,[25] to name just a few. Quantitative and direct measures of this backbonding have been thwarted by the difficulty in separating the σ and π bonding effects and by the problems in clearly relating the extent of backbonding to the particular physical observable that is measured. In this case the valence ionization the $^2T_{2g}$ state has pure π symmetry with respect to the carbonyls, and the delocalization of 20% of the six metal d electrons as indicated by the spin–orbit splitting corresponds to 0.1 electron charge assigned to each carbonyl π^* orbital in the positive ion. It should be remembered that there are many possible definitions for charges of atoms in molecules,[26,27] and any given number is most significant only in the context in which it is given.

What one really wishes to know is how the stability and geometry of a particular state of the positive ion compares to that in the neutral molecule. In other words, how does the electron density difference between the positive ion and the neutral molecule affect bond distances, bond angles, force constants, and other features of the molecular potential surface?

2.2. Vibrational Fine Structure and Bond Distances in Metal Carbonyls

The shape and structure of an ionization band is dependent on the potential surface of the positive ion state and its relationship to the initial state in the neutral molecule. This relationship is most often exploited in the study of small molecules and ions, where vibrational progressions are clearly observed with the electronic excitation, and Franck–Condon analysis of the vibrational intensity pattern yields the bond distances and angles in the ion. Carbon monoxide is commonly used to illustrate this principle,[2,4] and this example is relevant to our purposes. The first ionization of CO is dominated by transition to the ground vibrational level in the positive ion, showing that the potential well for the ground ion state is only slightly shifted with respect to the potential well of the neutral molecule. The 2200-cm^{-1} spacing[2] to the next (weak) component of the progression shows that the vibrational frequency in the positive ion is slightly greater than the 2157-cm^{-1} frequency in the neutral molecule. Thus, the ground ion state is just slightly more bonding than the neutral molecule. This is reasonable since the first ionization corresponds to removal of an electron from the weakly antibonding 5σ orbital to create the $^2\Sigma_g^+$ state of the positive ion. In comparison, the second ionization of CO (to the first excited electronic state of CO^+) displays a long vibrational progression of at least nine members, and the most probable (vertical) transition is to the third member. The spacing of the vibrational components is only about 1600 cm^{-1}, which is much less than the frequency in the neutral molecule. The substantial shift of the potential well and reduction in force constant follows from the reduced bond order when an electron is removed from the π bond to produce the $^2\Pi_u$ state of the positive ion.

The direct information on the positive ion state provided by this vibrational information is easy to appreciate. The interpretation does not require Koopmans' approximation or other such theoretical assistance. Unfortunately, conventional photoelectron spectroscopy studies of transition metal molecules generally have not shown evidence of vibrational structure in the ionization band envelopes. This is due in part to the low frequency of metal–ligand vibrations and the potentially large number of normal mode progressions that may overlap and obscure the structure. It is also due to the limitations of the instruments that are most commonly used for these studies. In recent years we have demonstrated the instrumentation requirements for observation of this fine structure.[16,28-31] Here we would like to illustrate some of the applications of these principles to the study of transition metal ions, and list a set of molecular, electronic, and vibrational characteristics for a metal complex that favor observation of this structure.

The first observation of individual components of a vibrational progression in an ionization of a metal complex is shown in the ionization band of $W(CO)_6$ in Figure 1. The broad, low shoulder on the high-ionization-energy side (~8.8 eV) of the $^2U'$ excited state is the second member of a short vibrational progression with a frequency of ~2100 cm^{-1} in the positive ion. The normal vibrational modes of the metal hexacarbonyls have been analyzed in detail.[23] This frequency relates to the $\nu_1(a_{1g})$ fundamental stretching mode, which has a frequency of 2126 cm^{-1} in the neutral molecule. The relative nuclear motions of this mode in the neutral molecule. The relative nulcear motions of this mode in the neutral molecule are shown in **1**. This mode is identified as the

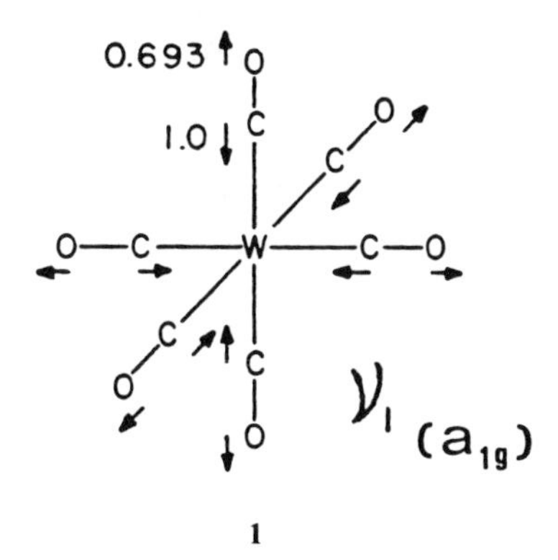

1

totally symmetric, primarily carbon–oxygen stretch. The observation of the short carbon–oxygen vibrational progression in the positive ion of the complex shows that some of the formally metal electron density that was removed in the ionization was actually delocalized to the carbonyls, as expected for backbonding to the antibonding carbonyl π^* orbitals.

Closer inspection of the $W(CO)_6$ ionization in Figure 1 reveals finer detail in the band profile. A detailed examination of the ionization to the $^2E''$ ground state is shown in Figure 2. The structure of this ionization is accurately modeled with a progression of symmetric, evenly spaced component bands with a nearly Gaussian intensity pattern. This progression corresponds to the $\nu_2(a_{1g})$ vibrational mode, which is identified as the totally symmetric metal–carbon stretching mode **2**.

2

The frequency spacing of the progression in the positive ion ($390\,cm^{-1}$) is considerably less than the frequency in the neutral molecule ($426\,cm^{-1}$). This shows that removal of an electron from the metal-based t_{2g} orbitals to produce the ground state ion significantly reduces the bond strength between the metal and the carbonyls. The intensity pattern of the progression, with the most intense ionization to the third or fourth excited metal-carbonyl vibrational level of the ion, shows that the metal-carbon bond distance in the ion is considerably longer than in the neutral molecule.

The magnitude of the change in metal-carbon distance between the neutral molecule and the positive ion is determined from a normal Franck-Condon analysis[16,32] of the vibrational progression. The results of the analysis are illustrated in Figure 3. This figure is constructed with the correct scale between the vibrational potential wells of the ground state and the positive ions and matches the vibrational quantum levels with the observed vibrational progressions in the ionization band, shown on edge in the center of the figure. The $\nu_2(a_{1g})$ normal mode potential well is plotted versus the internal coordinate representing the W—C bond distances. The simple result from this analysis is that the W—C bond distances in the positive ion are displaced 0.10 Å from the distances in neutral $W(CO)_6$.

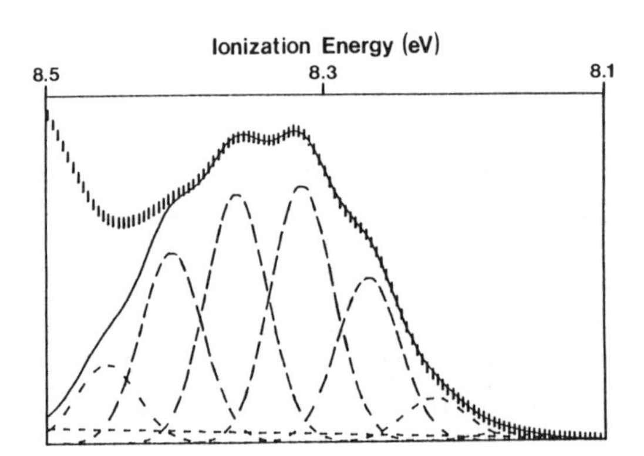

Figure 2. Detailed examination of ionization to the $^2E''$ electronic ground state of $W(CO)_6^+$. The ionization prole is fit with a progression of symmetric, evenly spaced vibrational components. The vertical dashes are the experimental data. The solid line is the fit sum from the dashed vibrational components and the baseline.

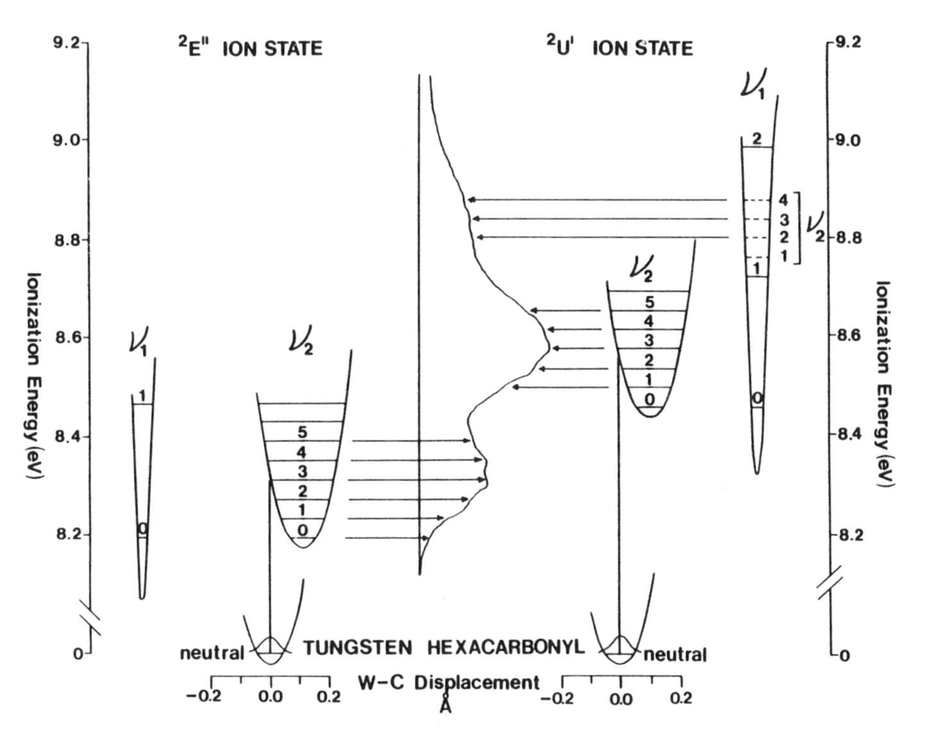

Figure 3. Illustration of the Franck-Condon evaluation of the $\nu_1(a_{1g}$, C—O stretch) and $\nu_2(a_{1g}$, W—C stretch) potential surfaces of the $^2E''$ and $^2U'$ states of the $W(CO)_6^+$ ion based on the ionization data and the normal mode vibrational analysis of the neutral molecule. The experimental ionization band is shown on edge in the center of the figure and the vertical transitions from the initial state are indicated.

The first ionization bands of $Cr(CO)_6$ and $Mo(CO)_6$ shown in Figure 1 are quite broad in comparison to the band of $W(CO)_6$, indicating long vibrational progressions. The separation of individual vibrational components in these progressions is not visually apparent when the data are plotted in this way. However, the data are more precise than the width of the line used to plot Figure 1. When the signal-to-noise is this high, the derivative of the data can be used to reveal more subtle features, as is commonly done in other spectroscopies.[33] Figure 4 shows how the data would appear if collected in derivative mode, as determined digitally from the data in Figure 1. The derivative spectrum shows evenly spaced inflections of the $\nu_2(a_{1g})$ vibrational mode with a frequency separation of about 330 cm^{-1}. This is again much less than the neutral molecule frequency of 379 cm^{-1}. The larger change in force constant and the greater length of the vibrational progression show that the chromium-carbonyl bond is more sensitive to the positive ion charge than the tungsten-carbonyl bond. Franck-Condon analysis of the $Cr(CO)_6$ data shows the change in chromium-carbonyl bond length to be at least 0.14 Å.

This trend in bond distances is supported by trends in M—CO and M—C (single bond) distances from crystal structure determinations. The W—C

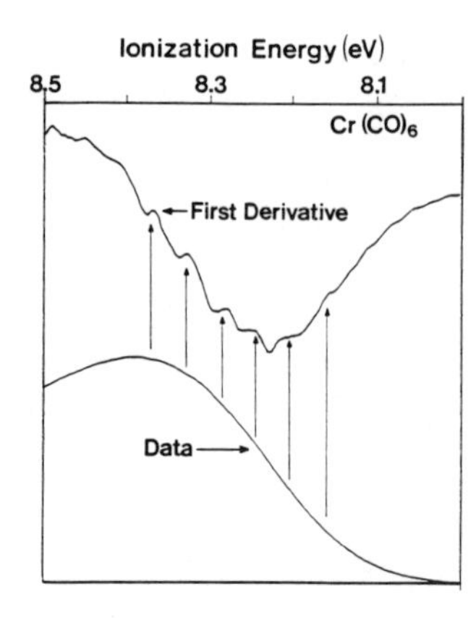

Figure 4. Detailed examination of the first ionization band of $Cr(CO)_6$. The noise in the data is less than the width of the line in the plot. The first derivative of the data allows visual observation of an evenly spaced progression on the leading edge of the ionization.

(single bond) distance is about 0.19 Å longer than the W—CO distance,[34] and the Cr—C (single bond) distance is about 0.28 Å longer than the Cr—CO distance.[35,36] In each case the M—CO distance in the ion, as determined from analysis of the vibrational progression in the ionization, is about halfway between the M—CO distance in the neutral molecule and the M—C single bond distance.

The intermediate bond length in the ion is reasonable since the ionization removes π bond order between the metal and the carbonyls. If the removal of one of the d^6 t_{2g} electrons is envisioned as loss of one-sixth of the total metal-Carbonyl π bond order in a ground state orbital description, then the bond lengthening appears disproportionately large for a single ionization. Studies of the vibrational structure in the ionizations of main-group molecules generally emphasize only this orbital bond order factor, as in the example of CO above. This example demonstrates the additional importance of the ion state description in transition metal systems. In this case the ionization increases the effective oxidation state of the metal, which contracts the metal orbitals and reduces the ability of the remaining t_{2g} electrons to backbond to the carbonyls. We have observed similar oxidation state effects in other systems.[37]

We have now gained experience in observing vibrational fine structure in a variety of other transition metal systems.[21,28,37–41] A set of molecular, electronic, and vibrational characteristics of these systems have emerged that indicate favorable circumstances for observation of these features. These criteria often assist in assigning the positive ion electronic state. The vibrational progression is most clearly observed when excitation of only a single normal mode in a particular frequency range dominates the band contour. Although large molecules have a large number of normal modes, there are effective "selection rules" that severely restrict the number of modes that can lead to observable progressions, particularly for transition metal complexes. The most important factor governing the occurrence of a progression is that a change in equilibrium bond distance must be caused by the particular ionization. The typical localization of metal-centered orbitals and normal vibrational modes dictates that most modes will not give extensive progressions. Even if the particular orbital is delocalized, local electronic symmetry and normal mode symmetry can assist the observation of vibrational fine structure because the totally symmetric mode normally dominates. Stevens *et al.*[42] have also observed this effect in the electron detachment spectra of $Ni(CO)_n^m$ ($n = 1 - 3$; $m = 0$ and -1), where only one vibrational mode, the totally symmetric C—O stretching mode, is significantly activated by the ionization process.

2.3. Jahn–Teller Effects and Angular Distortions in Metal Complex Positive Ions

The Jahn–Teller theorem has found many applications in the interpretation of the electronic structure and geometries of transition metal complexes. The theorem is most often illustrated with examples of splittings in the visible/UV spectra of transition metal complexes. Interestingly, discussion of Jahn–Teller effects in the photoelectron spectra of transition metal complexes is practically nonexistent,† even though Jahn–Teller splittings have been identified in the photoelectron spectra of small molecules (such as the $^2T_2$ ground ion state of methane[45–49] and the $^2E'$ ground ion state of cyclopropane[50,51]) and given much attention. Observation of Jahn–Teller splittings in the ionizations of metal complexes requires high-quality photoelectron spectra. We will illustrate here a "textbook" case of the Jahn–Teller effect in the photoelectron spectrum of $Fe(CO)_5$.[52] This example will demonstrate the factors that determine when the Jahn–Teller effect is important to the geometry of metal complex ions. This example is also significant because iron carbonyl ions and clusters have received extensive attention in gas phase and matrix isolation studies. The $Fe(CO)_5^+$ ion is an important member of the general class of metal pentacarbonyls that have been studied extensively both experimentally and theoretically. It is a reactive radical species that is isoelectronic with the d^7 [$Mn(CO)_5$] fragment.

The essential features of the neutral molecule electronic structure are provided entirely by the well-known symmetry description of the D_{3h} complex, shown in Figure 5. Attention is focused on the valence e' and e'' molecular orbitals. These orbitals are primarily metal in character and are completely occupied to constitute the formal d^8 configuration. Both are the correct symmetry for backbonding into empty carbonyl π^* orbitals, but only the e' orbitals, which are oriented in the equatorial plane of the trigonal bipyramid, are also the correct symmetry to interact with the carbonyl σ donor orbitals. This filled-filled σ interaction tends to destabilize the e' molecular orbitals relative to the e'' molecular orbitals and makes the e' the highest occupied orbitals in the complex.

This qualitative description of the electronic structure agrees with the photoelectron spectrum of the complex. Two distinct, predominantly metal-based valence ionization bands are observed in the region of 7- to 11-eV

† Jahn–Teller splitting has been assigned in the spectra of $M(NR_2)_4$ complexes.[43,44] The orbitals involved are primarily nitrogen lone pairs rather than metal *d*.

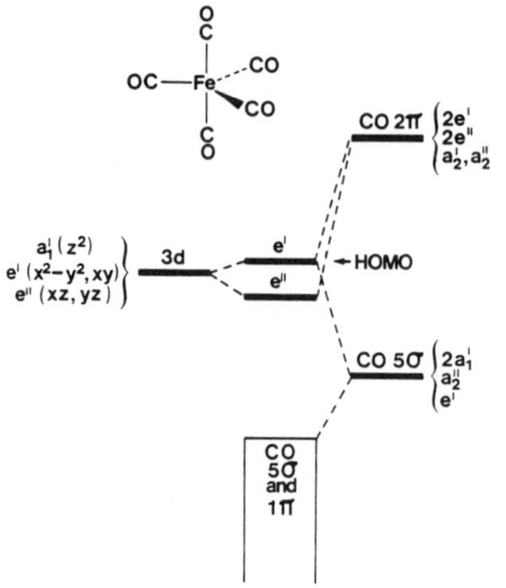

Figure 5. Orbital symmetry and approximate energy interaction diagram for D_{3h} $Fe(CO)_5$.

ionization energy, as shown in Figure 6. The band near 8.5 eV is assigned to production of the $^2E'$ ground ion state, and the band near 10 eV is assigned to production of the $^2E''$ excited ion state. Close examination of the profile of the ionization to the $^2E'$ state shows that it is really composed of two bands of rougly equal intensity and shape, as expected for a Jahn–Teller splitting of this state. The ionization corresponding to the $^2E''$ excited ion state is adequately modeled with a single band and does not show evidence of a Jahn–Teller splitting. Both the $^2E'$ and $^2E''$ states are doubly degenerate and susceptible by symmetry to Jahn–Teller splitting. The remainder of this section will discuss why one state displays the splitting and the other does not.

Symmetry considerations also determine, in addition to the qualitative orbital description, which normal mode vibrational distortions are able to couple appropriately with the degenerate electronic states of the ion. In this case both states couple with the same symmetry vibrations. The two normal vibrational modes found to be most active in the Jahn–Teller coupling are e' in symmetry. An elegant vibrational analysis of $Fe(CO)_5$ has been carried out by Jones *et al.*[53] The relative nuclear displacements of these modes, labeled

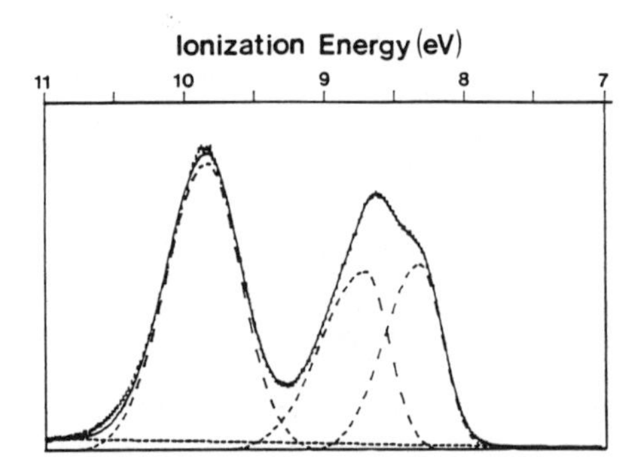

Figure 6. Ionization to the lowest positive ion states of $Fe(CO)_5^+$ at room temperature. The ionization between 8 and 9 eV requires two approximately equivalent asymmetric Gaussian bands to adequately model the profile, as expected for a Jahn–Teller splitting. The ionization near 10 eV is represented by a single asymmetric Gaussian band.

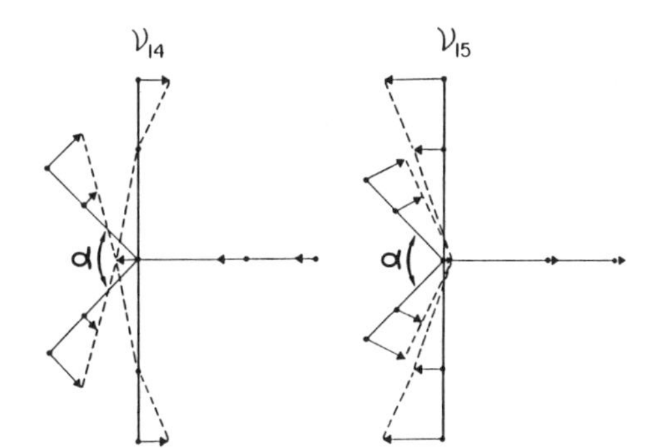

Figure 7. The relative nuclear displacements for the low frequency e' normal vibrational modes of $Fe(CO)_5$.

ν_{14} and ν_{15}, are shown in Figure 7. These vibrations basically represent in-plane bending of two of the equatorial carbonyls from the bond angle of 120°, combined with in-phase (ν_{14}) and out-of-phase (ν_{15}) bending of the two axial carbonyls. The mode ν_{15} has a low frequency of 74 cm^{-1} and corresponds to distortion along the path of a Berry pseudorotation process[54] toward a C_{4v} square-based pyramid geometry. The mode ν_{14} also has a low frequency of 105 cm^{-1}.

The angle α between two of the equatorial carbonyls defines the magnitude of the nuclear displacements. The potential surface for a degenerate e' vibration is represented as a cylindrically symmetric well about the D_{3h} geometry ($\alpha = 120°$). The most probable angle at a given temperature is not 120° because of the circular area of the well for angles away from the well minimum. (This is similar to the most probable distance of a hydrogen atom electron from the nucleus not being zero because of the volume in the shell $r + \delta r$.) This means that the most probable Franck–Condon transition does not occur from the minimum of the well, where the ion states are degenerate (and the area term is zero), but from some finite distortion away from $\alpha = 120°$, where the ion states are split. This is the semiclassical interpretation of the apparent splitting of the ionization in the photoelectron spectrum.

Figure 8 shows the relationship of the vibrational probability distribution to the energy dependencies of the ion states and the observed photoelectron spectrum. The bottom of the figure shows the vibrational probability distribution of α for the mode ν_{15} at 300 K. At this temperature the most probable angle is about 6° away from 120°. Theoretical calculations show that the $^2E'$ ground state of the positive ion has a strong dependence on this distortion, splitting into the lower-symmetry $^2A_1$ and $^2B_2$ states. The splitting of the $^2E''$ state is much less for an equivalent distortion. The calculated energy dependencies shown in Figure 8 model the edges of the ion state potential wells, and the mapping of the neutral molecule vibrational distribution onto these states gives an excellent account of both the splitting and the bandwidths observed in the photoelectron spectrum.

This analysis prompted us to consider the effects of increasing temperature on the photoelectron spectrum. Normally, higher temperatures simply increase

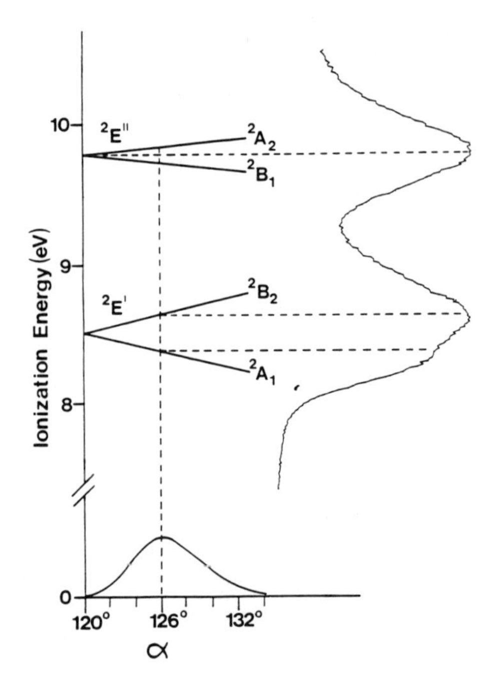

Figure 8. Relationship between the vibrational analysis, electronic structure calculations, and experimental photoelectron spectrum of $Fe(CO)_5$. The vibrational probability distribution for the angle α between equatorial carbonyls is shown at the bottom. The calculated energy dependencies of the positive ion states on the angle α are shown above, and the correspondence of vertical transitions to the observed spectrum (on edge at the right) is indicated.

the intensities of hot bands observed in the spectrum. In this case higher temperatures make the vibrational distribution in Figure 8 more diffuse, and the maximum probability of α occurs at a larger nuclear displacement from D_{3h}. This has the effect of broadening the ionization bands, generally shifting them to lower ionization energy because of more population in higher levels in the neutral molecule, and increasing the Jahn–Teller splitting. The spectrum collected at 473 K shows exactly these effects, and the 25% increase in the splitting of the $^2E'$ ion state is accurately predicted by the calculations. A splitting in the $^2E''$ state also begins to emerge.

The observed splitting of the $^2E'$ ionization indicates that the most stable geometry of d^7 $Fe(CO)_5^+$ in the gas phase is not D_{3h}. It is most likely that the geometry of $Fe(CO)_5^+$ is proceeding toward C_{4v} symmetry. Extended Hückel calculations slightly favor the C_{4v} geometry, although these calculations indicate that the D_{3h} and C_{4v} geometries are energetically very close for d^7 $M(CO)_5$ complexes.[55] An *ab initio* calculation on $Mn(CO)_5$, which is isoelectronic with $Fe(CO)_5^+$, indicates that the D_{3h} geometry is about 0.44 eV less stable than the C_{4v} geometry.[56] The observed Jahn–Teller splitting in the $^2E'$ ionization of $Fe(CO)_5$ provides an indication of the stability gained by $Fe(CO)_5^+$ when it distorts from D_{3h} symmetry. The ionization energy at the center of the Jahn–Teller-split bands represents the D_{3h} geometry. The adiabatic ionization energy represents the most stable position of the positive ion potential well. The difference between these two energies is the stabilization gained in relaxing to the favored geometry. It is not possible to derive an exact value for this stabilization from the experimental spectrum because of uncertainty in observation of the exact adiabatic ionization energy and because of contributions from changing bond distances, but it is possible to place a lower limit on this stabilization at about 0.4 eV, in good agreement with the *ab initio* calculation.

The identification of a degenerate electronic state in the ion is clearly not sufficient to determine that the geometry will be sensitive to Jahn–Teller distortion. The critical factors in the case of $Fe(CO)_5$ are first that the total energy of the neutral molecule is relatively insensitive to normal motions such as ν_{15}. Thus, this vibration is low in frequency and the absolute amplitude of the nuclear motion is relatively large. Second, the energies of the positive ion states are much more sensitive than the energy of the neutral molecule to these distortions from D_{3h} symmetry. The ground $^2E'$ state is most sensitive to these distortions because of the partial σ interaction between the equatorial metal orbitals and the carbonyls. The change in σ interaction with the low-frequency vibration is apparently crucial since a large number of degenerate ion states with predominantly π interactions with the ligands have shown no evidence of Jahn–Teller effects on the vibrational progressions or ionization band profiles.

3. ELECTRON DISTRIBUTION AND BONDING IN THE POSITIVE ION

As shown to this point, ionizations to positive ion states can give a measure of the electron delocalization, bonding, and structure of the metal complex ion as related to the neutral molecule. The ionizations can be identified as metal-based or ligand-based, and the number of metal-based ionizations, including degeneracy, determines the formal d electron count (and in turn the formal oxidation state) of the metal. In many cases the assignment of the formal d electron count and oxidation states based on the ionizations is quite clear and agrees directly with qualitative schemes of electron counting. For instance, the spin–orbit splitting in the ionization of $W(CO)_6$ shows that the metal should be considered d^6, and the Jahn–Teller splitting and intensity of the $Fe(CO)_5$ valence ionizations show that the metal should be considered d^8.

However, in other cases the qualitative electron counting scheme may be ambiguous in dividing electrons between the metal and the ligands, or the assignment of a formal (integral) oxidation state may be misleading in relation to the actual distribution of electron charge in the neutral molecule and distribution of positive potential in the complex ion. As mentioned in Section 1, in most metal complex ions the primary positive charge and the lowest empty orbital are considered localized on the metal center. This generalization may not be true if the associated neutral metal complex is electron rich, meaning that certain antibonding orbitals with the ligands are occupied, or if the ligands themselves have inherently low electronegativity or occupied antibonding orbitals. These factors have important implications for the stability, geometry, reactivity, and photochemistry of the complex and the ion.

A simple example is the case of nitrosyl bound to a metal atom. The nitrosyl has alternatively been counted as a NO^+ cation, a neutral $NO\cdot$ radical, and a NO^- anion. This qualitative electron counting problem has been avoided by simply assigning electrons to the metal–nitrosyl as a combined unit,[57] but this still leaves the question of distribution of charge in the molecule.

In this section we will show how examination of the photoelectron ionizations clarifies the classification of formal oxidation states, distribution of charge, and bonding in metal complexes. The question we will focus on is determination of the nature of the ion in its ground electronic state. This will help demonstrate some of the factors that need to be considered when comparing the chemistry of ions in the gas phase to the chemistry of neutral molecules.

These points are illustrated by the photoelectron study of the cyclopentadienyl dicarbonyl complexes of cobalt and rhodium.[39] The specific molecules included in the study (for reasons that will become apparent) are $CpCo(CO)_2$, $Cp^*Co(CO)_2$, $CpRh(CO)_2$, and $Cp^*Rh(CO)_2$ [where Cp is η^5-C_5H_5 and Cp* is η^5-$C_5(CH_3)_5$]. The geometry and basic orbital interaction diagram of these complexes is shown in Figure 9. The orbital interaction diagram is constructed from the e_1'' orbitals of the cyclopentadienyl ring (designated e_1^+ and e_1^- according to the symmetry with respect to the *xz* mirror plane of the molecule) and the valence orbitals of $Co(CO)_2^+$. The e_1'' orbitals

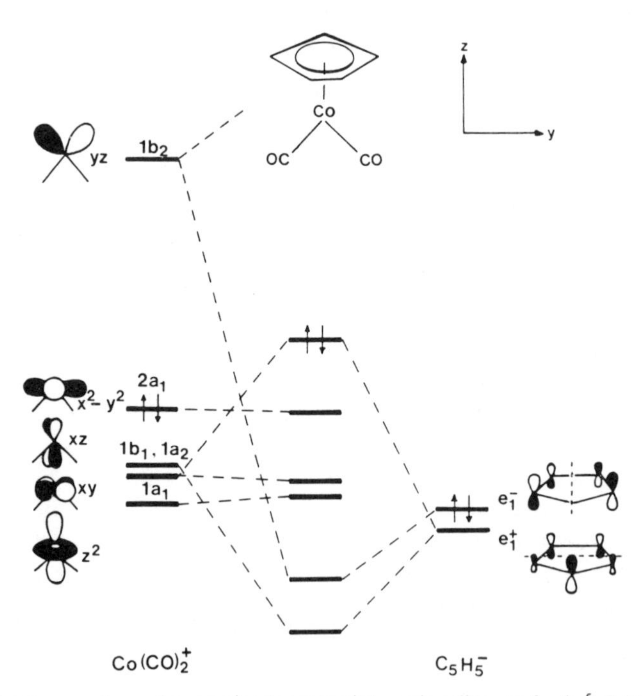

Figure 9. Orbital symmetry and approximate energy interaction diagram for $(\eta^5\text{-}C_5H_5)Co(CO)_2$.

of the cyclopentadienyl ring are considered filled, giving the cylopentadienyl ring a formal mononegative charge and leaving eight valence electrons for the metal dicarbonyl. Thus, ionization to six different ion states is expected in the low-valence region, with four having orgins traced back to the metal d orbitals and two having origins traced back to the π orbitals of the ring. The complex is electron rich with filled bonding and antibonding interactions taking place between the ring e_1^+ orbital and the metal d_{xz}. The antibonding combination of the ring e_1^+ and the metal d_{xz} is expected to be the highest occupied orbital in the complex.

The valence He I photoelectron spectra of these complexes are shown in Figure 10. The spectrum of $CpCo(CO)_2$ gives evidence of six valence ionizations as indicated by the above analysis. The He I band intensities give the first indication of the nature of the ionizations. The fifth ionization band (9.87 eV) is much more intense in this spectrum than the other ionization bands, showing that this ionization is largely associated with the carbon π character in the cyclopentadienyl ring. This band is labeled ring I. The sixth ionization band is intermediate in intensity, suggesting a mixing of ring and metal character. The position of this band suggests that it be labeled ring II. The first through fourth ionization bands are lowest in He I intensity and therefore presumably highest in metal character, and will be labeled metal I through metal IV. These labels are only provided for bookkeeping purposes to keep in mind their primary ancestry in the orbital correlation diagram. The

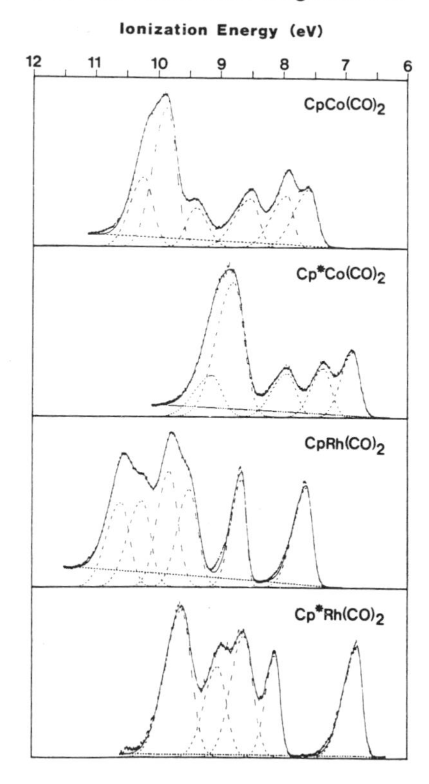

Figure 10. Valence He I photoelectron spectra of $CpCo(CO)_2$, $Cp^*Co(CO)_2$, $CpRh(CO)_2$, and $Cp^*Rh(CO)_2$ [where Cp is η^5-C_5H_5 and Cp* is η^5-$C_5(CH_3)_5$].

labels are not meant to imply the actual charge distribution and bonding in the positive ion state. It is unwise to attempt to derive this information from a single photoelectron spectrum, even with the aid of sophisticated calculations, on a molecule of this complexity.

The variety of closely spaced ionizations displayed by these complexes dictate that a broad arsenal of approaches and ionization techniques be applied to obtain a definitive interpretation of the electronic ground state of the ion. The relevant techniques are He I/He II intensity comparisons, ionization energy comparisons with ring methylation and metal substitution, correlation of core and valence ionization energy shifts, and examination of detailed ionization band profiles and fine structures. In other cases any one of these techniques might be sufficient to establish the significant electronic features. The principles of these approaches have been reviewed recently.[14] Here it is appropriate to briefly summarize the information obtained on these systems.

Table 1 lists the vertical ionization energies and relative integrated ionization band areas from the He I and He II photoelectron spectra of these complexes. The "heavy atom effect" (increase in He I photoelectron cross

Table 1. Photoelectron Energies and Relative Cross Sections for $CpM(CO)_2$ Complexes

Complex	Vertical ionization energy (eV)	Relative integrated peak area		Assignment
		He I	He II	
$CpCo(CO)_2$	7.59	1.00	1.00	Metal I
	7.95	0.84	1.09	Metal II
	8.51	0.89	1.06	Metal III
	9.41	0.62	0.63	Metal IV
	9.87	2.35	1.37	Ring I
	10.23	1.18	0.79	Ring II
$Cp^*Co(CO)_2$	6.88	1.00	1.00	Metal I
	7.37	0.84	1.62	Metal II
	7.95	0.85	1.49	Metal III
	8.79[a]	3.07	3.48	Metal IV, Ring I
	9.15	0.94	0.76	Ring II
CpRh(CO)	7.64	1.00	1.00	Metal I
	8.65	0.87	1.40	Metal II
	9.50	1.03	1.62	Metal III
	9.82	1.22	0.70	Ring I
	10.25	1.04	0.83	Metal IV
	10.59	1.02	0.71	Ring II
$Cp^*Rh(CO)_2$	6.80	1.00	1.00	Metal I
	8.12	0.69	1.56	Metal II
	8.62	1.27	1.00	Ring I
	9.05	0.95	1.35	Metal III
	9.63[a]	1.79	2.16	Metal IV, Ring II

[a] Two overlapping peaks under band.

section with metal atom substitution down a group[19]) is visually apparent in comparison of the cobalt and rhodium complex spectra in Figure 10. Table 1 shows that the area of the first ionization band (labeled metal I) is consistently greater than the areas of bands metal II through metal IV when they can be independently observed in the He I spectra, indicating that the first ionization has more ligand-type nature associated with it than bands metal II through metal IV. This does not indicate whether the ligand nature is associated with backbonding to the carbonyls or delocalization with the π system of the ring. The changes in intensities from the He I spectra to the He II spectra also indicate the intermediate metal–ligand nature of the first ionization. Metal-based ionizations have increasing cross section relative to organic ligand-based ionizations in going from He I to He II excitation.[9–11] In comparison to a relative area of 1.0 for the first ionization band (metal I) of these complexes, the bands metal II through metal IV increase in relative He II cross section, and the ring bands decrease in relative cross section.

The methylation of the cyclopentadienyl ring and the substitution of rhodium for cobalt are chemical approaches to perturbing the electronic structure of the complex. The effects of these perturbations on the core and valence ionization energies are shown in Table 2. Methylation of the ring has the greatest effect on the ionization states that mix with the e (C_{3v} symmetry) combination of the C—H bonds.[28] Thus, the ionizations ring I and ring II associated with the π orbitals of Cp show the largest shifts between the Cp and Cp* complexes. The metal II through metal IV ionizations shift about half as much. The metal core ionization shifts slightly more than the metal II through metal IV ionizations, consistent with Jolly's principle on the correlation of core and valence ionization energies for localized orbitals.[58] On the other hand, the first ionization (labeled metal I) shifts more with methylation than

Table 2. Shifts in Core and Valence Ionizations Energies (eV[a]) Resulting from Ring Methylation and Metal Subsitution

Ionization	$CpCo(CO)_2$ to $Cp^*Co(CO)_2$	$CpRh(CO)_2$ to $Cp^*Rh(CO)_2$	$CpCo(CO)_2$ to $CpRh(CO)_2$	$Cp^*Co(CO)_2$ to $Cp^*Rh(CO)_2$
Carbon core 1s (Cp)	−0.17	−0.23	−0.02 (9)	−0.08 (4)
Carbon core 1s (CO)	−0.65	−0.88	+0.3 (3)	+0.1 (2)
Oxygen core 1s	−0.51	−0.56	+0.07 (8)	+0.02 (7)
Metal core	−0.63[b]	−0.66[c]		
Valence band metal I	−0.71	−0.84	+0.05	−0.08
Valence band metal II	−0.58	−0.53	+0.70	+0.75
Valence band metal III	−0.56	−0.45	+0.99	+1.10
Valence band metal IV	−0.62[d]	−0.62[d]	+0.84	+0.84
Valence band ring I	−1.08[d]	−1.20	−0.05	−0.17
Valence band ring II	−1.08	−0.96[d]	+0.36	+0.48

[a] Shifts to lower binding energy are negative.
[b] Cobalt $2p_{3/2}$.
[c] Rhodium $3d_{5/2}$.
[d] Bands overlap in the pentamethylcyclopentadienyl complexes.

the metal core ionization. This directly indicates that the first ion state has significant overlap and mixing with the methyl *e* orbitals, meaning that this state is partially delocalized into the π system of the Cp ring.

The photoelectron spectra of the rhodium complexes in Figure 10 appear dramatically different from the spectra of the cobalt complexes. The changes are almost entirely due to the 0.7- to 1.1-eV shifts of ionizations metal II–IV, with a smaller shift in ring II and negligible shifts in metal I, ring I, and the core ionizations. The negligible shift of the first ionization emphasizes its ligand nature. The constant core ionizations indicate that the charge distribution and potential that the metal and ring orbitals experience is not significantly altered by rhodium is place of cobalt. This is reasonable in view of the very similar CO stretching frequencies,[39] proton NMR shifts,[39] and structural features[59] of these cobalt and rhodium analogues, all of which are sensitive to the neutral molecule electron distribution. It is surprising that the electron distribution in the neutral molecule is so constant if the metal levels are destabilized as much as 1 eV from cobalt to rhodium as indicated by a simplistic *ground state* orbital interpretation of the shifts of the metal II–IV ionizations.

The differences in ionization energies between $CpCo(CO)_2$ and $CpRh(CO)_2$ are traced to electron relaxation in the positive ions, and not features of the neutral molecule. It has been found both experimentally[60–63] and theoretically[64–67] that electron relaxation energies can be substantial for transition metal complexes, particularly for first-row metals to the right of the transition metal series and for d^8 configurations.[65,66] In addition, electron relaxation energies are expected to decrease on descending a column of the periodic table. *Ab initio* calculations show that cobalt 3*d* ionizations will have almost twice the relaxation energy of rhodium 4*d* ionizations.[65] This difference in relaxation energy accounts for the lower ionization energy of cobalt-based ionizations compared to rhodium-based ionizations. Although the neutral molecules may have similar descriptions and properties, these relaxation factors must be taken into account when comparing the gas phase ion properties and chemistry. Metal complexes that appear similar in neutral ground state properties can be much different in ion state properties.

Electron reorganization can also affect the charge potential throughout the molecule. If the molecular electron density is sufficiently fluid, as is often the case in organometallic systems, removal of an electron from a metal-based orbital to create the ion may be accompanied by extensive reorganization of electron density from the ligands to the metal. The flow of density toward the metal may be such that the net change in charge potential at the metal is very low, and the increase in positive charge potential is largely distributed among the ligands.

In the photoelectron spectra of the rhodium complexes the first ionization band is sufficiently well separated from the other ionizations to allow detailed examination of the ionization band profile. The close-up of the first ionization band of $Cp^*Rh(CO)_2$ is shown in Figure 11. Three components of a CO stretching vibrational progression are clearly observed. In comparison to the two vibrational components observed in the spectra of the d^6 metal hexacar-

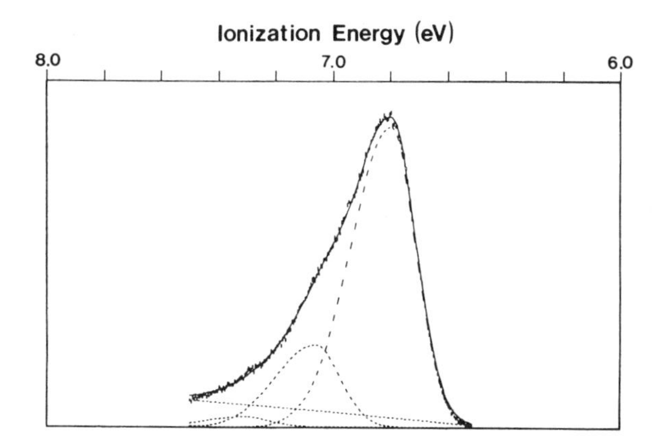

Figure 11. Detailed examination of the first ionization band of $[\eta^5\text{-}C_5(CH_3)_5]Rh(CO)_2$.

bonyls, the intensities of these three components indicate more extensive delocalization of the metal electrons into the carbonyls and a more substantial change in CO bond distance when an electron from the orbital correlating with the first ionization band is removed. The frequency spacing between vibrational components in the spectrum of $Cp^*Rh(CO)_2$ is $2270 \pm 80\ cm^{-1}$, which is considerably greater than the 2012 and 1947 cm^{-1} frequencies in the neutral molecule.[39] This is the most extensive delocalization into the carbonyl π^* orbitals that we have observed from vibrational structure in an ionization.

Thus, there are several independent indications from the spectroscopy that the unpaired electron in the ground electronic state of the ion is extensively delocalized throughout both the π system of the cyclopentadienyl ring and the π^* orbitals of the carbonyls. This electron would formally be assigned to the metal in normal electron counting schemes, but in fact only six electrons, corresponding to the ionizations metal II, metal III, and metal IV, can clearly be assigned to the metal. Although the simple electron counting schemes are usually supported by the ionizations, we have observed other important cases in which formal oxidation state and *d*-electron count descriptions are misleading in comparison to the actual electron distributions.[15,38]

4. IONIZATION ENERGY–BOND ENERGY RELATIONSHIPS

4.1. Fundamental Relationships

Knowledge of individual thermodynamic bond energies of molecules and fragments with metals is essential for understanding and systematizing the chemistry of transition metal systems. Unfortunately, the values of many fundamental thermodynamic quantities for key transition metal interactions are only roughly estimated or not known at all.[68] At the most basic level the ionization energy is a well-defined thermodynamic quantity that can be precisely measured. The ionization energy is related to other thermodynamic

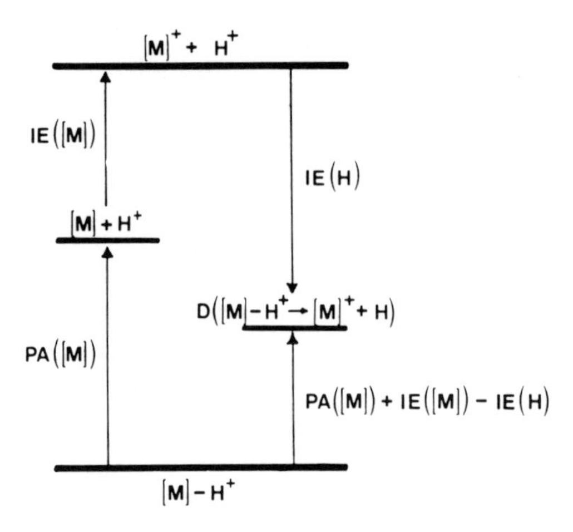

Figure 12. Energy cycle relating the proton affinity and ionization energy of a complex [M] to the homolytic dissociation of a hydrogen atom from the complex ion.

quantities through energy level diagrams, such as the familiar contribution of atomic ionization energies to the understanding of ionic bonds through the Born–Haber cycle. Gas phase ion experiments are able to obtain other thermodynamic information that goes hand in hand with molecular ionization energies. For example, mass spectrometry can measure the energy (appearance potential) for production of a molecular fragment ion from a molecule. Subtraction of the first ionization energy of the molecule from the appearance potential for creation of the fragment ion gives the dissociation energy of the bond in the molecular ion that produces the fragment. Another well-known and important example is the determination of gas phase proton affinities (the energy of $[M]{-}H^+ \rightarrow [M] + H^+$) by ion cyclotron resonance.[69] Knowledge of the proton affinity (PA) of a metal complex [M] and the ionization energy (IE) of the complex [M] is sufficient to determine the homolytic bond dissociation energy (D) of $[M]{-}H^+$ to $[M]^+$ and H•. The conservation of energy equation for these quantities is:

$$D\{[M]{-}H^+ \rightarrow [M]^+ + H\bullet\} = PA\{[M]\} + IE\{[M]\} - IE\{H\} \tag{4}$$

The energy cycle relating these quantities is shown in Figure 12.

The molecular ionization energies are also directly significant in relation to metal–ligand bond energies. Figure 13 shows the connection between ionization energies and bond dissociation energies from the neutral molecule and the ground state of the positive ion. The equation relating these quantities for a ligand L bound to a metal complex [M] to form an [M]—L bond is†:

$$IE\{[M]{-}L\} - IE\{[M]\} = D\{[M]{-}L \rightarrow [M] + L\} - D\{[M]{-}L^+ \rightarrow [M]^+ + L\} \tag{5}$$

Knowledge of any three of the quantities in this equation gives knowledge of the fourth. This equation is used extensively by Herzberg,[70] most often to

† If the ionization energies used in Equation 5 are not the adiabatic ionization energies, then the dissociation energies must be corrected for the vibrational excitation.

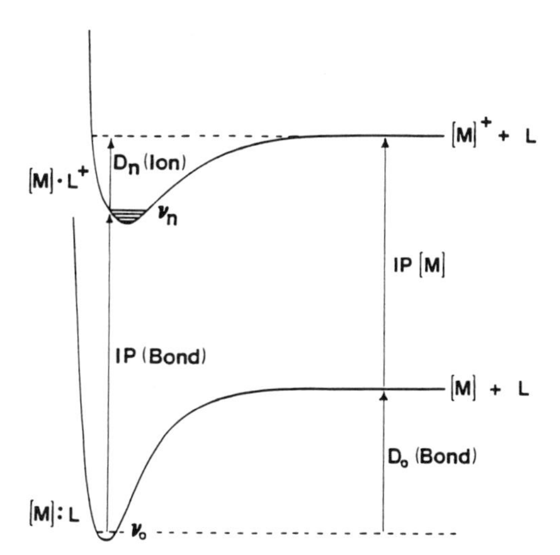

Figure 13. Relationship of the ionization energies of a [M]—L complex and the [M] fragment to the [M]—L bond potential wells in the neutral molecule and the ion.

obtain the depth of the well for a diatomic molecule positive ion from the ionization energy of the diatomic molecule, the ionization energy of one of the atoms, and the bond energy of the neutral molecule. If the bond dissociation energy of the positive ion state is known from gas phase ion experiments, the ionization energies can be used to extend this information to bond energies of neutral complexes.

The ionization energy of the coordinatively unsaturated metal complex [M] may be difficult to obtain directly (*vide infra*). Even if this value is not known, it is still possible to obtain a measure of the relative bond energies of different ligands coordinated to the same metal complex. The ionization energy of the common metal complex fragment [M] cancels in the comparison of the bonding of the two ligands, and the change in ionization energy between the two [M]—L complexes is related to the changes in [M]—L bond dissociation energies:

$$\Delta \text{IE}\{[\text{M}]\text{—L}\} = \Delta D\{[\text{M}]\text{—L} \rightarrow [\text{M}] + \text{L}\} - \Delta D\{[\text{M}]\text{—L}^+ \rightarrow [\text{M}]^+ + \text{L}\} \quad (6)$$

Thus, the ionization energies can be used to relate the changes in bond energies observed in positive ions to the changes in bond energies of neutral molecules. Knowledge of the relative changes in bond energies is sufficient for equilibrium processes. Furthermore, if the metal-ligand bond energy for one ligand of the series is known from independent information, then the bond energies for all ligands in the series can be obtained.

With the limited amount of data presently available on bond energies of transition metal-containing molecules and ions, it is more often the case that only the ionization energies are known. If this the case, it is possible to take advantage of simple relationships for the depth of the positive ion potential well in comparison to the neutral molecule potential well. For example, if the bond is entirely covalent, removal of an electron from the bond reduces the bond order by one-half, and the bond energy in the ion is expected to be approximately one-half the bond energy in the neutral molecule. The expression for the bond energy in terms of the ionization energies is then

$$\mathrm{IE}\{[\mathrm{M}]\text{—L}\} - \mathrm{IE}\{[\mathrm{M}]\} = \frac{1}{2} D\{[\mathrm{M}]\text{—L} \rightarrow [\mathrm{M}] + \mathrm{L}\} \tag{7}$$

Application of this approximation to the bond energy of the hydrogen molecule yields a predicted bond dissociation energy of 4.6 eV,† which compares with the known bond dissociation energy of 4.5 eV.[71] If the interaction between the atoms is *entirely* ionic, then removal of an electron from the anion completely removes the ionic bond. In this case the homolytic dissociation energy is directly equated to the left side of Equation 7. A generalized expression is:

$$\mathrm{IE}\{[\mathrm{M}]\text{—L}\} - \mathrm{IE}\{[\mathrm{M}]\} = (1 - K) D\{[\mathrm{M}]\text{—L} \rightarrow [\mathrm{M}] + \mathrm{L}\} \tag{8}$$

where K tends toward a value of one-half for ionization associated with a covalent bond and a value of zero for ionization associated with an ionic bond. If the depth of the potential well is not affected by the ionization, then the value of K is one. Stronger bonding in the positive ion state than in the neutral molecule is associated with values of K greater than one, and ionizations to repulsive states correspond to negative values of K.

If the only quantity that is known is the ionization energy of the neutral molecule, then Equations 6 and 8 can be combined to compare the bond energies of different ligands to a common metal fragment:

$$\Delta\mathrm{IE}\{[\mathrm{M}]\text{—L}\} = (1 - K)\Delta D\{[\mathrm{M}]\text{—L} \rightarrow [\mathrm{M}] + \mathrm{L}\} \tag{9}$$

This equation requires the assumption that the [M]—L bonds and ionizations for the two ligands fall in the same class, so that the value of K in each case is essentially the same.

4.2. $Mn(CO)_5$ Bound to H, CH_3, and $Mn(CO)_5$

As a simple example of the application and current limitations of these relationships, consider the Mn—H and Mn—CH_3 bond energies in $Mn(CO)_5H$ and $Mn(CO)_5CH_3$. These bond energies have been reported to be 2.2 eV and 1.6 eV, respectively, on the basis of microcalorimetric measurements.[72] The adiabatic ionization energies associated with the Mn—H and Mn—CH_3 bonds of these complexes are approximately 10.0 eV[73] and 9.3 eV,[74] respectively. The relative stabilities of the bonds and the ionizations are in the same order, with a difference of 0.6 eV from calorimetry and 0.7 eV from the ionizations.

† The assumption of removing half the bond energy with removal of one of the bond electrons is applied at the equilibrium internuclear separation in the neutral molecule. Therefore, the vertical ionization of the hydrogen molecule at 15.9 eV is used in Equation 7.

These values can be used in Equation 9 to determine K, which in turn can be supplied to Equation 8 to estimate the ionization energy of the $Mn(CO)_5^{\bullet}$ fragment. The estimated first ionization energy of the $Mn(CO)_5^{\bullet}$ fragment on this basis is 7.4 eV.† This is reasonable in comparison to the higher first ionization energy of $Mn_2(CO)_{10}$ (~8.05 eV vertical[73] and 7.7 eV adiabatic[72]) that is expected from the stabilization of bonding between the two $Mn(CO)_5$ fragments. It is also reasonable in relation to an extrapolation of the ionization information from $Fe(CO)_5$ (see Section 2.3) which produces the isoelectronic d^7 $Fe(CO)_5^+$. This information yields an ionization energy of the fragment between 7.6 and 7.2 eV for the expected angle of 100–105°[55,56] between the axial and equatorial carbonyls of C_{4v} $Mn(Co)_5^{\bullet}$.‡

An estimate of the ionization energy of the $Mn(CO)_5^{\bullet}$ fragment reported from other calorimetric data is 8.1 eV,[72] and a mass spectrometry appearance potential measurement indicates a value of 8.4 eV.[75] It is bothersome that these values are greater than the initial ionization energy of $Mn_2(CO)_{10}$. The strengths and limitations of various methods for determining individual thermodynamic bond energies have been discussed elsewhere.[76,77] The interpretation of calorimetric data is subject to large uncertainties in many cases because the reported values are the result of small differences in large numbers and may include additional assumptions necessary to relate the data to individual bond energies. The relation between spectroscopic dissociation energies and thermodynamic bond energies should also be kept in mind. Mass spectrometric determinations of bond energies are also subject to errors of measurement and interpretation.[77b] For example, we have found that the reported appearance potential for the ground ion state of $W(CO)_6$ actually corresponds to ionization to the $^2U'$ excited state of the ion.[16] Likewise, the reported appearance potential of 8.6 eV and the homolytic bond dissociation energy of $Mn_2(CO)_{10}$ state rather than to the adiabatic ionization energy of 7.7 eV.[72] This 0.7-eV error is also contained in the 9.3-eV value reported for the appearance potential of the $Mn(CO)_5^+$ ion from $Mn_2(CO)_{10}$.[75] Using the corrected appearance potential of 8.6 EV and the homolytic bond dissociation energy of $Mn_2(CO)_{10}$ to $2Mn(CO)_5^{\bullet}$ (1.1 eV based on gas phase equilibrium measurements[78]) yields a value of 7.5 eV for the ionization energy of the $Mn(CO)_5^{\bullet}$ fragment, which is in good agreement with the values indicated above from the ionization energies.

The value of K obtained from Equation 9 in the analysis of the Mn—H and $Mn—CH_3$ bonds is near zero, and is actually slightly negative. This indicates substantial weaking of the bonds in the positive ions. The general trend of weaker bonds with increasing oxidation state is a characteristic feature of organometallic chemistry. Stevens and Beauchamp[69] suggest that metal-hydrogen bonds are substantially weaker in higher oxidation states of the same

† Either the [Mn]—H or the $[Mn]—CH_3$ ionization and bond dissociation energies can be used in Equation 8. The same answer is obtained.

‡ In Section 2 the adiabatic ionization energy of $Fe(CO)_5$ was shown to be 0.4 eV away from the D_{3h} ionization (8.5 eV) when α is 5.9°. The value of α is expected to be 15–20° for $Mn(CO)_5^{\bullet}$,[55,56] and the adiabatic ionization energy is scaled directly to the expected α.

metal, and reports a value of 1.0 eV for the dissociation of a hydrogen atom from the $Mn(CO)_5^+$ ion, in comparison to 2.2 eV for dissociation from the neutral molecule. The value of K obtained above suggests an even greater bond weakening. The assumption of equivalent K values for the different bonds and ionizations is not completely correct[79] and contributes to this difference. If this approximation is lifted, then an addition assumption must be made regarding the nature of the bonds, or additional information from other sources must be incorporated.

The point to be made is that there are close and direct relationships between the factors that influence ionization energies and the factors that influence bond energies. With the present limitations in obtaining thermochemical information, these relationships are helpful in providing a check on the conclusions that are drawn from the data. There is much ambiguity in the literature on the relative strengths of fundamental metal-ligand interactions, and these additional experiments are helping to resolve these differences.

The principle of additivity of ligand electronic effects provides another framework for comparing the ionization energy of the neutral molecule [M]—L with the ionization energy of the coordinatively unsaturated fragment [M].[20,21] This principle states that it is possible to define quantitatively the individual ligand overlap/bonding and charge potential effects on a metal-based ionization energy. The principle has been applied successfully to a wide range of metal electronic environments. Equation 5 shows that if addition of the ligand L to the metal complex [M] does not change the first ionization energy of the species (left side of Equation 5 equals zero), then that ligand will be bound just as strongly to the ion as to the neutral complex. If addition of the ligand to [M] increases the first ionization energy of the complex, such as the stabilization resulting from backbonding to a carbonyl, then the increase in ionization energy is related one-to-one to a stronger bond in the neutral molecule compared to the ground state of the positive ion.

These same principles apply to higher ionization energies and excited states of the positive ion, and contribute to understanding the relative inportance of individual bond interactions to stabilization of a complex. This will be illustrated in the final section.

4.3. Electronic Factors of Carbon–Hydrogen Bond Activation

One of the most remarkable and important properties of transition metals is their ability to activate H—H, C—H, and C—C bonds. The primary electronic structure and bonding factors that contribute to the activation can be viewed in terms of two limiting case descriptions,[80] which are shown diagramatically for C—H activation in Figure 14. The first occurs when the C—H σ (bonding) orbital donates electron density into an empty metal level, thus weakening the C—H bond order (Figure 14A). We refer to this interaction as σ activation. The second, which we call σ^* activation, occurs when filled metal levels donate into the empty C—H σ^* antibonding orbital, again weakening the C—H bond order (Figure 14B). Filled-filled orbital interactions can

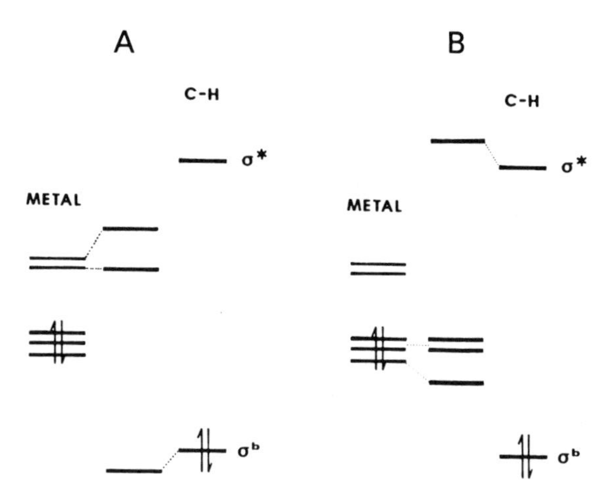

Figure 14. The σ activation (A) and σ^* activation (B) modes and their effect on the energy of metal-based ionizations.

also be important in determining the pathway and energetics at different stages of the process. These interactions have been the subject of numerous theoretical investigations.[84] The challenge remains to obtain experimental information relating to the relative energy contributions of these different interactions at different stages of the activation process.

A specific example of a C—H bond activation process that is particularly well known is the β-hydride elimination reaction. A metal with a bound alkyl group that contains a β-hydrogen atom can often undergo a facile rearrangement through a three-center C—H—M intermediate to form a metal hydride:

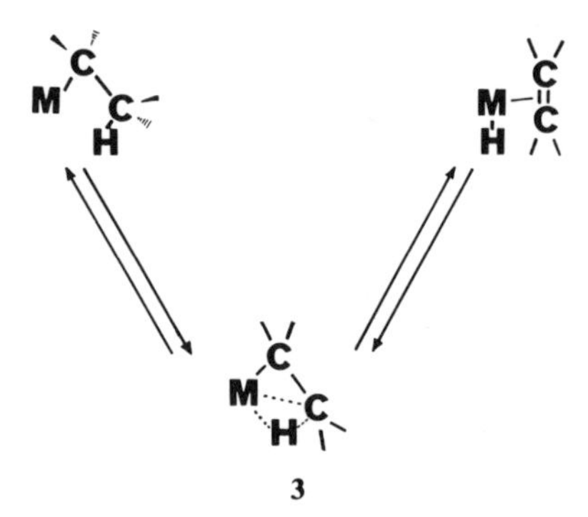

It is important to know the electronic factors that favor the metal alkyl, the metal hydride, and intermediates at any point along the path between the two. Stable complexes representative of many geometry points along the reaction coordinate are known.

A particularly interesting molecule with regard to this question is (cyclohexenyl)manganese tricarbonyl (**4**).[81]

(CO)$_3$Mn H H

4

One of the ring C—H bonds of this molecule is positioned on a coordination site of the metal, and the neutron diffraction study shows that this C—H bond is lengthened by nearly 10% in comparison to normal ring carbon-hydrogen bonds. Variable temperature NMR studies[82] show that there are two types of fluxionality for this molecule (Figure 15). In one the C—H interaction with the metal is disrupted and there is only bonding of the carbon atoms of the ring with the metal. The other fluxional process apparently proceeds through formation of the metal hydride. Thus, the ground state of this molecule is stable as an intermediate along the β hydride elimination path, and the electronic factors contributing to its stability in this form can be examined.

Theoretical calculations show that both the σ activation and σ^* activation interactions are possible for this molecule. The interactions of the C—H σ bond with an empty metal level and the C—H σ^* orbital with occupied metal levels are shown in Figure 16. Thus, this molecule provides a special opportunity to experimentally evaluate the σ and σ^* factors in a metal carbonyl. This is accomplished by examining the effects of the C—H bond interaction with the metal on the ionizations of the complex. As shown in Figure 14, if the σ^* activation mode is significant, the effect will be manifested in the first ionization band of the complex. The σ activation mode, on the other hand, will affect the C—H bond ionization region.

The foundation for the analysis is provided by our previous photoelectron studies of (η^5-cyclopentadienyl)manganese tricarbonyl[28] and (η^5-cyclohexadienyl)manganese tricarbonyl.[83]

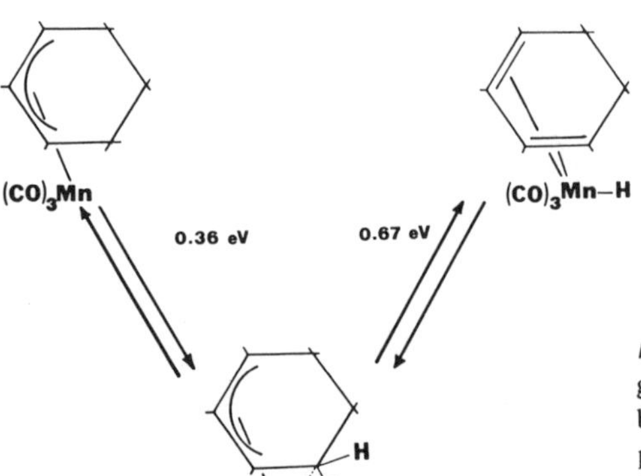

Figure 15. Fluxional processes and energies of (cyclohexenyl)manganese tricarbonyl determined from variable temperature NMR studies.[82] Note the similarity to the β-hydride elimination process shown in the text.

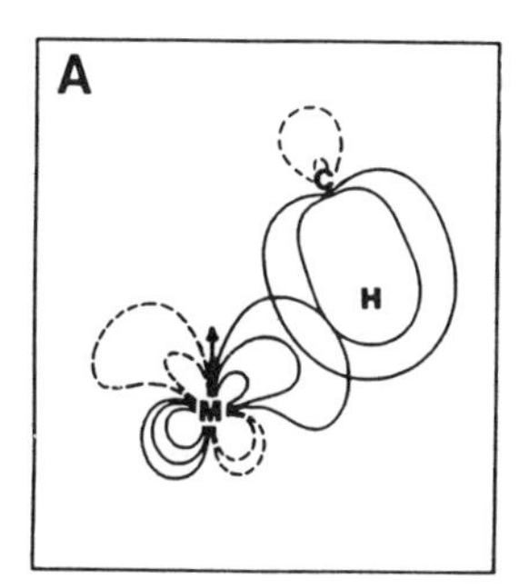

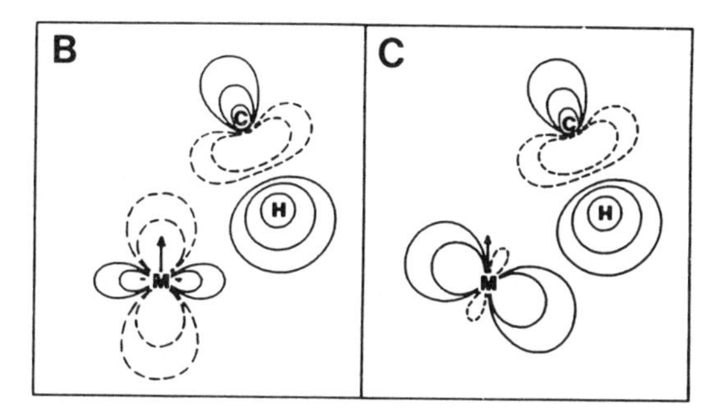

Figure 16. Calculated orbital contours illustrating the interactions of C—H orbitals with $Mn(CO)_3^+$ orbitals. The plotting plane includes one CO *trans* to the C—H and the principal z axis of the C_{3v} $Mn(Co)_3^+$ is indicated by the small arrow. The outermost contour value is 0.1 $e^-(au)^3$ and each successive inner contour value increases by a factor of 2. In (A) the filled C—H σ orbital donates to an empty metal-based orbital. In (B) and (C) the empty C—H σ^* orbital accepts density from occupied metal-based orbitals.

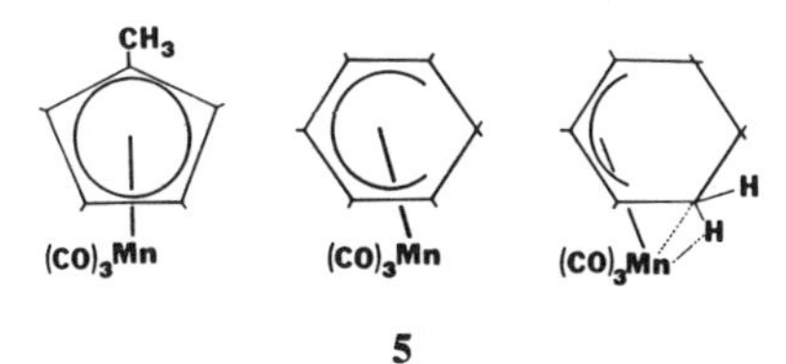

5

These studies show that the manganese tricarbonyl portion of the molecule establishes a pseudo-octahedral electronic symmetry about the d^6 metal center and the first ionization band is essentially threefold degenerate. The six-membered cylohexadienyl ring removes the near degeneracy of the two ring valence π ionizations, but does not disrupt the electronic symmetry at the metal. Figure 17 compares the He I and He II photoelectron spectra of the cyclohexadienyl complex with those of the cyclohexenyl complex that has the C—H—Mn interaction. In both cases the ring valence π ionizations split to about 8.5 and 10.5 eV. The He I/He II comparison confirms that the metal-based ionizations occur near 8 eV and remain degenerate, showing no evidence of C—H σ^* interaction. The only significant changes in ionizations are observed in the σ ionization region,[80] corresponding to donation of the C—H σ bond to the metal levels in an "agostic" interaction.[84]

It is important to note that the first ionization, aside from not showing any splitting, also does not shift (7.95 ± 0.06 eV) between the manganese complexes. Normally, an additional two-electron donation to a metal center will shift the metal ionizations to lower energy because of the increase in negative charge potential in the vicinity of the metal. However, in this case the negative charge potential donated from the C—H bond is compensated by the positive proton charge of hydrogen in the electron cloud. Thus, the electron density on manganese is not destabilized, and the bonding capabilities of the metal to the carbonyls and the ring are not significantly altered. Also,

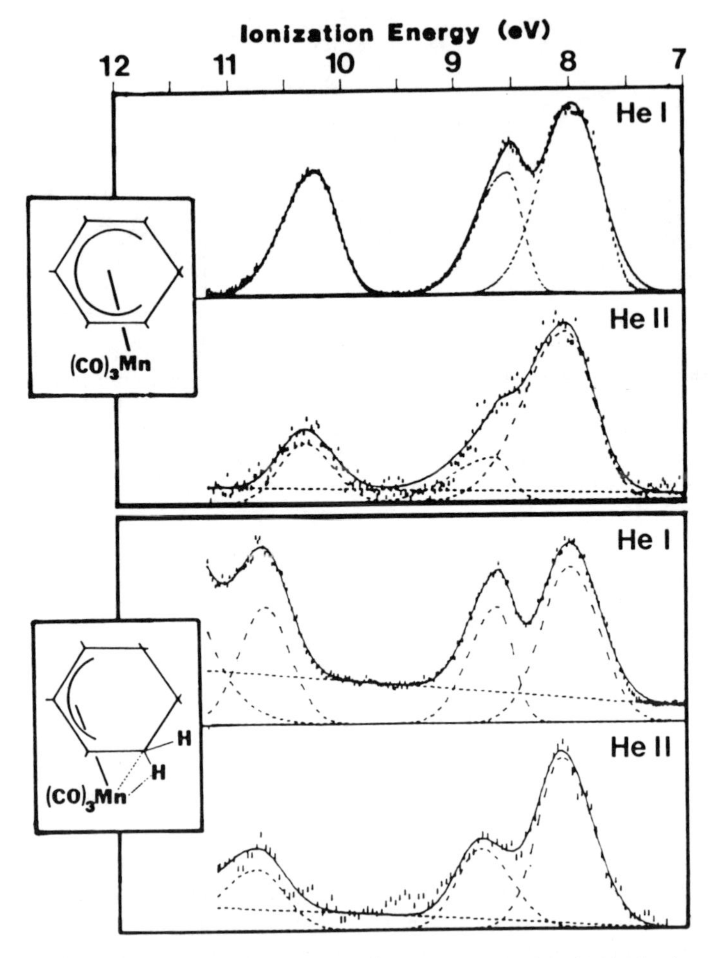

Figure 17. Valence photoelectron spectra of (η^5-cyclohexadienyl)$Mn(CO)_3$ (top) and (cyclohexenyl)$Mn(CO)_3$ (bottom). The ionization characteristics of these two molecules are very similar in this low-energy region.

because of the reasons summarized in the previous section, the lack of shift of the first ionization means that the agostic interaction will be as strong in the ground electronic positive ion state as in the neutral state of this metal carbonyl.

Figure 18 compares the ionization energy correlations for general agostic addition of a C—H bond to a metal center, β-hydride elimination, and oxidative C—H bond addition. The energies are shown to scale based on known ionization energies of relevant metal carbonyl complexes.[80] Formation of a metal hydride is generally unfavorable relative to the strength of the C—H bond unless compensated by additional stabilization elsewhere in the complex. The (cyclohexenyl)manganese tricarbonyl complex stops at the stage of agostic addition and stabilization of the three-center C—H—Mn interaction because

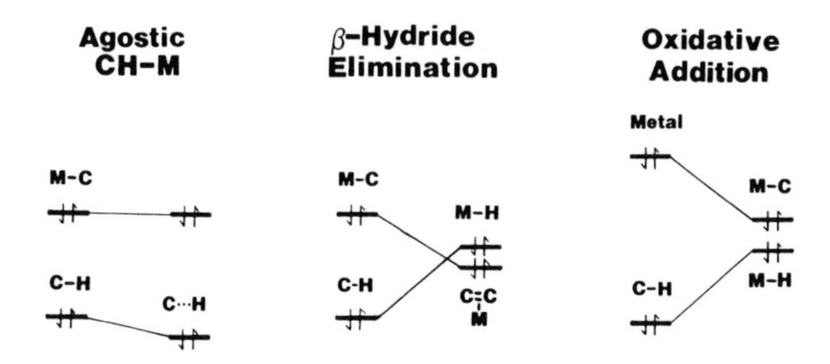

Figure 18. Ionization energy correlations for agostic C—H—M interaction, β-hydride elimination, and C—H oxidative addition. The relative ionization energies are to scale based on photoelectron studies of representative manganese carbonyl complexes with bound hydride, methyl, and olefin groups.

additional stabilization of the metal-carbon bonding does not occur to balance breaking of the carbon-hydrogen bond. If formation of a metal-olefin bond from a metal-carbon σ bond accompanies the C—H addition, the process is more likely to be favored. Likewise, oxidative addition can be favored if metal density is sufficiently stabilized in formation of M—C and M—H bonds. The ionization energies in Figure 18 show a general balance of bond stabilization and destabilization in these cases.

Much additional information is needed for application of the principles illustrated in Figure 18 to many specific reactions with transition metals. Experiments in gas phase transition metal chemistry are increasingly contributing valuable new information on metal reactions and bond energies. The photoelectron information is especially helpful in relating gas phase positive ion studies to neutral molecules and in identifying the most significant electronic structure and bonding factors. Most important, the ionization studies lead to models of transition metal behavior that help to inspire new progress in the understanding of this fascinating chemistry.

ACKNOWLEDGMENTS

We would like to thank Ann Copenhaver and Gary Darsey for helpful discussions, and Julia St. Lawrence for assistance in preparing the figures. This work has been supported by the U.S. Department of Energy (Division of Chemical Sciences, Office of Basic Energy Sciences, Office of Energy Research), the National Science Foundation, the Petroleum Research Fund, Research Corporation, and the Department of Chemistry, University of Arizona.

REFERENCES

1. Turner, D.W.; Baker, A.D.; Brundle, C.R. *Molecular Photoelectron Spectroscopy*; Wiley: London, 1970.

2. Ghosh, P.K. *Introduction to Photoelectron Spectroscopy*; Wiley: New York, 1983.
3. Brundle, C.R.; Baker, A.D. *Electron Spectroscopy: Theory, Techniques, and Applications*, Vol. 1; Academic: London, 1977.
4. Siegbahn, K.; Nordling, C.; Johansson, G.; Hedman, J.; Heden, P. F.; Hamrin, K.; Gelius, U.; Bergmark, T.; Werme, L.O. Manne, R.; Baer, Y. *ESCA Applied to Free Molecules*; North Holland: Amsterdam, 1969.
5. Hendrickson, D.N. In *Physical Methods in Chemistry*; Drago, R.S., Ed.; Saunders: Philadelphia, 1977.
6. Rabalais, J.W. *Principles of Ultraviolet Photoelectron Spectroscopy*; Wiley: New York, 1977.
7. Briggs, D. *Handbook of X-ray and Ultraviolet Photoelectron Spectroscopy*; Heyden: London, 1977.
8. Herring, F.G.; McLean, R.A.N. *Inorg. Chem.* 1972, *11*, 1667.
9. Green, J.C. *Struct. Bonding (Berlin)* 1981, *43*, 37.
10. Cowley, A.H. *Prog. Inorg. Chem.* 1979, *26*, 45.
11. Solomon, E.I. *Comments on Inorg. Chem.* 1984, *3*, 227.
12. Koopmans, T. *Physica (Utrecht)* 1934, *1*, 104.
13. Richards, W.G. *Int. J. Mass Spectrom. Ion Phys.* 1969, *2*, 419.
14. Lichtenberger, D.L.; Kellogg, G.E. *Acc. Chem. Res.* 1987, *20*, 379.
15. Lichtenberger, D.L.; Kellogg, G.E.; Pang, L.S.K. In *New Developments in the Synthesis, Manipulation, and Characterization of Organometallic Compounds*; Wayda, A.L.; Darensbourg, M.Y., Eds.; *ACS Symp. Ser.* 1987, *357*, 265.
16. Hubbard, J.L.; Lichtenberger, D.L. *J. Am. Chem. Soc.* 1982, *104*, 2132.
17. Griffith, J.S. *The Theory of Transition Metal Ions*; University Press: Cambridge, 1961.
18. Higginson, B.; Loyd, D.R.; Burroughs, P.; Gibson, D.M.; Orchard, A.F. *J. Chem. Soc., Faraday Trans. 2* 1973, *69*, 1659.
19. Böhm, M.C. *J. Chem. Phys.* 1983, *78*, 7044.
20. Bursten, B.E.; Darensbourg, D.J.; Kellogg, G.E.; Lichtenberger, D.L. *Inorg. Chem.* 1984, *23*, 4361.
21. Lichtenberger, D.L.; Kellogg, G.E.; Landis, G.H. *J. Chem. Phys.* 1985, *83*, 2759.
22. Cotton, F.A.; Kraihanzel, C.S. *J. Am. Chem. Soc.* 1962, *84*, 4432.
23. Jones, L.H.; McDowell, R.S.; Goldblatt, M. *Inorg. Chem.* 1969, *8*, 2349.
24. Svec, H.J.; Michels, G.D.; Flesch, G.D. *Inorg. Chem.* 1980, *19*, 479.
25. Barber, M.; Connor, J.A.; Hillier, I.H.; Saunders, V.R. *J. Chem. Soc., Chem. Commun.* 1971, 682.
26. Cusachs, L. Ch.; Politzer, P. *Chem. Phys. Lett.* 1968, *1*, 529.
27. Bader, R.F.W.; Beddall, P.M.; Cade, P.E. *J. Am. Chem. Soc.* 1971, *93*, 3095.
28. Calabro, D.C.; Hubbard, J.L.; Blevins, C.H., II; Campbell, A.C.; Lichtenberger, D.L. *J. Am. Chem. Soc.* 1981, *103*, 6839.
29. Hubbard, J.L. *Diss. Abstr. Intl. B* 1983, *43*, 2203.
30. Kellogg, G.E. *Diss. Abstr. Intl. B* 1986, *46*, 3838.
31. Lichtenberger, D.L.; Kellogg, G.E.; Kristofzski, J.G.; Page, D.; Turner, S.; Klinger, G.; Lorenzen, J. *Rev. Sci. Instrum.* 1986, *57*, 2366.
32. Eland, J.H.D.; Danby, C.J. *Int. J. Mass. Spectrom. Ion Phys.* 1968, *1*, 111.
33. Cahill, J.E.; Padera, F.G. *Am. Lab.* 1980, *12*, 101.
34. Forder, R.A.; Gale, G.D.; Prout, K. *Acta Cryst. Sect. B* 1975, *B31*, 307.
35. Beach, N.A.; Gray, H.B. *J. Am. Chem. Soc.* 1968, *90*, 5713.
36. Kruasse, J.; Marx, G.; Schodl, G. *J. Organomet. Chem.* 1970, *21*, 159.
37. Lichtenberger, D.L.; Blevins, C.H., II *J. Am. Chem. Soc.* 1984, *106*, 1636.
38. Pang, L.S.K. *Diss. Abstr. Intl. B.* 1986, *46*, 3839.
39. Lichtenberger, D.L.; Calabro, D.C.; Kellogg, G.E. *Organometallics* 1984, *3*, 1623.
40. Lichtenberger, D.L.; Hubbard, J.L. *Inorg. Chem.* 1985, *24*, 3835-3841.
41. Blevins, C.H. II *Diss. Abstr. Intl. B* 1984, *45*, 1186.
42. Stevens, A.E.; Feigerle, C.S.; Lineberger, W.C. *J. Am. Chem. Soc.* 1982, *104*, 5026.
43. Cowley, A.H. *Prog. Inorg. Chem.* 1979, *26*, 45.

44. Flamini, A.; Semprini, E.; Stefani, F.; Cardaci, G.; Bellachioma, G.; Andreocci, M. *J. Chem. Soc., Dalton Trans.* 1978, 695.
45. Baker, A.D.; Baker, C.; Brundle, C.R.; Turner, D.W. *Int. J. Mass Spectrosc. Ion Phys.* 1968, *1*, 285.
46. Pullen, B.P.; Carlson, T.A.; Moddeman, W.E., Schweitzer, G.K.; Bull, W.E.; Grimm, F.A. *J. Chem. Phys.* 1970, *53*, 768.
47. Brundle, C.R.; Robin, M.B.; Basch, H. *J. Chem. Phys.* 1970, *53*, 2196.
48. Dixon, R.N. *Mol. Phys.* 1971, *20*, 113.
49. Rabalais, J.W.; Bergmark, T.; Werme, L.O.; Karlsson, L.; Siegbahn, K. *Phys. Scr.* 1971, *3*, 13.
50. Basch, H.; Robin, M.B.; Kuebler, N.A.; Baker, C.; Turner, D.W. *J. Chem. Phys.* 1969, *51*, 52.
51. Rowland, C.G. *Chem. Phys. Lett.* 1971, *9*, 169.
52. Hubbard, J.L.; Lichtenberger, D.L. *J. Chem. Phys.* 1975, *75*, 2560.
53. Jones, L.H.; McDowell, R.S.; Goldblatt, M.; Swanson, B.I. *J. Chem. Phys.* 1972, *57*, 2050.
54. Berry, R.S. *J. Chem. Phys.* 1960, *32*, 933.
55. Eilian, M.; Hoffmann, R. *Inorg. Chem.* 1975, *14*, 1058.
56. Demuynck, J.; Strich, A.; Veillard, A.; *Nouv. J. Chim.* 1977, *1*, 217.
57. Enemark, J.H.; Feltham, R.D. *Coord. Chem. Rev.* 1974, *13*, 339.
58. Jolly, W.L. *Acc. Chem. Res.* 1983, *16*, 370.
59. Lichtenberger, D.L.; Blevins, C.H., II; Ortega, R.B. *Organometallics* 1984, *3*, 1614.
60. Connor, J.A.; Derrick, L.M.R.; Hall, M.B.; Hillier, I.H.; Guest, M.F.; Higginson, B.R.; Lloyd, D.R., *Mol. Phys.* 1974, *28*, 1193.
61. Evans, S.; Guest, M.F.; Hillier, I.H.; Orchard, A.F. *J. Chem. Soc., Faraday Trans. 2*, 1974, 417.
62. Hillier, I.H.; Guest, M.F.; Higginson, B.R.; Lloyd, D.R. *Mol. Phys.* 1974, *27*, 215.
63. Guest, M.F.; Hiller, I.H.; Higginson, B.R.; Lloyd, D.R. *Mol. Phys.* 1975, *29*, 113.
64. (a) Coutiere, M.M.; Demuynck, J.; Veillard, A. *Theor. Chim. Acta* 1972, *27*, 281; (b) Bagus, P.S.; Walgren, U.I.; Almlof, J. *J. Chem. Phys.* 1976, *64*, 2324.
65. Calabro, D.C.; Lichtenberger, D.L. *Inorg. Chem.* 1980, *19*, 1732, and references therein.
66. Böhm, M.C. *J. Chem. Phys.* 1983, *78*, 7044.
67. Böhm, M.C. *Z. Naturforsch. A* 1982, *37A*, 1193.
68. Bruno, J.W.; Marks, T.J.; Morss, L.R. *J. Am. Chem. Soc.* 1983, *105*, 6824, and references therein.
69. Stevens, A.E.; Beauchamp, J.L. *J. Am. Chem. Soc.* 1981, *103*, 190.
70. Herzberg, G. *Molecular Spectra and Molecular Structure. I. Spectra of Diatomic Molecules*, 2nd ed.; Van Nostrand: New York, 1950.
71. Vedeneyev, V.I.; Gurvich, L.V.; Kondrat'yev, V.N.; Medvedev, V.A.; Frankevich, Y.L. *Bond Energies, Ionization Potentials, and Electron Affinities*; Arnold: London, 1966.
72. Connor, J.A.; Zafarani-Moattar, M.T.; Bickerton, J.; El Saied, N.I.; Suradi, S.; Carson, R.; Takhin, G.A.; Skinner, H.A. *Organometallics* 1982, *1*, 1166.
73. Evans, S.; Green, J.C.; Green, M.L.H.; Orchard, A.F.; Turner, D.W. *Discuss. Faraday Soc.* 1969, *47*, 112.
74. Lichtenberger, D.L.; Fenske, R.F. *Inorg. Chem.* 1974, *13*, 486.
75. Bidinosti, D.R.; McIntyre, N.S. *J. Chem. Soc., Chem. Commun.* 1966, 555.
76. (a) Halpern, J. *Acc. Chem. Res.* 1982, *15*, 238, and references therein; (b) Halpern, J. *Pure Appl. Chem.* 1979, *51*, 2171.
77. (a) Pilcher, G.; Skinner, H.A. In *the Chemistry of the Metal-Carbon Bond*; Hartley, F.R.; Patai, S., Eds.: Wiley: New York, 1982; pp. 43–90; (b) Connor, J.A. *Top. Curr. Chem.* 1977, *71*, 71; (c) Kochi, J.K. *Organometallic Mechanisms and Catalysis*; Academic: New York, 1978; Chapter 11.
78. Bidinostic, D.R.; McIntyre, N.S. *Can. J. Chem.* 1970, *48*, 593.
79. Calhorda, M.J.; Martinho Simoes, J.A. *Organometallics*, 1987, *6*, 1188.
80. Lichtenberger, D.L.; Kellogg, G.E. *J. Am. Chem. Soc.* 1986, *108*, 2560, and references therein.
81. Brookhart, M.; Green, M.L.H. *J. Organomet. Chem.* 1983, *250*, 395.
82. Brookhart, M.; Lamanna, W.; Humphrey, M.B. *J. Am. Chem. Soc.* 1982, *104*, 2117.
83. Whitesides, T.H.; Lichtenberger, D.L.; Budnik, R.A. *Inorg. Chem.* 1975, *14*, 68.
84. Brookhart, M.; Green, M.L.H. *J. Organomet. Chem.* 1983, *250*, 395.

9

Chemistry and Photochemistry of Bare Metal Cluster Ions in the Gas Phase

Steven W. Buckner and Ben S. Freiser

1. INTRODUCTION

The last few years have seen cluster research develop into one of the most intensely studied areas of chemistry and physics. Theoretical and experimental techniques have been applied to study the formation, properties, and structures of clusters. Much of the interest has been due to the possibility of bridging the gap between the gas and condensed phases with clusters. In addition, clusters are of importance because of their role in astrophysics and homogeneous catalysis and their possible use in the future development of microelectronics.

Gas phase ionic clusters have received a great deal of attention because of the ease with which they can be size selected and manipulated. Many of the techniques which have been used successfully to study organic and inorganic ion–molecule reactions in the gas phase have been applied to the study of cluster ions. Specifically, ion beam and ion trapping techniques have proven to be particularly well suited for studying the size-dependent properties of clusters in the transition to bulk. Techniques for generating neutral clusters generally produce a distribution of sizes and, in addition, fragmentation often occurs during the ionization process prior to detection. Also, as cluster size increases, the effects of charge on the properties of the cluster should diminish. These are a few of the main reasons why charged clusters are gaining popularity.

As an indication of the rapid growth of this area, a host of review articles on clusters have recently appeared.[1] This chapter will focus on the bimolecular and unimolecular reactions of bare metal cluster ions in the gas phase. The

Steven W. Buckner and Ben S. Freiser • Department of Chemistry, Purdue University, West Lafayette, Indiana 47907.

chemistry of neutral clusters will be discussed only when it is directly relevant to the ionic cluster studies.

2. BIMOLECULAR REACTIONS OF METAL CLUSTER IONS

2.1. Dissociative Chemisorption

In one of the earliest studies on cluster reactivity, Smalley and co-workers reported[2] on the reactivity of neutral cobalt and niobium clusters, Co_n and Nb_n ($n = 1$–20), with H_2 in a fast-flow reaction tube attached to a supersonic expansion source for the production of clusters[3] (see Figure 1). These clusters showed a remarkable size-dependent reactivity toward H_2 (or D_2). Cobalt clusters with 3 to 5 atoms and 10 or more atoms are highly reactive with H_2 via dissociative chemisorption, whereas other size clusters are almost unreactive. Niobium clusters with 1, 2, 8, 10, and 16 atoms are unreactive with H_2 while all other clusters (up to $n = 20$) rapidly undergo dissociative chemisorption. In contrast, the reactions of H_2 with Fe_n ($n = 1$–25) did not show these drastic oscillations in reactivity, but a strong correlation between Fe_n reactivity and the ionization potential of the cluster (see Figure 2) was observed.[4] The authors proposed a model whereby electrons are donated from the cluster to the σ^* orbital in H_2, which facilitates H—H bond cleavage and, therefore, dissociative chemisorption onto the cluster.

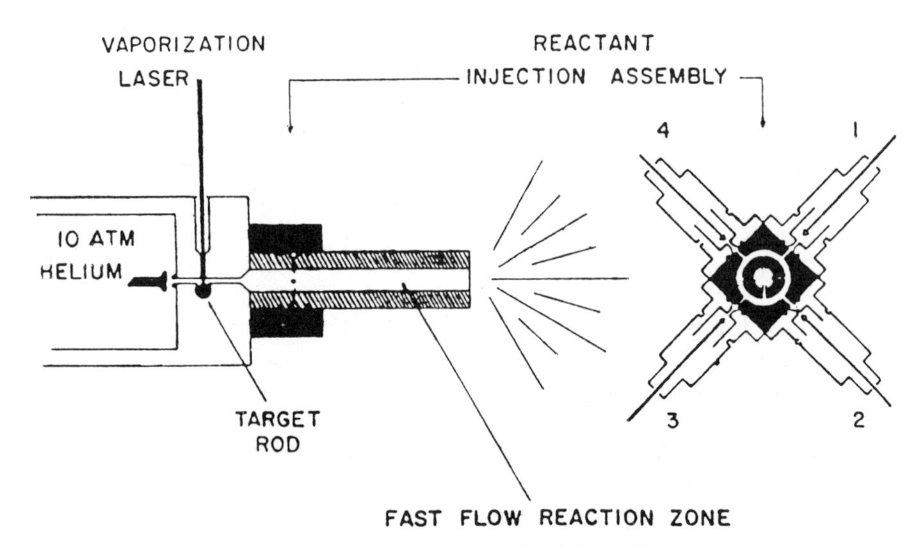

Figure 1. Schematic of fast-flow reactor used in studies of cluster reactivities. A laser pulse (30–40 mJ, 6 ns) is fired at a metal target. A pulse (1–2 atm) of helium is simultaneously passed over the metal surface and the clusters are formed and rapidly thermalized in a near-sonic expansion. The reactant gas is introduced into the cluster beam downstream where reaction occurs. (Reprinted from Reference 3 with permission.)

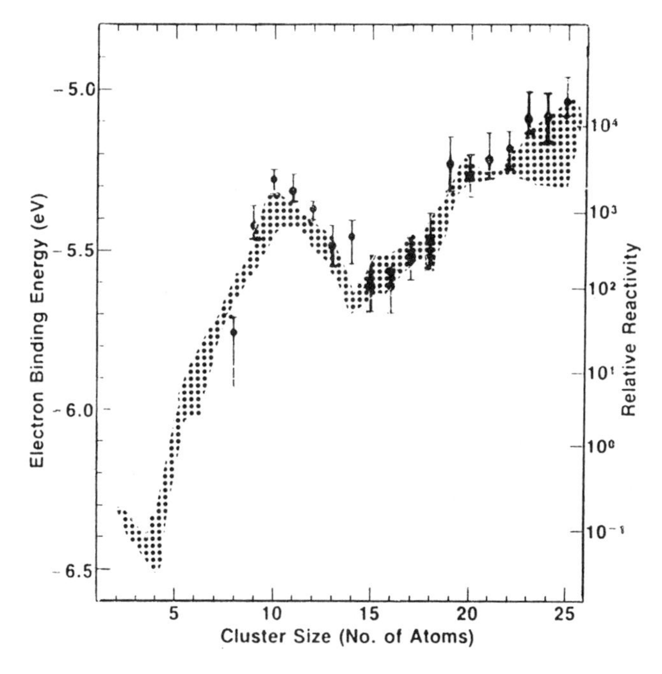

Figure 2. Comparison of measured ionization thresholds (left-hand scale) with intrinsic relative reactivities of Fe clusters with H_2 and D_2 (right-hand scale). The gray band represents the uncertainty in the ionization thresholds, and the vertical lines represent uncertainty in the reactivities. (Reprinted from Reference 4 with permission.)

In 1985, Smalley and co-workers interfaced a supersonic expansion source for the production of metal cluster ions to a Fourier transform mass spectrometer (FTMS).[5] This important achievement was particularly impressive considering the low-pressure requirements of an FTMS and the high gas loads in supersonic cluster sources. The first application of this new technique was a study of the reactivity of ionic niobium clusters with H_2 which revealed a close similarity to the size-dependent reactivity of the neutral cluster species.[6] Nb_7^+ was reported to react rapidly with H_2 via dissociative chemisorption, chemisorbing up to four molecules of H_2, whereas Nb_8^+ and Nb_9^+ were far less reactive, only adding one molecule of H_2 slowly. The authors suggest that these data indicate that the electron donation model may be incorrect and that size or structure may be the determining factor in cluster reactivity. Without data on the second ionization potential of the clusters, however, the electrostatic model cannot be completely ruled out, but a model which considers surface structure appears essential.

A model proposed by Phillips could reconcile the data.[7] Structural factors are considered important, and nearest-neighbor surface structure is proposed as necessary for dissociative chemisorption of H_2. He also proposed that clusters with non-nearest-neighbor surface structure should have high ionization potentials, consistent with the Fe_n data.

2.2. Reactions with Organic Compounds

Before discussing the reactivity of bare metal cluster ions with organic compounds, a brief review of the extensive literature on bare atomic metal ion reactivity with organic molecules will be useful. Atomic metal ions have been observed to be very reactive, activating C—X (X = O, N, halogens, S) bonds in polar compounds and C—C and C—H bonds in hydrocarbons.† Two proposed mechanisms for metal ion reactions with alkanes are shown in Schemes 1 and 2. As shown in the schemes, initial C—C and C—H insertion can lead to dehydrogenation and alkane elimination. For example, initial oxidative addition of the metal ion to a C—H bond (Scheme 1), followed by a β-hydride or a β-alkyl shift results in reductive elimination of H_2 or an alkane. Likewise, oxidative addition of a metal ion to a C—C bond (Scheme 2), followed by a β-hydride or a β-alkyl shift and reductive elimination of hydrogen or an alkane has also been proposed. Of course, which mechanism is applicable has been found to vary with the metal ion, and β-hydride shifts are more common than β-alkyl shifts.

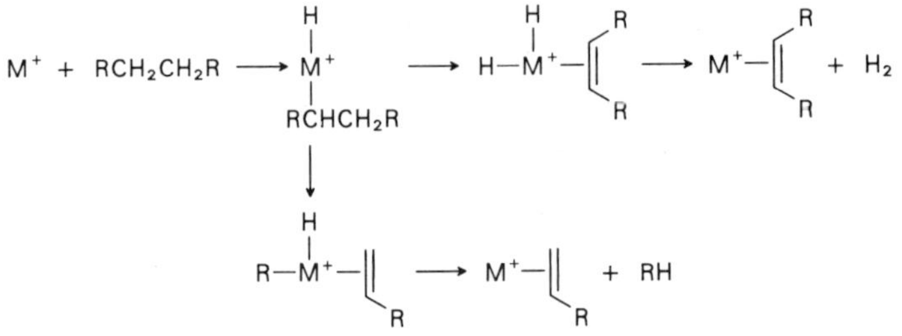

M^+ + RCH_2CH_2R ⟶ H–M^+–$RCHCH_2R$ ⟶ H–M^+–(RCH=CHR) ⟶ M^+–(RCH=CHR) + H_2

↓

R–M^+(H)–(CH_2=CHR) ⟶ M^+–(CH_2=CHR) + RH

Scheme 1

M^+ + RCH_2CH_2R ⟶ R–M^+–CH_2CH_2R ⟶ H–M^+(R)–(CH_2=CHR) ⟶ M^+–(CH_2=CHR) + RH

↓

R–M^+(R)–(CH_2=CH_2) ⟶ M^+–(CH_2=CH_2) + RR

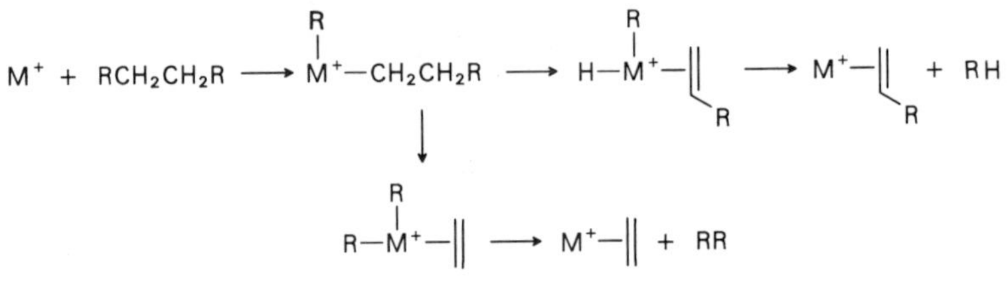

Scheme 2

Fe^+, Co^+, and Ni^+ are selective toward C—C bond activation.[9] Other first-row metal ions Mn^+, Cr^+, and Cu^+ are unreactive with alkanes,[10,11] whereas the early transition metal ions Sc^+,[12] Ti^+,[13] and V^+[14] predominantly activate C—H bonds. The bond energy of H· and CH_3· to the first-row metal ions has been correlated with the energy to promote the metal ion into a $3d^n4s^1$ electronic configuration.[15] This strongly suggests that a 4s electron on the metal is necessary to form a σ bond to a hydrogen atom or an alkyl group.

† For a review of gas phase metal ion chemistry, see Reference 8.

One of the first and most straightforward methods for generating metal cluster ions in a mass spectrometer was by electron impact on multinuclear carbonyl complexes, as in reactions 1 and 2:

$$Co_2(CO)_8 \xrightarrow{e^-} Co_2^+ + 8CO + 2e^- \quad (1)$$

$$Mn_2(CO)_{10} \xrightarrow{e^-} Mn_2^+ + 10CO + 2e^- \quad (2)$$

This method is fairly limited, however, in that only relatively small cluster ions can be formed, and appropriate volatile polynuclear metal compounds are necessary to generate the metal cluster ions.

The fragments generated upon electron impact of volatile metal carbonyls undergo subsequent reactions with the parent neutral to form large carbonyl-containing cluster ions.[16] The reactivity of metal cluster ions with metal carbonyl compounds is discussed in Chapter 3 of this book.

In the first study of the chemistry of bare metal cluster ions, the reactions of Mn_2^+ and Co_2^+ with alcohols and alkyl halides were investigated by Ridge.[17a] Mn_2^+ was observed to react by three pathways which included splitting of the cluster, abstraction of a halide atom, and formation of Mn_2HX^+ (X = Br, OH). This reactivity was attributed to a weak Mn—Mn bond. Co_2^+ was observed to react via dehydration, dehydrohalogenation, and formation of Co_2HX^+. A complete study of the kinetics of the reactions of Mn_2^+ with a variety of *n*-donor bases was carried out.[17b] The main reaction pathway, displacement of Mn to form MnB^+ (B = reactive base), was used in conjunction with previously determined values for $D^\circ(Mn^+—Mn)$ and relative reaction rates for various bases with Mn_2^+ to construct a scale of absolute Mn^+ affinities.

Both Mn_2^+ and Co_2^+ are unreactive toward alkanes. This is in contrast to Co^+ which, as previously mentioned, reacts rapidly with alkanes, activating C—C and C—H bonds. In a subsequent report by Freas and Ridge, a CO ligand was observed to have a dramatic effect on the reactivity of Co_2^+ toward alkanes.[18] The reactivity of Co_2CO^+ is different from that of Co^+, $CoCO^+$, or Co_2^+. Co_2CO^+ is very selective toward C—H bonds. The enhanced reactivity of Co_2CO^+ relative to Co_2^+ was explained in terms of the polarizing effect of CO on the Co—Co bond, which concentrates positive charge on the unbound Co and increases its reactivity. A further discussion of the reactivity of fragments from metal carbonyls with organic compounds can be found in Chapter 3 of this book.

In 1984 a method was developed by Jacobson and Freiser for the generation of mixed-metal cluster ions in a Fourier transform mass spectrometer and was demonstrated for $CoFe^+$.[19] In this method, atomic transition metal ions, M^+, are generated from laser desorption of a pure metal target followed by reaction of M^+ with a volatile metal carbonyl, as in reaction 3:

$$M^+ + Fe(CO)_5 \longrightarrow MFe(CO)_x^+ + (5 - x)CO \quad (3)$$

Collisional activation of the product cluster carbonyl ions results in sequential loss of the remaining CO ligands to produce the bare metal cluster ion:

$$MFe(CO)_x^+ \xrightarrow{-CO} MFe(CO)_{x-1}^+ \xrightarrow{-CO} \cdots \xrightarrow{-CO} MFe^+ \quad (4)$$

In general, the collision energies of the cluster carbonyl ions are in the 25- to 75-eV laboratory energy range for maximal production of the bare cluster ions. Figure 3 demonstrates this methodology for generating $NbFe^+$ and illustrates the power of FTMS for multistep ion manipulation. Since the first report, this technique has been used extensively to study the chemistry and photochemistry of dimeric and trimeric mixed metal cluster ions, as briefly discussed below. A summary of the reactions of a number of heterodinuclear cluster ions appears in Tables 1 and 2.

Ligand displacement reactions were used to determine the $CoFe^+$ bond strength.[19] $CoFe^+$ reacts with benzene to form $Co(benzene)^+$, exclusively, but is unreactive with acetonitrile. From these results, 71 ± 3 kcal/mol $> D^\circ(Co^+—benzene) > D^0(Co^+—Fe) > D^\circ(Co^+—CH_3CN) > 61 \pm 4$ kcal/mol, and a value of $D^\circ(Co^+—Fe) = 66 \pm 7$ kcal/mol was reported.

The reactions of $CoFe^+$ with hydrocarbons were reported.[20] Like Co_2^+, $CoFe^+$ is unreactive with alkanes, and two possible explanations for the lack of reactivity were considered by the authors. First, reaction with alkanes could be endothermic due to the weak binding of alkenes to $CoFe^+$, but this was discounted because the binding of olefins to $CoFe^+$ was believed to be similar to that of Co^+ and Fe^+. The second explanation involved the energetics of the C—H bond insertion step. As discussed above, the correlation between $D^\circ(M^+—X)$ ($X = H, CH_3$) and the electronic promotion energy of the metal ion ($3d^n \rightarrow 3d^{n-1}4s^1$) indicates that the $M^+—H$ and $M^+—CH_3$ bonds are primarily of s character. Since the $Co^+—Fe$ bond presumably involves the $4s$ electrons, the $CoFe^+—H$ and $CoFe^+$—alkyl bonds were proposed to be weaker and the initial C—H insertion less thermodynamically favorable than for the atomic metal ions, Co^+ and Fe^+. Recent results, however, indicate that $D^\circ(Fe_2^+—H)$ is comparable to $D^\circ(Fe^+—H)$. Specifically, an endothermic reaction threshold experiment with Fe_2^+ and ethane yielded $D^\circ(Fe_2^+—H) = 52 \pm 16$ kcal/mol,[21] and deprotonation reactions of Fe_2H^+ indicated $PA(Fe_2) > 241.8$ kcal/mol from which $D^\circ(Fe_2^+—H) > 63$ kcal/mol is derived.[22,23] Also, Armentrout has observed Co_2^+ and Co^+ to react comparably with D_2 in an ion beam apparatus and has suggested that $D^\circ(Co_2^+—D)$ and $D^\circ(Co^+—D)$ are quite similar.[24] Reaction of the dimers requires formation of two bonds for insertion into a C—H or C—C bond, however, and this may account for the reduced reactivity. Other factors may be responsible for the inert behavior of these cluster ions with simple alkanes, including kinetic barriers arising from electronic restructuring.

$CoFe^+$ is also unreactive with ethene, propene, isobutene, and butadiene but reacts with larger alkenes mainly via dehydrogenation.[20] Reactivity with the larger alkenes is attributed to both the presence of an allylic C—H bond which is weak and, therefore, more susceptible to oxidative addition and to

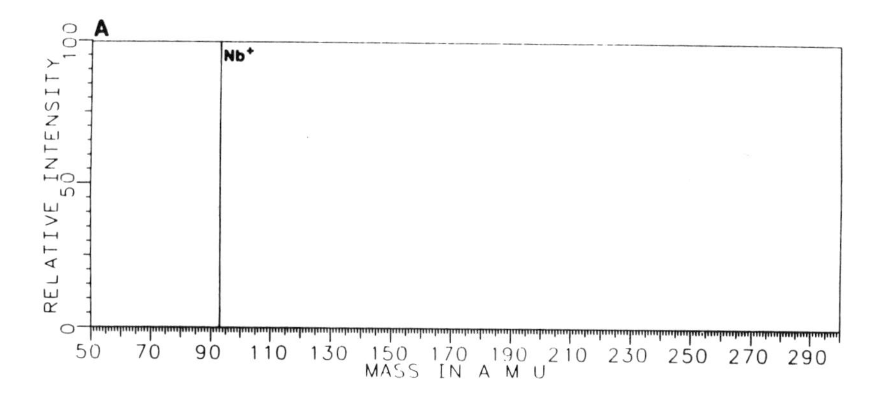

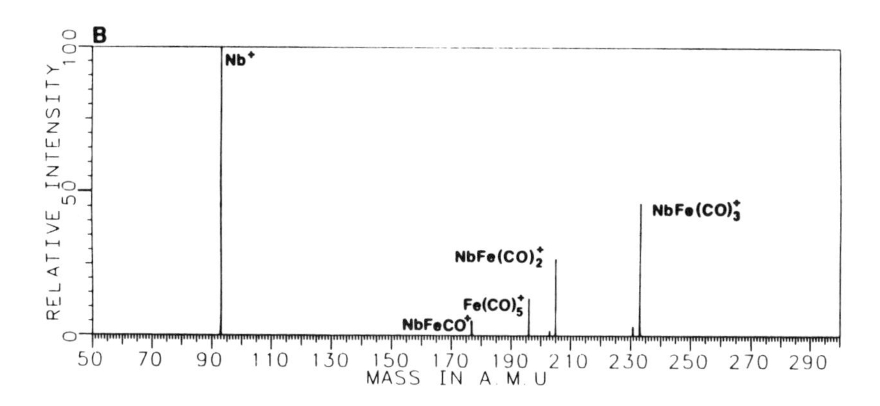

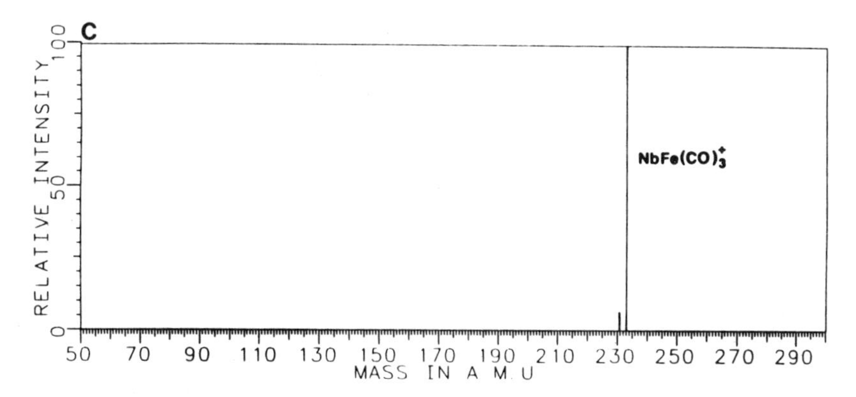

Figure 3. (A) Laser desorption of Nb^+ from a pure niobium foil; (B) reaction of Nb^+ with $Fe(CO)_5$; (C) isolation of $NbFe(CO)_3^+$; (D) collisional activation of $NbFe(CO)_3^+$; (E) isolation of $NbFe^+$; (F) reaction of $NbFe^+$ with cyclohexene; (G) isolation of $NbFeC_6H_6^+$ [and $Nb(C_6H_6)_2^+$]; (H) collisional activation of $NbFeC_6H_6^+$. (See Reference 39.)

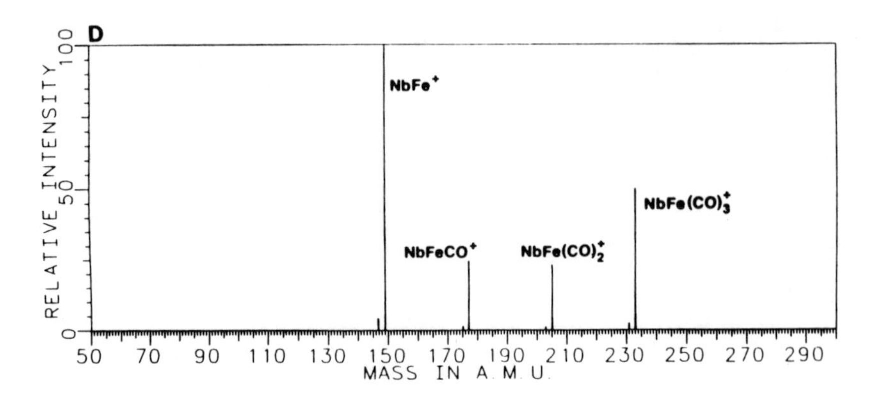

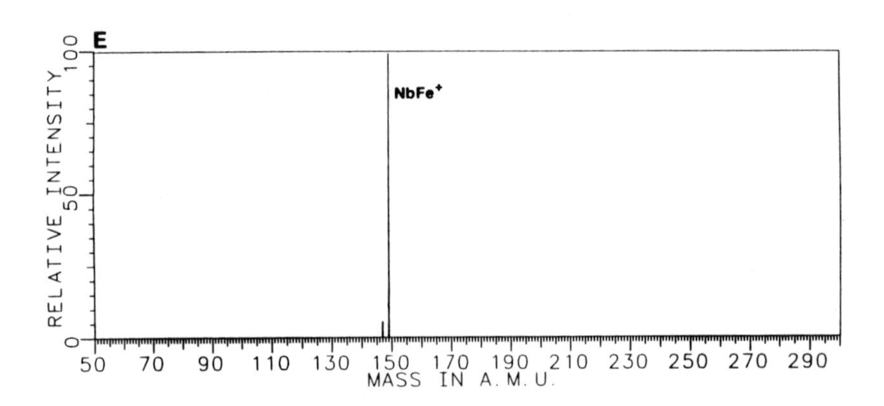

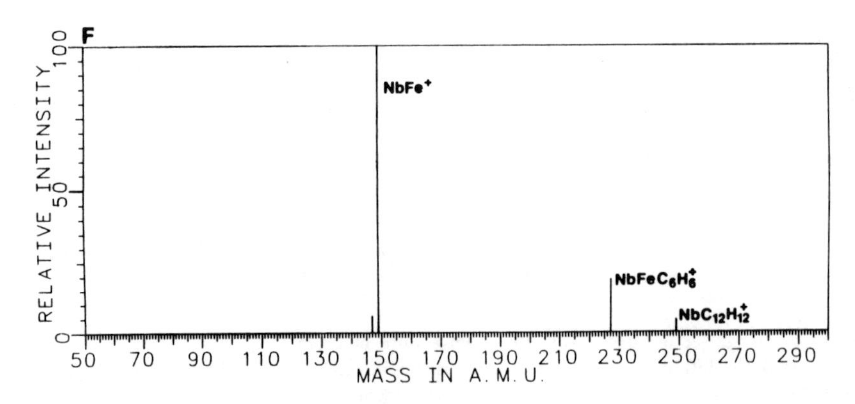

Figure 3. (continued)

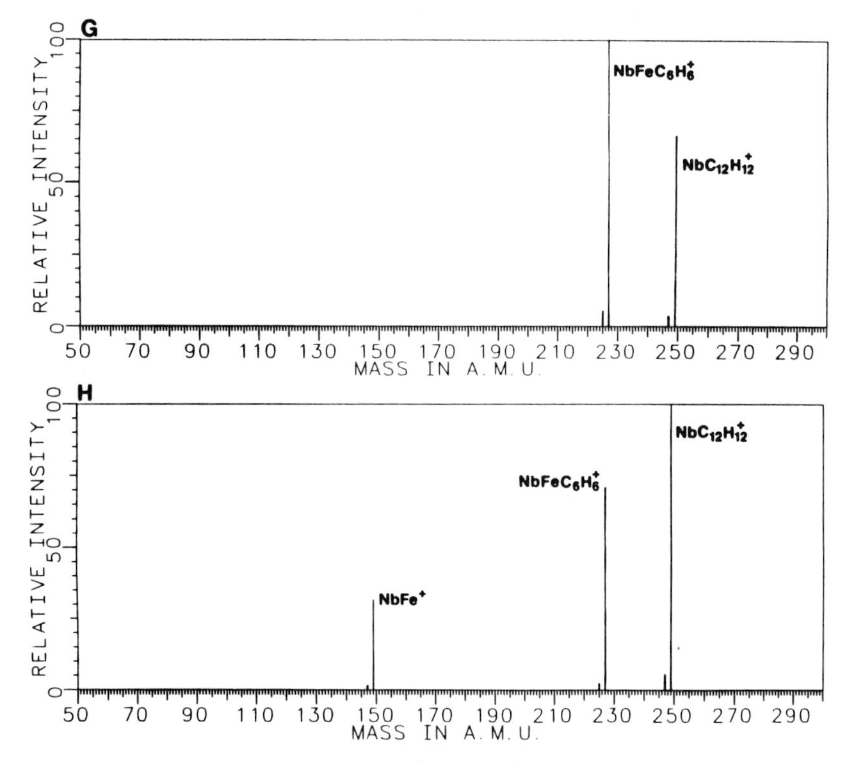

Figure 3. (continued)

Table 1. Primary Product Distributions (%) for Reactions of MM'^{+} with Allcanes and Various Alkenes

Reactant	Products	$NbFe^{+a}$	VFe^{+b}	$ScFe^{+a}$	$CuFe^{+c}$	$CoFe^{+d}$
Alkanes		NR[e]	NR	NR	NR	NR
Ethene		NR	NR	NR	NR	NR
Propene	$MM'C_3H_4^+ + H_2$	100	NR	NR	NR	NR
1-Butene	$MM'C_4H_6^+ + H_2$	82	NR	100	100	100
	$MM'C_4H_4^+ + 2H_2$	18				
1-Pentene	$MM'C_5H_8^+ + H_2$		NR	100	100	100
	$MM'C_5H_6^+ + 2H_2$	100				
1-Hexene	$MM'C_6H_{10}^+ + H_2$		NR	100	34	58
	$MM'C_6H_8^+ + 2H_2$	69			66	42
	$MM'C_6H_6^+ + 3H_2$	31				
Cyclopentene	$MM'C_5H_6^+ + H_2$	68	NR	100	100	100
	$MM'C_5H_4^+ + 2H_2$	32				
Cyclohexene	$MM'C_6H_6^+ + 2H_2$	100	100	62	100	88
	$MC_6H_6^+ + FeH_2 + H_2$			38		12

[a] Reference 39.
[b] Reference 25.
[c] Reference 28.
[d] Reference 20.
[e] NR: No reaction.

Table 2. *Primary Product Distributions (%) and Pertinent Thermochemical Limits for Reactions of MM'^{+} with Various Hydrocarbons*[a]

Reactant	Product	$LaFe^+$	$RhFe^+$	$RhCo^+$	Implications (if observed)
CH_4		NR[b]	NR	NR	
C_2H_6	$MM'C_2H_4^+ + H_2$	100	100	100	$D°(MM'^+$—ethene$) > 33$ kcal/mol
C_3H_8	$MM'C_3H_6^+ + H_2$	100	92	83	$D°(MM'^+$—propene$) > 30$ kcal/mol
	$MM'C_3H_4^+ + 2H_2$		8	17	$D°(RhM^+$—allene$) > 69$ kcal/mol
n-C_4H_{10}	$MM'C_4H_8^+ + H_2$	20			$D°(LaFe^+$—butene$) > 30$ kcal/mol
	$MM'C_4H_6^+ + 2H_2$	80	100	100	$D°(MM'^+$—butadiene$) > 56$ kcal/mol
n-C_5H_{12}	$MM'C_5H_8^+ + 2H_2$	100	69	10	
	$MM'C_5H_6^+ + 3H_2$		6	20	$D°(RhM^+$—cyclopentadiene$) > 67$ kcal/mol
	$MC_5H_8^+ + H_2 + (M' + H_2)$[c]			24	
	$MC_5H_6^+ + 2H_2 + (M' + H_2)$[c]		10	37	$D°(Rh^+$—cyclopentadiene$) > 114$ kcal/mol
	$MC_5H_5^+ + 3H_2 + M'H$		15	9	$D°(Rh^+$—cyclopentadienyl$) > 189$ kcal/mol
n-C_6H_{14}	$MM'C_6H_{10}^+ + 2H_2$	100	40	27	
	$MM'C_6H_8^+ + 3H_2$		44	73	
	$MM'C_6H_6^+ + 4H_2$		16		$D°(RhFe^+$—benzene$) > 60$ kcal/mol
c-C_3H_6	$MM'C_3H_4^+ + H_2$	62	63	59	$D°(MM'^+$—allene$) > 33$ kcal/mol
	$MM'C_2H_2^+ + CH_4$	38	37		$D°(MM'^+$—acetylene$) > 24$ kcal/mol
	$MM'CH_2^+ + C_2H_4$			41	$D°(RhCo^+$—$CH_2) > 93$ kcal/mol

c-C_4H_8	$MM'C_4H_6^+ + H_2$	75			
	$MM'C_4H_4^+ + 2H_2$	25	90	100	
	$MM'C_2H_4^+ + C_2H_4$		10		$D^\circ(RhFe^+—ethene) > 45$ kcal/mol
c-C_5H_{10}	$MM'C_5H_6^+ + 2H_2$	100	14	21	$D^\circ(MM'^+—cyclopentadiene) > 50$ kcal/mol
	$MC_5H_6^+ + H_2 + (M' + H_2)$[c]		14	65	$D^\circ(Rh^+—cyclopentadiene) > 97$ kcal/mol
	$MC_5H_5^+ + 2H_2 + M'H$		72	14	$D^\circ(Rh^+—cyclopentadienyl) > 162$ kcal/mol
c-C_6H_{12}	$MM'C_6H_8^+ + 2H_2$		35		
	$MM'C_6H_6^+ + 3H_2$	100	65	100	$D^\circ(MM'^+—benzene) > 49$ kcal/mol
i-C_4H_{10}	$MM'C_4H_8^+ + H_2$	100	29		
	$MM'C_4H_6^+ + 2H_2$		71	100	
neo-C_5H_{12}	$MM'C_5H_8^+ + 2H_2$	57			
	$MM'C_5H_6^+ + 3H_2$	43			
	$MM'C_4H_6^+ + CH_4 + H_2$		100	100	
Ethylene oxide	$MM'O^+ + C_2H_4$	[d]	38	55	$D^\circ(MM'^+—O) > 85$ kcal/mol
	$MM'CH_2^+ + CH_2O$		62	45	$D^\circ(MM'^+—CH_2) > 78$ kcal/mol
O_2	$M^+ + M'O_2$		100	100	
	$MO^+ + M'O$	100			

[a] Data taken from Ref. 29.
[b] No reaction.
[c] Parentheses used where neutral composition is not conclusive.
[d] Reaction of $LaFe^+$ with ethylene oxide was not investigated due to rapid reaction between La^+ and ethylene oxide.

the presence of a double bond which coordinates to the cluster and provides additional energy to the collision complex prior to C—H insertion.

The reactivity of $CoFe^+$ with alkenes is very different from that of Co^+ or Fe^+, which react with butene and larger alkenes predominantly by C—C cleavage. For $CoFe^+$, C—C cleavage is only observed in the reactions with 3,3-dimethyl-1-butene, which contains no β-hydrogens. Dehydrocyclization occurs readily upon collisional activation of $CoFe(olefin)^+$, when the olefin is a linear C_5 or C_6 alkadiene. Branched alkenes, however, were not observed to undergo skeletal rearrangements.

The secondary reactions of $CoFe^+$ with alkenes show more extensive dehydrogenations than the primary reactions. It appears that an alkene ligand activates the cluster ion toward hydrocarbons in analogy to the effect of CO on the reactivity of Co_2^+ discussed above. Coupling of the ligands could also account for the more extensive neutral losses in the ligated cluster ion.

The method of reactions 3 and 4 was used to prepare VFe^+.[25] V^+ was produced by laser desorption. VFe^+ was observed to be unreactive with all alkanes studied (C_1-C_6) and all alkenes studied (C_2-C_6) except for cyclic C_6 and C_7 alkenes.

VFe^+ reacts with cyclohexene to produce $VFe(benzene)^+$:

$$VFe^+ + c\text{-}C_6H_{10} \rightarrow VFe(benzene)^+ + 2H_2 \tag{5}$$

$VFe(benezene)^+$ reacts with cyclohexene as in reactions 6 and 7:

$$VFe(benzene)^+ + c\text{-}C_6H_{10} \longrightarrow VFe(C_6H_6)_2^+ + 2H_2 \tag{6}$$

$$\longrightarrow V(C_6H_6)_2^+ + (Fe + 2H_2) \tag{7}$$

Collisional activation of $VFeC_6H_6^+$ yields VFe^+, exclusively. Collisional activation of $VFe(C_6H_6)_2^+$ produces $V(C_6H_6)_2^+$. This can be compared to $CuFe(C_6H_6)_2^+$ which forms $CuFeC_6H_6^+$ upon collisional activation[28] and $CoFe(C_6H_6)_2^+$ which forms $Co(C_6H_6)_2^+$ and $CoFeC_6H_6^+$ upon collisional activation.[20] A diagram indicating the energetics of the VFe^+ and benzene system is shown in Figure 4.

VFe^+ reacts with cycloheptatriene to form $VFeCH_2^+$ and benzene. This indicates $D^\circ(VFe^+—CH_2) > 70$ kcal/mol, in comparison with $D^\circ(V^+—CH_2) = 80 \pm 8$ kcal/mol[14b] and $D^\circ(Fe^+—CH_2) = 82 \pm 5$ kcal/mol.[27]

The gas phase chemistry of $CuFe^+$ has recently been studied.[28] Although Fe^+ is very reactive with hydrocarbons, Cu^+ is unreactive with alkanes. Cu^+ has recently been observed to react with larger alkenes, but the mechanism probably does not involve oxidative addition to C—H bonds, due to the d^{10} configuration of Cu^+.[10] This interesting mix of metals produces a cluster ion with reactivity trends similar to $CoFe^+$ (see Table 1). $CuFe^+$ activates C—H bonds in alkenes containing a linear C_4 or larger backbone.

Recently, Huang *et al.* reported on the activation of C—H and C—C bonds in alkanes by the heterodinuclear cluster ions $RhFe^+$, $RhCo^+$, and

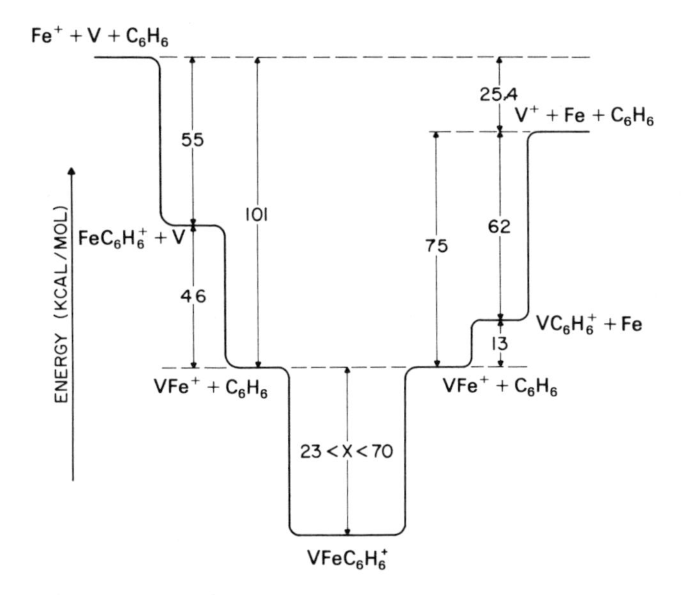

Figure 4. Energetics of [V, Fe, benzene]$^+$ system. Heats of formation determined from photodissociation and ion-molecule reactions. (See References 25 and 52b.)

$LaFe^+$.[29] All three cluster ions react with alkanes larger than methane mainly via dehydrogenation (see Table 2). In this respect, the reactivities of $RhFe^+$, $RhCo^+$, and $LaFe^+$ are more like those of Rh^+ and La^+ than those of Fe^+ and Co^+. $RhFe^+$ and $RhCo^+$ show similar patterns of reactivity with multiple dehydrogenations of alkanes larger than ethane. *n*-Pentane and cyclopentane react with both $RhFe^+$ and $RhCo^+$ to split the cluster and form both $Rh(c\text{-}C_5H_5)^+$ and $Rh(c\text{-}C_5H_6)^+$, implying $D°[Rh^+—(c\text{-}C_5H_5)] >$ 189 kcal/mol. $LaFe^+$ shows less extensive dehydrogenation in its reactions with alkanes. The authors point out that cluster ions containing a second- or third-row metal bound to a first-row metal are much more reactive than cluster ions containing two first-row metals. Reduced overlap in the metal–metal bond for orbitals of different size and energy is proposed to account for the differences in reactivity, but high-level calculations are necessary before an understanding of the bonding and novel reactivity of these clusters can be obtained.

The trinuclear species $FeCo_2^+$ was studied by Jacobson and Freiser and observed to activate C—H bonds in alkanes.[30] The reactivity of $FeCo_2^+$ was thus significantly different from that of Co^+, Fe^+, Co_2^+, and $CoFe^+$. $FeCo_2^+$ was formed by collisional activation of the products from reactions 8 and 9:

$$Fe^+ + Co_2(CO)_8 \longrightarrow FeCo_2(CO)_6^+ + 2CO \quad (8)$$

$$ \longrightarrow FeCo_2(CO)_5^+ + 3CO \quad (9)$$

Thermal dehydrocyclization of hexane to benzene occurs in the reaction with $FeCo_2^+$. Collisional activation of $FeCo_2(alkene)^+$ species produced in reactions with alkanes yields cleavage of the carbon skeleton and extensive dehydrogenation.

Fast atom bombardment (FAB) is a well-known technique for production of metal cluster ions from metal surfaces.[31] Recently, a number of groups have exploited this technique to study the chemistry of metal cluster ions.

Freas *et al.* have used an 8–10-keV Xe sputtering source coupled with a double-focusing mass spectrometer to generate and study a number of interest-

Table 3. Fragment Ions in the CID/MIKES Spectra of Copper/Isobutane Cluster Ions[a]

Parent ion		Daughter ions		
m/z	Stoichiometry	m/z	Stoichiometry	Percent relative abundance[a]
119	$[Cu(C_4H_8)]^+$	117	$[CuC_4H_6]^+$	88
		115	$[CuC_4H_4]^+$	12
		103	$[CuC_3H_4]^+$	10
		63	Cu^+	100
121	$[Cu(C_4H_{10})]^+$	119	$[CuC_4H_8]^+$	12
		105	$[CuC_3H_6]^+$	12
		78	$[CuCH_3]^+$	2
		63	Cu^+	100
		57	$[C_4H_9]^+$	3
		43	$[C_3H_7]^+$	1
		41	$[C_3H_5]^+$	1
		39	$[C_3H_3]^+$	1
179	$[Cu(C_4H_{10})_2]^+$	121	$[CuC_4H_{10}]^+$	100
		119	$[CuC_4H_8]^+$	6
		105	$[CuC_3H_6]^+$	9
		91	$[CuC_2H_4]^+$	<1
		63	Cu^+	34
		57	$[C_4H_9]^+$	5
		43	$[C_3H_7]^+$	<1
		41	$[C_3H_5]^+$	<1
		39	$[C_3H_3]^+$	<1
247	$[Cu_3(C_4H_{10})]^+$	245	$[Cu_3C_4H_8]^+$	<1
		189	$[Cu_3]^+$	100
		184	$[Cu_2C_4H_{10}]^+$	4
		182	$[Cu_2C_4H_8]^+$	1
		126	$[Cu_2]^+$	16
		121	$[CuC_4H_{10}]^+$	1
		63	Cu^+	1
305	$[Cu_3(C_4H_{10})_2]^+$	247	$[Cu_3C_4H_{10}]^+$	100
		189	$[Cu_3]^+$	95
		184	$[Cu_2C_4H_{10}]^+$	15
		126	$[Cu_2]^+$	29
		121	$[CuC_4H_{10}]^+$	2
		63	Cu^+	2
363	$[Cu_3(C_4H_{10})_3]^+$	305	$[Cu_3(C_4H_{10})_2]^+$	100

[a] Data taken from Ref. 33.

ing species.[32] Copper cluster ions, Cu_n^+ ($n = 1, 3, 5$), were observed to form adducts with isobutane in the high-pressure (0.1–0.2 torr) ion source yielding $[Cu_n(C_4H_{10})_m]^+$ ($n = 1, 3, 5$; $m = 1, 2, 3$).[33] The structures of these species were probed by collision-induced dissociation mass-analyzed ion kinetic energy spectroscopy (CID/MIKES) to determine if the isobutane physisorbs on the cluster or whether it undergoes dissociative chemisorption. Table 3 lists the fragment ions in the CID/MIKES spectra for various copper–isobutane cluster ions. Dehydrogenation, loss of isobutane, and loss of neutral copper atoms dominate the spectra. No C—C cleavage products were observed, indicating that the isobutane is simply physisorbed onto the copper cluster ions. The authors noted that loss of Cu and Cu_2 in the $[Cu_3(C_4H_{10})_n)]^+$ ($n = 1, 2$) spectra indicates a linear or bent structure for the Cu_3 center as opposed to a triangular structure.

In a further study, Freas and Campana used collision-induced dissociation to study the reactivity of oxygen-deficient, $(Co_xO_{x-1})^+$, and oxygen-equivalent, $(Co_xO_x)^+$, cobalt cluster ions with isobutane.[34] The cobalt–oxygen–isobutane cluster ions were formed in a high-pressure source by condensation of the isobutane onto the cobalt–oxygen cluster ions. Table 4 shows fragment ions from the CID of the cluster ions. Oxygen-equivalent cluster ions were found to lose primarily isobutane to regenerate the cobalt–oxygen cluster ions, while oxygen-deficient ions yielded dehydrogenation products, $[Co_xO_{x-1}(C_4H_4)]^+$. These results were attributed to structural differences in the oxygen-equivalent and oxygen-deficient cluster ions. Calculations showed that the oxygen-equivalent cluster ions consist of closed structures, whereas the oxygen-deficient cluster ions consist of structures with terminal cobalt atoms (see

Table 4. Collision-Induced Neutral Fragments of Cobalt–Oxygen–Isobutane Cluster Ions

	Percent relative abundances[a,b]								
Parent ion	H	H_2	O	H_2O	CH_4	C_2H_4	C_3H_4	C_4H_8	C_4H_{10}
$[Co_2OC_4H_8]^+$	3	3	12	11		3	5	100	
$[(Co_2O)HC_4H_8]^+$		7	31					100	
$[Co_2OC_4H_{10}]^+$		90	10						100
$[Co_3O_2C_4H_8]^{+c}$	<1	2		2		6	29	100	
$[(Co_3O_2)HC_4H_8]^+$	3	1	8	3			100	83	
$[Co_2O_2C_4H_{10}]^+$	5	2		2					100
$[Co_3O_3C_4H_{10}]^+$	2	<1							100
$[CoC_4H_8]^{+d}$	3	15			22			100	
$[CoC_4H_{10}]^+$	2	33			100				55
$[Co_2C_4H_{10}]^+$									100
$[Co_3C_4H_{10}]^+$		10							88

[a] Relative to 100% for the most abundant fragmentation. No correction has been made for unicluster dissociations. Fragmentations corresponding to cobalt-containing moieties are not shown.
[b] Data taken from Ref. 34.
[c] The loss of CH_3 was observed also (relative abundance = 6).
[d] The loss of C_3H_7 was observed also (relative abundance = 8).

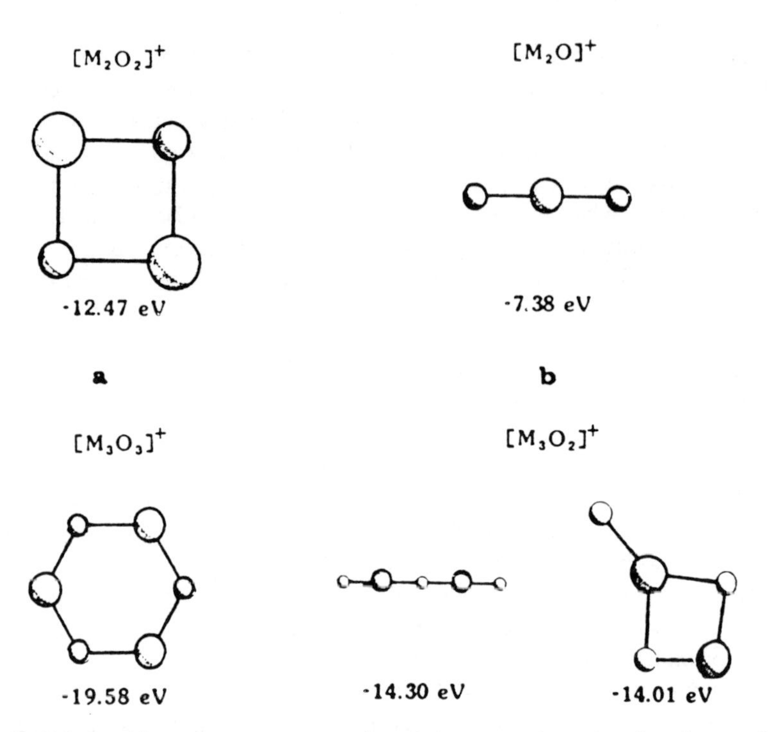

Figure 5. Calculated lowest-energy structures for cobalt–oxygen cluster ions based on a pairwise potential mode. (Reprinted from Reference 34 with permission.)

Figure 5). Thus, oxygen-deficient cluster ions are more reactive because of positionally accessible cobalt atoms, analogous to surface steps or defects.

Hanley and Anderson have also developed an instrument for studying the chemistry of transition metal cluster ions generated by sputtering from metal foils.[35] As the authors point out, ions produced by sputtering can have high translational and internal energies. By allowing the cluster ions which are produced to interact with argon (~10 collisions), the authors demonstrate that the ions can be significantly translationally cooled. The reactions of cooled and uncooled Co_n^+ with ethylene were compared. Uncooled Co_n^+ showed significant fragmentation of the cluster ion. Furthermore, under these conditions, endothermic products were observed from the reaction of sputtered Co^+ with ethylene (reaction 10). These processes were substantially reduced for cooled Co_n^+. The x subscript on H in reactions 10–18 is used because the instrument lacked sufficient resolution to differentiate the number of hydrogen atoms.

$$(Co^+)^* + C_2H_4 \rightarrow CoC_2H_x^+ \tag{10}$$

Hanley and Anderson also studied the reactions of uncooled vanadium, chromium, manganese, and iron cluster ions with C_2H_4. Vanadium cluster ions were observed to be the most reactive, proceeding mainly by dehydrogenation of ethylene [reactions 11 and 12 (n = 1-4)]:

$$(V_n^+)^* + C_2H_4 \rightarrow V_nC_2H_x^+ \tag{11}$$

$$(V_nO^+)^* + C_2H_4 \rightarrow V_nOC_2H_x^+ \tag{12}$$

V_3^+, however, yielded a unique C—C cleavage product:

$$(V_3^+)^* + C_2H_4 \rightarrow V_3CH_x^+ \tag{13}$$

Manganese, iron, and chromium cluster ions were found to be unreactive with ethylene, except for Cr_2^+, which undergoes reaction 14:

$$(Cr_2^+)^* + C_2H_4 \rightarrow Cr_2C_2H_x^+ \tag{14}$$

Several reactions involving the atomic metal species were also reported:

$$(Cr^+)^* + C_2H_4 \rightarrow CrC_2H_x^+ \tag{15}$$

$$(Mn^+)^* + C_2H_4 \rightarrow MnC_2H_x^+ \tag{16}$$

$$(MnO^+)^* + C_2H_4 \rightarrow MnOC_2H_x^+ \tag{17}$$

In a subsequent study, aluminum cluster ions were found to react with ethylene by reaction 18, exclusively.[36]

$$Al_n^+ + C_2H_4 \rightarrow Al_nCH_x^+ + CH_{4-x} \qquad (n = 1, 2, 3) \tag{18}$$

The ion kinetic energy was varied by accelerating the ions at different potentials after formation in an ion beam apparatus. Reactions with n = 1 or 2 show a decrease in reaction cross section with increasing cluster ion kinetic energy, indicative of an exothermic process. The dimer is ~10 times more reactive than Al^+. Al_3^+ shows an increase in reaction cross section for up to ~1-eV acceleration energy, indicating an endothermic process. The reaction cross section for the trimer is ~2.5% of the maximum for the dimer.

Magnera *et al.* have used a FAB sputtering source to generate metal cluster ions. The chemistry of Ni_n^+ (n = 1-9), Pd_n^+ (n = 1-4), and Pt_n^+ (n = 1, 2) with n-butane was studied at collision energies of ~1 eV.[37] Table 5A summarizes the results, which show that dehydrogenation and dealkylation both occur for all three metals. As cluster ion size increases, the extent of dehydrogenation increases, and C—C cleavage products are suppressed. As in the study by Hanley and Anderson, the effects of collisional cooling on the reactivity of the cluster ions was tested. The ions were passed through a collision chamber where they underwent 10 to 100 collisions with argon. Ions which were cooled exhibited a greater tendency to dehydrogenate butane and the probability of C—C cleavage was reduced (see Table 5B). The authors concluded that C—H

Table 5A. Relative Product Abundances for Reactions of Cluster Ions with n-Butane[a,b]: $M_n^+ + C_4H_{10} \rightarrow X + Y^+$

M_n^+	X: H_2 Y^+: $M_n^+C_4H_8$	$2H_2$ $M_n^+C_4H_6$	$3H_2$ $M_n^+C_4H_4$	CH_4 $M_n^+C_3H_6$	C_2H_6 $M_n^+C_2H_4$	C_2H_8 $M_n^+C_2H_2$	$M + C_2H_6$ $M_{n-1}^+C_2H_4$
Ni^+	6.0			1.1	9.8		
Ni_2^+	0.8	4.1	0.03		0.7	0.6	2.3
Ni_3^+	0.8	1.7	0.5				
Ni_4^+	1.3						
Ni_5^+		1.6					
Ni_6^+		0.2	0.7				
Ni_7^+			0.3				
Ni_8^+			0.6				
Ni_9^+			0.3				
Ni_{10}^+			0.5				
Pd^+	0.2			0.49	1.6		
Pd_2^+		16.0					
Pd_3^+		14.0					
Pd_4^+		0.3	0.3				
Pt^+		11		<0.5	1		
Pt_2^+			18			2.0	3.2

[a] Parent intensity, 100.0.
[b] Data taken from Ref. 37.

Table 5B. Reactions of Ni_n^+ with n-Butane: Internal Ion Energy Effects[a]

M_n^+		Product relative abundance[b]				
		$M_n^+C_4H_8$	$M_n^+C_4H_6$	$M_n^+C_4H_4$	$M_{n-1}^+C_2H_4$	M_{n-1}^+
Ni_2^+	Collision cooled	6.7	8.1		1.7	0.3
	Directly sputtered	0.8	4.1	0.02	2.3	3.8
Ni_3^+	Collision cooled		3.0			
	Directly sputtered	0.8	1.4	0.5		2.0

[a] Data taken from Ref. 37.
[b] Parent intensity: 100.0.

insertions are characteristic of the thermal reactivity of the cluster ions, while C—C cleavage pathways are due, at least in part, to energetic effects.

Weil and Wilkins reported on the gas phase reactivity of Au^+ with alcohols.[38] Au^+ was generated by laser desorption of gold oxide. The authors also noted the production of Au_2^+ and Au_3^+. Subsequent to this, Freiser and co-workers generated Zn_2^+ from laser desorption of ZnO, and Ag_3O^+ and Ag_3^+ from laser desorption of AgO.[39] Charge-transfer reactions of Zn_2^+ indicated $IP(Zn_2) = 9.0 \pm 0.2$ eV, from which $D^o(Zn^+—Zn) = 0.56 \pm 0.2$ eV was derived. Surprisingly, laser desorption of an AgO/ZnO mixture produced an enhancement of the silver cluster ions, with Ag_n^+ ($n = 1$–11) observed as well as various silver oxide cluster ions (see Figure 6). Since Zn is inert in the gas phase and

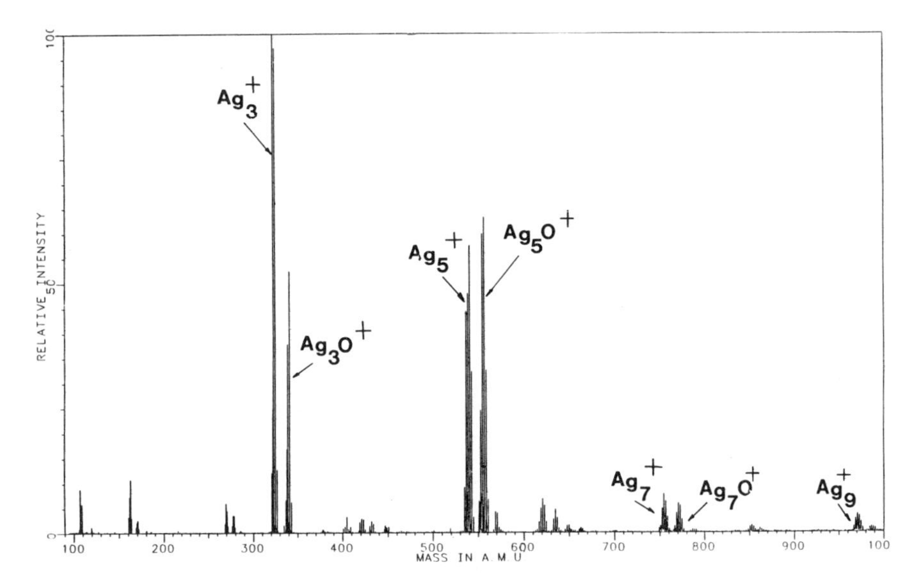

Figure 6. Direct laser desorption of an AgO/ZnO mixture in an FTMS. (See Reference 39.)

Zn^+ only reacts via charge transfer,[40] zinc may act as a third body to stabilize collisions in the laser-desorbed plasma and thus increase cluster size. The spectra were found to be fairly insensitive to the AgO/ZnO ratio. Charge transfer was observed in the reactions of directly laser-desorbed Ag_3^+ and Ag_5^+ with $Fe(c\text{-}C_5H_5)_2$. Addition of excess argon to the trapping cell, which allowed the ions to undergo ~20 thermalizing collisions prior to reaction with ferrocene, suppressed charge transfer for both cluster ions. Following thermalization, the silver cluster ions can be kinetically excited by the application of a radio frequency pulse resonant with the cyclotron frequency of the ions, after which rapid charge transfer is once again observed. This indicates that some portion of the cluster ion population produced by direct laser desorption is kinetically excited. Thermalized Ag_n^+ ($n = 3, 5$) is unreactive with small alkanes, alkenes, and alcohols, but $Ag_nL_2^+$ ($n = 3, 5$; $L =$ *sec*-butylamine) reacts with *sec*-butylamine via dehydrogenation and deamination to produce, presumably, $Ag_nL_2(\text{butadiene})^+$.

2.3. Oxide Chemistry

Many of the previously described techniques have been used to study the reactivity of transition metal cluster ions with oxygen-containing compounds. These studies have revealed a rich chemistry which often differs greatly from that of the bare atomic metal ions.

In one of the earliest studies on the reactivity of transition metal cluster ions, Armentrout *et al.* reported on the reactivity of Mn_2^+ with O_2.[41] Mn_2^+ was prepared by electron impact on $Mn_2(CO)_{10}$ with the electron energy kept just above the appearance potential for Mn_2^+ in order to minimize excess internal energy. The reaction studies were carried out in an ion beam apparatus.

Three ionic products result from the reaction of Mn_2^+ with O_2:

$$Mn_2^+ + O_2 \rightarrow Mn_2O^+ + O \quad (19)$$

$$\rightarrow Mn^+ + MnO_2 \quad (20)$$

$$\rightarrow MnO^+ + MnO \quad (21)$$

Kinetic energy studies indicated that all three reactions are exothermic, but that production of MnO^+ in reaction 21 exhibits an activation barrier. Reaction 19 implies $D°(Mn_2^+—O) > 119$ kcal/mol, which agrees well with a result obtained in an ion beam reaction of Mn_2^+ with CO of $D°(Mn_2^+—O) \sim 127$ kcal/mol. This value is almost twice the value for MnO^+ of $D°(Mn^+—O) \sim 70$ kcal/mol.

Using the techniques illustrated by Equations 3 and 4, Jacobson and Freiser studied the oxide chemistry of a number of groups 8 and 9 transition metal cluster ions.[42] The thermochemical data derived from the reactions in this study are listed in Table 6. The bare metal ions, Fe^+ and Co^+, do not abstract oxygen from ethylene oxide or O_2 due to their low oxide bond strengths, $D°(Fe^+—O) = 68 \pm 3$ kcal/mol and $D°(Co^+—O) = 65 \pm 3$ kcal/mol.[43] The cluster ions, however, are very reactive toward oxygen and ethylene oxide. $FeCo^+$, Fe_2^+, and Co_2^+ react with ethylene oxide by sequentially abstracting two oxygen atoms:

$$MM'^+ + \text{(ethylene oxide)} \xrightarrow{-C_2H_4} MM'O^+ + \text{(ethylene oxide)} \xrightarrow{-C_2H_4} MM'O_2^+ \quad (M, M' = Fe, Co) \quad (22)$$

All three cluster ions are also observed to react with O_2 (reactions 23-26):

$$M_2^+ + O_2 \rightarrow M^+ + MO_2 \quad (23)$$

$$\rightarrow M_2O^+ + O \quad (24)$$

$$CoFe^+ + O_2 \rightarrow Co^+ + FeO_2 \quad (25)$$

$$\rightarrow CoFeO^+ + O \quad (26)$$

Reactions 24 and 26 imply $D°(MM'^+—O) > 119$ kcal/mol, almost twice the value of the atomic metal ion values. The authors note that reaction 24 is very slow for Co_2^+, indicating that the reaction may be slightly endothermic. This was confirmed when Armentrout and co-workers[44] found $D°(Co_2^+—O) \sim 117$ kcal/mol using an ion beam apparatus.

$FeCo_2^+$ and Co_3^+ abstract up to three oxygen atoms sequentially from ethylene oxide:

$$MM_2'^+ + 3\ \text{(ethylene oxide)} \rightarrow MM_2'O_3^+ + 3C_2H_4 \quad (27)$$

Table 6. Summary of New Thermochemical Data for Small Fe- and Co-Containing Clusters[a]

Cluster	ΔH_f (kcal/mol)	D°(A—B) A—B	D°(A—B) kcal/mol
Ions			
Fe_2O^+	$<257 \pm 7$	Fe_2^+—O	>119
		FeO^+—Fe	$>117 \pm 8$
Co_2O^+	$<258 \pm 7$	Co_2^+—O	>119
		CoO^+—Co	$>120 \pm 8$
$CoFeO^+$	$<256 \pm 7$	$CoFe^+$—O	>119
		CoO^+—Fe	$>117 \pm 8$
$Fe_2O_2^+$	215 ± 18	Fe_2^+—2O	221 ± 17
$Co_2O_2^+$	216 ± 18	Co_2^+—2O	221 ± 17
$CoFeO_2^+$	214 ± 18	$CoFe^+$—2O	221 ± 17
$FeCo_2^+$	$>315 \pm 19$	$FeCo^+$—Co	$<102 \pm 19$
Co_3^+	$> 318 \pm 19$	Co_2^+—Co	$<102 \pm 19$
$FeCo_2O^+$		$FeCo_2^+$—O	>85
Co_3O^+		Co_3^+—O	>85
$FeCo_2O_2^+$		$FeCo_2O^+$—O	>85
$Co_3O_2^+$		Co_3O^+—O	>85
$FeCo_2O_3^+$		$FeCo_2O_2^+$—O	>85
$Co_3O_3^+$		$Co_3O_2^+$—O	>85
$FeCo_2O_4^+$		$FeCo_2O_3^+$—O	<25
$Co_3O_4^+$		$Co_3O_3^+$—O	<25
Neutrals			
FeO_2[b]	$<34 \pm 7$	Fe—2O	$>185 \pm 7$
CoO_2[b]	$<34 \pm 7$	Co—2O	$>185 \pm 7$

[a] Data taken from Ref. 42.
[b] IP > 181 kcal/mol (7.86 eV).

This reaction implies that $D^\circ(MM_2'O_n^+—O) > 85$ kcal/mol ($n = 0$–2). These clusters also react with O_2 as in reactions 28–36:

$$FeCo_2^+ + O_2 \longrightarrow FeCo^+ + CoO_2 \quad (28)$$
$$\longrightarrow Co_2^+ + FeO_2 \quad (29)$$
$$\longrightarrow FeCoO^+ + CoO \quad (30)$$
$$\longrightarrow Co_2O^+ + FeO \quad (31)$$
$$\longrightarrow FeCoO_2^+ + Co \quad (32)$$
$$\longrightarrow Co_2O_2^+ + Fe \quad (33)$$

$$Co_3^+ + O_2 \longrightarrow Co_2^+ + CoO_2 \quad (34)$$
$$\longrightarrow Co_2O^+ + CoO \quad (35)$$
$$\longrightarrow Co_2O_2^+ + Co \quad (36)$$

The reactivity of several cluster ions containing carbonyl ligands was also investigated. $Fe_2(CO)_4^+$, $CoFe(CO)_3^+$, and $Co_2(CO)_x^+$ ($x = 1$–3) react with O_2 via displacement of all carbonyls to form $MM'O_2^+$. $Co_2(CO)_4^+$ and $Co_2(CO)_5^+$ react with O_2 slowly via displacement of three carbonyls, and $Co_2(CO)_y^+$ ($y =$ 6–8) is unreactive with O_2. Jacobson and Freiser attribute this trend to coordinative saturation effects. If initial O_2 dissociation on the cluster ion is followed by loss of CO ligands, then coordinative saturation may prevent O_2 dissociation.

The trimeric carbonyl cluster ions react with O_2, as in reactions 37–45. $FeCo_2O_3^+$ and $Co_3O_3^+$ were observed to be unreactive with O_2, but $M'M_2O_4^+$ reacts with O_2 as in reaction 46, indicating $D^\circ(M'M_2O_3^+—O) < 25$ kcal/mol.

$$FeCo_2(CO)_5^+ + O_2 \rightarrow FeCo_2(CO)(O)_2^+ + 4CO \quad (37)$$

$$Co_3(CO)_6^+ + O_2 \rightarrow Co_3(CO)_3(O)_2^+ + 3CO \quad (38)$$

$$\rightarrow Co_3(CO)_2(O)_2^+ + 4CO \quad (39)$$

$$FeCo_2(CO)(O)_2^+ + O_2 \rightarrow FeCo_2O_4^+ + CO \quad (40)$$

$$\rightarrow FeCo_2O_3^+ + CO_2 \quad (41)$$

$$Co_3(CO)_2(O)_2^+ + O_2 \rightarrow Co_3O_4^+ + 2CO \quad (42)$$

$$\rightarrow Co_3O_3^+(CO + CO_2) \quad (43)$$

$$Co_3(CO)_3(O)_2^+ + O_2 \rightarrow Co_3O_4^+ + 3CO \quad (44)$$

$$\rightarrow Co_3O_3^+ + (2CO + CO_2) \quad (45)$$

$$M'M_2O_4^+ + O_2 \rightarrow M'M_2O_3^+ + O_3 \quad (46)$$

Collisional activation of the product oxides indicated that the species are very stable. $M'MO^+$ species were not observed to fragment in collisions up to ~100 eV (lab energy), suggesting structure **1**. $M'MO_2^+$ fragmented to form M^+ and $M'MO^+$ in very low efficiency, suggesting structure **2**.

$M'M_2O_3^+$ species do not undergo dissociation in collisions up to ~75 eV (lab energy). $Co_3O_4^+$, however, undergoes CID to form $Co_3O_3^+$ and a small amount of $Co_3O_2^+$. Similary, $FeCo_2O_4^+$ undergoes CID to form $FeCo_2O_3^+$ with small amounts of $FeCoO_2^+$ and $Co_2O_2^+$ also observed.

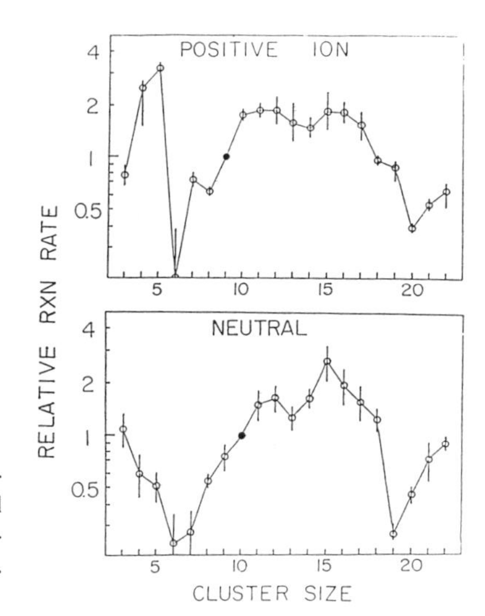

Figure 7. Relative reaction rates of N_2 chemisorption by cobalt cluster ions (top trace) and neutral cobalt clusters (bottom trace). Reaction was performed in the fast-flow reactor shown in Figure 1. (Reprinted from Reference 45 with permission.)

Smalley and co-workers studied the reactivity of Nb_n^+ and Co_n^+ ($n = 3$–22) with CO, CO_2, and N_2.[45] As discussed above for H_2, studies of the reactivities of the neutral clusters and their ionic counterparts with these simple molecules showed virtually the same size dependence. Figure 7, for example, shows the similarities in the reactivities of Co_n^+ and Co_n with N_2. The niobium cluster ions showed a similar behavior to that of Co_n^+. CO reacts faster than N_2 and shows a slow monotonic rise in reactivity with increasing cluster size, in contrast to the sharply size-dependent N_2 data. The Co_n^+ and Nb_n^+ data for reaction with CO_2 were exactly analogous to those for the neutral clusters. The apparent lack of charge dependence, coupled with the distinct size dependence of cluster reactivity, agrees with the Phillips model previously discussed (see Section 2.1). Niobium cluster ions were observed to react with CO_2 by abstracting an oxygen atom, indicating $D°(Nb_n^+—O) > 125$ kcal/mol for $n = 3$–7,[45,23] with larger cluster ions forming the adduct $Nb_nCO_2^+$.

Photodissociation was used to "desorb" CO and N_2 from $Nb_nN_2^+$ and Nb_nCO^+. The energy required to desorb the ligand from the cluster ion, which was always observed as the lowest-energy dissociation channel, was found to vary along with the cluster ion reactivity, with more tightly bound ligands corresponding to more reactive cluster ions.

Woste and co-workers have reported on the reactivity of Ni_n^+ with CO.[46] Ni_n^+ was produced via a sputtering source, and reaction was carried out in an ion beam scattering cell. Both $Ni_n(CO)_x^+$ and $Ni_n(C)(CO)_x^+$ complexes were observed. Kinetic measurements indicated that the reactions involve three-body collisions, except for the reaction of Ni_3^+. The maximum number of carbonyl ligands adsorbed by the cluster ions was observed to correlate well with recently proposed electron counting rules. The observed number of CO ligands for Ni_n^+ ($n = 2$–13) is shown in Table 7. For example, Ni_4^+ becomes saturated with 10 CO ligands suggesting an 18-electron configuration about each metal

Table 7. Maximum Number of CO Ligands as a Function of Cluster Size Following One of the Three Reaction Channels[a]:

$$Ni_n^+ + CO \rightarrow \begin{cases} Ni_n(CO)_k^+ \\ Ni_nC(CO)_l^+ \\ Ni_{(n-1)}(CO)_m^+ \end{cases}$$

Cluster	k	l	m	Cluster	k	l	m
Ni_2^+	9	5	8	Ni_8^+	16	14	17
Ni_3^+	8	7	9	Ni_9^+	17	12	18
Ni_4^+	10	7	11	Ni_{10}^+	18	16	19
Ni_5^+	12	9	13	Ni_{11}^+	19	—[b]	20
Ni_6^+	13	11	11	Ni_{12}^+	20	20	21
Ni_7^+	15	13	14	Ni_{13}^+	22	20	22

[a] Data taken from Ref. 46(b).
[b] Not observed.

atom for a tetrahedral arrangement of the Ni atoms, as in Figure 8. The proposed structures for Ni_n^+ (n = 2–13) are shown in Figure 9. The $Ni_nC(CO)_l^+$ species have, in general, two less carbonyl ligands than the corresponding $Ni_n(CO)_k^+$ species with the same number of nickel atoms. The authors propose that the C atom resides in the middle of the cluster ion and that the four electrons, previously provided by the two carbonyl ligands, now come from the C atom.

Anderson and co-workers have studied the reactions of Al_n^+ (n = 1–5) with O_2 and D_2O in an ion beam apparatus.[36,47] Al_2OD^+ and $Al_2OD_2^+$ are the major exothermic product ions observed for each of the cluster ions (n = 2–5) with D_2O. The other observed product channel for the larger cluster ions (n = 3–5) is collision-induced dissociation to form smaller cluster ions, with the expected endothermic channel behavior. Al^+ reacts with O_2 via an endothermic process forming AlO^+. The threshold at 3.0 eV disagrees somewhat with a previously reported threshold of 3.6 eV. The lower value is attributed to reactions occurring outside the scattering cell. Al_2^+ forms Al_2O^+

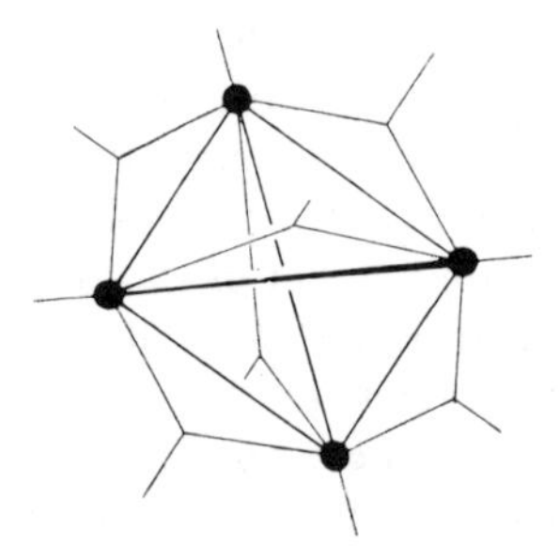

Figure 8. Proposed structure for $Ni_4(CO)_{10}^+$. (Reprinted from Reference 46 with permission.)

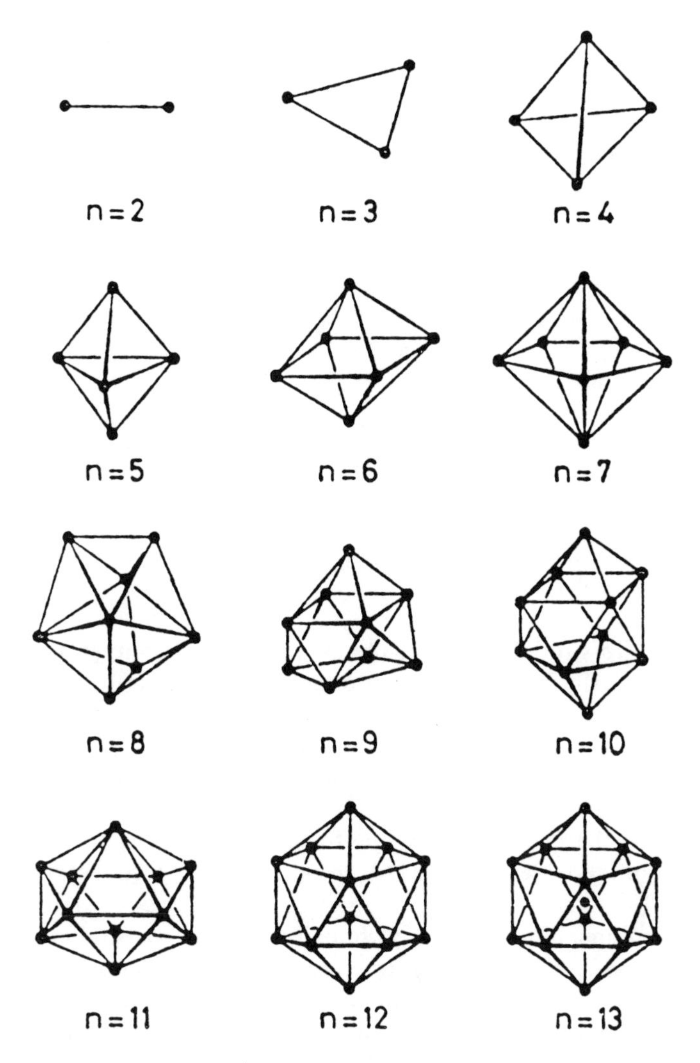

Figure 9. Proposed structures for Ni_n^+ (n = 2–13). (Reprinted from Reference 46 with permission.)

from O_2 in an exothermic process, indicating $D^o(Al_2^+—O) > 119$ kcal/mol. For larger cluster ions (n = 3–5), Al_nO^+ (n = 3–5) is not observed at energies up to 5 eV. Instead, Al_2O^+ is observed, as well as smaller cluster ion fragments. These results are in agreement with data obtained by Jarrold and Bower,[48] who used a low-energy ion beam apparatus with center-of-mass collision energies of 1 eV to study the reactions of Al_n^+ (n = 4–25) with O_2. Argon was used in place of O_2 to ensure that products did not occur simply by CID processes. No reactions were observed with Ar at 1-eV center-of-mass collision energies. A histogram showing the reactions of Al_n^+ with O_2 is shown in Figure 10. No oxygen abstraction is observed. Loss of Al_5 is a major product channel for reactions of Al_n^+ with n = 7–12 and 19. For reactions of Al_n^+ with n = 13–25,

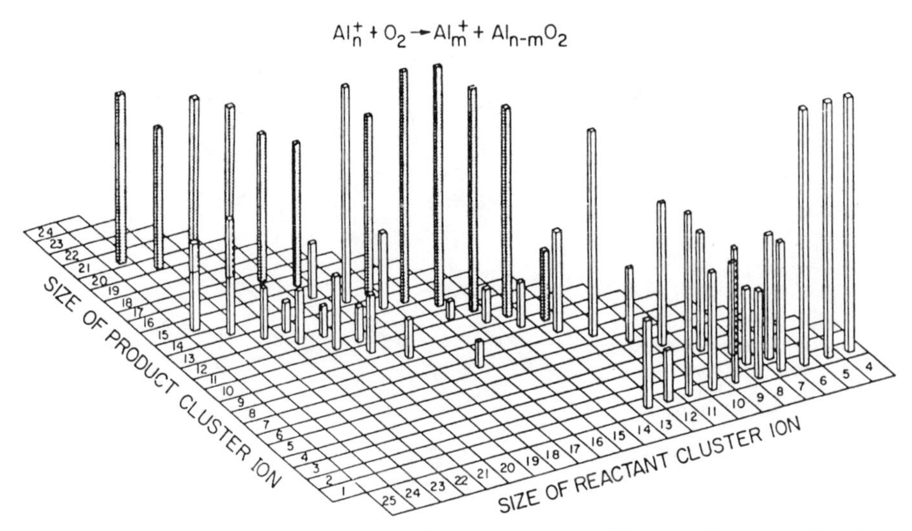

Figure 10. Histogram showing the products from reaction of Al_n^+ (n = 4–25) with O_2. (Reprinted from Reference 48 with permission.)

excluding 19, loss of Al_4 becomes the major reaction channel. For Al_n^+ with $n > 16$, loss of Al_{10} is also a major product. For the neutral products, which are surely oxygenated since no reaction occurs with Ar, structures could not be conclusively determined [i.e., Al_4O_2, Al_5O_2, and $Al_{10}O_2$] from the reaction studies. Further studies by these authors on the CID of various oxygenated aluminum cluster ions ($Al_nO_m^+$; n = 3–26, m = 1, 2) showed loss of Al_2O to be the main pathway due to the high stability of Al_2O.[49] This indicates neutral losses for the $Al_n^+ + O_2$ reactions such as (Al_4O_2) may consist of ($2Al_2O$).

As a continuation of the cobalt oxide cluster ion studies previously described, Freas *et al.* studied the structure and collision-induced dissociation of cobalt oxide cluster ions.[50] Cobalt clusters were produced from a pure metal foil by FAB with an 8-keV beam. The cobalt cluster ions were reacted with O_2 in a high-pressure source. As before, oxygen-deficient cluster ions, $[Co(CoO)_x]^+$, and oxygen-equivalent cluster ions, $[Co_xO_x]^+$, were the most abundant, with $[(CoO)_xO]^+$ also observed in low abundance. $[(CoO)_xH]^+$ and $[(CoO)_xOH]^+$ were also observed, resulting from the reaction of cobalt oxide cluster ions with water or hydrocarbon contaminants in the source. Proton transfer to neutral cobalt oxide clusters could also account for these species.

CID fragments for the $[Co_xO_x]^+$ species are shown in Table 8. For larger cluster ions, the main fragmentation pathway is loss of $(CoO)_x$ to form $(CoO)_y^+$. Smaller cluster ions form $Co(CoO)_y^+$ products also. These results are consistent with a $(CoO)_x^+$ structure, with each CoO unit being ionic in nature, as opposed to $Co_xO_x^+$. For smaller ions, the positive charge becomes more important and formation of $Co(CoO)_y^+$, where the extra Co can accommodate the positive charge, becomes more favorable. In contrast, the oxygen-deficient cluster ions

Table 8. Fragment ions and Neutral Losses[a] from the Collision-Induced Dissociation of $[(CoO)_x]^{+b}$

	Relative abundance[c]								
Parent ion	Co^+	$[CoO]^+$	$[Co_2]^+$	$[Co(CoO)]^+$	$[(CoO)_2]^+$	$[Co(CoO)_2]^+$	$[(CoO)_3]^+$	$[Co(CoO)_3]^+$	$[(CoO)_4]^+$
$[(CoO)_5]^+$	18	—	—	31	28	8	64	8	100
$[(CoO)_4]^+$	15	4	4	33	25	12	100		
$[(CoO)_3]^{+d}$	59	12	9	100	85	18			
$[(CoO)_2]^+$	100	15	11	56					
$[CoO]^+$	100								
	Relative abundance[e]								
	O	O_2	CoO	CoO_2	$(CoO)_2$	$(CoO)_2O$	$(CoO)_3$	$(CoO)_3O$	$(CoO)_4O$
$[(CoO)_5]^+$	—	—	100	8	64	8	28	31	18
$[(CoO)_4]^{+f}$	—	—	100	12	25	33	4	15	
$[(CoO)_3]^{+g}$	18	12	85	100	12	59			
$[(CoO)_2]^+$	56	11	15	100					
$[CoO]^+$	100								

[a] The neutral fragment stoichiometries are inferred empirically from the ionic fragments.
[b] Data taken from Ref. 50.
[c] The values are relative to 100 for the most abundant fragment ion of each parent ion.
[d] The formation of $[Co_3O]^+$ was observed (relative abundance = 12).
[e] The values are relative to 100 for the most abundant fragment lost.
[f] The loss of Co_2O_4 was observed (relative abundance = 4).
[g] The loss of CoO_3 was observed (relative abundance = 9).

Table 9. Fragment Ions and Neutral Losses[a] from the Collision-Induced Dissociation of $[Co(CoO)_x]^{+b}$

	Relative abundance[c]					
Parent ion	Co^+	$[CoO]^+$	$[Co_2]^+$	$[Co(CoO)]^+$	$[(CoO)_3]^+$	$[(CoO)_4]^+$
$[Co(CoO)_4]^+$	45	—	—	79	79	100
$[Co(CoO)_3]^{+d}$	71	—	—	94	100	
$[Co(CoO)_2]^{+e}$	52	4	4	100		
$[Co(CoO)]^+$	100	11	15			
	Relative abundance[f]					
	Co	CoO	Co(CoO)	$(CoO)_2$	$(CoO)_3$	$(CoO)_4$
$[Co(CoO)_4]^+$	100	—	79	—	79	45
$[Co(CoO)_3]^+$	100	29	—	94	71	
$[Co(CoO)_2]^{+g}$	48	100	4	52		
$[Co(CoO)]^{+h}$	11	100				

[a] The neutral fragment stoichiometries are inferred empirically from the ionic fragments.
[b] Data taken from Ref. 50.
[c] The values are relative to 100 for the most abundant fragment ion of each parent ion.
[d] The formation of $[(CoO)_2O]^+$ was observed (relative abundance = 29).
[e] The formation of $[(CoO)_2]^+$ was observed (relative abundance = 48).
[f] The values are relative to 100 for the most abundant fragment lost.
[g] The loss of (CoO)O was observed (relative abundance = 4).
[h] The loss of O was observed (relative abundance = 15).

fragment by loss of Co and Co(CoO), as shown in Table 9. The authors use this to formulate the oxygen-deficient cluster ions as $[Co(CoO)_x]^+$. A simple pair potential model was used to calculate geometries for the cluster ions. The lowest-energy structures for the oxygen-equivalent species were found to be highly symmetric with many nearest neighbors. In contrast, the oxygen-deficient cluster ions were found to have low symmetry with protruding Co atoms.

Armentrout and co-workers have developed a cluster ion source similar in design to that of Smalley *et al.* but which utilizes an 8-kHz Cu vapor laser in a continuous flow of He.[(51)] This allows continuous generation of cold cluster ion beams of very high intensity. The ions can then be mass selected by a magnetic sector. When the He flow is seeded with a small amount (~0.5%) of O_2, oxygenated cluster ions can be generated. For iron cluster ions there is a large depletion of the bare cluster ion signal with a commensurate increase in the oxygenated cluster ion signal. Specifically, Fe_2^+ is observed to form $Fe_2O_3^+$ and $Fe_2O_5^+$. The authors suggest that $Fe_2O_5^+$ consists of a strongly bound $Fe_2O_3^+$ unit weakly bound to an O_2 molecule, since a chemically bonded $Fe_2O_5^+$ would place the iron atoms in extremely high oxidation states.

3. UNIMOLECULAR REACTIONS OF METAL CLUSTER IONS

3.1. Photodissociation

Direct measurement of the absorption spectra of ions in the gas phase is difficult or impossible due to the inability to generate these species in high enough concentrations. Photodissociation, however, provides a means of obtaining this information indirectly.[52] A photodissociation spectrum is obtained by monitoring the following reaction as a function of wavelength:

$$AB^+ + h\nu \rightarrow A^+ + B \tag{47}$$

A considerable data base is available on the photodissociation of organic ions and, more recently, photodissociation studies of organotransition metal ions and transition metal cluster ions have begun to appear.

In general, both spectroscopic and thermodynamic information can be obtained from photodissociation experiments but, for species containing transition metals, the density of low-lying electronic states makes the thermodynamic information much more accessible.[52b]

Particularly for transition metal cluster ions, the density of low-lying electronic states can be extremely high. Even for small clusters, electronic structure reminiscent of band structure in bulk metals is observed. For example, a recent HF-CI calculation of the electronic structure of Fe_2 indicated 112 electronic states at or below 0.6 eV[53] [though experimental evidence from Lineberger and co-workers suggests this value is high (*vide infra*)]. For this type of system, photon absorption occurs over a broad wavelength range which includes the dissociation limit and, therefore, photodissociation thresholds (i.e., the longest wavelength at which photodissociation is observed) can yield excellent thermochemical information.

In one of the first studies in this area, Helm and Moller reported on the formation and photodissociation of Cs_n^+ ($n = 2$–9).[54] Cs_n^+ was produced by a liquid metal ion source (LMIS). Figure 11 shows the ion and light sources, ion optics, and detection system. Photoproducts were energy analyzed to determine parent/daughter relationships according to

$$E^{\text{lab}}(Cs_m^+) = E^{\text{lab}}(Cs_n^+)[m/n] + \Delta E \tag{48}$$

where m is the number of atoms in the daughter (Cs_m^+) and n is the number of atoms in the parent (Cs_n^+), and ΔE, the kinetic energy release, is very small for high-energy parent ion beams. Initial study of Cs_2^+ indicated a lower limit for $D°(Cs^+—Cs)$ of 0.59 ± 0.06 eV from the kinetic energy release thresholds for the photodissociation fragments. Figure 12 shows the photodissociation spectrum for Cs_n^+ with all parent/daughter relationships indicated.

After initial study of Mn_2^+ by ion beam methods (*vide infra*), a photodissociation study of Mn_2^+ was reported by Jarrold *et al.*[55] Mn_2^+ was produced

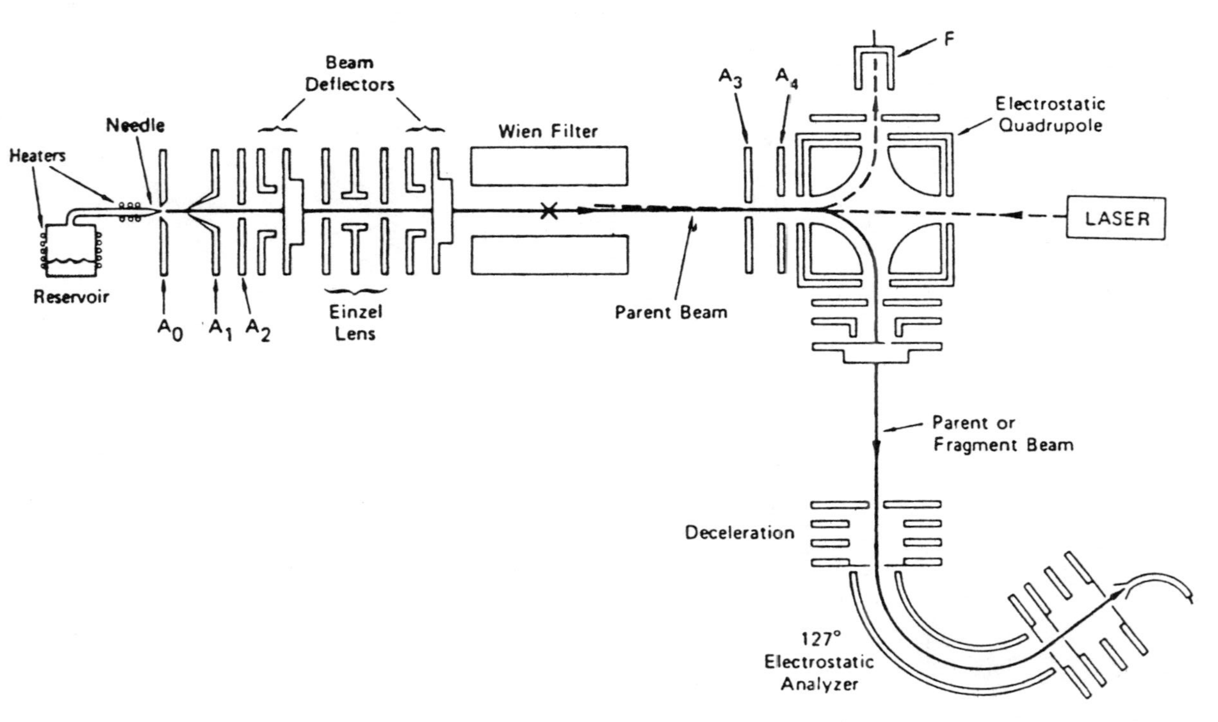

Figure 11. Schematic of instrument used for production and photodissociation of Cs_n^+. The kinetic energy release of the photofragments is determined by the electrostatic sector prior to the detector. (Reprinted from Reference 54 with permission.)

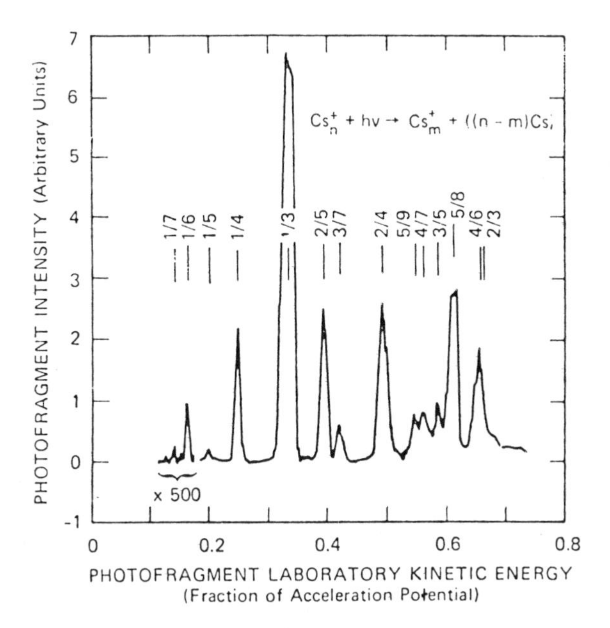

Figure 12. Kinetic energy spectrum of the photofragments from photodissociation of Cs_n^+ at 752.5 nm with the daughter/parent relationships indicated. (Reprinted from Reference 54 with permission.)

by electron impact on $Mn_2(CO)_{10}$, and photodissociation was carried out using a crossed high-energy ion/laser beam apparatus, as shown in Figure 13. Relative kinetic energies of the photoproducts were obtained using the electrostatic sector. N_2 was added to the ion source to collisionally cool the Mn_2^+. Figure 14 shows plots of the relative kinetic energy of the photoproduct, Mn^+, from photodissociation of Mn_2^+. The products can be formed with excess kinetic or electronic energy. The low relative kinetic energy peak was attributed to formation of Mn^+ in an excited electronic state, and the higher-energy peak was attributed to ground state product. The broadness of the peaks is due to initially vibrationally excited Mn_2^+. Thus, the thresholds represent lower limits for the Mn_2^+ bond strength. The relative kinetic energy threshold for photodissociation at 488 nm implies $D°(Mn^+—Mn) > 1.15$ eV, whereas the relative kinetic energy threshold for photodissociation at 514 nm implies $D°(Mn^+—Mn) > 1.39$ eV. These values are somewhat larger than $D°(Mn^+—Mn) = 0.85 \pm 0.2$ eV previously reported. This difference will be discussed later.

Hettich and Freiser reported on the photodissociation of a series of dimeric cluster ions of the type MFe^+(M = Sc, Ti, V, Cr, Fe, Co, Ni, Cu, Nb, Ta), prepared in reactions analogous to reactions 3 and 4.[56] The ions were trapped in the FTMS cell for ~3-10 s while illuminated by light from an arc lamp. The photoproducts, which included both M^+ and Fe^+ in each case, were monitored as a function of wavelength in the region 230-600 nm. A high-energy peak at ~250 nm was observed in each of the spectra and was assigned to a $\sigma \rightarrow \sigma^*$ transition due to 4*s*-4*s* electron overlap. For the early transition metals

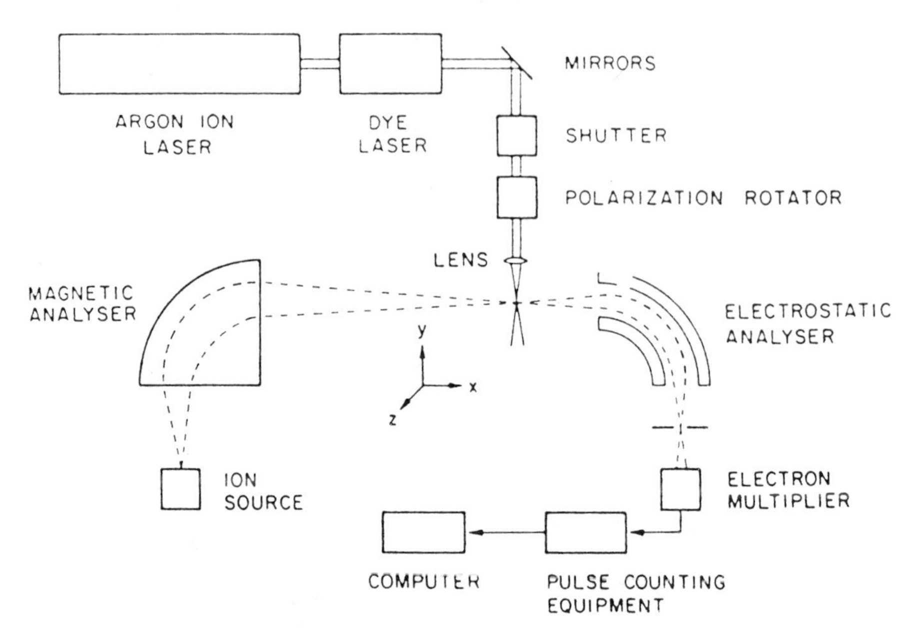

Figure 13. Crossed laser/ion beam apparatus used for photodissociation of Mn_2^+. (Reprinted from Reference 55 with permission.)

(Sc to V) one or more low-energy peaks in the 300–450-nm range are observed presumably due to more extensive *d–d* bonding. Cutoffs in the photodissociation spectra were attributed to thermodynamic as opposed to spectroscopic factors, given the high density of low-lying electronic states in metal cluster species. Ion-molecule reactions were used to bracket the bond strengths of MFe^+ (using similar techniques to those discussed above in the determination of the $CoFe^+$ bond strength) and in all cases showed good agreement with the photodissociation results, supporting the assumption that the cutoffs were due to thermodynamic factors. Figure 15 shows the photodissociation spectrum for VFe^+. Table 10 lists the thermochemical data derived for MFe^+ from photodissociation and ion-molecule bracketing experiments. Using previous, predominantly theoretical, data for D°(M—Fe), the ionization potentials for MFe were determined using the following equation:

$$\mathrm{IP(MFe)} = D^\circ(\mathrm{M{-}Fe}) + \mathrm{IP(M)} - D^\circ(\mathrm{M^+{-}Fe}) \tag{49}$$

With the exception of $ScFe^+$, the bond energies for the cluster ions greatly exceed the bond energies for the neutral clusters. It was suggested that the positive charge may reduce the internuclear distance and allow more *d–d* bonding.

Smalley and co-workers reported on the photodissociation of niobium, nickel, and iron cluster ions produced by a supersonic expansion source.[57] For Ni_2^+ and Nb_2^+, well-resolved vibronic bands were observed for the cooled

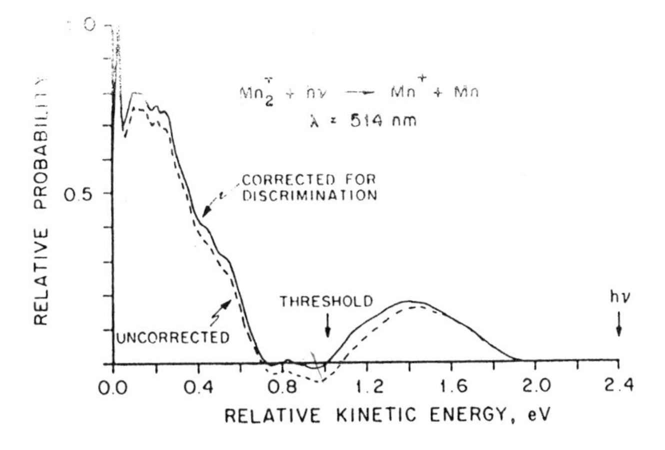

b)

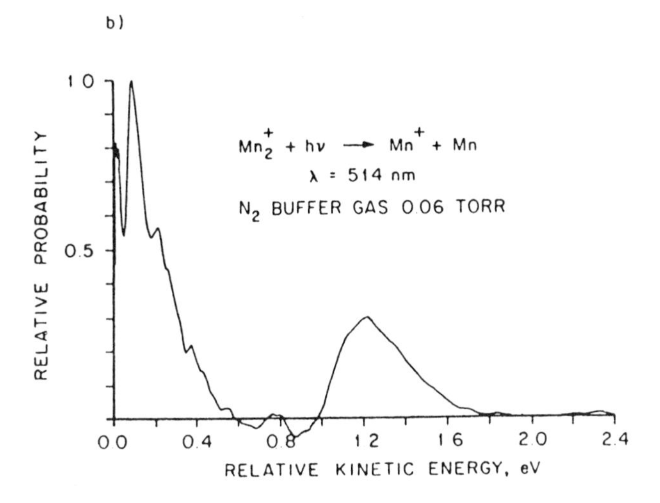

c)

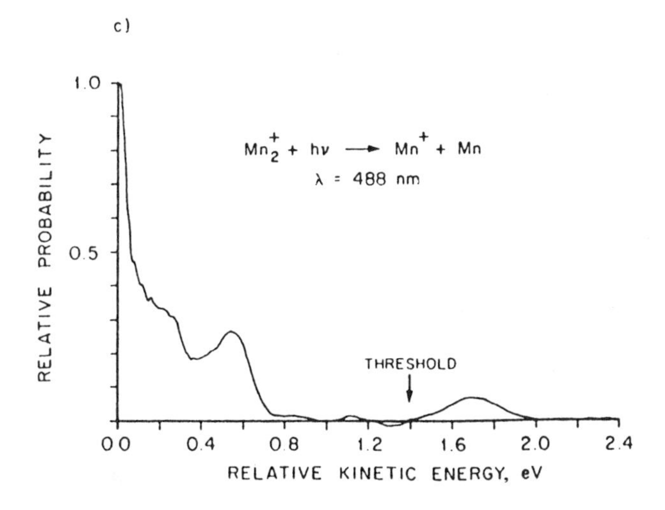

Figure 14. Product relative kinetic energy distributions for the photoproducts from photodissociation of Mn_2^+. (A) Results from photodissociation at 514 nm. Mn_2^+ formed by electron impact with no collisional cooling. (B) Results at 514 nm following collisional cooling of Mn_2^+ with N_2. (C) Results at 488 nm with no cooling. (Reprinted from Reference 55 with permission.)

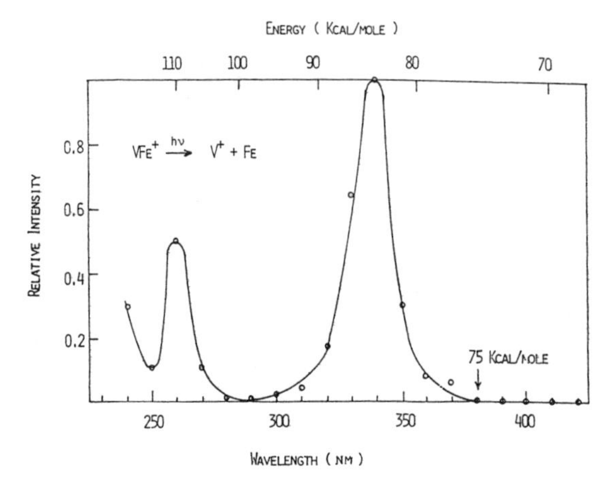

Figure 15. Photodissociation spectrum of VFe^+ in the 240- to 420-nm range. The threshold at 380 nm is attributed to thermodynamic factors and a value of $D^\circ(V^+—Fe) = 75 \pm 5$ kcal/mol is implied. (Reprinted from Reference 25 with permission.)

Table 10. Thermodynamic Data for Metal Clusters[a]

AB	D°(A—B)[b] (kcal/mol)	IP(AB) (eV)	PDS[e] $D^\circ(A^+—B)$ (kcal/mol)	I-M[c] $D^\circ(A^+—B)$ (kcal/mol)	PDS[e] $D^\circ(B^+—A)$[d] (kcal/mol)
ScFe	55	6.8	48 ± 5	49 ± 6[f]	79 ± 5
TiFe	42	6.0	60 ± 6	>49	84 ± 6
VFe	44	5.4	75 ± 5	>62[g]	101 ± 5
CrFe	32	6.0	50 ± 7	—[h]	75 ± 7
Fe_2	18	6.0	62 ± 5	58 ± 7[i]	62 ± 5
CoFe	30	6.5	62 ± 5	62 ± 6[i]	62 ± 5
NiFe	39	6.6	64 ± 5	<68 ± 5[f]	69 ± 5
CuFe	30	6.7	53 ± 7	>52	56 ± 7
NbFe	59[j]	6.5	68 ± 5	>60[f]	91 ± 5
TaFe	60[j]	7.4	72 ± 5	>60[f]	72 ± 5

[a] Data taken from Ref. 56.
[b] Taken from Weltner, W., Jr.; Van Zee, R.J. *Ann. Rev. Phys. Chem.*, 1984, *35*, 291.
[c] I-M = Ion-molecule reactions.
[d] Calculated from $D^\circ(A^+—B)$, IP(A), and IP(B).
[e] PDS = photodissociation.
[f] Lech, L.M.; Jacobson, D.B.; Buckner, S.W.; Freiser, B.S.; Unpublished results.
[g] Taken from Ref. 25.
[h] Not available.
[i] Taken from Ref. 19.
[j] Taken from Miedema, A.R.; *Faraday Symp. Chem. Soc.* 1980, *14*, 136.

species. For Fe_2^+ and larger (n = 2–10) iron and niobium cluster ions, broad absorptions with very few spectral features were observed. In these cases, as for the MFe^+ ions, photodissociation cutoffs were attributed to thermodynamic and not spectroscopic factors.

The bond dissociation energy of Fe_2^+ was determined to be between 2.43 and 2.92 eV, in agreement with the result obtained by Hettich and Freiser,[56] and $D^\circ(Fe_2^+—Fe)$ was bracketed between 1.17 and 2.18 eV. For Fe_6^+, the only one-photon photodissociation (2.33-eV light) process observed was loss of one atom to form Fe_5^+. Multiphoton absorption by Fe_6^+ produces Fe_n^+ ($n = 4$–1) with one Fe atom lost per photon, except for Fe^+ which is produced via a two-photon process from Fe_2^+. No one-photon dissociation processes were observed for Fe_n^+ ($n = 2$–6) with 1.17-eV light, indicating $1.17\ \text{eV} < d^\circ(Fe_n^+\text{—}Fe) < 2.33\ \text{eV}$ ($n = 3$–6). For all Fe_n^+ and Nb_n^+ ($n = 2$–10) the predominant one-photon photodissociation process is loss of one atom.

On a similar apparatus, electron photodetachment of the cluster anions Cu_n^-, Ag_n^-, Ni_n^-, and Nb_n^- was observed.[58] One modification to the source involved production of low-energy electrons at the point where the clusters emerge from the supersonic nozzle by firing an ArF laser at the tip of the nozzle. The free electrons combine with the metal clusters to produce the cluster anions.

Interestingly, photodissociation does not compete with photodetachment for Cu_n^-, Ni_n^-, and Nb_n^-. However, photofragmentation does occur for Ag_n^- ($n < 15$) except for Ag_2^- and Ag_{10}^-. Photoproducts from Ag_{10}^-, Ag_7^-, and Ag_3^- are shown in Figure 16.

Electron affinities for copper clusters were determined by measuring the onset of photodetachment of Cu_n^- by two methods—varying the photodetachment wavelength and varying the laser fluence at constant wavelength. Both

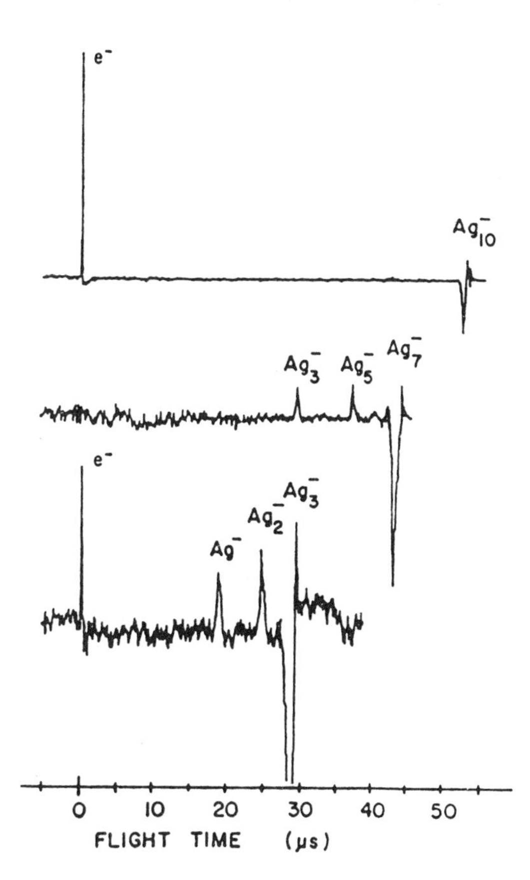

Figure 16. Mass spectra of the photoproducts from photodissociation and photodetachment of Ag_n^- by 532-nm light. The vertical scale shows the difference signal between light on and light off. Ag_{10}^- (top spectrum) only undergoes photodetachment while Ag_7^- and Ag_3^- undergo photodissociation as well. (Reprinted from Reference 58 with permission.)

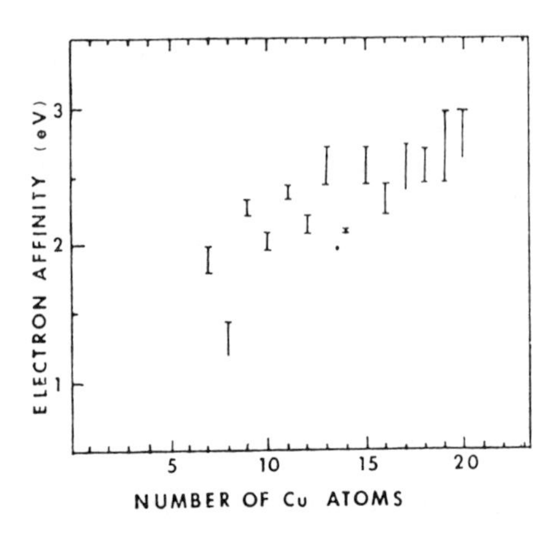

Figure 17. Electron affinities of Cu_n (n = 7–20) determined by photodetachment of Cu_n^- using laser fluence-dependent measurements. (Reprinted from Reference 58 with permission.)

techniques were shown to give similar results but, due to experimental considerations, the constant-wavelength technique was preferred.

Figure 17 shows the variation of electron affinity of Cu_n with cluster size, which the authors point out is very similar to the behavior of the ionization potentials for Cu_n as a function of size. The odd clusters have higher electron affinities and lower ionization potentials than the even clusters. The explanation for this behavior is that the even clusters have singlet ground states with highly bonding doubly occupied HOMOs, while the odd clusters have doublet ground states with singly occupied HOMOs of non-bonding character. Thus, the LUMOs of the even clusters should be strongly antibonding while for the odd clusters, the extra electron in the anion would fill up the half-filled nonbonding orbital.

A method for probing the electronic structure of neutral metal clusters is negative ion photoelectron spectroscopy.[59] In this technique, an anion absorbs a photon, and the photodetached electron is energy analyzed at very high resolution (~10 meV). Leopold, Lineberger, and co-workers have successfully interfaced two sources for generation of metal cluster anions with a negative ion photoelectron spectrometer.[60] The first source used multinuclear carbonyl complexes [e.g., $Re_2(CO)_{10}$, $Fe_2(CO)_9$, $Co_2(CO)_8$] seeded in a He flow which passed through a 2.45-GHz microwave cavity discharge to produce the bare metal cluster anions (e.g., Re_2^-, Fe_2^-, Co_2^-). The photoelectron spectrum of Re_2^- is shown in Figure 18. From the well-resolved vibrational structure in this spectrum, force constants of 6.4 ± 0.8 mdyn/Å and 5.6 ± 0.5 mdyn/Å were obtained for Re_2 and Re_2^-, respectively. These high force constants indicate multiple bonding for both Re_2 and Re_2^-. One interesting aspect of the photoelectron spectrum of Re_2^-, which was also observed for that of Fe_2^- and Co_2^-, is its simplicity. As previously mentioned, high-level calculations on Fe_2 and Co_2 indicated high-density manifolds of low-lying excited electronic states. These spectra would be expected to be very complex based on these calculations. The spectra also indicate extensive d–d bonding for Fe_2 and Co_2, in contrast to the $4s\sigma$ bonds predicted by these calculations.

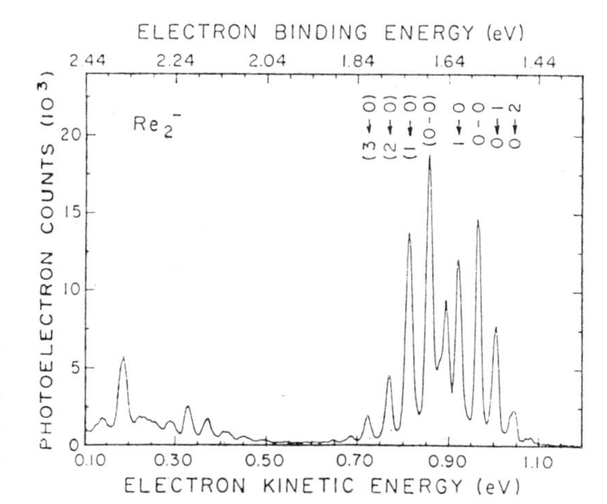

Figure 18. Photoelectron spectrum of Re_2^-. Labels indicate $Re_2 \leftarrow Re_2^-$ vibrational quantum numbers for transitions to the ground and excited Re_2 states. (Reprinted from Reference 60 with permission.)

The second source used was a cold cathode dc discharge placed in a flowing afterglow source. Using this source, photoelectron spectra for Cu_n^- (n = 1–10) were obtained. Accurate vertical and adiabatic electron affinities were obtained and found to be in qualitative agreement with trends observed for larger clusters by Smalley and co-workers. This technique shows great promise in understanding the electronic structure of metal clusters.

Fayet and Woste have carried out photodissociation on a number of sputtered metal cluster ions.[61] The experimental apparatus used in their studies is shown in Figure 19. Photochemical investigations were performed on Ag_n^+ and Ag_n^-. Spectra for $Ag_7^{\pm}$ appear in Figure 20. Photodissociation as well as photodetachment occur, in agreement with results from Smalley's lab. All silver cluster ions fragment preferentially to form daughter cluster ions with odd numbers of atoms, in agreement with the higher stability of odd cluster ions.

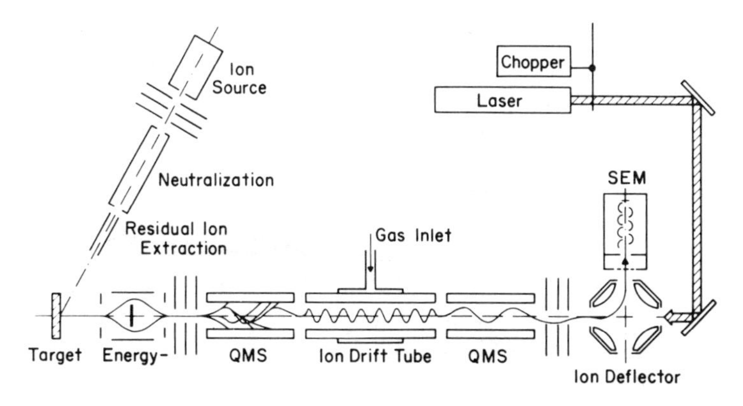

Figure 19. Photodissociation apparatus used for dissociating positive and negative cluster ions produced by a sputtering source. (Reprinted from Reference 61 with permission.)

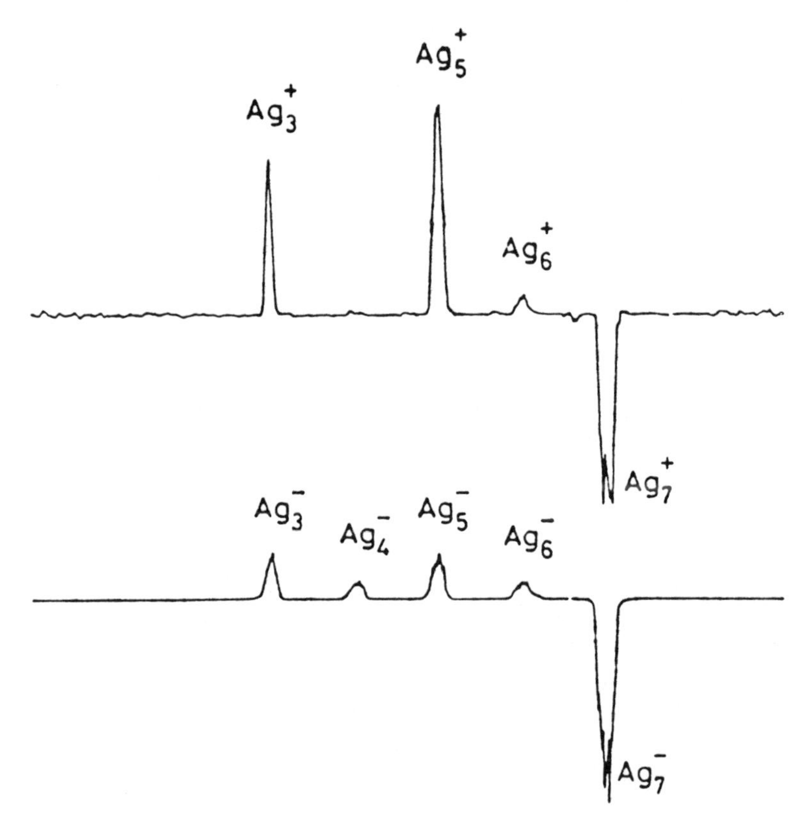

Figure 20. Photodissociation spectra for $Ag_7^\pm$ with 488-nm laser light. The vertical scale represents the difference signal between light on and light off. (Reprinted from Reference 61 with permission.)

3.2. Collision-Induced Dissociation and Unimolecular Dissociation

In one of the earliest studies on cluster ions, Armentrout and co-workers used collision-induced dissociation to determine the bond strength of Mn_2^+.[62] Mn_2^+ was produced by electron impact on $Mn_2(CO)_{10}$ near the appearance potential of Mn_2^+ to minimize excess internal energy. Figure 21 shows the variation of Mn^+ abundance with Mn_2^+ kinetic energy for the reaction:

$$Mn_2^+ + Ar \rightarrow Mn^+ + Mn + Ar \tag{50}$$

The appearance energy of 0.85 eV yields a strict lower limit of $D°(Mn^+—Mn) >$ 0.85 eV, but the authors suggested that this value could be the true bond strength (within ±0.2 eV) as Mn_2^+ was produced at the appearance potential for Mn_2^+. The results described above from Bowers' lab[55] of $D°(Mn^+—Mn) >$ 1.35 eV, however, indicate that Mn_2^+ might still have had internal energy. Using a modified apparatus to cool the ions, which included a side-on EI source equipped with a high-pressure drift cell to thermalize the ions, a threshold for

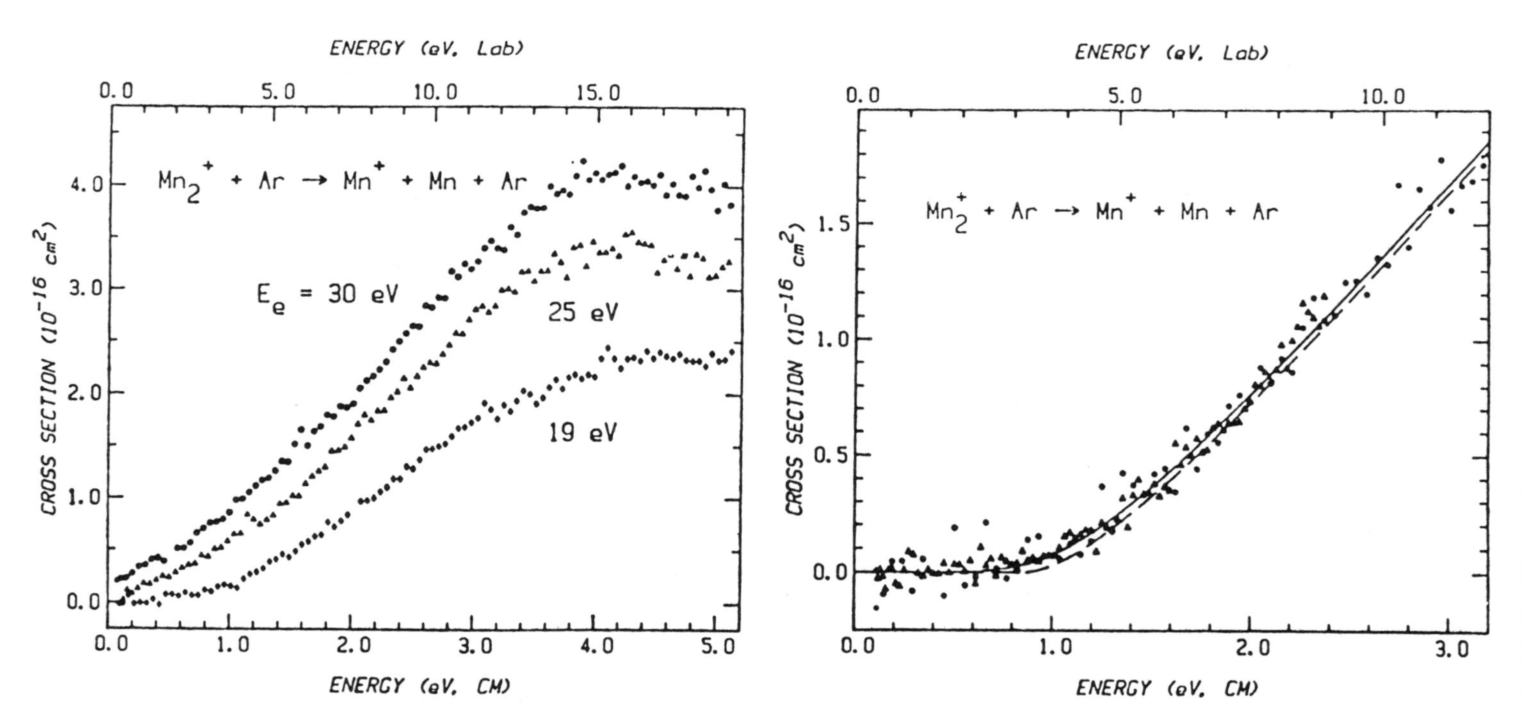

Figure 21. Collision-induced dissociation cross sections as a function of energy for dissociation of Mn_2^+. The curves represent production of Mn_2^+ at different electron energies. The right-hand panel shows the cross section near the threshold with Mn_2^+ produced near its appearance potential. (Reprinted from Reference 62 with permission.)

dissociation of Mn_2^+ was observed at 1.1 ± 0.2 eV, close to the value obtained by photodissociation. Similar experiments on Co_2^+ suggested $D^\circ(Co^+—Co) > 2.8 \pm 0.3$ eV.

Begemann *et al.* have studied the collision-induced dissociation and unimolecular decomposition of a large number of sputtered cluster ions, including Al_n^+, Cu_n^+, W_n^+, Mo_n^+, and Pb_n^+.[63] The main process was loss of one atom from the cluster, with loss of two atoms observed for Al_n^+ and Cu_n^+.

The mass spectra of sputtered cluster ions have been used to infer the stability of various-sized cluster ions by comparison of the observed intensities. Cluster sizes with large intensity are presumed to be stable. The probabilities for the cluster ions in Begemann's study were observed to mirror the ion abundances of the cluster ions in the mass spectra. The authors suggest that the decay probability and the ion abundances in the mass spectra should be a function of the cluster stability and structure.

In a modified apparatus similar to the one used for reaction studies, Anderson and co-workers studied the collision-induced dissociation of Al_n^+ (n = 2-7).[64] Threshold energies for dissociation of Al_n^+ are shown in Table 11. Using these thresholds, the stabilities for Al_n^+ are:

$$Al_7^+ \gg Al_6^+ > Al_{3,5}^+ > Al_2^+ > Al_4^+$$

Formation of Al^+ was the major reaction pathway for all cluster ions except for Al_3^+, where Al_2^+ and Al^+ were produced in a 3:1 ratio. The authors explained the anomalous behavior of Al_3^+ using spin conservation rules. Al_3^+ and Al^+ have singlet ground states, whereas Al_2 has a triplet ground state with an excited singlet state at 0.38 eV. If the observed threshold represents production of excited singlet Al_2, an allowed process, then the threshold would be high by 0.38 eV, and the actual order of stabilities would be:

$$Al_7^+ \gg Al_6^+ > Al_5^+ > Al_{2,3}^+ > Al_4^+$$

Jarrold *et al.* also studied the CID of Al_n^+ (n = 3-26).[65] All of their experiments were carried out at a center-of-mass collision energy of 5.25 eV (Ar

Table 11. Threshold Energies (in eV) for $Al_n^+ + Xe \rightarrow Al_m^+$ [a]

Parent cluster	Product channel					
	Al^+	Al_2^+	Al_3^+	Al_4^+	Al_5^+	Al_6^+
Al_2^+	0.90 ± 0.30	—	—	—	—	—
Al_3^+	1.30 ± 0.35	1.12 ± 0.35	—	—	—	—
Al_4^+	0.85 ± 0.40	2.55 ± 0.35	2.03 ± 0.30	—	—	—
Al_5^+	1.30 ± 0.35	3.35 ± 0.65	4.30 ± 0.45	2.57 ± 0.35	—	—
Al_6^+	1.55 ± 0.30	4.90 ± 0.55	5.67 ± 0.55	5.30 ± 0.35	2.40 ± 0.30	—
Al_7^+	2.55 ± 0.70	N.O.[b]	N.O.[b]	N.O.[b]	7.25 ± 0.40	4.47 ± 0.40

[a] Data taken from Ref. 64.
[b] Product channel not observed.

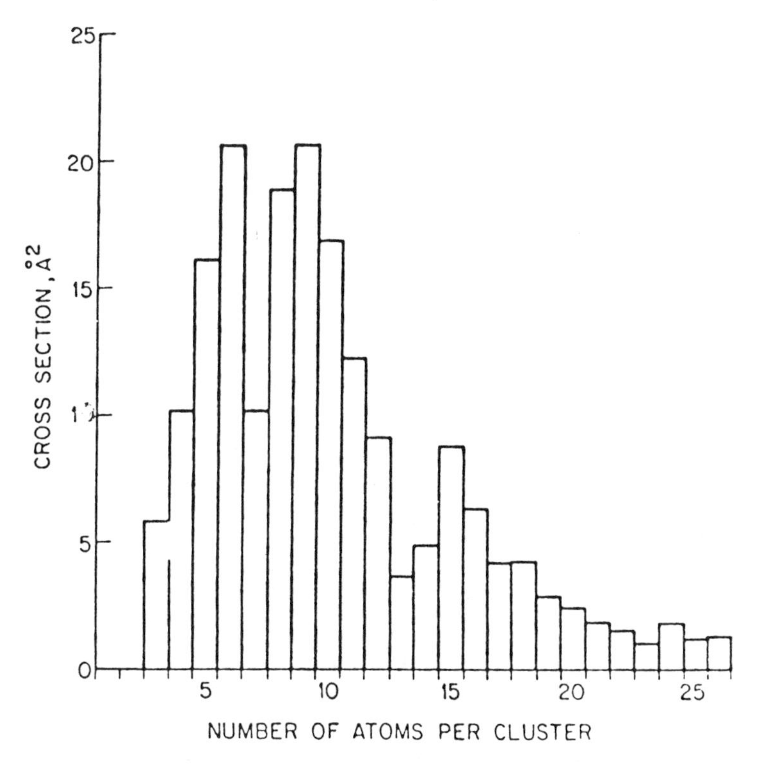

Figure 22. Histogram showing total cross section for collision-induced dissociation of Al_n^+ (n = 3-26). Variations in cross section indicate relative cluster ion stability; see text. (Reprinted from Reference 65 with permission.)

collision gas). Figure 22 shows a histogram of the total cross sections for CID of Al_n^+. This plot indicates that Al_7^+, Al_{13}^+, Al_{14}^+, and Al_{23}^+ are particularly stable due to their significantly lower cross sections. This is in agreement with the findings of Anderson and co-workers, who found Al_7^+ to be highly stable.

3.3. Multiply Charged Clusters

Small multiply charged molecules have been observed in the gas phase in cases where high binding energy exceeds the coulomb repulsion, but they are relatively rare.[66] Recently, Sattler *et al.* reported the formation of Pb_n^{2+} by electron impact on the corresponding neutral clusters, for $n > 30$.[67] For $n < 30$, the Coulomb repulsion is too great and the charges migrate to opposite sides of the cluster which undergoes a "coulomb explosion" or charge splitting reaction:

$$Pb_n^{2+} \rightarrow Pb_{n-x}^+ + Pb_x^+ \qquad n < 30 \tag{51}$$

For multiply charged van der Waals clusters composed of inert atoms, "atom boil off" or loss of neutral atoms is observed instead of charge splitting for clusters larger than a critical size.

Table 12. Stability of M_3^{2+}: Predictions and Experimental Observation[a]

M	r_a (nm)	Λ (eV)	E_b (eV)	E_c (eV)	$R = E_c/E_b$	M_3^{2+} detected
Ni	0.124	4.44	2.39	2.90	1.21	Yes
Cu	0.127	3.50	1.97	2.83	1.44	No
Ag	0.144	2.96	1.65	2.45	1.49	No
Au	0.144	3.78	2.30	2.45	1.07	Yes
Ga	0.122	2.78	1.40	2.95	2.11	No
In	0.162	2.60	1.04	2.22	2.14	No
Sn	0.158	3.12	1.93	2.28	1.18	Yes
W	0.141	8.66	4.68[b]	2.55	0.545	Yes
Bi	0.182	2.15	2.04	1.98	0.975	Yes
Sb	0.161	2.70	3.04	2.24	0.737	Yes

[a] Data taken from Ref. 68.
[b] E_b for W estimated from the binding energy to the surface, Λ.

M_3^{2+} was observed for Ni, Au, and W but not for Cu and Ag.[68] Rough estimates for the cluster binding energy (E_b) and coulomb repulsion (E_c) were used to define the ratio $R = E_c/E_b$. Clusters with a value of R below ~1.3 were found to be stable (see Table 12). Finally, recent experimental[69,70] and theoretical[71] studies have demonstrated the existence of stable M_2^{2+}. Specifically, Mo_2^{2+} [69] and $LaFe^{2+}$ [70] have been observed, with the latter being sufficiently long lived to study its reaction chemistry in the gas phase.

4. CONCLUSION

The interfacing of new sources for production of metal cluster ions with well-developed mass spectrometric techniques is proving to be a powerful combination. The use of ionic species in the study of gas phase clusters greatly increases the ability to investigate their size and energy-dependent properties. These techniques have been demonstrated on a few systems, but a complete picture of the behavior of this novel phase of matter has only begun to emerge and awaits much further experimental and theoretical study.

ACKNOWLEDGMENTS

Metal cluster ion research in our laboratory is supported by the Division of Chemical Sciences in the Office of Basic Energy Research in the United States Department of Energy (DE-FG02-87ER13766) and the National Science Foundation (CHE-8612234).

REFERENCES

1. (a) Muetterties, E.L.; Rhodin, T.N.; Band, E.; Brucker, C.F.; Pretzer, W.R. *Chem. Rev.* 1979, *79*, 91; (b) *Faraday Symp. Chem. Soc.* 1980, *14*; (c) *Ber. Bunsenges. Phys. Chem.* 1984, *88*(3); (d) *Surf. Sci.* 1985, *156*; (e) Castleman, A.W., Jr.; Keesee, R.G. *Annu. Rev. Phys. Chem.* 1986, *37*, 525; (f) Morse, M.D. *Chem. Rev.* 1986, *86*, 1049.

2. Geusic, M.E.; Morse, M.D.; Smalley, R.E. *J. Chem. Phys.* 1985, *82*, 590.
3. Geusic, M.E.; Morse, M.D.; O'Brien, S.C.; Smalley, R.E. *Rev. Sci. Instrum.* 1985, *56*, 2123.
4. Whetten, R.L.; Cox, D.M.; Trevor, D.I.; Kaldor, A. *Phys. Rev. Lett.* 1985, *54*, 1494.
5. Alford, J.M.; Williams, P.E.; Trevor, D.J.; Smalley, R.E. *Int. J. Mass Spectrom. Ion Proc* 1986, 72, 33.
6. Alford, J.M.; Weiss, F.D.; Laaksonen, R.T.; Smalley, R.E. *J. Phys. Chem.* 1986, *90*, 4480.
7. Phillips, J.C. *J. Chem. Phys.* 1986, *84*, 1951.
8. Allison, J. In *Progress in Inorganic Chemistry*; Lippard, S.J., Ed.; Wiley-Interscience: New York, 1986; Vol. 34, p. 628.
9. (a) Allison, J.; Freas, R.B.; Ridge, D.P. *J. Am. Chem. Soc.* 1979, *101*, 1332; (b) Freas, R.B.; Ridge, D.P. *J. Am. Chem. Soc.* 1980, *102*, 7129.
10. Peake, D.A.; Gross, M.L. *J. Am. Chem. Soc.* 1987, *109*, 600.
11. Armentrout, P.B.; Halle, L.F.; Beauchamp, J.L. *J. Am. Chem. Soc.* 1981, *103*, 6501.
12. Tolbert, M.; Beauchamp, J.L. *J. Am. Chem. Soc.* 1984, *106*, 8117.
13. Byrd, G.D.; Burnier, R.C.; Freiser, B.S. *J. Am. Chem. Soc.* 1982, *104*, 3565.
14. (a) Elkind, J.L.; Armentrout, P.B. *J. Phys. Chem.* 1985, *87*, 5626; (b) Aristov, N.; Armentrout, P.B. *J. Am. Chem. Soc.* 1984, *106*, 4065; (c) Elkind, J.L.; Armentrout, P.B. *J. Am. Chem. Soc.* 1986, *108*, 2765.
15. Armentrout, P.B.; Halle, L.F.; Beauchamp, J.L. *J. Am. Chem. Soc.* 1981, *103*, 6501.
16. (a) Foster, M.S.; Beauchamp, J.L. *J. Am. Chem. Soc.* 1975, *97*, 4808; (b) Foster, M.S.; Beauchamp, J.L. *J. Am. Chem. Soc.* 1975, *97*, 4814; (c) Weddle, G.H.; Allison, J.; Ridge, D.P. *J. Am. Chem. Soc.* 1977, *99*, 105; (d) Wronka, J.; Ridge, D.P. *J. Am. Chem. Soc.* 1984, *106*, 67; (e) Fredeen, D.A.; Russell, D.H. *J. Am. Chem. Soc.* 1985, *107*, 3762; (f) Meckstroth, W.K.; Ridge, D.P.; Reents, W.D., Jr. *J. Phys. Chem.* 1985, *89*, 612.
17. (a) Ridge, D.P. In *Lecture Notes in Chemistry*; Hartmann, H.; Wanczek, K.P., Eds.; Springer-Verlag: Berlin, 1982; Vol. 31, p. 140; (b) Larsen, B.S.; Freas, R.B.; Ridge, D.P. *J. Phys. Chem.* 1984, *88*, 6014.
18. Freas, R.B.; Ridge, D.P. *J. Am. Chem. Soc.* 1984, *106*, 825.
19. Jacobson, D.B.; Freiser, B.S. *J. Am. Chem. Soc.* 1984, *106*, 4623.
20. Jacobson, D.B.; Freiser, B.S. *J. Am. Chem. Soc.* 1985, *107*, 1581.
21. Forbes, R.A.; Lech, L.M.; Freiser, B.S. *Int. J. Mass Spectrom Ion Proc.* 1987, *77*, 107.
22. Jacobson, D.B., private communication.
23. Rosenstock, H.M.; Draxl, D.; Steiner, B.W.; Herron, J.T. *J. Phys. Chem. Ref. Data* 1977, *6*, Suppl. 1.
24. Armentrout, P.B. In *Structure/Reactivity and Thermochemistry of Ions*; Ausloos, P.; Lias, S.G., Eds.; Reidel: Holland, 1987; p. 97.
25. Hettich, R.L.; Freiser, B.S. *J. Am. Chem. Soc.* 1985, *107*, 6222.
26. Halle, L.F.; Armentrout, P.B.; Beauchamp, J.L. *Organometallics* 1982, *1*, 963.
27. Hettich, R.L.; Freiser, B.S. *J. Am. Chem. Soc.* 1986, *108*, 2537.
28. Tews, E.C.; Freiser, B.S. *J. Am. Chem. Soc.* 1987, *109*, 4433.
29. Huang, Y.; Buckner, S.W.; Freiser, B.S. In *The Physics and Chemistry of Small Clusters*; NATO ASI Series B, Physics Vol. 158, Jena, P.; Rao, B.K.; Khanna, S.N., Ed.; Plenum: New York; 1987; p. 891.
30. Jacobson, D.B.; Freiser, B.S. *J. Am. Chem. Soc.* 1984, *106*, 5351.
31. Blaise, G. In *Materials Characterization Using Ion Beams*; Thomas, J.P.; Cachard, A., Eds.; Plenum: New York, 1978.
32. Freas, R.B.; Ross, M.M.; Campana, J.E. *J. Am. Chem. Soc.* 1985, *107*, 6195.
33. Freas, R.B.; Campana, J.E. *J. Am. Chem. Soc.* 1985, *107*, 6202.
34. Freas, R.B.; Campana, J.E. *J. Am. Chem. Soc.* 1986, *108*, 4659.
35. Hanley, L.; Anderson, S.L. *Chem. Phys. Lett.* 1985, *122*, 410.
36. Hanley, L.; Anderson, S.L.; *Chem. Phys. Lett* 1986, *129*, 429.
37. Magnera, T.F.; David, D.E.; Michl, J. *J. Am. Chem. Soc.* 1987, *109*, 936.
38. Weil, D.A.; Wilkins, C.L. *J. Am. Chem. Soc.* 1985, *107*, 7316.
39. Freiser, B.S.; Buckner, S.W.; Gord, J.R.; Lech, L.M. Unpublished results.
40. Bartmess, J.E.; Kester, J. G. *Inorg. Chem.* 1984, *23*, 1877.
41. Armentrout, P.B.; Loh, S.K.; Ervin, K.M. *J. Am. Chem. Soc.* 1984, *106*, 1161.

42. Jacobson, D.B.; Freiser, B.S. *J. Am. Chem. Soc.* 1986, *108*, 27.
43. Jackson, T.C.; Jacobson, D.B.; Freiser, B.S. *J. Am. Chem. Soc.* 1984, *106*, 1252.
44. Armentrout, P.B. Private communication.
45. Brucat, P.J.; Pettiette, C.L.; Yang, S.; Zheng, L.-S.; Craycraft, M.J.; Smalley, R.E. *J. Chem. Phys.* 1986, *85*, 4747.
46. (a) Fayet, P.; Woste, L. *Z. Phys. D* 1986, *3*, 177; (b) Fayet, P.; McGlinchey, M.J.; Woste, L.H. *J. Am. Chem. Soc.* 1987, *109*, 1733.
47. Hanley, L.; Ruatta, S.; Anderson, S. In *The Physics and Chemistry of Small Clusters*; Jena, P., Ed.; Plenum: New York; 1987; p. 781.
48. Jarrold, M.F.; Bower, J.E. *J. Chem. Phys.* 1986, *85*, 5373.
49. Jarrold, M.F.; Bower, J.E. *J. Chem. Phys.* 1987, *87*, 1610.
50. Freas, R.B.; Dunlap, B.I.; Waite, B.A.; Campana, J.E. *J. Chem. Phys.* 1987, *86*, 1276.
51. Loh, S.K.; Hales, D.A.; Armentrout, P.B. *Chem. Phys. Lett.* 1986, *129*, 527.
52. (a) Dunbar, R.C. In *Gas Phase Ion Chemistry*; Bowers, M.T., Ed.; Academic Press: New York, 1984; Vol. 3, Chapter 20; (b) Hettich, R.L.; Jackson, T.C.; Stanko, E.M.; Freiser, B.S. *J. Am. Chem. Soc.* 1986, *108*, 5086.
53. Shim, I.; Gingerich, K.A. *J. Chem. Phys.* 1982, *77*, 2490.
54. (a) Helm, H.; Moller, R. *Phys. Rev. A* 1983, *27*, 2493; (b) Helm, H.; Moller, R. *Rev. Sci. Instrum*, 1983, *54*, 837.
55. Jarrold, M.K.; Illies, A.J.; Bowers, M.T. *J. Am. Chem. Soc.* 1985, *107*, 7339.
56. Hettich, R.L.; Freiser, B.S. *J. Am. Chem. Soc.* 1987, *109*, 3537.
57. Brucat, P.J.; Zheng, L.-S.; Pettiette, C.L.; Yang, S.; Smalley, R.E. *J. Chem. Phys.* 1986, *84*, 3078.
58. (a) Zheng, L.-S.; Brucat, P.J.; Pettiette, C.L.; Yang, S.; Smalley, R.E. *J. Chem. Phys.* 1985, *83*, 4273; (b) Zheng, L.-S.; Karner, C.M.; Brucat, P.J.; Yang, S.H.; Pettiette, C.L.; Craycraft, M.J.; Smalley, R.E. *J. Chem. Phys.* 1986, *83*, 1681.
59. Mead, R.D.; Stevens, A.E.; Lineberger, W.C. In *Gas Phase Ion Chemistry*; Bowers, M.T. Ed.; Academic Press: New York, 1984; Vol. 3, Chapter 23.
60. (a) Leopold, D.G.; Miller, T.M.; Lineberger, W.C. *J. Am. Chem. Soc.* 1986, *108*, 178; (b) Leopold, D.G.; Lineberger, W.C. *J. Chem. Phys.* 1986, *85*, 51; (c) Leopold, D.G.; Ho, J.; Lineberger, W.C. *J. Chem. Phys.* 1987, *86*, 1715.
61. Fayet, P.; Woste, L. *Surf. Sci.* 1985, *156*, 134.
62. Ervin, K.; Loh, S.K.; Armentrout, P.B. *J. Phys. Chem.* 1983, *87*, 3593.
63. (a) Begemann, W.; Dreihofer, S.; Meiwes-Broer, K.H.; Lutz, H.O. *Z. Phys. D* 1986, *3*, 183; (b) Begemann, W.; Meiwes-Broer, K.H.; Lutz, H.O. *Phys. Rev. Lett.* 1986, *56*, 2248.
64. Hanley, L.; Ruatta, S.A.; Anderson, S.L. *J. Chem. Phys.* 1987, *87*, 260.
65. Jarrold, M.F.; Bower, J.E.; Kraus, J.S. *J. Chem. Phys.* 1987, *86*, 3876.
66. Koch, W.; Maquin, F.; Stahl, D.; Schwartz, H. *Chimia* 1985, *33*, 376.
67. Sattler, K.; Muhlback, J.; Echt, O.; Pfau, P.; Recknagel, E. *Phys. Rev. Lett.* 1981, *47*, 160.
68. Jentsch, Th.; Drachsel, W.; Block, J.H. *Chem. Phys. Lett.* 1982, *93*, 144.
69. Tsong, T.T. *J. Chem. Phys.* 1986, *85*, 639.
70. Huang, Y.; Freiser, B.S. *J. Am. Chem. Soc.* 1988, *110*, 4435.
71. Liu, F.; Press, M.R.; Khanna, S.N.; Jena, P. *Phys. Rev. Lett.* 1987, *59*, 2562.

10

Photodissociation of Metal-Containing Gas Phase Ions

Robert C. Dunbar

1. INTRODUCTION

As this volume makes clear, chemical study of charged molecules has played a major role in the rapid progress of gas phase inorganic chemistry. A constant difficulty in studying gas phase ions is the relative scarcity of powerful tools for characterizing the molecules, since most of the array of methods which make conventional neutral-molecule chemistry a tractable field of study, such as NMR, ESR, chromatography, optical activity, and IR/Raman, are not now feasible for gaseous ions. Photodissociation holds the outstanding promise of filling some of this gap†: Through photodissociation spectroscopy, it supplies optical spectroscopic information corresponding to that provided by visible/UV spectroscopy (and perhaps eventually to that provided by IR as well), while the observation of the photofragmentation patterns is effectively a way of performing mass spectrometric analysis on the ionic molecules. It is the rapidly growing application of these methods to the new frontier of gas phase metal ion chemistry that inspires this chapter.

Ion photodissociation is at once a spectroscopic and a photochemical approach to studying charged molecules under the uniquely interesting conditions of the low-pressure gas phase, in which the opportunity an individual molecule has to react with its surroundings is controlled or eliminated. Photodissociation as a spectroscopic method is based on the idea that photon absorption must always precede photodissociation, so that any feature in the

† For recent reviews, see Reference 1.

Robert C. Dunbar • Department of Chemistry, Case Western Reserve University, Cleveland, Ohio 44106.

photodissociation spectrum (as we designate the spectrum of the extent of photodissociation as a function of wavelength) must be a feature in the optical absorption spectrum, while (with certain exceptions which we believe are understood and predictable) absorption of photons by the molecule is expected to lead to dissociation, so that features in the absorption spectrum are normally reflected in the photodissociation spectrum. Thus, there is often a one-to-one correspondence between the photodissociation spectrum and the true optical spectrum. Most of what is known about the spectroscopy of polyatomic gas phase ions comes from applying this idea.

The interest in photodissociation as a photochemical process is twofold. First, the dissociation of the ion after the deposition of a known internal energy by photon impact is a revealing method of characterizing the ion and is useful as an analytical and structural probe in the familiar senses of mass spectrometry. Second, ions are particularly easy to manipulate and detect, and it is rather easy to study photodissociation of isolated gas phase molecules regarded as the initial photochemical process initiating subsequent photochemistry. Viewing the photodissociation event in isolated molecules is an attractive prospect, but it has been extraordinarily difficult to realize in photochemical study of neutral molecules, which are difficult to manipulate in isolation and to detect individually. The realization that for ions such studies are possible and even easy has opened new avenues in the study of photochemical and photophysical processes.

This is a fortunate time for a survey of this field, because there is considerable and growing interest and excitement about gas phase metal chemistry and the information about it coming from spectroscopic and photochemical approaches, but the field is so new that the literature is not yet unmanageably large. A more-or-less complete survey of the literature matching our title theme through 1986 will be feasible. Before the more detailed discussion of specific experiments, an overview of the perspectives and possibilities of the various research groups may be helpful. The field divides naturally into ion cyclotron resonance (ICR) ion trap work and ion beam work.

Ion Traps. Freiser's group[2-9] has been very active in ICR studies of metal-containing ions. They have pioneered the production of metal-containing ions by laser surface ionization combined with gas phase ion–molecule reactions, have exploited their capability for taking photodissociation spectra at modest optical resolution, and have been most interested in using photodissociation threshold characteristics to get at bond-breaking thermochemistry. Our own group[10,11] has recently emphasized laser photodissociation, which offers possibilities of photodissociation spectroscopy at high optical resolution and opens various interesting possibilities of multiphoton chemistry and time resolution of photochemical events. Brauman's group[12,13] has been most concerned with anions, for which the usual photochemical process is electron detachment, but these interests have led them to some interesting dissociation chemistry as well. Beauchamp's group[14] has combined their extensive interests in IR-multiphoton ion chemistry and in ion chemistry of transition metal ions, in a ground-breaking investigation of IR-multiphoton processes in some of these ions, with the emphasis on deeper understanding of the mechanistic details of dissociation.

Ion Beams. Commerical double-focusing mass spectrometers have given several groups a convenient entry to this field. Bowers' group[15] has been interested in highly detailed study of the photofragmentation processes in simple ions, which has led them to some metal ion work. They have combined information on kinetic energy release, angular dependence, and wavelength dependence in search of a deeper understanding of the fragmentation process. The Stanford Research Institute group[16] has used their home-built beam in similar approaches on simple ions. Russell's group[17,18] (using their own cluster study technology), and the Naval Research Laboratory (NRL) group[21] have all been interested in the identity of the photoproducts from some larger cluster-type ions, while Fukuda and Campana[22,23] have had a similar goal with some organometallic complexes.

The information coming from photodissociation is wide-ranging. The chapter will be organized according to the different kinds of information which have been, or can be, obtained for metal-containing ions.

2. EXPERIMENTAL METHODS

2.1. Ion Traps

All of the trapped-ion work described here uses the ICR ion trap.[24] The trap is a Penning ion trap, which uses a magnetic field to prevent ion loss in the x and y directions and an electrostatic potential well to prevent ion loss in the z direction. Ion motion in the x–y plane is the familiar cyclotron motion, whose frequency is $qB/2\pi m$ (where B is the magnetic field strength, q is the charge of the ion, and m is its mass.) Ion detection is accomplished by exciting the ions at their cyclotron frequency to give them cyclotron velocity in the x–y plane and detecting the rf signal at the cyclotron frequency which the moving ions induce on the receiver plates. A variety of detection schemes are used, including bridge detectors and Fourier transform detectors, but in all cases the result is an ICR signal whose amplitude is proportional to the abundance of ions at a given mass. Photodissociation experiments are, in principle, a straightforward proposition of irradiating the ICR cell with light and observing the decrease in parent ion abundance and/or the increase in fragment ion abundance.

The ICR instrument can be operated at pressures in the 10^{-8}-torr region, which allows trapping for periods of seconds (needed for many photodissociation studies with low light intensity or low photodissociation cross section) and also makes possible impressive mass resolution ($M/\Delta M$ up to at least several million in favorable cases). The ICR cell may also be run with a higher pressure of neutrals to allow ion–molecule reaction chemistry to occur. Ions may be formed by electron impact ionization of volatile neutral compounds or by laser vaporization/ionization by pulsed laser ablation of a metal surface.

Freiser's group has carried ion production technolgy to an impressive level, employing laser vaporization/ionization to create bare metal ions, followed by sequences of ion–molecule reactions and cyclotron ejection of unwanted ions to clothe the metal ions with the desired ligands.[7]

2.2. Ion Beams

For ion beam studies, commercial double-focusing mass spectrometers offer an impressive set of capabilities, with only the provision of a window between the first and second mass analyzer being necessary to do photodissociation. The hardware and software of these instruments is already arranged for collisional dissociation of mass-selected parent ions, and all of the existing collision-induced dissociation (CID) capabilities, such as linked scans and investigation of kinetic energy release, carry over when photon activation is substituted for collisional activation. A persistent problem with such instruments is that the ions emerging from the standard source are not well thermalized and may have substantial internal excitation. This can give quite different photodissociation results from those obtained with thermalized ions.

2.3. Clusters

The study of cluster species is a rapidly emerging discipline with its own technology. A highly successful instrumental configuration is that used by Smalley's group in their photodissociation studies,[19] in which laser vaporization from a metal surface introduces metal atoms directly into the helium flow of a supersonic expansion nozzle. Multiphoton ionization near the expansion nozzle creates large numbers of ions, and the supersonic expansion cools them to extremely low internal temperatures. Smalley's group uses time-of-flight mass analysis for detection, although in principle any mass filter could serve as the detector.

3. SPECTROSCOPY

3.1. What Color Are the Ions?

As suggested above, the photodissociation spectrum of an ion can be regarded as an approximation to the optical absorption spectrum. Thus, one hopes to learn the usual spectroscopic things about the ions, which is to say the location of excited states and, from the peak intensities, some symmetry information about them. Considered as a structural tool, the spectrum can give information about the chromophoric features in the ion and is a powerful and often unambiguous way of distinguishing isomeric structures. As one expects for transition metal-containing molecules, these ions show a most promising array of strong spectral peaks at visible/near-UV wavelengths, and it is both a current frustration and a future challenge that for most of them theory is not yet able to give much guidance in the interpretation of the spectra.

The equivalence of the photodissociation and absorption spectra of an ion depends on the assumption that each photon absorbed by the ion leads to dissociation. This may not be true if the ion loses its energy by fluorescence or by collisional energy transfer faster than it dissociates. In the case where the energy of the photon is sufficient to bring about a dissociation, experience has shown that this is almost never a problem. Of hundreds of ions which have been examined for fluorescence,[25] only a small handful fluoresce in preference to internal conversion of the photon energy to vibrational excitation, which in turn leads to dissociation much faster than infrared fluorescent relaxation or collisions at the pressures used. The main cause for concern is the case where the photon energy is insufficient for dissociation, so that the energized ion simply remains intact until the energy is lost by radiation or by collisions. The wavelength beyond which one-photon dissociation is impossible is the thermochemical threshold wavelength. As will be illustrated below for ferricenium ion, it is possible to use multiphoton dissociation to do photodissociation spectroscopy even beyond the thermochemical threshold, which relieves an important limitation of the technique.

3.2. Organometallic Ions with Reasonably Complete Ligand Shells

3.2.1. Metal Carbonyl Anions

Early photodissociation spectroscopy was done on a number of metal carbonyl anions. The first published work fitting the title of this chapter was an ICR study by Richardson *et al.*[12] on $Fe(CO)_4^-$ and $Fe(CO)_3^-$, which was further elaborated in a later publication by Rynard and Brauman.[13] The spectra obtained in these studies are shown in Figure 1. The photodissociation spectrum of $Fe(CO)_4^-$ shows peaks of unknown identity at 750 and 320 nm. The peak near 320 nm also appears as a well-defined broad peak at 310 nm in the solution spectrum of $Fe(CO)_4^-$ formed by pulse radiolysis of iron pentacarbonyl in THF.[26] Photodissociation of $Fe(CO)_3^-$ shows a broad feature between 440 and 920 nm, which was attributed to an allowed $A_1 \rightarrow A_1$ transition with a very broad Franck–Condon envelope. Superimposed on this is a remarkably narrow peak at 627.5 nm, in which nearly resolved vibrational structure is clearly visible in the high-resolution laser spectrum. This peak was assigned to an allowed $A_1 \rightarrow B_1$, B_2 transition with a narrow Franck–Condon envelope.

Dunbar and Hutchinson[10] reported an early spectroscopic study of the photodissappearance of a series of $M(CO)_n^-$ molecules. [It was believed that the photodisappearance reflected dissociation rather than electron photodetachment, although this is not known for certain except for $Fe(CO)_4^-$.] For all of the ions [except $Mn(CO)_5^-$ which showed no photodisappearance] a sharp increase in photodissociation was observed at a wavelength which depended on both the metal and the number of carbonyls. As shown in Figure 2, the energy of this onset increases regularly going to the right in the periodic table, and it decreases regularly with an increase in the number of ligands. Both of these trends were considered consistent with the assignment of the

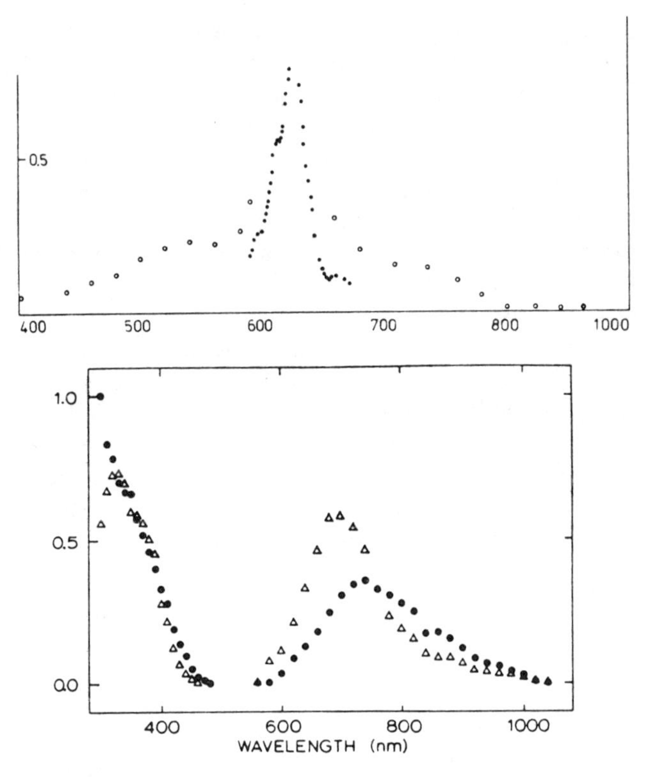

Figure 1. Photodissociation spectra of $Fe(CO)_3^-$ (top) and $Fe(CO)_4^-$ (bottom) obtained with monochromator and laser light sources in the ICR spectrometer. (Reprinted with permission from References 12 and 13.)

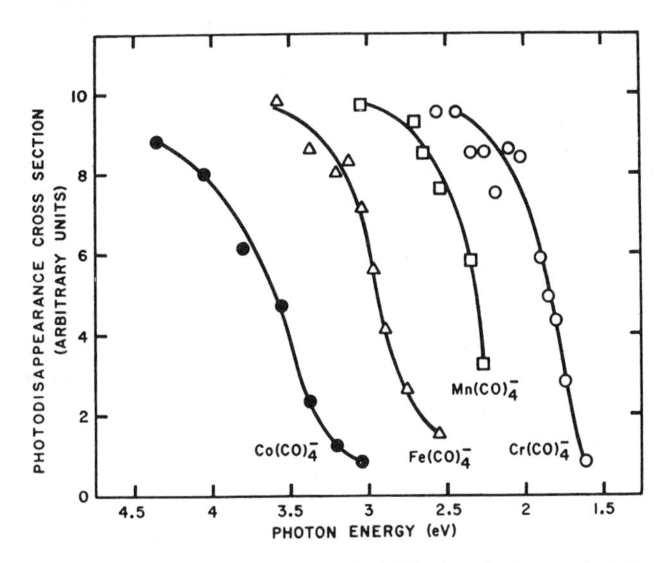

Figure 2. Photodisappearance spectra for various $M(CO)_4^-$ ions in the wavelength regions where the spectra rise sharply. (Reprinted with permission from Reference 10.)

strong photodissociation transition as a metal-to-ligand charge-transfer transition.

3.2.2. Ferricenium Ion

Some of the principles involved in interpreting the photodissociation spectroscopy of gas phase organometallic ions can be illustrated for the case of ferricenium ion, recently under study in our laboratory. Unlike most metal-containing ions, this ion has been rather extensively studied in condensed phases,[27-29] where its stability is surprisingly high for an odd-electron cation. A schematic molecular orbital diagram (Figure 3, reproduced from Reference 28) shows the hole in the half-filled HOMO which is the important feature in the spectroscopic behavior. The visible-wavelength optical transitions which lead to the strong color typical of odd-electron cations usually arise from inner-electron promotion into this hole. In this case the HOMO is largely derived from the metal 3*d* orbitals, and the optical transition which makes ferricenium much more colorful than ferrocene is an $e_{1u} \rightarrow e_{2g}$ electron promotion; the e_{1u} orbital is largely ligand derived, so this can be described as a ligand-to-metal charge-transfer band. It is parity and symmetry allowed and appears at red wavelengths with a modest extinction coefficient ε_{max} of around 500 $l\text{-mol}^{-1}\text{-cm}^{-1}$.

Photodissociation of gas phase ferricenium ion at visible wavelengths is a two- or three-photon process, since the energy required to cleave one of the ligands

$$Fe(Cp)_2^+ \rightarrow FeCp^+ + C_5H_5 \qquad (1)$$

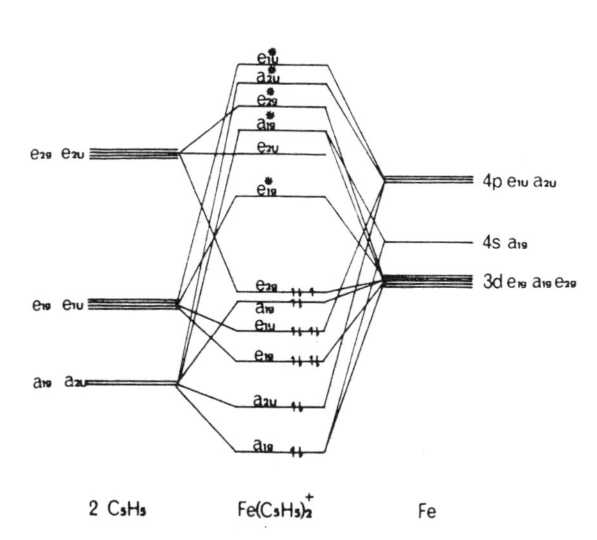

Figure 3. Molecular orbital diagram for ferricenium ion. (Reprinted with permission from Reference 28.)

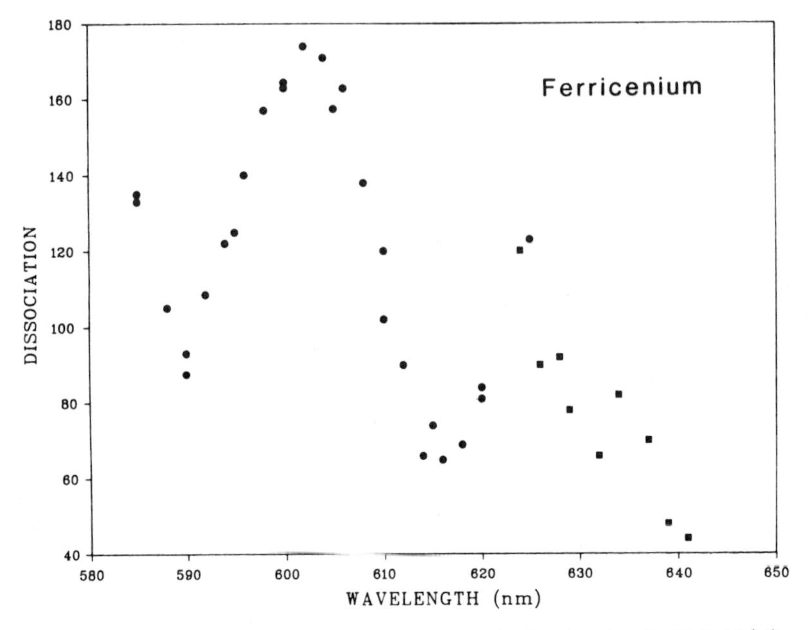

Figure 4. ICR laser photodissociation spectrum of ferricenium ion, $Fe(C_5H_5)_2^{+}$.[11]

is (according to the available data†) around 6 eV. (These electron impact appearance energy numbers are not very reliable, but it seems quite sure that the dissociation energy is not less than perhaps 4 eV.) Two-photon and multiphoton dissociations of organic cations have been extensively studied using the sequential absorption of two or more photons from a cw laser,[1] and it has been found possible to observe the photodissociation spectrum for such dissociations at quite high optical resolution with a tunable laser.[31] The ferricenium ion has a lower photodissociation intensity than has been the case in previous successful studies, but it is possible to work with it. The photodissociation spectrum as it has been obtained so far is shown in Figure 4.[28] The gas phase photodissociation spectrum is similar in form to the 290 K liquid phase spectrum, but the solution spectrum appears to be substantially red-shifted relative to the gas phase spectrum. The broadness and lack of resolved vibrational structure in both the gas phase photodissociation spectrum and the solution spectrum, as compared with the well-resolved vibrational structure in the glassy-matrix spectrum[29] at 6 K, can be attributed to hot bands and congestion at the higher temperature. Multiphoton dissociation thus gives the possibility of using photodissociation spectroscopy even far beyond the one-photon dissociation threshold wavelength.

† Ion thermochemical values from Reference 30.

3.2.3. Other Organometallics

Freiser's group has taken an interesting spectrum of the $V(C_6H_6)_2^+$ ion.[7] It has a very strong peak at about 310 nm, which there has been no effort to assign. With a peak cross section of $2.3 \times 10^{-17}\ cm^2$ ($\varepsilon_{max} \sim 10^4$) and an oscillator strength of the order of 0.1, it seems too intense to be a *d*–*d* transition, making it presumably a charge-transfer band, but it has not been analyzed beyond this.

Some time ago Burnier and Freiser obtained a fine spectrum of the $Ni(C_5H_5)NO^+$ ion.[3] Well-defined peaks were seen at 300 nm and 530 nm. The 300-nm peak corresponded very well to the energy of the $^2E_2$ ligand-localized orbital seen in the photoelectron spectrum of the neutral compound. The dipole-allowed ligand-to-metal charge-transfer transition involving this orbital is expected to be near 4.1 eV, in excellent agreement with the photodissociation spectrum. On the other hand, the 530-nm transition did not correspond to any peak in the photoelectron spectrum, suggesting that it corresponds to a transition carrying an electron from the HOMO to an empty higher orbital not seen in the photoelectron spectrum.

Fukuda and Campana have looked at photodissociation of a number of organometallic cations as discussed below for diketonates[22] and porphyrins.[23] This work is not very informative about the spectroscopy of the ions, since only 488-nm irradiation was used. The authors observed that the photodissociations were very intense, similar to strongly allowed cases like bromobenzene ion, leading again to the conclusion that these are charge-transfer rather than *d*–*d* transitions. The diketonate cations have been suggested to have a half-empty ligand molecular orbital,[32] and the almost universal strong photodissociation observed in this whole series of transition metal complexes seems very likely to correspond to a *d* orbital-to-ligand charge-transfer transition, which could well be highly allowed with an extinction coefficient in the 10^4 range.

Freiser *et al.*[2] reported an interesting study of the photodissociation spectroscopy of a number of organic molecules bound to Li^+. Since the lithium basically served as a charge probe of the molecule, and played no essential spectroscopic role, we will not describe these results under the heading of this chapter.

3.3. Organometallic Ions with Major Coordinative Unsaturation

The Freiser group has reported photodissociation spectra of a large number of somewhat exotic ions having one or two transition metal atoms accompanied by an insufficient number of ligands to provide coordinative saturation,[4,7-9] of which the spectrum of $FeCH_3^+$ shown in Figure 5 may serve as an example. Table 1 collects the peak wavelengths and (where they were measured) the intensities of these peaks. The measured intensities are all large, and it is reasonable to presume that these peaks are generally metal-to-ligand

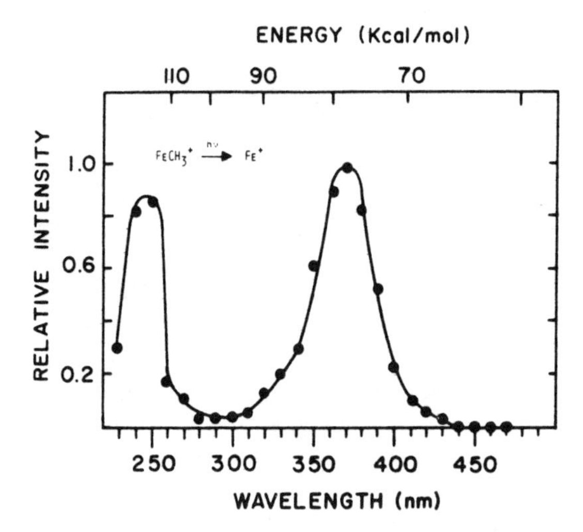

Figure 5. ICR photodissociation spectrum of $FeCH_3^+$. (Reprinted with permission from Reference 7.)

or ligand-to-metal charge-transfer transitions, but little effort has yet been made to identify the transitions responsible for these peaks.

A convincing example of the application of photodissociation spectra to distinguish isomeric ions came from the study of two $FeC_4H_6^+$ species formed in different ways[(7)]:

$$Fe^+ + \text{(butene)} \longrightarrow Fe^+(\text{butadiene}) + H_2 \quad (2)$$

1

$$\text{(trimethylenemethane)}\text{—}Fe(CO)_3 \xrightarrow{\text{E.I.}} |||\text{—}Fe^+\text{—}|| + 3CO \quad (3)$$

2

(The indicated structures **1** and **2** are speculative, although supported by the fragmentation patterns in collision-induced dissociation and photodissociation.) Ion **1** shows a prominent photodissociation peak at 300 nm, while ion **2** shows peaks at 250 and 340 nm. These entirely dissimilar photodissociation spectra give the strongest possible indication that these two ions do indeed have different structures, although giving no indication of what the structures are.

3.4. Diatomic Ions

3.4.1. FeM^+

Among the small ions studied by Freiser's group, the mixed-metal diatomic ions[5,6] have shown an interesting extent of electronic structure in their photodisappearance spectra. It has been considered likely that the enormous number of low-lying electronic states with their accompanying vibrational levels will

Table 1. Photodissociation Peaks[a] Observed by Freiser and Co-workers for ML^+

Ion	Peaks (cross sections)	Reference
$V(C_6H_6)^+$	$\geqslant$430 (0.004), 300 (0.02)	7
$V(C_6H_6)_2^+$	310 (0.23)	7
FeO^+	350 (0.04), 260 (0.07)	7
FeS^+	320 (0.30)	7
$FeOH^+$	335, 290	4
$FeCO^+$	345, 280	4
$FeCH_2^+$	330 (0.06), 260 (0.12)	8
$FeCH_3^+$	370 (0.22), 250 (0.19)	7
$FeC_2H_4^+$	330, 285	4
$FeC_4H_6^+$		
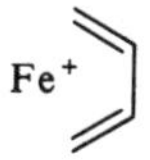 Fe^+	290 (0.02)	7
$Fe^+C_4H_6$	340 (0.14)	7
$Fe(C_6H_6)^+$	440 (0.005), 260 (0.03)	7
CoS^+	390 (0.05), 300 (0.09)	7
$CoOH^+$	340	4
$CoCH_2^+$	370 (0.01), 320 (0.05)	8
$CoCH_3^+$	350 (0.21)	7
$Co(C_6H_6)^+$	~390 (0.006), 260 (0.02)	7
NiS^+	420 (0.08), 280 (0.10)	7
$NiC_4H_8^+$		
$\|-Ni^+-\|$	310 (0.08)	7
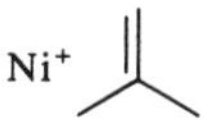 Ni^+		7
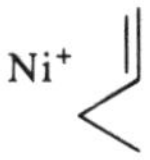 Ni^+	320 (0.05)	7
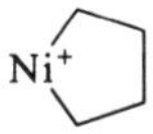 Ni^+	320 (0.12)	7
$Ni(C_5H_5)NO^+$	530, 300	3
$NbCH_2^+$	390 (0.06), 240 (0.20)	9
$RhCH_2^+$	240 (0.06)	9
$LaCH_2^+$	320 (0.02), 260 (0.19)	9

[a] Peak wavelength in nm, with cross sections in Å^2.

lead to a bandlike optical spectrum with substantial absorption intensity at all visible wavelengths. This has been called the "vibronic soup" limit for the spectrum.[19] (For instance, there are of the order of 100 spectroscopically known electronic states of $Fe + Fe^+$ within 1 eV of the ground state,[33] and there are calculated[34] to be about 200 states within 0.5 eV of the ground state for Fe_2.) The FeM^+ ions do indeed appear to absorb significant amounts of light and photodissociate to some extent at all energetically possible wavelengths, but superimposed on this background are prominent peaks, as

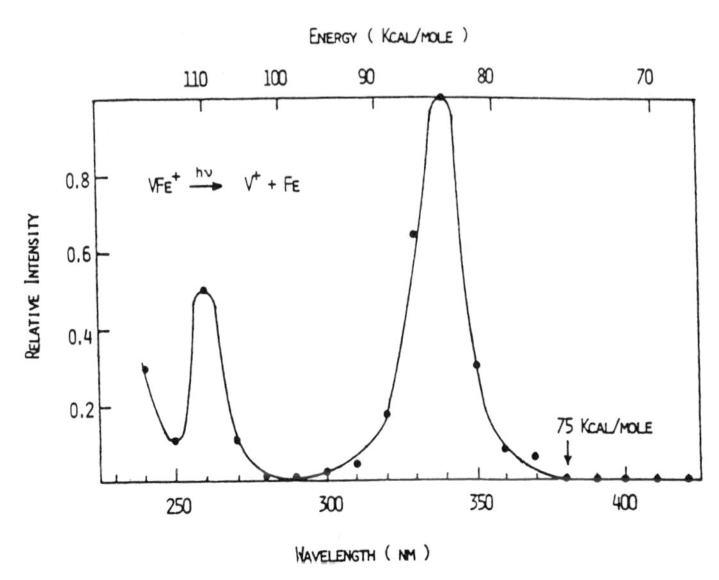

Figure 6. ICR photodissociation spectrum of FeV^+. (Reprinted with permission from Reference 5.)

illustrated for FeV^+ in Figure 6. The peaks appear to be of the order of 30 nm wide, but since this is near the optical resolution of the monochromator light source used, the peak shapes are poorly resolved, and there could be unresolved finer structure. It is evident that for these diatomics the vibronic soup model is not fully descriptive and that there are a few strong optical transitions which dominate the spectrum. The stronger peaks typically were measured to have cross sections of the order of 3×10^{-17} cm^2 ($\varepsilon \sim 10^4$), which corresponds roughly to an oscillator strength of 0.1, so these peaks are quite strongly allowed (if they correspond, as seems likely, to single electronic states and not to the chance coincidence of many weakly allowed transitions to different excited states). Figure 6 is typical of the FeM^+ dimers, which nearly all showed two or three of these spectral peaks in the 240 to 450-nm range. The identity of these particularly optically favored excited electronic states is not clear with the present state of theory of these molecules.

3.4.2. M_2^+

Diatomic metal ions have been attractive targets for high-resolution spectroscopy because one can at least hope for spectra sufficiently simple to understand in a complete way. The possiblity of producing extremely cold ions by supersonic beam expansion, as illustrated below for Nb_2^+, is important, in offering simplification of the spectrum by suppressing the congestion due to hot bands. However, the enormous number of low-lying excited states in many metal diatomics noted above promises to make many spectra terribly complex, and the observation of a simple, interpretable photodissociation spectrum is always noteworthy.

In order for sharp spectral structure to be present, the optical transition must be a bound–bound transition, while if the molecule is to dissociate, it must ultimately find its way to a dissociative state in a predissociation process. Two ions which have turned out to have such a predissociated bound–bound transition are Nb_2^+ and Cs_2^+.

3.4.2a. Nb_2^+. Nb_2^+ was studied by Smalley's group as part of their study of niobium cluster ions.[19] It shows an apparently complex and as yet unassigned spectrum of vibronic transitions in the 18,000-cm^{-1} region. The crucial importance of cooling of the ions after ionization in the supersonic expansion is clearly seen in Figure 7, which shows that the internal energy imparted by vertical photoionization of cooled neutral Nb_2 (Figure 7a) is large enough to wash out this structure, probably by congestion of hot bands and rotational levels. The structure clearly seen in Figures 7b and 7c must, from its sharpness, be due to predissociated bound–bound transitions. The spectrum has a density of vibronic transitions approaching a peak each 10 cm^{-1}, which must reflect overlapping and interaction of a number of electronic states. Even this spectrum of a very cold diatomic metal ion approaches a degree of congestion that would require Doppler-free techniques for resolution, and it is easy to justify the reasoning of Smalley's group's publications that the normal situation for

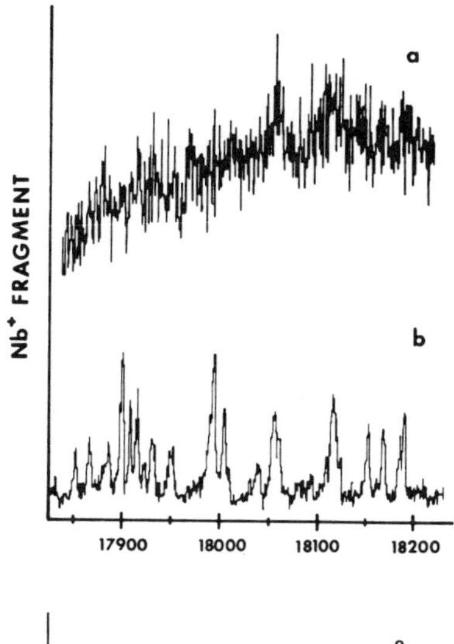

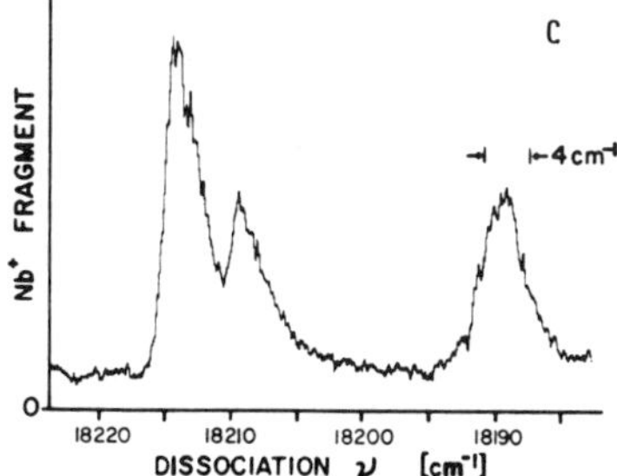

Figure 7. Photodissociation spectra of Nb_2^+. (a) Ions formed by photoionization of neutral Nb_2 cooled by supersonic beam expansion. (b), (c) Ions formed by photoionization of Nb_2 in the nozzle of the supersonic beam source and cooled in the subsequent expansion. (Reprinted with permission from Reference 19.)

metal clusters will be a "vibronic soup" limit in which optical absorption and predissociation will be observed at any wavelength where it is energetically allowed. For Fe_2^+, in fact, Brucat *et al.* did not resolve any sharp spectral structure even for the cooled ions.[19]

3.4.2b. Cs_2^+. The fast ion beam instrument at Stanford Research Institute has been extensively used for extreme-high-resolution, Doppler-tuned spectroscopy of simple ions, and for various wavelength-tuned laser photodissociation studies.[1b] Their tuned laser photodissociation study[16] of Cs_2^+ showed a broad bound-free transition around 8000 nm and, more interestingly, as shown in Figure 8, a sharp bound-bound transition at 8000 nm, assigned as a $X^2\Sigma^+g \rightarrow {}^2\Sigma_u$ transition. Two spin-orbit components were observed, separated by 3000 cm^{-1}. At a laser resolution of 4 cm^{-1}, weak vibrational structuring of these two peaks was observed, but the compact appearance of the peaks led the authors to conclude that the Franck-Condon envelope is very narrow and that the excited state has a similar bond length to the ground state.

3.5. Cluster Ions

In a recent study Tecklenburg and Russell[17,18] have obtained some photodissociation spectroscopic information on $M_n(CO)_m^+$ clusters (M = Fe, Mn, Co) from laser photodissociation at a number of visible wavelengths in their ion beam instrument. The iron and cobalt clusters in general had their most intense photodissociation in the green around 500 nm, while the manganese clusters in general were most intense in the red. Particularly interesting[18] is the spectrum of $Fe_3(CO)_{12}^+$, which shows a prominent peak near 616 nm. This peak is very similar to the peak observed by Tyler *et al.*[35] in the single-crystal spectrum of neutral $Fe_3(CO)_{12}$. The neutral-molecule peak

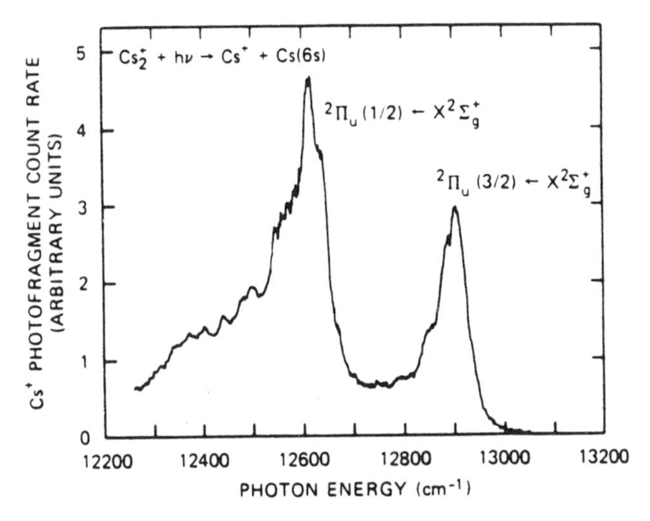

Figure 8. Ion beam laser photodissociation spectrum of Cs_2^+. (Reprinted with permission from Reference 16.)

has a modest ε_{max} of about 3000, and Tyler *et al.* confidently assign it as a metal d–d transition, ruling out a charge-transfer assignment. Removal of an electron to form the cation apparently has little effect on the position of this absorption transition.

4. DISSOCIATION PRODUCTS

4.1. What Do the Fragments Tell Us?

The structural point of view regards the fragmentation pattern from an ion activated by the energy of a photon as a clue to the structure of the original ion. We will see this theme below in the Freiser group's study of isomeric complexes of uncertain structure and in Fukuda and Campana's search for characteristic fragments from organometallic complexes. From a thermochemical point of view, as emphasized by Freiser's group, the formation of a particular fragment at a given photon energy sets an upper limit on the corresponding bond strength. In the fast-expanding study of cluster ions, the fragmentation patterns following photoexcitation offer clues to the forces holding the clusters together and to the patterns of stability for both ionic and neutral fragments.

4.2. Ion Structures

4.2.1. Fragmentation Patterns

A popular theme in analytical mass spectrometry is the MS/MS idea,[36] in which an ion from the primary mass spectrum of the sample is mass selected and activated by energy deposition, after which its fragmentation pattern is acquired as a secondary mass spectrum to give clues about its structure. Photon impact is an excellent way to accomplish the energy deposition step, so that photodissociation can be approached as a way of doing MS/MS analysis.

4.2.1a. Diketonates. Fukuda and Campana have done some photodissociation work with metal complexes at 488 nm with the primary motivation of seeing how informative and useful the photon-induced fragmentation patterns would be in the context of analytical applications of this sort. They studied the photodissociation products of a variety of ionized metal chelates with diketonate ligands,[22] of the general form

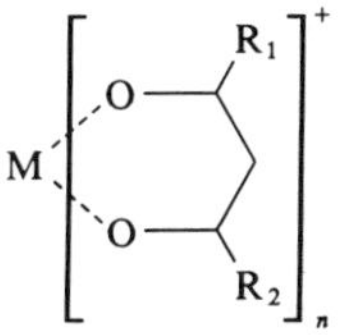

$$\left[M \left(\begin{array}{c} O-C(R_1) \\ \quad CH \\ O-C(R_2) \end{array} \right) \right]_n^+ \qquad (n = 2 \text{ or } 3)$$

where R_1 and R_2 represent various alkyl and fluorinated alkyl groups, and M represents a number of a number of divalent and trivalent metals, both transition metals (Mn, Fe, Co, Ni, VO, Cr, Cu, Zn) and non-transition metals (Al, Ga).

No photodissociation was observed for the non-transition metal Al and Ga complexes, which was attributed to lack of optical absorption at 488 nm, but all the other parent ions [except $VO(tfa)_2^+$] photodissociated with large cross sections which would be typical for allowed optical absorptions. In many cases more than one fragment ion was formed in significant abundance, with branching to two or three products being typical. The fragments may be considered in three classes: (1) loss of one complete ligand; (2) loss of a CH_3 or CF_3 group from one ligand; and (3) other fragmentations involving rearrangement, multiple bond fission, and so forth. Type (2) products were predominant in ML_3^+ complexes, while type (1) products were large or predominant in ML_2^+ complexes. When CF_3 was present, it was lost in preference to CH_3 (and with higher kinetic energy release), presumably reflecting the weaker $C—CF_3$ bond. Various type (3) products appeared with no obvious pattern.

The most prominent difference from electron impact fragmentation was the fact that ML_2^+ usually photodissociated by CH_3 (or CF_3) loss, in contrast to the usual loss of a complete ligand in electron impact. Fukuda and Campana's interpretation of this , which seems entirely reasonable, is based on the more nearly monoenergetic nature of photodissociation: It has been shown that photodissociation, which involves parent ions having a relatively narrow range of internal energies, can give strikingly different fragmentation products from techniques like electron impact or photoionization which excite the ions to a broad range of internal energies.[37]

4.2.1b. Porphyrins. A photodissociation study of a number of porphyrin ions[23] offers an outstanding illustration of the potential applications of photon energy deposition for fragmentation analysis of large molecules in mass spectrometry, although little was learned about the spectroscopy or photochemistry of the ions. Fukuda and Campana looked at a number of on-hand porphyrins, of which a typical example would be

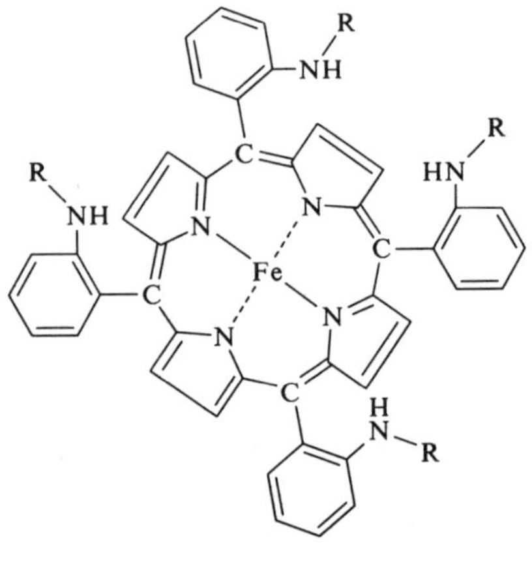

Using cw multiple-wavelength radiation from 515 to 458 nm, photodissociation fragmentations were observed reflecting loss of various neutral fragments. From an analytical point of view, it was most interesting that many side-chain cleavages were induced, giving structural information about the original molecule.

4.2.2. Isomeric Ions

Using fragmentation patterns to distinguish isomeric ions must always be approached with caution, since ion fragmentation patterns are so extremely sensitive to internal energy differences in the parent ions, and it is difficult to ensure thorough thermalization of the ions being compared. Hettich *et al.* have given some convincing evidence from fragmentation patterns as a complement to photodissociation spectroscopic evidence in suggesting the independence of at least four $NiC_4H_8^+$ isomers.[7] Each of the isomers, which they assign as structures **3–6**,

$$Ni^+ + \text{(n-butane)} \xrightarrow[-H_2]{} \|\text{—}Ni^+\text{—}\| \quad \mathbf{3} \tag{4}$$

$$Ni^+ + \text{(neopentane)} \xrightarrow[-CH_4]{} Ni^+\text{—(isobutene)} \quad \mathbf{4} \tag{5}$$

$$Ni^+ + \text{(n-hexane)} \xrightarrow[-C_2H_6]{} Ni^+\text{—(butene)} \quad \mathbf{5} \tag{6}$$

$$Ni^+ + \text{(cyclopentanone)} \xrightarrow[-CO]{} \text{(nickelacyclopentane)}^+ \quad \mathbf{6} \tag{7}$$

shows a unique pattern of photofragments. The independence of **4**, which is unique in giving loss of CH_4, and **5**, which gives a characteristic loss of CH_3, is strongly indicated. The distinction of **6** from **3**, based either on the photodissociation spectra or the characteristic loss of H_2 from **6**, is on less firm ground, although other studies suggest the independence of these two isomers.

4.3. Thermochemistry and Bond Strengths

4.3.1. Thresholds in the Photodissociation Spectrum

A principal focus of the Freiser group's extensive photodissociation studies of metal-containing cations has been the determination of bond strengths and dissociation energetics. The approach is to determine the longest

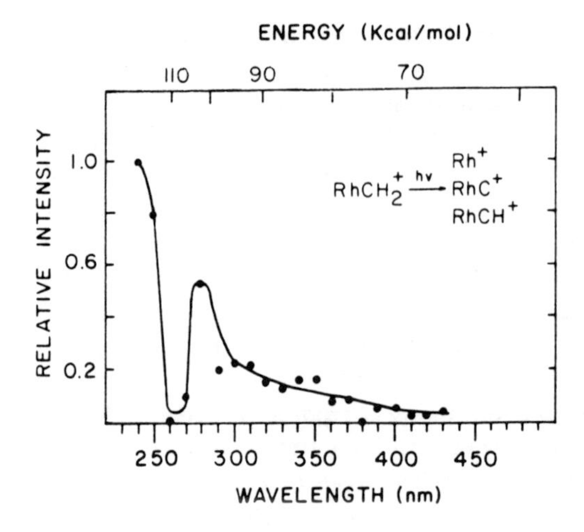

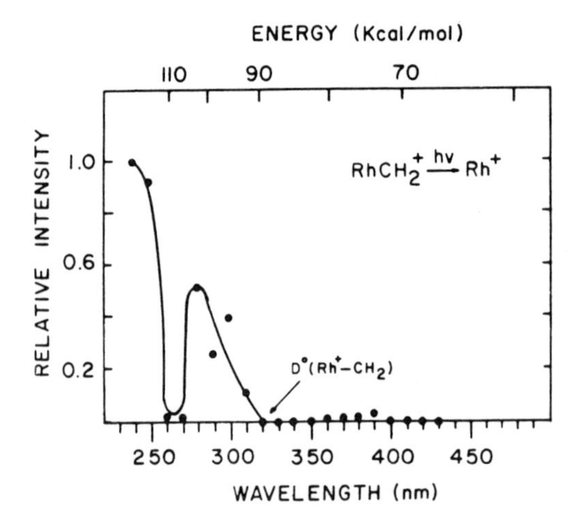

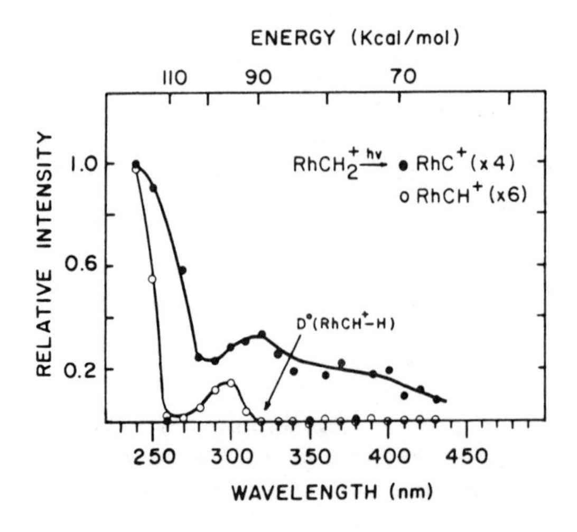

Figure 9. Photodisappearance spectrum of $RhCH_2^+$ ions (top) photoproduction spectrum of the daughter ions Rh^+ (middle) and RhC^+, and $RhCH^+$ (bottom). (Reprinted with permission from Reference 9.)

wavelength at which photodissociation of the bond of interest can be observed. An exceptionally favorable example of this[9] is shown for $RhCH_2^+$ in Figure 9: The photodisappearance spectrum of the parent ion, shown in the top panel, extends beyond 400 nm, with only a small region around 260 nm where the cross section is not substantial. The thresholds at 320 nm (89 kcal) for breaking the $Rh^+—CH_2$ bond and also near 320 nm for breaking the $RhCH^+—H$ bond are clearly evident, while $RhC^+—H_2$ bond breaking occurs out to at least 600 nm (45 kcal). The bond strengths derived from the thresholds, $D(Rh^+—CH_2) = D(RhCH^+—H) = 89$ kcal/mol and $D(RhC^+—H_2) \leq$ 45 kcal/mol, are conservatively precise within ±5 kcal, and (especially when corroborated by other evidence) are very convincing.

Several nontrivial assumptions are involved in this analysis, and (as the authors of these papers have been careful to make clear) the interpretation must be cautious. (1) The observed threshold must not be determined spectroscopically; that is, the drop in photodissociation at the threshold must be due to the insufficiency of the photon energy, and not due to a drop in photoabsorption rate. This is assured in cases like that shown in Figure 9 for $Rh^+—CH_2$ cleavage, where another photodissociation process continues to longer wavelength, but is problematic where, as, for instance, for $Fe^+—S$, (1) only one dissociation channel is observed which tails off from small values to zero at the threshold. (2) The thermochemical threshold may be red-shifted by internal energy in the parent ion. This may be tested and alleviated by using a high pressure of neutral collision gas, but for larger molecules, such as $V(C_6H_6)_2^+$, the internal energy of even thermalized ions may be 5 kcal or more. (3) Two-photon dissociation below the one-photon threshold may obscure the threshold if high sensitivity and high light intensity are used. Intensity-dependence or pressure-quenching studies can diagnose this problem.[38] (4) In cases of product branching, as in Figure 9, the apparent threshold for a higher-energy product (Rh^+) may actually be determined by its unsuccessful competition with a lower-energy, higher-cross-section product (RhC^+) in the threshold wavelength region.

With these potential problems in mind, the photodissociation threshold approach to measuring bond energies in cations seems highly promising. It has been shown capable in favorable nonmetal cases of yielding accuracy rivaling that of the most precise photoionization techniques. This brief discussion may serve to highlight the promise of the approach. Detailed presentation and discussion of the thermochemical information obtained, in the context of values obtained from other methods, is appropriate for other chapters in this volume and will not be pursued here.

4.3.2. Kinetic Energy Release

The study of Mn_2^+ by Bowers'group[15] in the ion beam is a convincing source of information on the strength of the $Mn^+—Mn$ bond. Using fixed laser wavelength (at a variety of wavelengths in the visible), they observed the kinetic energy distribution of the photodissociation product ions, as illustrated

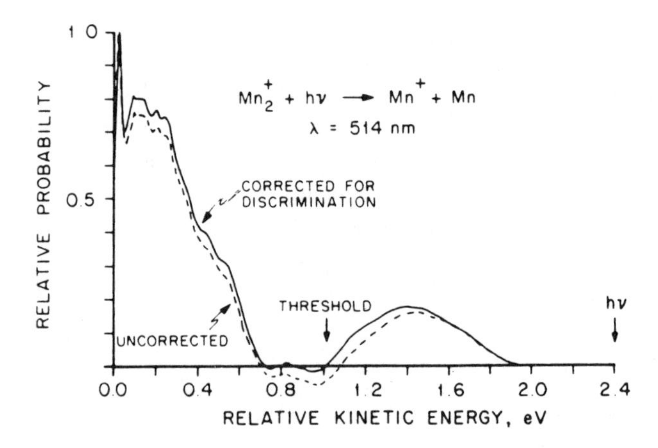

Figure 10. Kinetic energy release distribution for Mn_2^+ photodissociated in the ion beam. (Reprinted with permission from Reference 15.)

in Figure 10. The band of kinetic energy releases in the 0-to-0.7-eV region is assigned to production of Mn^+ products in an excited state. The band between 1.0 and 1.9 eV is assigned as reflecting production of ground state products, giving the bond strength according to the energy conservation relation

$$\begin{aligned} &(\text{Photon energy}) + (\text{Initial energy of } Mn_2^+) \\ &\quad = (Mn_2^+ \text{ bond energy}) + (\text{Kinetic energy released}) \\ &\quad\quad + (\text{Excitation energy of } Mn^+ + Mn) \end{aligned} \tag{8}$$

The Mn_2^+ parents have a spread of initial vibrational energies, so that (assuming zero excitation of products) the low end of the kinetic energy release peak, marked "Threshold" on the figure, gives a lower bound on the Mn_2^+ bond strength. The most stringent such lower bound was obtained at 515 nm, which gave $D(Mn^+{-}Mn) \geq 1.39$ eV.

4.4. Cluster Ions

The study of cluster ions has taken on enormous momentum in just a few years,[39] with a considerable amount of work on both clusters of metal atoms and also clusters of metal atoms mixed with ligands like CO. Since thinking about these molecules is so new, photodissociation is being used in the hope of casting light on very basic questions about the structures, bonding types, bond strengths, and dissociation dynamics. These studies are still largely at the level of empirical observation of how many and which pieces of the cluster break off after absorption of a photon.

4.4.1. Bare Metal Clusters

The development by Smalley's group at Rice University,[40] among others, of highly effective ways of generating and observing cluster ions of large size has resulted in a recent explosion of interest in the chemistry and properties of these entities. Other chapters in this volume cover much of this work, but some of the photodissociation studies most closely tied to the themes of this chapter will be of interest here.

The most ambitious photodissociation study of ionic clusters of pure metals is described in two papers from Smalley's group on cationic and anionic clusters. In the cation study,[19] iron, nickel, and niobium clusters were generated from a source in which metal is laser vaporized into a helium stream and expanded in a supersonic beam expansion nozzle to yield a pulsed, high-ion-density ($\sim 10^5$ ions per shot) cloud of cold cluster ions and molecules. The ions were mass selected for the subsequent photodissociation study, or alternatively the neutral clusters were photoionized with an excimer laser pulse and then similarly mass selected. The photodissociation spectrum of Nb_2^+ gave highly resolved vibronic spectral structure, as shown in Figure 7, which served as a diagnostic of the ion temperature: From the rotational envelopes of the vibronic peaks it was found that photoionization of the neutral clusters after the expansion nozzle gave hot ions, but that expansion of the ions formed directly by the initial metal vaporization pulse gave cold ions having a rotational temperature near 20 K.

The larger cluster ion reported in most detail was Fe_6^+. Photodissociation yielded the series of daughters Fe_5^+, Fe_4^+, and Fe_3^+. The plot of ion abundance versus laser fluence shown in Figure 11 is revealing. The Fe_5^+ daughter shows the fluence dependence (linear at low fluence, leveling off at higher fluence) expected for one-photon dissociation. The Fe_4^+ daughter shows the initially quadratic curve expected for two-photon dissociation, and similarly the subsequent smaller fragments show correspondingly higher power-law

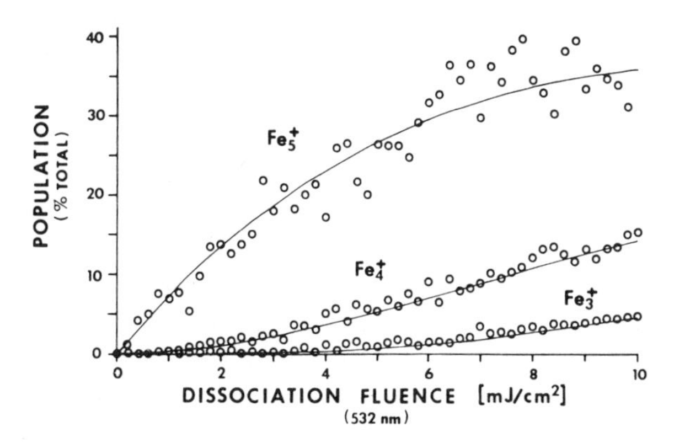

Figure 11. Plots of daughter ion production versus laser fluence for photodissociation of Fe_6^+. (Reprinted with permission from Reference 19.)

dependences on fluence at the limit of low fluence. The solid lines shown on the figure are in fact the one-parameter fit to the simple kinetic scheme

$$Fe_6^+ \rightarrow Fe_5^+ \rightarrow Fe_4^+ \rightarrow Fe_3^+ \tag{9}$$

in which only the photon absorption rate k is varied to give the best fit. Brucat *et al.*[19] point out that the excellence of this fit does not prove the kinetic scheme and that for instance an alternative kinetic scheme in which all the photons are absorbed by Fe_6^+, which then dissociates to varying extents depending on the number of photons absorbed, is kinetically indistinguishable from the kinetic scheme in Equation 9. This kinetic ambiguity makes it hard to draw conclusions about the dissociation processes of the smaller ions in this sequence, but the conclusion is clear that the loss of a single Fe atom from Fe_6^+ is a single-photon process, which places an upper limit of 2.33 eV on the dissociation energy for $Fe_6^+ \rightarrow Fe_5^+ + Fe$. Similar results for the cluster ions Fe_n^+ and Nb_n^+ with $3 \leq n \leq 10$ led to the same 2.33-eV upper bound for Fe or Nb atom loss from all of the clusters, while the lack of any one-photon dissociation at 1.17-eV photon energy gave a lower bound on the atom loss energy. Brucat *et al.* also concluded that for all of these cluster ions the observed single-neutral-atom loss was indeed the lowest-energy dissociation process.[19]

In contrast to Nb_2^+, Fe_2^+ showed no sharp spectral structure, yielding $Fe^+ + Fe$ at all wavelengths studied. The fluence dependence curve at 2.92-eV photon energy was typically one-photon, while that at 2.43 eV had two-photon shape, leading to the assignment of the dissociation energy between these two values, implying an Fe_2 bond energy between 0.83 and 1.32 eV.

The anions studied by Zheng *et al.*[20] are somewhat more complicated because of the competing possibility of electron photodetachment. The most important experimental difference from the cation study described above was the use of an ArF excimer laser pulse striking the nozzle region near the time of emergence of the vaporized metal cloud, to produce a plasma which then cooled on expansion and from which the desired cluster negative ions were extracted. Cluster anions of most metals, including Ni, Nb, and Cu, show electron detachment as the only photon-initiated process. Only Ag_n^- shows dissociation in competition with detachment, a distinction which Zheng *et al.* attribute to the exceptional circumstance in silver anions of the dissociation energy being lower than or comparable to the electron detachment energy.

In contrast to the cations, photodissociation of the Ag_n^- ions showed loss of several metal atoms at a time, by what was tentatively assigned (from fluence-dependence measurements) as a one-photon process. For instance, Ag_7^- yielded comparable amounts of Ag_5^- and Ag_3^-. The evident preference for odd anion clusters is consistent with the observation in Cu, Si, Ge, and GaAs anion clusters that the odd anions have higher electron affinity than the even ones. Whether the neutral fragments appear through the boiling off of successive single Ag atoms or through the fissioning of a single neutral Ag_m cluster cannot be determined from the data.

4.4.2. Metal Carbonyl Clusters

Tecklenburg and Russell[17] have given a graphic illustration of the effect of excess internal energy on dissociation patterns. $Fe_3(CO)_{12}^+$ was formed by electron impact, extracted promptly from the source, and photodissociated in the MS50 ion beam spectrometer at 515 nm. The parent ion extraction time scale is clearly too short for any appreciable radiative ion cooling, but long enough for most dissociation to be complete, so the parent ion population must consist of those parent ions having energy below or not far above the first dissociation threshold. Loss of three carbonyls to give $Fe_3(CO)_9^+$ predominated and was approximately five times greater than loss of two carbonyls. However, when the ions were cooled (to an unknown degree) by extraction from a high-pressure argon-bath-gas source, loss of two carbonyls to give $Fe_3(CO)_{10}^+$ was approximately three times greater than loss of three carbonyls.

These same authors observed loss of various numbers of carbonyls, as well as metal–metal bond cleavages, for a variety of $Fe_n(CO)_m^+$ and $Co_n(CO)_m^+$ ions. They found apparent wavelength thresholds suggesting about 1 eV per bond broken is required in fragmenting these clusters, but this threshold information only gives lower limits to the true fragmentation thresholds, since the extent of internal excitation of the parent ions is unknown.

The NRL group[21] has also investigated the photofragmentation of some metal carbonyl cluster ions, $Mn_n(CO)_m^+$, $Cr_n(CO)_m^+$, and $Cr_n(CO)_m^-$. The ions (n from 1 to about 4) were photodissociated at 515 nm in the VG ZAB mass spectrometer. With only the exception of $Mn_3(CO)_{10}^+$, the ions with an even number of electrons showed no dissociation; since it is certain that the photon energy is sufficient to remove at least one CO ligand, this is presumably due to low photoabsorption cross section, and one may speculate that this in turn reflects a filled HOMO and the unavailability of inner-orbital transitions to a half-filled HOMO (as was discussed in connection with ferricenium ion in Figure 3). The odd-electron clusters, on the other hand, gave ready photodissociation. With few exceptions photodissociation cleaved CO ligands, without removing metal atoms from the clusters. This was constrasted with collision-induced dissociation, which frequently (for Mn clusters, at least) cleaved metal–metal bonds. In the Mn cluster case, three to five carbonyls were cleaved; since the laser-intensity dependence indicated a single-photon process, this suggests that the 56-kcal/mol photon energy is sufficient to break off several carbonyls, showing a binding of the order of 10–20 kcal/mol per carbonyl.

5. *MECHANISTIC ASPECTS*

5.1. *Probing the Details of the Photodissociation Process*

Because it is so convenient to observe ions under collisionless conditions, they offer an attractive opportunity to study the initial photochemical events following photon absorption. Some of the questions to ask concern the time it takes for dissociation to occur, the flow and distribution of energy within the excited molecule, the path taken on the various potential surfaces during dissociation, the distribution of energy into the degrees of freedom of the fragmentation products, and so forth. We consider here a few studies which may shed light on some of these questions.

5.2. Competitive Fragmentation

Photodissociation is a relatively low energy process, and often the only fragmentation observed is to the obvious lowest-energy fragment. Much more interesting are the cases where competition to form several different fragments occurs, because this competition can be revealing about the events occurring in the ion between photon absorption and ultimate dissociation.

Typical of the Freiser group's observations on coordinatively unsaturated ML^+ and FeM^+ ions has been competitive formation of two or more products. An example among many is the fragmentation[8] of $FeCH_2^+$ to $FeCH^+$, FeC^+, and Fe^+. All three photofragment ions are observed at short wavelengths. $FeCH^+$ disappears abruptly at 350 nm, and Fe^+ drops to zero near 400 nm; these thresholds are interpreted as thermochemical thresholds and used in the assignment of bond strengths. There seems not yet to have been an effort to fit the fragmentation patterns into an RRKM[41] or phase space[42] theory of fragmentation, and it is unclear whether the different fragment ions reflect dissociation from different electronic states or competitive dissociation from a common excited state.

Collision-induced dissociation in the ICR ion trap is likely to be a multistep process, in which energy is pumped into the ion by several successive collisions before fragmentation. Photodissociation involves a one-step excitation (leaving multiphoton dissociation aside for the moment). Whether because of this difference or some other factors, the photodissociation product pattern often differs from the CID pattern, as for instance with $NiC_4H_8^+$ formed by reacting Ni^+ with neopentane.[7] This ion (whose structure is not certain) yields Ni^+ in CID, but yields equal amounts of Ni^+ and $NiC_3H_4^+$ in photodissociation. Photodissociation seems often to give more different fragments than CID, which would be consistent with the idea that the energy deposited in the parent ion is higher in the photodissociation case. However, to illustrate one counter-example, $FeCH_2^+$ yields only Fe^+ with CID, which is 16 kcal more endothermic than FeC^+ production.[8] Thus, CID does not necessarily yield the lowest-energy fragment as this simple idea would predict, and there are evidently further factors governing this very interesting contrast.

It is interesting that the mixed diatomic metal ions FeM^+ seem to dissociate according to the energetics of the products, with the more stable product ion predominating.[5,6] One is accustomed to this behavior in polyatomic ion fragmentations, where vibrational predissociation through the extremely dense manifold of vibrational states makes a statistical description appropriate (as in RRKM theory) and gives essentially thermodynamic control of the products.

However, one thinks of diatomic molecule fragmentation as proceeding through specific pathways involving a few electronic excited states, so that a highly nonstatistical and nonthermodynamic control of products is normal. In the case of the metals, however, it appears that the density of electronic states is so high that statistical ways of analyzing the behavior may be fruitful, the possible argument being that the lower-energy product channel has a much greater number of final electronic states at a given energy and is statistically more likely.

5.3. Infrared Laser Photochemistry

Multiphoton excitation by infrared laser gives a means of depositing energy in the ion which differs in interesting ways from visible/UV excitation. Actually, there is a strong distinction between pulsed and continuous IR excitation: The pulsed laser deposits a large number of IR photons during a time (typically a few nanoseconds) which is shorter than many dissociation processes, and has much in common with other energizing methods (visible/UV excitation, fast collisional excitation, surface collision excitation) which put the ion promptly above the dissociation threshold. Pulsed IR photodissociation, while well explored for hydrocarbon ions,[(43)] has yet to have an impact on metal-containing ion work. On the other hand, cw IR excitation has unique interest because the rate of photon input is slow compared with dissociation: The ion will steadily absorb photons until it just exceeds the threshold of the easiest dissociation and will then dissociate. Since the energy of an IR photon is ~2.5 kcal, this experiment automatically reflects dissociation of a narrowly energy-selected ion population lying not more than 2.5 kcal above the lowest threshold. One might speak of the IR dissociation reaction as being nearly adiabatic on the dissociation potential surface.

Hanratty *et al.*[(14)] gave an elegant demonstration of these ideas in a study of the IR dissociation of cobalt–pentene ion association complexes using CO_2-laser light. They investigated the products resulting from dissociation of $Co(1\text{-}C_5H_{10})^+$ and $Co(2\text{-}C_5H_{10})^+$, as well as Co^+ adducts with branched-chain pentenes and cylopentane. The cyclopentane adduct gave no dissociation, which was correlated with the lack of IR absorption features in neutral cyclopentane in the CO_2-laser region from 900 to 1100 cm^{-1}, suggesting that this association complex may simply not absorb radiation fast enough. The branched-chain pentene adducts behaved similarly to the 2-pentene adduct, so the real interest of this study was the 1-pentene and 2-pentene results.

Each complex gave only a single dissociation product, in line with the above discussion. In the 1-pentene case, the photodissociation proceeded by loss of C_2H_4 to give $Co(C_3H_6)^+$, while the 2-pentene adduct photodissociated by loss of CH_4 to give $Co(C_4H_6)^+$. In each case, the reaction is pictured as proceeding by initial C—C bond insertion to form an allyl complex, followed by further rearrangement and final elimination of the alkane or alkene neutral fragment, as shown in Equations 10 and 11:

$$\mathrm{Co}\text{-(1-pentene)}^{+\bullet} \xrightarrow{\mathrm{C-C}} \mathrm{CH_3{-}Co^+}\text{(allyl)} \longrightarrow \mathrm{CH_3}(\mathrm{H})\mathrm{Co^+}\text{(butadiene)} \xrightarrow{-\mathrm{CH_4}} \mathrm{Co^+}\text{(butadiene)}$$

$$\Delta H = -52\ \mathrm{kcal/mol} \qquad (10)$$

$$\mathrm{Co}\text{-(2-pentene)}^{+\bullet} \xrightarrow{\mathrm{C-C}} \text{(ethyl)}\mathrm{{-}Co^+}\text{(allyl)} \longrightarrow \|\mathrm{{-}Co^+{-}}\| \xrightarrow{-\mathrm{C_2H_4}} \mathrm{Co^+{-}}\|$$

$$\Delta H = -29\ \mathrm{kcal/mol} \qquad (11)$$

It is interesting to compare these results with other (nonadiabatic) fragmentation results, as in Table 2. For the 1-pentene adduct, C_2H_4 loss is clearly predominant even for the more energetic ions formed in the other methods, but in the 2-pentene case, even the very low energy metastable-ion fragmentation does not discriminate between CH_4 loss and H_2 loss, which must have very similar activation energies. Many of the other fragments noted in Table 2 resulting from more energetic fragmentation methods can be accounted for by paths initiated by C—H bond insertion, which is apparently more energetic than the C—C bond insertion postulated in Equations 10 and 11.

It is striking that the 1-pentene adduct dissociation does not yield the lowest-energy product, which is CH_4 loss. The dissociation apparently follows the potential surface to the product ion of lowest activation energy, as in Figure 12. What seems significant in this study is the apparently totally selective dissociation along the most favorable reaction path, in contrast to the product branching observed with all the prior techniques.

Table 2. $CoC_5H_{10}^+$ Fragment Distributions for Different Fragmentation Methods

		Fragmentation method			
Isomer	Fragment	Ion beam[a]	Metastable[b]	CID[c]	IR[d]
1-C_5H_{10}	H_2	0.11	0.02	0.02	0
	CH_4	0.13	0.01	0.03	0
	C_2H_4	0.58	0.95	0.72	1.0
	C_3H_6	0.18	0.01	0.10	0
2-C_5H_{10}	H_2	0.28	0.47	0.35	0
	CH_4	0.33	0.43	0.38	1.0
	C_2H_4	0.29	0.09	0.16	0
	C_3H_6	0.10	0.01	0.02	0

[a] Endothermic Co^+-pentane reaction at 0.5 eV kinetic energy; from Reference 45.
[b] Metastable-ion decomposition; from Reference 46.
[c] High-energy collision-induced dissociation; from Reference 46.
[d] Continuous IR-laser dissociation; from Reference 14.

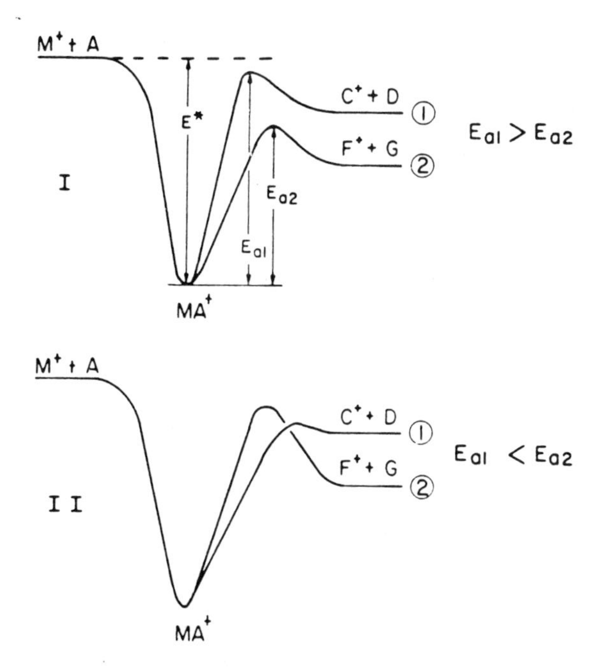

Figure 12. Hypothetical schematic potential surfaces for $Co(1\text{-}C_5H_{10})^+$ dissociation, illustrating how the adiabatic traversal of the path with lowest activation energy leads to a product higher in energy than the lowest-energy product in case II, while in case I the lowest-activation-energy path also leads to the lowest-energy product. (Reprinted with permission from Reference 14.)

5.4. Angular Dependence

As has been clear above, a prominent challenge in interpreting photodisappearance spectra of metal-containing ions is the appearance of many spectral peaks which are hard to assign to particular excited states. The theoretical methods which have been so helpful in assignment of hydrocarbon ion spectra are not advanced enough to be useful in these electronically complex ions.

A source of information which should be useful for assignments is the polarization, or angular dependence, of the photodissociation at a given wavelength. This is most simply illustrated for a diatomic, although the principles carry over to polyatomics. Consider a dimer ion $A{-}B^+$, which dissociates to $A^+ + B$, with the A^+ ion flying off with significant velocity. Assume that the optical transition is polarized parallel to the A—B bond, so that light will be adsorbed preferentially when it is linearly polarized along the bond. Then if a randomly oriented sample of ions is irradiated with light linearly polarized in the z direction, there will be preferential photodissociation and preferred direction of flight of the A^+ products along the z direction. The opposite will of course be true if the optical transition is polarized perpendicular to the A—B bond. Comparing the number of products emitted in the z direction with the number emitted in the x or y direction will thus give the same information about the polarization of the optical transition as would be obtained in the ideal experiment in which the ions were aligned in the z direction and their absorption spectrum measured with linearly polarized light. This photodissociation angular dependence will depend on two conditions: (1) the dissociation must be accompanied by observable kinetic energy release and (2) the dissociation must occur sufficiently soon after photon absorption so that the orientation of the ion is not meanwhile randomized by rotation. (Since free rotation of a molecule never fully randomizes its initial orientation, some angular dependence remains even for slow photodissociation, but the angular asymmetry drops off rapidly as the dissociation time becomes comparable to the rotational reorientation time.) The observation of a large angular dependence thus gives information about the optical polarization, the kinetic energy release, and the dissociation rate following photon absorption.

It was shown[44] that ICR photodissociation spectroscopy can be used to observe the dissociation angular dependence, since product ions with sufficient energy can escape along the magnetic field direction but are trapped efficiently if their velocity is perpendicular to the field. Richardson *et al.*[12] used this approach in finding that the photodissociation $Fe(CO)_4^- \rightarrow Fe(CO)_3^- + CO$ is anisotropic (polarized perpendicular to the direction of preferred CO emission). This confirms the expected nontetrahedral geometry of $Fe(CO)_4^-$, but the polarization information has not been interpreted in terms of specific electronic transitions.

A dissociation which releases kinetic energy can also be examined for angular dependence in an ion beam instrument by comparing the product-ion kinetic energies obtained with the laser polarized parallel and perpendicular to the beam direction. For instance, if the optical transition is polarized parallel to the diatomic axis, kinetic energy will be released along the beam direction if both molecule and laser are oriented along the beam axis, but will be released in the perpendicular direction if molecule and laser are perpendicularly oriented. In this case, then, the kinetic energy distribution will be broadened for the laser-parallel orientation relative to the laser-perpendicular orientation.

Jarrold *et al.*[15] used this capability in examining Mn_2^+ photodissociation at visible wavelengths in their ion beam instrument. Kinetic energy releases of the order of 1 eV were ample for good measurement. The asymmetry parameter β was found to be about +0.5, independent of the kinetic energy release value. This indicates a parallel transition, but the magnitude of β is no larger than would be expected for a rotationally averaged dissociation, compared with the value of +2.0 expected for a prompt parallel-polarized photodissociation. The firm conclusion can be drawn that this is not a direct photodissociation transition to an unbound continuum state of Mn_2^+, which would necessarily give a β value near 2 or −1. Beyond this, the interpretation is ambiguous: The low β value might arise from a rotationally averaged parallel transition, or it might reflect a near cancellation of superimposed, comparable-strength parallel and perpendicular transitions.

ACKNOWLEDGMENT

The support of the National Science Foundation and of the donors of the Petroleum Research Fund, administered by the American Chemical Society, during preparation of this chapter and during the research of our group described here is gratefully acknowledged.

REFERENCES

1. (a) Dunbar, R.C. In *Gas Phase Ion Chemistry*; Bowers, M.T., Ed.; Academic: New York, 1984; Vol. 3, Chapter 20; (b) Moseley, J.T. In *Photodissociation and Photoionization*; Lawley, K.P., Ed.; Wiley: New York, 1984; p. 245.
2. Freiser, B.S.; Staley, R.H.; Beauchamp, J.L. *Chem. Phys. Lett.* 1976, *39*(1), 49.
3. Burnier, R.C.; Freiser, B.S. *Inorg. Chem.* 1979, *18*, 906.
4. Cassady, C.J.; Freiser, B.S. *J. Am. Chem. Soc.* 1984, *106*, 6176.
5. Hettich, R.L.; Freiser, B.S. *J. Am. Chem. Soc.* 1985, *107*, 6222.
6. Hettich, R.L.; Freiser, B.S. *J. Am. Chem. Soc.* 1987, *109*, 3537.
7. Hettich, R.L.; Jackson, T.C.; Stanko, E.M.; Freiser, B.S. *J. Am. Chem. Soc.* 1986, *108*, 5086.
8. Hettich, R.L.; Freiser, B.S. *J. Am. Chem. Soc.* 1986, *108*, 2537.
9. Hettich, R.L.; Freiser, *B.S. J. Am. Chem. Soc.* 1987, 3543.
10. Dunbar, R.C.; Hutchinson, B.B. *J. Am. Chem. Soc.* 1974, *96*, 3816.
11. Faulk, J.; Dunbar, R.C. To be published.
12. Richardson, J.H.; Stephenson, L.M.; Brauman, J.I. *J. Am. Chem. Soc.* 1974, *96*, 3671.
13. Rynard, C.M.; Brauman, J.I. *Inorg. Chem.* 1980, *19*, 3544.
14. Hanratty, M.H.; Paulsen, C.M.; Beauchamp, J.L. *J. Am. Chem. Soc.* 1985, *107*, 5074.
15. Jarrold, M.F.; Illies, A.J.; Bowers, M.T. *J. Am. Chem. Soc.* 1985, *107*, 7339.
16. Helm, H.; Cosby, P.C.; Huestis, D.L. *J. Chem. Phys.* 1983, *78*, 6451.
17. Tecklenburg, R.E., Jr.; Russell, D.H. Presented at the 34th Annual Conference on Mass Spectrometry and Allied Topics, Cincinnati, OH., June 8-13, 1986.
18. Tecklenburg, R.E.; Russell, D.H.; *J. Am. Chem. Soc.* 1987, *109*, 7654.
19. Brucat, P.J.; Zheng, L.-S.; Pettiette, C.L.; Yang, S.; Smalley, R.E. *J. Chem. Phys.* 1986, *84*, 3078.
20. Zheng, L.-S.; Karner, C.M.; Brucat, P.J.; Yang, S.H.; Pettiette, C.L.; Craycraft, M.J. Smalley, R.E. *J. Chem. Phys.* 1986, *85*, 1681.
21. Freas, R.B.; Ross, M.M.; Campana, J.E.; Ridge, D.P. Presented at the 34th Annual Conference on Mass Spectrometry and Allied Topics, Cincinnati, OH., June 8-13, 1986.
22. Fukuda, E.K.; Campana, J.E. *Int. J. Mass Spectrom. Ion Proc.* 1985, *65*, 321.
23. Fukuda, E.K.; Campana, J.E. *Anal. Chem.* 1985, *57*, 949.
24. See, for instance, Hartmann, H.; Wanczek, K.-P., Eds. *Ion Cyclotron Resonance Spectrometry II, Lecture Notes in Chemistry*, Vol. 31; Springer-Verlag: New York, 1982.
25. See, for instance, Klapstein, D.; Maier, J.P.; Misev, L. In *Molecular Ions: Spectroscopy, Structure and Chemistry*; Miller, T.A.; Bondybey, V.E., Eds.; North Holland: New York, 1983; Chapter 7.
26. Reed, D.T.; Meckstroth, W.K.; Ridge, D.P. *J. Phys. Chem.* 1985, *89*, 4578.
27. Prins, R. *J. Chem. Soc., Chem. Commun.* 1970, 280.
28. Rowe, M.D.; McCaffery, A.J. *J. Chem. Phys.* 1972, *59*, 3786.
29. Rowe, M.D.; Gale, R.; McCaffery, A.J. *Chem. Phys. Lett.* 1973, *21*(2), 360.
30. Rosenstock, H.M.; Draxl, K.; Steiner, B.W.; Herron, J.T. *J. Phys. Chem. Ref. Data* 1977, *6*, Suppl. 1.
31. Dunbar, R.C.; Klein, R. *J. Am. Chem. Soc.* 1976, *98*, 7994.
32. Westmore, J.B. *Chem. Rev.* 1976, *76*, 695.
33. Moore, C.E. *Atomic Energy Levels*; National Bureau of Standards: Washington, DC, 1971; Vol. II, NSRDS-NBS 37.

34. Moskovits, M.; Hulse, J.E. *J. Chem. Phys.* 1977, *66*, 3988.
35. Tyler, D.R.; Levenson, R.A.; Gray, H.B. *J. Am. Chem. Soc.* 1978, *100*, 7888.
36. McLafferty, F.W., Ed. *Tandem Mass Spectrometry*; Wiley: New York, 1983.
37. (a) Chen, J.H.; Dunbar, R.C. *Int. J. Mass Spectrom. Ion. Proc.* 1986, *72*, 115; (b) Chen, J.H.; Dunbar, R.C. *Int. J. Mass Spectrom. Ion Proc.* 1987, *76*, 1.
38. Dunbar, R.C.; Honovich, J.P. *Int. J. Mass Spectrom. Ion Proc.* 1984, *58*, 25.
39. Castleman, A.W., Jr.; Keesee, R.G. *Chem. Rev.* 1986, *86*, 589.
40. Zheng, L.S.; Brucat, P.J.; Petteiette, C.L.; Yang, S.; Smalley, R.E. *J. Chem. Phys.* 1985, *83*, 4273.
41. See, for instance, (a) Forst, W. *Theory of Unimolecular Reactions*; Academic: New York, 1983; (b) Robinson, T.J.; Holbrook, K.H. *Unimolecular Reactions*; Wiley-Interscience: New York, 1972.
42. See, for instance, Chesnavich, W.J.; Bowers, M.T. *Prog. React. Kinet.* 1982, *11*, 137.
43. See, for instance, Jasinski, J.M.; Rosenfeld, R.N.; Brauman, J.I. *J. Am. Chem. Soc.* 1982, *104*, 652.
44. (a) Dunbar, R.C.; Kramer, J. M. *J. Chem. Phys.* 1973, *95*, 6511; (b) Orth, R.; Dunbar, R.C.; Riggin, M. *Chem. Phys.* 1977, *19*, 279.
45. Armentrout, P.B.; Halle, L.F.; Beauchamp, J.L. *J. Am. Chem. Soc.* 1981, *103*, 6624.
46. Hanratty, M.A.; Illies, A.J.; Bowers, M.T.; Beauchamp, J.T. *J. Am. Chem. Soc.* (to be submitted).

11

Photodissociation of Gas Phase Metal Clusters

Veronica Vaida

1. INTRODUCTION

Over the past decade, research involving transition metal complexes emphasized the synthesis and structural characterization of novel metal-containing compounds. However, the 1980s are seeing a rapidly developing field at the interface of organometallic chemistry and chemical physics, addressing questions of bonding, electronic structure, and reactivity of transition metal complexes. Interest in these issues is reinforced by the promise of metal-containing species for practical applications in areas such as catalysis[1-5] and microelectronics.[6,7]

The excited state chemistry of organometallic compounds is much more efficient than that of inorganic coordination complexes. The high quantum yield for dissociation of organometallic species leads to "unpleasant" spectroscopic consequences. Short excited state lifetimes caused by photoreaction preclude rotational and vibrational structure in the electronic spectra of these compounds and cause unmeasurably small quantum yields for emission.[8] UV-visible absorption and emission studies have been possible for only a few organometallic compounds such as $W(CO)_5L$ and $W(CO)_4L'$ (L = acetylpyridine, L′ = *o*-phenanthroline)[9,10] and $(bipy)Re(CO)_3Cl$[11] which have relatively low quantum yields for dissociation. In the case of highly dissociative organometallic compounds, much of the data used to test chemical theories of excited state bonding and reactivity come from photochemical experiments.[8]

Only recently, techniques have been developed with the potential for providing the thermochemical, photochemical, and mechanistic data necessary for a comprehensive study of organometallic species. This has come about as

Veronica Vaida • Department of Chemistry and Biochemistry, University of Colorado, Boulder, Colorado 80309-0215.

experimental chemical physics has added to its repertoire sophisticated and sensitive time- and energy-resolved spectroscopic probes. Recently, methods able to handle many-electron systems, necessary for the study of metal-containing complexes, have become available to theoretical chemistry. As a consequence, at the interface of organometallic chemistry and experimental and theoretical chemical physics, one can attempt the characterization of potential energy surfaces for transition metal complexes, which ultimately leads to an understanding of their dynamics and chemistry.

Most of the information to date on the photochemistry of organometalic complexes comes from condensed phase work.[8] Primarily, two classes of photoreactions are known: (1) cleavage of the metal–metal bonds with fragmentation to species of low nuclearity, and (2) dissociative ligand loss with substitution of CO. Condensed phase data have been crucial to the development of the field to date; however, they contain the effects of the interaction of the sample with a solvent or matrix, and the characterization of these effects is beyond the current capabilities of theoretical pictures. Photochemical experiments on isolated transition metal molecules under collision-free conditions provide a data base directly relevant to theory.[12] In what follows, the photodissociation of organometallic complexes in gas phase is reviewed. The gas phase photochemistry is compared with the vast amount of photochemical data available from condensed phase experiments.[8] The results are viewed in light of the one-electron, electronic structure picture generally used in understanding and predicting the photochemistry of transition metal complexes.[8,13–15]

2. TECHNIQUES

Chemical physics has developed in recent years sophisticated spectroscopic techniques for the study of photodissociative molecules.[16–20] Only a small subset of these techniques are applicable to the study of transition metal compounds as these low-vapor-pressure samples require very sensitive detection. In addition, the fragments themselves are often reactive, nonfluorescent species, elusive to much of the battery of available spectroscopic probes. In spite of the difficulties, studies are already under way to characterize the nature of photofragments formed on dissociation of transition metal complexes,[12,21,22] the energy content of the organometallic photoproducts,[23] and their structure[24–29] and chemistry.[30–32] Ultraviolet photolysis of organometallic complexes in gas phase followed by detection of the reactive intermediates by schemes based on UV absorption[33] (see Figure 1), chemical trapping,[34,35] infrared absorption[27–29,36] (see Figure 1), and photofragment mass analysis[12,21,37–50] (see Figure 2) have proven to be sensitive and specific. These methods, which have yielded much of the information available to date about the photofragmentation of oganometallic complexes, are briefly outlined below.

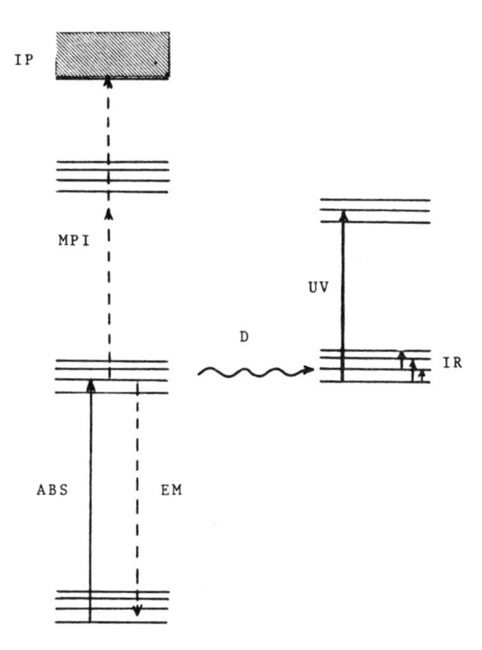

Figure 1. Summary of photodissociation experiments using spectroscopic detection of fragments. ABS = absorption, MPI = multiphoton ionization, EM = emission, D = dissociation, UV = electronic absorption spectroscopy, IR = infrared absorption spectroscopy. Dashed lines indicate processes which for organometallic compounds have unmeasurably small quantum yields.

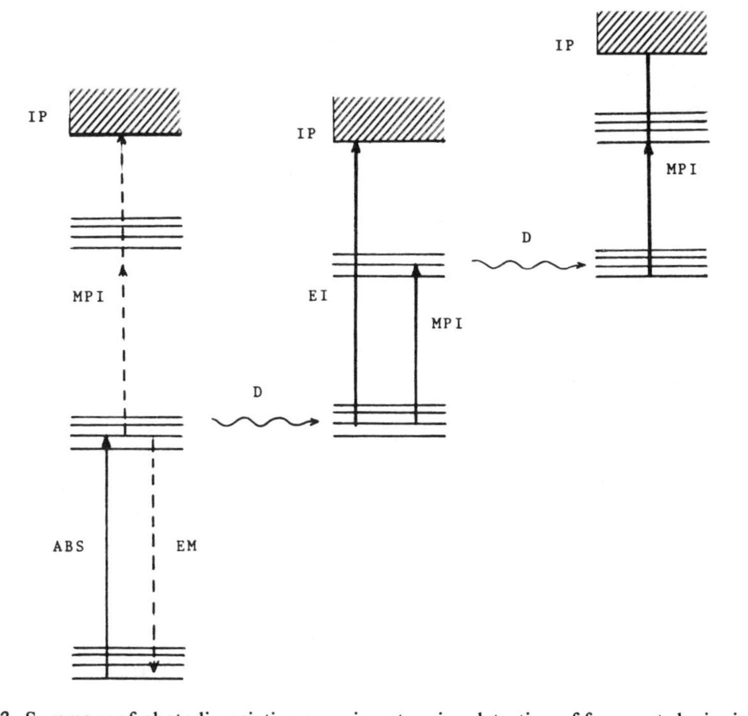

Figure 2. Summary of photodissociation experiments using detection of fragments by ionization. ABS = absorption, MPI = multiphoton ionization, EM = emission, D = dissociation, EI = electron impact ionization, IP = ionization potential. Dashed lines indicate processes which for organometallic compounds have unmeasurably small quantum yields.

2.1. Detection of Organometallic Photoproducts by Chemical Trapping

The earliest experiments attempting to explore the primary photochemistry of organometallic complexes in gas phase relied on UV laser photolysis of gaseous mixtures of $Fe(CO)_5$[34] or $Cr(Co)_6$[35] and PF_3. In these experiments, the primary photofragments react with PF_3 and these stable products are analyzed gas chromatographically.

A similar experiment led to investigations of gas phase reactions catalyzed by photolytically generated unsaturated transition metal carbonyl fragments. Gas phase mixtures of the metal carbonyl with olefins were photolysed, and then the reaction products were analyzed gas chromatographically. Olefin isomerization and hydrogenation was observed to occur in gas phase, catalyzed by photofragments of $Fe(CO)_5$[32] and $Cr(CO)_6$.[31]

2.2. Detection of Organometallic Photoproducts by Ionization Techniques

Ionization techniques have provided, to date, most of the information available about the gas phase photodissociation of organometallic compounds. All ion-counting techniques are extremely sensitive and have the special advantage of allowing the combination of optical and mass spectroscopies for unambiguous identification of signal carriers. Metal carbonyls with relatively high vapor pressure, high ultraviolet absorption cross sections, and high dissociation quantum yields[8] have been particularly amenable to such studies.[12,21] Figure 2 outlines the mechanism of laser photodissociation (LPD)/ionization experiments. Two alternatives are shown: one relying on extremely sensitive multiphoton ionization (MPI) of photofragments,[12] the other employing electron impact ionization (EI) of photofragments.

The mechanism for LPD/MPI in metal carbonyls consists of several steps. First, excitation of the parent to an excited electronic state occurs. During the lifetime of this excited state, the parent efficiently dissociates into neutral fragments. By absorbing more photons, the fragments can either further dissociate, to the limit of a metal atom, or reach their own ionization limits, resulting in detected ion fragments. Because the excitation of the neutral fragment is the slowest, and hence rate-limiting step in the overall process, spectroscopic information on the neutral fragment can be obtained whenever the laser frequency matches a molecular or atomic transition. This process contrasts with the one observed for less reactive inorganic compounds in which the parent is known to ionize first, with further ion dissociation occurring at higher laser fluxes.[51-54]

LPD/MPI has been established as a sensitive probe of the photochemical pathways for the highly dissociative metal carbonyls.[12,21,38-50] However, there are several drawbacks associated with using these techniques. In metal carbonyls, the MPD/MPI process favors dissociation all the way to the bare metal clusters or atoms, while metal-carbonyl fragments are rarely observed. Thus, the MPI method cannot identify the primary, partially decarbonylated intermediate. In addition, multiphoton ionization cross sections are intractable.

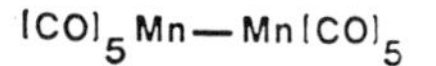

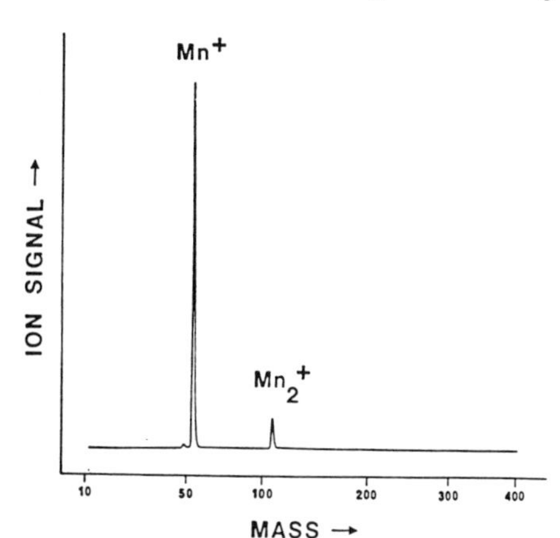

Figure 3. Multiphoton ionization (MPi) mass spectrum of gas phase $Mn_2(CO)_{10}$ following excitation at 337 nm.[12,45] Copyright by the American Chemical Society.

As a consequence, no meaningful information can be gleaned from intensities and, hence, no quantitative data can be obtained concerning the competition of reactive pathways in the excited state.

Many experiments have used one laser to carry the molecule all the way from fragmentation through ionization. With two lasers, however, the excited state chemistry can be probed unambiguously by tuning the first laser specifically to coincide with a wavelength region of photochemical interest. The resulting fragments can then be detected with the focused output of a second laser and the ion signal detected through typical mass spectral means. A companion technique gives spectroscopic information on the fragments. If the MPI laser is a tunable dye laser, then the scanning laser will yield enhancements in the ion signal whenever it coincides with an energy level of the fragment [i.e., resonance-enhanced MPI (REMPI)]. Figures 3 and 4 illustrate the two different types of information obtainable from MPI for the molecule $Mn_2(CO)_{10}$. Figure 3 shows the fragments that result from multiphoton ionization of the fragments on photolysis at 337 nm. Figure 4 shows the spectral structure of the fragments when the laser is scanned. The atomic

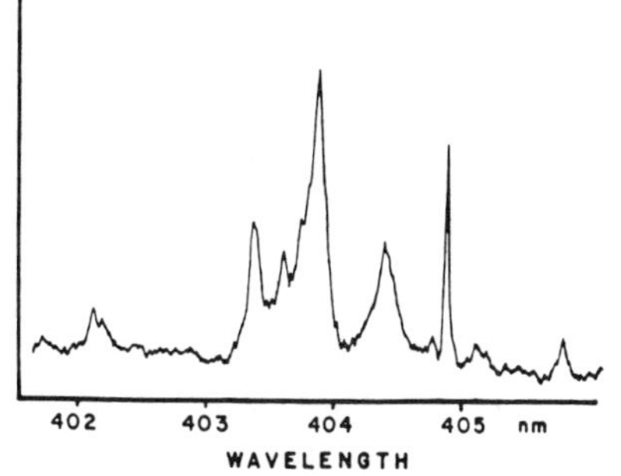

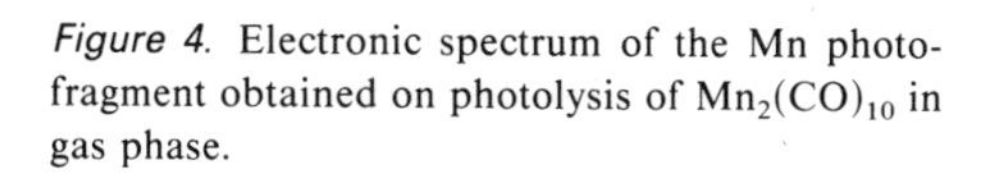
Figure 4. Electronic spectrum of the Mn photofragment obtained on photolysis of $Mn_2(CO)_{10}$ in gas phase.

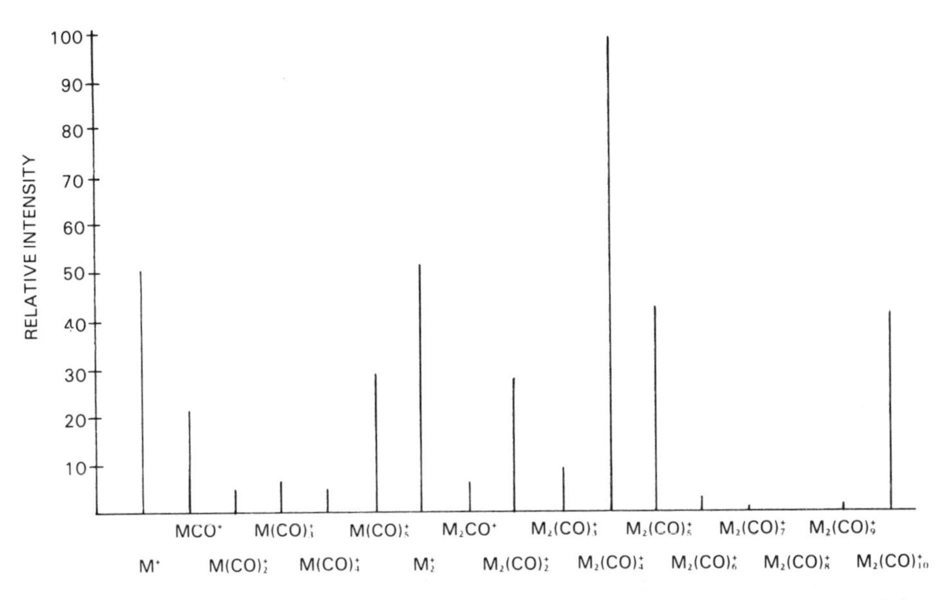

Figure 5. Electron impact ionization (electron energy ≈70 eV) mass spectrum of $Mn_2(CO)_{10}$.[58]

transitions in Mn are easily observable and clearly identifiable[55] by monitoring the Mn^+ signal.†

Alternatively, the MPI spectra of the departing ligand can be monitored in an LPD/MPI experiment. This has been done for both CO and NO ligands, and clear-cut information was obtained about the departing ligands energy content.[23,56] The technical details of these experiments have been described in the literature.[12,21,38-50]

The outcome of a laser photolysis experiment can also be probed by electron impact ionization and mass detection of the photofragments.[57,58] Electron impact ionization occurs at high energy (>20 eV), above a compound's ionization potential. Under these conditions, the parent and daughter ions are observed. The EI mass spectrum of $Mn_2(CO)_{10}$ is shown in Figure 5, for comparison with the MPI mass spectrum of Figure 3.

Making use of EI for detection, the LPD/EI experiments obtain the relevant signal from comparing at each fragment mass the electron impact ion count with and without the laser tuned to an electronic state of the molecule. In this experiment it is possible to detect primary photofragments even when they are highly dissociative unsaturated transition metal carbonyls. This is possible because electron impact ionization avoids the repulsive electronic state of these fragments. Relative cross sections for EI ionization can be more easily quantitated than cross sections for MPI. Consequently, LPD/EI studies promise to identify the primary photofragments and quantitatively indicate the branching ratios when more than one photolytic pathway is possible. While

† The Mn_2^+ signal did not show any resonance enhancement for these experimental conditions (i.e., this particular laser flux and wavelength).

this experiment yields more direct and more quantitative information about the nature of photoproducts than that obtained using the MPI techniques, it is considerably less sensitive.

2.3. Detection of Organometallic Photoproducts by Spectroscopic Techniques

Spectroscopic techniques for gas phase detection of organometallic photoproducts are being developed. To date, these experiments are more difficult due to the low volatility of organometallic compounds and the intrinsically lower sensitivity of spectroscopic methods as compared with ion-counting detection methods. Potentially, spectroscopic detection can provide not only clear-cut photoproduct identification but also characterization of the energetics and structures of these unsaturated species. Figure 1 outlines two photodissociation experiments that employ spectroscopic product analysis.

Early work relied on *emission* from photolytically generated transition metal fragments.[50,59,60] As the primary, dissociative, unsaturated carbonyl fragments do not fluoresce, the emission comes from metal atoms generated ultimately on photolysis. Like MPI detection, these laser-induced fluorescence (LIF) probes are sensitive but rather indirect reflections of photochemical processes in organometallic compounds.

Alternatively, gas phase organometallic photofragments have been detected using *ultraviolet absorption.*[33] Electronic absorption studies of transition metal complexes are possible because the high absorption cross section compensates for the low vapor pressure of the compound. The UV spectra of organometallic transients contain, unfortunately, no structural information and can only be interpreted based on data from matrix isolation experiments.[36] Vibrational spectra have more easily accessible structural information than UV-visible spectra, due to the measurable IR shifts associated with even minor structural changes. Extremely interesting are techniques being developed that are aimed at obtaining structural information about photofragments resulting from gas phase photodissociation processes. To date, the most successful methods are the ones using *infrared spectroscopy* for the characterization of gas phase photofragments.[27–29,36] Intrinsically insensitive, IR absorption can be applied to the study of carbonyl fragments because of the unusually large IR cross sections for CO bound to transition metals. These techniques are being improved to allow for the study of less volatile species and the detection of noncarbonyl ligands.

The techniques developed for IR detection of organometallic photoproducts in gas phase rely on initiation of the photochemical event with a UV laser, followed by kinetic measurements at fixed IR wavelength. The monitoring IR wavelength is changed synchronously with the next UV pulse. The infrared sources currently used for gas phase experiments are line tunable CO lasers[27–29,61] and conventional globar and monochromator arrangements.[62]

In principle, this experiment provides both structural and kinetic information about primary photofragments and promises the most direct probe of photodissociation mechanism. In practice, the low sensitivity of the technique coupled with the low vapor pressure of organometallic compounds has limited its application.

3. EXAMPLES

In what follows, both qualitative and quantitative aspects of the excited state chemistry of organometallic species in gas phase will be discussed. The questions addressed are (1) the proper description of excited electronic states, (2) the effect of electronic excitation on product distribution, and (3) the structure and reactivity of the unsaturated metal fragments in gas phase.

3.1. The Photofragmentation of Group 6 Complexes

The group 6 transition metal carbonyls have been the object of extensive photochemical investigations in condensed phase.[8,36,63] Qualitatively different products are obtained on photolysis of $M(CO)_6$ (M = Cr, Mo, W) in vapor and condensed phases.[64] Single-photon excitation of these compounds in solution or low-temperature matrices leads to the dissociation of one metal–CO bond. Photochemical studies of the group 6 carbonyls in condensed phase have focused on the mechanism[8] and the dynamics[64,65] of this process and the effect of the environment on the structure of the photoproducts formed.[63]

In gas phase, extensive fragmentation occurs, leading to unsaturated metal carbonyls. The primary photoproducts and gas phase photodissociation mechanism have been studied using chemical trapping,[31,34,35] multiphoton ionization,[21,40,64] emission,[66] time-resolved UV–visible absorption,[33] and time-resolved IR spectroscopy.[29,67] The most extensive study of the gas phase photochemistry of $Cr(CO)_6$ to date has been performed using flash photolysis followed by time-resolved IR absorption spectroscopy.[67] In this experiment, the CO product vibrational energy distribution is measured using time-resolved CO laser absorption spectroscopy following excitation of $M(CO)_6$ at 351 and 249 nm. The data are compared with phase space model calculations of CO product vibrational energy distributions as a function of available energy. From the comparison of experiment and calculated results, a model of the photodissociation mechanism can be constructed. When possible, IR spectra of $M(CO)_n$ are obtained and interpreted using extensive spectroscopic information available for the group 6 unsaturated carbonyls trapped in low-temperature matrices.[63]

Excitation at 28,490 cm^{-1} (351 nm) of $Cr(CO)_6$ populates the $\tilde{a}^3T_{1g}$ state. The measured CO product vibrational energy distributions indicate that photodissociation occurs from the excited electronic state to give $Cr(CO)_5(\tilde{a}^3E)$ and CO $(X^1\Sigma^+)$. Excitation at 40,161 cm^{-1} (249 nm) of the predominantly $^1A_{1g} \rightarrow d^1T_{1u}$ $(M \rightarrow \pi^*CO)$ transition deposits 115 kcal/mol into the molecule. Consistent with available energetic information,[68] the IR absorption spectra[67] indicate that photolysis at this energy generates $Cr(CO)_5$ with a large amount of internal energy so that it can in turn lose another CO.

$$Cr(CO)_6 \rightarrow [Cr(CO)_5]^* + CO \qquad \text{(at 249 nm)}$$

$$[Cr(CO)_5]^* \rightarrow Cr(CO)_4 + CO$$

The CO product is formed translationally and rovibrationally excited.

Excitation of the ${}^3T_{1g}$ excited state of $W(CO)_6$ at 28,490 cm^{-1} (351 nm) leads to intersystem crossing to the ${}^1A_{1g}$ ground state, where dissociation occurs. The much larger spin-orbit coupling constant for tungsten than for chromium appears to be responsible for the dissociation of $W(CO)_6$ on the ground electronic surface while dissociation of $Cr(CO)_6$ proceeds on the excited electronic surface.

3.2. The Photofragmentation of Group 7 Complexes

The first metal-metal bond to be characterized[70] is the formally single Mn—Mn bond in $Mn_2(CO)_{10}$. This compound has often been used as the model for developing electronic structure pictures.[8,69] Extremely efficient photofragmentation is responsible for the structureless electronic spectra and lack of emission following excitation of this molecule. As a consequence, experimental data on the photofragmentation rather than the spectroscopy of $Mn_2(CO)_{10}$ are needed to check theoretical models. Most of the photochemical experiments in the past explored the reactions of the lowest excited singlet state in the near ultraviolet. According to the simple one-electron electronic structure description (see Figure 6), excitation into this strong $\sigma \rightarrow \sigma^*$ band should lead to a reduction of the metal-metal bond order to zero and therefore result in homolytic cleavage into $M(CO)_5$ units. While condensed phase experiments did at first seem to agree with this prediction,[8,13,69-71] more recent time-resolved studies painted a more complex picture[73,74]: in addition to metal-metal bond cleavage, CO loss and isomerization are important photochemical pathways in condensed phase. The photochemistry of this molecule in gas phase is required to check theoretical predictions as these cannot include solvent or matrix effects. A clear picture of the photodissociation of $Mn_2(CO)_{10}$ in gas phase is just emerging.

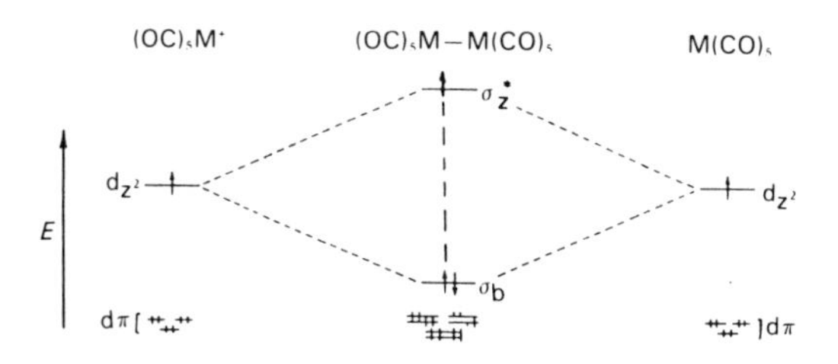

Figure 6. One-electron electronic structure description[8] of the S_1 electronic state of $Mn_2(CO)_{10}$.

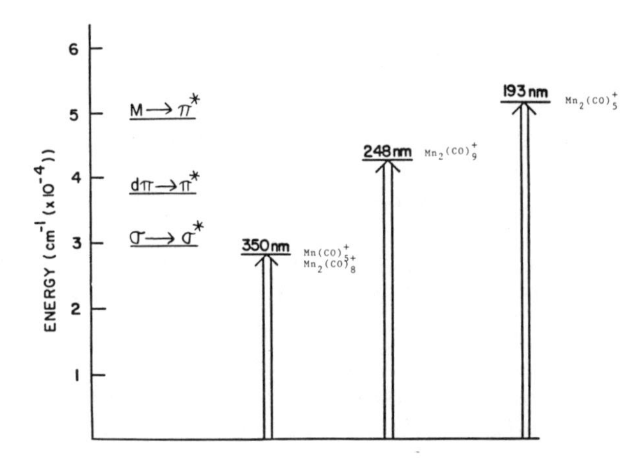

Figure 7. Energetic constraints of the photodissociation of $Mn_2(CO)_{10}$. The energies of the excited electronic states are shown schematically. The arrows indicate the photolysis energies employed. The primary photofragments observed at each energy are listed.

The gas phase photodissociation of this molecule has been investigated by photodissociation followed by electron impact ionization (LPD/EI).[58] Figure 7 shows on a unified energy scale the absorption spectrum of the molecule and the photolysis energy. At each photolysis energy, Figure 7 displays the photofragments obtained by monitoring the electron impact signal with the laser on and off. This experiment shows that excitation of S_1 at 350 nm leads to production of equal amounts of fragments that retain the metal-metal bond [e.g., $Mn_2(CO)_8^+$] and those that are a result of homolysis [e.g., $Mn(CO)_5^+$]. At 350 nm the molecule receives sufficient energy for cleavage of more than one bond. For this reason, $Mn_2(CO)_8^+$ and not $Mn_2(CO)_9^+$ is the predominant binuclear fragment detected mass spectroscopically.

Excitation at 249 nm populates π^* states. The photofragments produced at this energy are exclusively a result of metal-CO bond cleavage. At very low laser fluxes, the fragment seen is $Mn_2(CO)_9^+$, with other $Mn_2(CO)_n^+$ fragments formed very efficiently at higher laser fluxes. The same is true on excitation at 193 nm (the $M \rightarrow \pi^*$ transition), where only $Mn_2(CO)_n^+$ fragments are seen. Excitation of the higher energy $d\pi \rightarrow \pi^*$ and $M \rightarrow \pi^*$ transitions is therefore seen to enhance the metal-CO channel at the expense of the Mn—Mn homolysis. These conclusions regarding the gas phase fragmentation of $Mn_2(CO)_{10}$ are confirmed by photolysis experiments employing IR spectroscopic detection[74] and by LPD/MPI studies.[45] In the latter, excitation by a low-flux N_2 laser (337 nm) of the lowest singlet state of $Mn_2(CO)_{10}$ leads to dissociation which is probed by MPI of the ensuing fragments. The resulting mass spectrum is shown in Figure 3 and consists of Mn^+ and Mn_2^+ with no carbonylated species detectable. The lack of carbonylated fragments in this study is in keeping with results on all other metal carbonyl complexes and the expectation of high dissociation yields for most such fragments. These results indicate that $\sigma \rightarrow \sigma^*$ excitation of this compound does lead to both homolytic cleavage of the metal-metal bond and ligand loss with retention of the metal-metal bond.

These indirect LPD/MPI measurements are very sensitive and could be applied to study $Re_2(CO)_{10}$.[45] Two pathways, Re—Re and Re—CO bond cleavage, are possible on excitation of the σ^* state. The intensities of the Re_2^+/Re^* as compared with the Mn_2^+/Mn^+ fragments suggest a stronger Re=Re bond than Mn—Mn bond. However, intensities in MPI cannot be interpreted and a quantitative study of the photodissociation of $Re_2(CO)_{10}$ awaits more direct experiments.

3.3. The Photochemistry of Group 8 Complexes

The most extensively studied group 8 compound in gas phase is $Fe(CO)_5$. Its photochemistry has been investigated using a range of gas phase techniques. The most clear-cut data concerning the primary photochemistry of this molecule came from laser photodissociation followed by IR detection of photofragments.[36,75] Photolysis of $Fe(CO)_5$ at 249 nm yields as primary photoproducts $Fe(CO)_2$, $Fe(CO)_3$, and $Fe(CO)_4$. These are identified by comparison of their IR spectra with those of the same fragments obtained in low-temperature matrices.[30] The relative yields of the photoproducts are wavelength dependent, with $Fe(CO)_2$ increasing with higher-energy photoexcitation. The loss of more than one CO group on photolysis of $Fe(CO)_5$ is possible as indicated by the specific bond dissociation energies,[76,77] which are consistent with loss of up to three CO groups on photolysis of $Fe(CO)_5$ at 249 nm in gas phase. In contrast with this gas phase result, one-photon excitation in condensed phase of $Fe(CO)_5$ generates only $Fe(CO)_4$,[78,79] indicating rapid loss to the solvent of internal energy deposited in the molecule on excitation.

The gas phase IR spectrum of $Fe(CO)_4$ could be interpreted to obtain structural information about this species.[75] Two C—O vibrations were observed, leading to a distorted, nontetrahedral gas phase structure of the unsaturated molecule, consistent with that observed for $Fe(CO)_4$ in a low-temperature matrix.[80]

The photodissociation of $Fe(CO)_5$ and several other group 8 complexes was investigated using the sensitive, even if less direct, MPI and LIF techiques. In these experiments, excitation leads to complete stripping of all ligands from the metal fragment. For example, analysis of the spectrum of the Fe fragment generated on photolysis of $Fe(CO)_5$[50] indicated a highly nonstatistical distribution of excited atomic energy levels. Analysis of the departing CO by MPI techniques favored a dissociation model involving an unbound lowest ligand field state densely mixed with higher-lying ligand field and charge-transfer states.[23]

Some group 8 cluster carbonyls allow nonmetal atoms to form symmetric bridges across metal atoms of the cluster. Notable examples are $Fe_2S_2(CO)_6$ and $Co_2(CO)_9CCH_3$. The interesting photochemistry of these two cluster complexes of group 8 transition metals and heteroatoms has been investigated[81,84] in gas phase and is discussed below.

The gas phase chemistry of $Fe_2(SCH_3)_2(CO)_6$ and $Fe_2S_2(CO)_6$ was investigated using LPD/MPI techniques.[81] The energetics and fragmentation of the

Fe/S clusters could be studied to explore the influence of the bridging ligands on metal–metal bonding as well as the bonding between the metal atoms and the terminal ligands. The photofragments detected following LPD/MPI of these iron–sulfur cluster carbonyls at 390 nm are Fe^+, FeS^+, Fe_2^+, FeS_2^+, Fe_2S^+, and $Fe_2S_2^+$. In LPD/MPI the parent compound undergoes single-photon excitation to a dissociative electronic state. The photofragments generated undergo further single/multiphoton excitation resulting in the removal of the terminal ligands (CO, CH_3) rather than ionization. The final step involves MPI, producing the observed fragment ions. The predominance of Fe^+ and FeS^+ can be accounted for by the existence of several dissociative channels following terminal ligand loss.

It has been established for numerous transition metal carbonyls that LPD/MPI favors dissociation all the way to bare metal clusters.[8] The $Fe_2S_2(Co)_6$ and $Fe_2(SCH_3)_2(CO)_6$ systems are interesting in this respect because on dissociation they produce the metal/heteroatom clusters FeS, FeS_2, Fe_2S, and Fe_2S_2.

The bonding of a carbon atom to three cobalt atoms in a pyramidal core gives methinyltricobaltenneacarbonyls, $Co_3(CO)_9CY$, unique properties and unusual stability.[82,83] The gas phase chemistry of a representative molecule of this class, $Co_3(CO)_9CCH_3$, has been investigated using photofragmentation of the parent compound followed by MPI detection of photofragments.[84] For the photolytic wavelength of 450 nm, extensive destruction of the Co_3 core occurs with only the smallest cobalt fragments Co^+, CoC, and Co_2^+ being observed. This is in keeping with the antibonding description for the electronic state involved and the interpretation of the solution photochemistry experiment. However, when the photolysis wavelength is changed to 337 nm, fragments such as Co_2^+ and Co_2CCH^+ predominate in the mass spectrum, with fragments containing the Co_3C unit also being present. The much higher degree of retention of the Co_3C central core in this case suggests a photolytic event destabilizing the ligand–core structure. This result contrasts with the solution experiment in which no marked wavelength dependence was observed.

In keeping with other molecules that have been studied with the MPI technique, the presumption is that the ion clusters observed are created as neutral species in the photofragmentation event and are ionized only in the final multiphoton step. Primary fragments cannot be identified in the MPD/MPI technique since the remaining carbonyl ligands are lost in the multiphoton-detection step. The ionic fragments that are ultimately observed merely identify the major branching patterns in the photofragmentation by identifying which bonds within the core structure have been preserved. The patterns of fragmentation are clearly quite distinct between the two wavelengths employed in the study of $Co_3(CO)_9CCH_3$, with the results at 337 nm favoring a greater degree of retention of the Co_3C core. As with other metal compounds, no fragments retaining carbonyl ligands were observed, presumably because such unsaturated carbonyls are themselves very labile and decarbonylation competes efficiently with MPI.

4. CONCLUSIONS

This review focuses on the primary photochemistry of transition metal complexes. Employing sensitive and specific chemical physics techniques, it is possible to obtain quantitative information about the gas phase photofragmentation of transition metal complexes. The examples reviewed here illustrate the questions under study in this field, namely (1) the elucidation of the primary photochemistry, (2) identification of primary photoproducts, and (3) the structure and reactivity of intermediates.

ACKNOWLEDGMENTS

Financial support from the NSF Grant CHE8607697 is gratefully acknowledged.

REFERENCES

1. Masters, C. *Homogeneous Transition Metal Catalysis*; Chapman and Hall: New York, 1981.
2. Moggi, L.; Juris, A.; Sandrini, D.; Manfrin, M.F. *Rev. Chem. Intermed.* 1981, *4*, 171.
3. Wrighton, M.S.; Ginley, D.S.; Schroeder, M.A.; Moise, D.L. *Pure Appl. Chem.* 1975, *41*, 671.
4. Bartholomew, C.H.; Agrawal, P.K.; Katzer, J.R. *Adv. Catal.* 1982, *31*, 135.
5. Schroeder, M.A.; Wrighton, M.S. *J. Am. Chem. Soc.* 1976, *98*, 551.
6. Mayer, T.M.; Fisanick, G.J.; Eichelberger, T.S., IV *J. Appl. Phys.* 1982, *53*, 8462.
7. Osgood, R.M.; Deutsch, T.F. *Science*, 1985, *227*, 709.
8. Geoffroy, G.L. Wrighton, M.S. *Organometallic Photochemistry*; Academic: New York, 1979.
9. (a) Lees, A.J.; Adamson, A. *J. Am. Chem. Soc.* 1982, *104*, 3840; (b) Manuta, D.M.; Lees, A.J. *Inorg. Chem.* 1983, *22*, 572.
10. Zink, J.I. *Coord. Chem. Rev.* 1985, *64*, 93.
11. (a) Bradley, P.G.; Kress, N.; Homberger, B.A.; Dallinger, D.F.; Woodruff, W.H. *J. Am. Chem. Soc.* 1981, *103*, 7441; (b) Smothers, W.K.; Wrighton, M.S. *J. Am. Chem. Soc.* 1983, *105*, 1067.
12. Hollingsworth, W.E.; Vaida, V. *J. Phys. Chem.* 1986, *90*, 1235.
13. Wrighton, M.S.; Graff, J.L.; Luong, J.C.; Reichel, C.L.; Robbins, J.L. In *Reactivity of Metal-Metal Bonds*; Chisholm, M.H., Ed.; American Chemical Society: Washington, D.C., 1981.
14. Hoffman, R. *Science* 1981, *211*, 995.
15. Hirst, D.M. *Adv. Chem. Phys.* 1982, *50*, 517.
16. Vaida, V. *Acc. Chem. Res.* 1986, *19*, 114.
17. Imre, D.; Kinsey, J.; Field, R.; Katayama, D. *J. Phys. Chem.* 1982, *86*, 2564.
18. Leone, S.R. *Adv. Chem. Phys.* 1982, *50*, 255.
19. Simons, J.P. *J. Phys. Chem.* 1984, *88*, 1287.
20. Bersohn, R. *J. Phys. Chem.* 1984, *88*, 5145.
21. Gedanken, A.; Robin, M.B.; Kuebler, N.A. *J. Phys. Chem.* 1982, *86*, 4096.
22. Gobeli, D.A.; Yang, J.J.; El-Sayed, M.A. *Chem. Rev.* 1985, *85*, 529.
23. Whetten, R.L.; Fu, K.-J.; Grant, E.R. *J. Chem. Phys.* 1983, *79*, 4899.
24. Poliakoff, M.; Ceulemans, A. *J. Am. Chem. Soc.* 1984, *106*, 50.
25. Poliakoff, M.; Turner, J.J. *J. Chem. Soc., Dalton Trans.* 1974, 2276.
26. Duncan, I.R.; Harter, P.; Shields, C.J. *J. Am. Chem. Soc.* 1984, *106*, 7248.
27. Ouderkirk, A.J.; Werner, P.; Schultz, N.L.; Weitz, E. *J. Am. Chem. Soc.* 1983, *105*, 3354.
28. Seder, T.A.; Church, S.P.; Ouderkirk, A.J.; Weitz, E. *J. Am. Chem. Soc.* 1985, *107*, 1432.

29. Fletcher, T.R.; Rosenfeld, R.N. *J. Am. Chem. Soc.* 1983, *105*, 6358.
30. Ouderkirk, A.J.; Weitz, E. *J. Chem. Phys.* 1983, *79*, 1089.
31. Tumas, W.; Gitlin, B.; Rosan, A.M.; Yardley, J.T. *J. Am. Chem. Soc.* 1982, *104*, 55.
32. Whetten, R.L.; Fu, K.-J.; Grant, E.R. *J. Chem. Phys.* 1982, *77*, 3769.
33. Breckenridge, W.H.; Sinai, M.J. *J. Phys. Chem.* 1981, *85*, 3557.
34. Nathanson, G.; Gitlin, B.; Rosan, A.M.; Yardley, J.T. *J. Chem. Phys.* 1981, *74*, 361.
35. Yardley, J.T.; Gitlin, B.; Nathanson, G.; Rosan, A.M. *J. Chem. Phys.* 1981, *74*, 370.
36. Poliakoff, M.; Weitz, E. *Adv. Organomet. Chem.* 1986, *25*, 277.
37. Duncan, M.A.; Dietz, T.G.; Smalley, R.E. *Chem. Phys.* 1979, *44*, 415.
38. (a) Gerrity, D.P.; Rothberg, L.J.; Vaida, V. *Chem. Phys. Lett.* 1980, *74*, 1; (b) Rőthberg, L.J.; Gerrity, D.P.; Vaida, V. *J. Chem. Phys.* 1981, *74*, 2218.
39. Engelking, P.C. *Chem. Phys. Lett.* 1980, *74*, 207.
40. Fisanick, G.J.; Gedanken, A.; Eichelberger, T.S., IV; Keubler, N.A.; Robin, M.B. *J. Chem. Phys.* 1981, *75*, 5215.
41. Leutwyler, S.; Even, U. *Chem. Phys. Lett.* 1981, *84*, 188.
42. Lichtin, D.A.; Bernstein, R.B.; Vaida, V. *J. Am. Chem. Soc.* 982, *104*, 1830.
43. Vaida, V.; Cooper, N.J.; Hemley, R.J.; Leopold, D.G. *J. Am. Chem. Soc.* 1981, *103*, 7022.
44. Leopold, D.G.; Vaida, V. *J. Am. Chem. Soc.* 1983, *105*, 6809.
45. Leopold, D.G.; Vaida, V. *J. Am. Chem. Soc.* 1984, *106*, 3720.
46. (a) Nagano, Y.; Achiba, Y.; Kimura, K. *J. Chem. Phys.* 1986, *84*, 1063; (b) Nagano, Y.; Achiba, Y.; Kimura, K. *J. Phys. Chem.* 1986, *90*, 1288.
47. Huang, S.K.; Gross, M.L. *J. Phys. Chem.* 1985, *89*, 4422.
48. Stiller, S.W.; Johnston, M.V. *J. Phys. Chem.* 1985, *89*, 2717.
49. Stuke, M. *Appl. Phys. Lett.* 1984, *45*, 1175.
50. Karny, Z.; Naaman, R.; Zare, R.N. *Chem. Phys. Lett.* 1978, *59*, 33.
51. Wheeler, R.G.; Duncan, M.A. *J. Phys. Chem.* 1986, *90*, 1610.
52. Yu, C.F.; Youngs, F.; Tsukiyama, K.; Bersohn, R.; Press, J. *J. Chem. Phys.* 1986, *85*, 1382.
53. Chen, C.J.; Osgood, R.M. *J. Chem. Phys.* 1984, *81*, 318, 327.
54. Mitchell, S.A.; Hackett, P.A. *J. Chem. Phys.* 1983, *79*, 4815.
55. Moore, C.E. *Natl. Stand. Ref. Data Ser.*, No. 35; National Bureau of Standards: Washington, D.C., 1971.
56. Georgiou, S.; Wight, C.A. *Chem. Phys. Lett.* 1986, *132*, 511.
57. Freedman, A.; Bersohn, R. *J. Am. Chem. Soc.* 1978, *100*, 4116.
58. Prinslow, D.A.; Vaida, V. *J. Am. Chem. Soc.* 1987, *109*, 5097.
59. Callear, A.B.; Oldman, R.J. *Nature (London)* 1966, *210*, 730.
60. Trainor, D.W.; Mani, A.S. *Appl. Phys. Lett.* 1978, *33*, 81.
61. Moore, B.D.; Simpson, M.B.; Poliakoff, M.; Turner, J.J. *J. Chem. Soc., Chem. Commun.* 1984, 972.
62. Hermann, H.; Grevels, F.W.; Henne, A.; Schaffner, K. *J. Phys. Chem.* 1982, *86*, 5151.
63. (a) Perutz, R.N.; Turner, J.J. *J. Am. Chem. Soc.* 1975, *97*, 4800; (b) Burdett, J.K.; Graham, M.A.; Perutz, R.N.; Poliakoff, M.; Rest, A.J.; Turner, J.J.; Turner, R.F. *J. Am. Chem. Soc.* 1975, *97*, 4805; (c) Perutz, R.N.; Turner, J.J. *J. Am. Chem. Soc.* 1975, *97*, 4791; (d) Perutz, R.N.; Turner, J.J. *Inorg. Chem.* 1975, *14*, 262.
64. Welch, J.A.; Peters, K.S.; Vaida, V. *J. Phys. Chem.* 1982, *86*, 1941.
65. (a) Kelly, J.M.; Long, C.; Bonneau, R. *J. Phys. Chem.* 1983, *87*, 3344; (b) Kelly, J.M.; Bent, D.V.; Hermann, H.; Schulte-Frohlinde, D.; Koerner von Gustorf, E. *J. Organomet. Chem.* 1974, *69*, 4791.
66. Gerrity, D.P.; Rothberg, L.J.; Vaida, V. *J. Phys. Chem.* 1983, *87*, 2222.
67. (a) Fletcher, T.R.; Rosenfeld, R.N. *J. Am. Chem. Soc.* 1985, *107*, 2221; (b) Fletcher, T.R.; Rosenfeld, R.N. *ACS Symp. Ser.* 1987, *333*, 99; Suslick, K.S., Ed.; American Chemical Society: Washington, DC (in press).
68. Lewis, K.E.; Golden, D.M.; Smith, G.P. *J. Am. Chem. Soc.* 1984, *106*, 3905.
69. (a) Levenson, R.A.; Gray, H.B.; Ceaser, G.P. *J. Am. Chem. Soc.* 1970, *92*, 3653; (b) Levenson, R.A.; Gray, H.B. *J. Am. Chem. Soc.* 1975, *97*, 6042.

70. Wegman, R.W.; Olsen, R.J.; Gard, D.R.; Faulkner, L.R.; Brown, T.L. *J. Am. Schem. Soc.* 1981, *103*, 6089.
71. Waltz, W.L.; Hackelberg, O.; Dorfman, L.M.; Wojcicki, A. *J. Am. Chem. Soc.* 1978, *100*, 7259.
72. Yesaka, H.; Kobayashi, T.; Yasufuku, K.; Nagakura, S. *J. Am. Chem. Soc.* 1983, *105*, 6249.
73. (a) Rothberg, L.J.; Cooper, N.J.; Peters, K.S.; Vaida, V. *J. Am. Chem. Soc.* 1982, *104*, 3536; (b) Hepp, A.F.; Wrighton, M.S. *J. Am. Chem. Soc.* 1983, *105*, 5934; (c) Fox, A.; Poe, A. *J. Am. Chem. Soc.* 1979, *102*, 2498; (d) Hughey, J.L., IV; Anderson, C.P.; Meyer, T.J. *J. Organomet. Chem.* 1977, *C49*, 125; (e) Sweany, R.L.; Brown, T.L. *Inorg. Chem.* 1977, *16*, 421.
74. Seder, T.A.; Church, S.P.; Weitz, E. *J. Am. Chem. Soc.* (in print).
75. Ouderkirk, A.J.; Seder, T.A.; Weitz, E. *Proc. SPIE* 1984, *458*, 148.
76. Smith, G.P.; Laine, R.M. *J. Phys. Chem.* 1981, *85*, 1620.
77. Engelking, P.C.; Lineberger, W.C. *J. Am. Chem. Soc.* 1979, *101*, 5569.
78. Poliakoff, M. *J. Chem. Soc., Dalton Trans.* 1974, 210.
79. Poliakoff, M.; Turner, J.J. *J. Chem. Soc., Faraday Trans. 2*, 1974, *70*, 93.
80. Poliakoff, M.; Turner, J.J. *J. Chem. Soc., Dalton Trans.* 1973, 1351.
81. Prinslow, D. Ph.D. Thesis, University of Colorado, 1988.
82. Seyferth, D. *Adv. Organomet. Chem.* 1976, *14*, 97.
83. Penfold, B.R.; Robinson, B.H. *Acc. Chem. Res.* 1973, *6*, 73.
84. Hollingsworth, W.E.; Vaida, V. *J. Am. Chem. Soc.* (in press).

12

Tandem Mass Spectrometry and High-Energy Collisional Activation for Studies of Metal Ion–Molecule Reactions

Denise K. MacMillan and Michael L. Gross

1. INTRODUCTION

The study of gas phase ion–molecule reactions has been greatly facilitated in recent years by the rapid development of tandem mass spectrometry (MS/MS). Originally promoted as a method for analysis of complex mixtures, MS/MS is now also used to determine ion structure, as well as reaction cross sections, thermodynamics, and kinetics.[1] MS/MS enables an ion to be selected quickly and accurately, with greater sensitivity and specificity than is possible in traditional mass spectrometry, and then allows the reactions in which the ion participates to be followed. The availability of new instrumentation[2] and the development of new methods[3] for MS/MS have made tandem mass spectrometry especially useful for the study of gas phase ion chemistry.

The emphasis of this chapter will be principally collisionally activated dissociation (CAD), the MS/MS technique used primarily for structure determination. CAD may be applied to the study of a wide variety of ions, but this chapter will concentrate on its use in the study of the reactions of metal ions with organic molecules. The chapter begins with an explanation of MS/MS and CAD and a description of the instrumentation currently being used. Examples of metal ion studies performed by using various CAD techniques

Denise K. MacMillan and Michael L. Gross • Department of Chemistry, University of Nebraska, Lincoln, Nebraska 68588.

follow. The final portion of the chapter is a review of the application of high-energy collisional activation (CA) for studies of the reactivity of gas phase transition metal ions and organic molecules. Although low-energy CA of ions formed in reactions conducted in an ion cyclotron resonance or a Fourier transform (FT) mass spectrometer cell has been used extensively, particularly by Jacobson and Freiser,[4] this approach will not be covered here.

2. REVIEW OF TANDEM INSTRUMENTS

2.1. Sector Instruments

A sector MS/MS instrument employs a minimum of two independent mass analyzers, MS I and MS II, to produce the mass spectrum of a specific mass ion. To conduct a fragment ion scan (i.e., to obtain the mass spectrum of the product ions of a given precursor ion), the desired precursor ion is selected by MS I from the multitude of ions introduced into the analyzer from the ionization source or reaction chamber. For the CAD experiment, the precursor ion is then directed into a collision chamber filled with a neutral target gas, such as helium or argon, to a pressure of 10^{-4} to 10^{-3} torr. Inelastic collisions with the target gas increase the precursor ion's internal energy, inducing dissociation to characteristic fragment ions. A discussion of collision energy follows in Section 2.5. The fragment ions are mass analyzed by MS II after leaving the collision cell. The array of fragment ions is called a fragment ion spectrum or a CAD or MS/MS spectrum of the precursor.

2.1.1. Forward Geometry

The collision cell is positioned in a field-free region (FFR), according to the instrument's geometry. In the forward-geometry sector instrument, the electric sector (E) precedes the magnetic sector (B). The collision cell can be located between the two sectors or between the source and the electric sector. The chain of events occurring in a typical forward-geometry MS/MS or CAD experiment is depicted in Figure 1. A fragment ion scan performed on an EB instrument requires that both E and B be scanned such that the ratio between E and B is a constant over the scan. Unit mass resolution is obtained for the

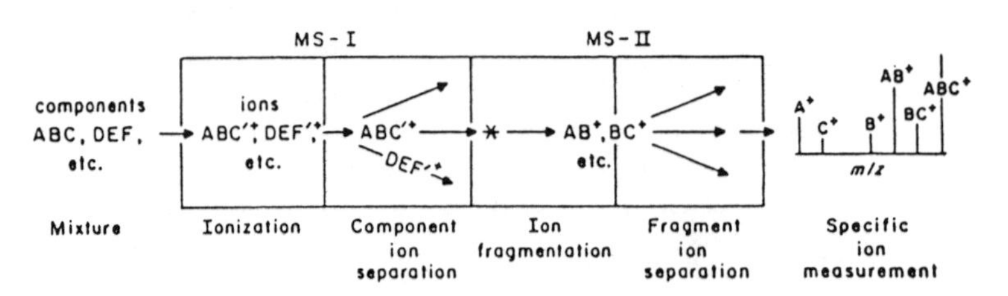

Figure 1. MS/MS analysis of complex mixtures. (Reprinted with permission from *Accounts of Chemical Research*, copyright 1980, American Chemical Society.)

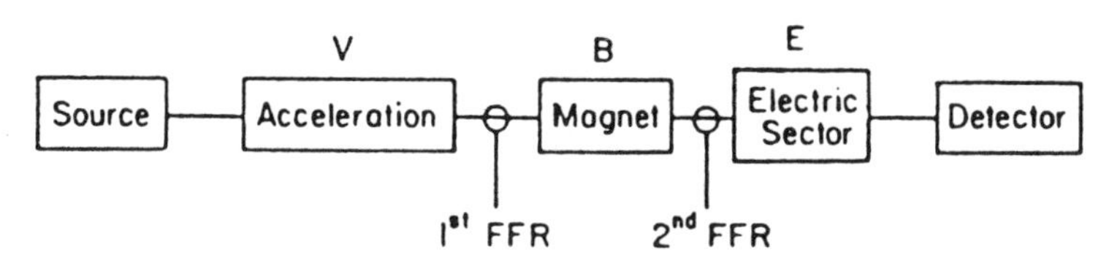

Figure 2. Block diagram of a sector mass spectrometer. (Reprinted with permission from J. Throck Watson, *Introduction to Mass Spectrometry*, copyright 1985, Raven Press, New York.)

fragment ions, but the spectrum is subject to artifact peaks.[1] Another approach is to link a scan of the accelerating voltage (V) and the electrostatic analyzer (ESA) voltage such that $V^{1/2}/E$ is a constant, where V and E are the accelerating and ESA voltages, respectively. This approach is seldom used because the scan causes the ion source focusing properties to change.

2.1.2. Reverse Geometry

For the reverse-geometry instrument, the momentum-analyzing magnetic sector is positioned prior to the energy-analyzing electric sector, as diagrammed in Figure 2. The collision cell is again between the two sectors or prior to the first sector. This instrument was developed largely through the efforts of Beynon, Cooks, and their students.[5] Freas and Ridge[6] first reported high-energy CAD of transition metal complexes on a reverse-geometry mass spectrometer in 1980. Fragment ion spectra are usually produced with a BE instrument by operating MS I in a static mode and MS II in a scanning mode although linked scans are also possible. Resolution for B is no better than 2000 and for E, a few hundred, at best. The upper mass limit of a double-focusing sector instrument manufactured prior to 1984 is ca. 3000 u, whereas the limit of contemporary instruments has been raised to greater than 10,000 u.[13]

2.1.3. Triple-Sector Instruments

Three-[7] and four-sector[8] instruments are also used in MS/MS. These instruments greatly improve upon some aspects of the resolution of double focusing instruments. An EBE instrument (Figure 3) usually contains two collision cells: one that precedes the first electric sector and a second cell following the magnetic sector. CAD spectra from an instrument of this design have been taken for ions separated at a mass resolution of 100,000 for collisional activation[2a] and 15,000 for metastable ion spectra.[7a]

A different three-sector instrument, of BEB configuration, appears to be especially versatile. The BEB instrument can be operated by coupling the first two sectors or the second two sectors. The BE-B configuration enables high precursor ion resolution but only low resolution of the daughter ions. The B-EB configuration permits only unit resolution of the precursor ion but high resolution of the fragment ions. Few applications of this instrument have been reported, however.

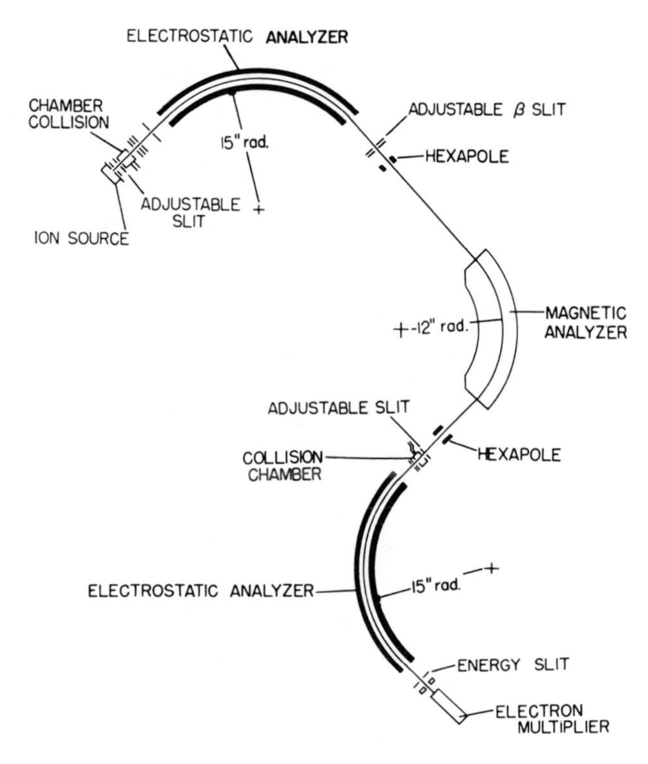

Figure 3. Ion Optics of the Kratos MS-50 triple analyzer. (Reprinted with permission from *Int. J. Mass Spectrom. Ion Phys.*, copyright 1982, Elsevier.)

2.1.4. Four-Sector Instruments

Four-sector instruments furnish medium to high resolution for operations of both MS I and MS II. Bursey and Hass[8] achieved fragment ion resolution of approximately 5500 with the BEEB configuration.

Instruments with a mass range of up to 15,000 u are now available commercially. With a BEEB instrument having a collision cell placed between the two electric sectors, Biemann *et al.*[9] have demonstrated medium mass resolution in the spectra of both the precursor ion and fragment ions for analytes such as fibrinopeptide A, with a mass of 1536 u, and renin substrate, with a mass of 1718 u. However, four-sector instruments have played little role in studies of ion chemistry thus far. Nevertheless, there will be numerous instances in advanced ion chemistry studies in which medium resolution for both MS I and MS II is required.

2.2. Triple Quadrupoles

The most common type of tandem mass spectrometer, the triple quadrupole (QQQ), was first used by Yost and Enke,[10] after McGilvery and

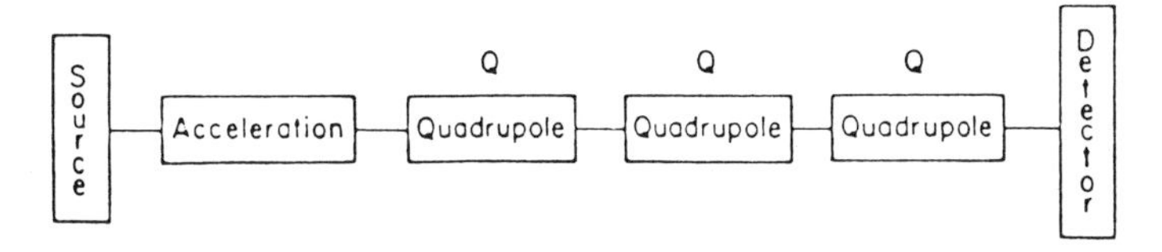

Figure 4. Block diagram of a triple quadrupole mass spectrometer. (Reprinted with permission from J. Throck Watson, *Introduction to Mass Spectrometry*, copyright 1985, Raven Press, New York.)

Morrison[11] developed it for ion photodissociation studies (see Figure 4). A QQQ instrument employs the first (Q1) and third (Q3) quadrupoles as mass analyzers and the middle quadrupole (Q2) as a collision cell. Q2 does not act as a mass filter, but instead allows all ions to pass through to Q3. Activation is done at low energy (ca. 50 eV, or less, in the laboratory frame compared to a few thousand eV in sector instruments).

The triple quadrupole offers sensitivity and selectivity advantages over sector MS/MS instruments. Collisional activation in Q2 has been shown to be 10% to 50% efficient.[10] CAD efficiency, the measure of the fraction of fragment ions collected from precursor ions, has been demonstrated at 11% in sector instruments,[12] and is generally less than that reported for triple quadrupoles.

Unit resolution, the ability to differentiate between M^+ and $(M + 1)^+$, is possible for both mass analyzers in the QQQ. QQQ instruments, which operate at low ion kinetic energy and select for mass-to-charge (m/q) ratio rather than energy-to-charge ratio (mv^2/q) or momentum-to-charge ratio (mv/q), are not affected by translational energy broadening of the product ion beam as are electrostatic and magnetic analyzers.

The advantages of a QQQ are diminished by its lower mass limit and less reproducible mass spectral outputs compared to a sector instrument. Theoretically, an electrostatic analyzer does not have an upper mass limit. Present-day quadrupoles transmit one-third the mass of the sector MS/MS instruments, so that analysis by a QQQ is constrained to molecules of low to mid-range mass. This is not a serious limitation for ion chemistry except for cluster chemistry. The major obstacle for widespread use of quadrupoles is the difficulty in reproducing low-energy CAD spectra.

2.3. *Hybrid Instruments*

Hybrid mass spectrometers combine sector and quadrupole mass analyzers. The first instrument of this type, introduced by Cooks and coworkers,[2b] consisted of a magnetic sector followed by a quadrupole mass filter. BEQ and EBQ hybrids are now commercially available. These instruments allow high-resolution precursor ion selection with unit resolution of fragment ions. McLuckey and Glish[14] recently reported high-resolution precursor and fragment ion selection using a QEB hybrid mass spectrometer. A custom-built BEQQ hybrid has been used by Cooks and co-workers[3] to develop new scan modes. The BEQQ permits high-resolution selection of the precursor ion,[15] with a resolution comparable to that in some of the first triple-sector instruments.[7a] Fragment ion spectra do not contain the diffuse peaks caused by translational energy broadening because the final mass analysis is performed by a quadrupole. Q1 serves as a collision chamber, effectively containing the collisionally activated ions that tend to scatter during CAD. These instruments, like QQQ tandem instruments, have had little impact in ion chemistry studies. Those hybrids having a Q as a final analyzer will suffer the same problems as QQQ instruments. The Q-EB-related instruments, on the other hand, may present opportunities for ion chemistry studies.

2.4. Fourier Transform Instruments

Fourier transform mass spectrometry (FTMS) can be adapted to perform MS/MS experiments. Radically different from conventional mass spectrometers, an FTMS instrument accomplishes mass analysis by using pulsed radio frequency (rf) beams and a trapped ion analyzer cell. Although the steps in an MS/MS experiment are separated physically in sectors and quadrupoles, the steps occur in the same cell, separated in time, in an FTMS instrument. Comisarow and Marshall[16] are credited with the first demonstration of FTMS, basing their technique on ion cyclotron resonance (ICR), a valued method for studying gas phase ion-molecule reactions.[17] CAD experiments were performed as early as 1968 in the older ICR mass spectrometers.[18]

Today, FTMS-CAD is a useful tool in the study of gas phase metal ion chemistry. Among the extensive reports of Freiser and his group are studies of mixed clusters of two metal ions[19] and the reactions of Fe^+, Co^+, and Ni^+ with linear alkanes.[20] The chemistry of larger clusters, such as Si_{39}^+[99a] and Nb_9^+,[99b] has been investigated with an FT-ICR instrument. Other illustrations of the applicability of FTMS-CAD in metal ion research are the studies of Allison's group on the reactions of Co^+ with thiols, ketones, and carboxylic acids[21a] and with amines[21b] and of metal ions with polyethers.[21c] Weil and Wilkins have used FTMS to study the group 11 metal ion reactions with alcohols.[22] The reactivity and specificity of Fe^+, Cr^+, and Mo^+ with aliphatic alcohols was recently reported by Gross *et al.*[23] Fredeen and Russell[24] have used the technique for studies of clusters containing metal ions. This listing accounts for most of the major contributors to metal ion chemistry, but only a few recent examples of work done in this broad field are cited.

2.5. Collision Energy Considerations

Collision energy differentiates the four main groupings of MS/MS instruments and the types of reactions that can be studied with each. Sector instruments analyze ions possessing high kinetic energies, in the range of 3 to 8 keV, whereas the ions in quadrupoles have kinetic energies near 50 eV or less. FTMS-CAD is usually low energy also, although ions with kinetic energies in the keV range can be a studied with larger cells and higher magnetic fields.[25] The amount of energy involved in the collision influences the amount of energy deposited in the ion of interest and the process that will follow. The quantity of transferred energy is also related to the relative energies of the colliding particles and their respective properties. The terms low and high energy refer to the energy in the laboratory frame of reference (E_{lab}). Only a portion of E_{lab} is available to excite the collision target because of conservation of energy and momentum.[26] The available energy, E_{CM}, is the collision energy in a coordinate system moving with the center of mass of the colliding entities. For a stationary target gas particle,

$$E_{CM} = E_{LAB}(m_G/(m_G + m_P)] \quad (1)$$

where m_G is the mass of the target gas and m_P is the mass of the precursor ion. For combinations of a light target gas and heavy precursor ion, E_{CM} becomes very small. Energies in the keV range in the laboratory frame are greatly diminished when one considers the center-of-mass frame.

Collisions between the ion of interest, M^+, and the target gas, such as an organic molecule, A, result in formation of an activated adduct, $[MA^+]^*$. The adduct may be more stable than the reactants, as in an association reaction (Equation 2), or it may be more highly energized. The highly energized adduct will fragment into characteristic fragment ions, F_1^+ and F_2^+, and neutral molecules, M_1 and M_2, as shown in Equation 3.

$$M^+ + A \rightarrow [MA^+]^* \quad (2)$$

$$M^+ + A \rightarrow [MA^+]^* \begin{cases} \rightarrow F_1^+ + M_1 \\ \rightarrow F_2^+ + M_2 \end{cases} \quad (3)$$

Collisionally activated decompositions can be studied at high or low E_{LAB} but associations to form stable adducts only occur in the low-E_{LAB} region. Charge stripping (Equation 4) and charge inversion (Equation 5) occur in the high-E_{LAB} region,

$$M^+ \rightarrow M^{2+} + e^- \quad (4)$$

$$M^- \rightarrow M^+ + 2e^- \quad (5)$$

whereas charge exchange between the sample and the collision target (Equation 6) occurs at both high and low energy.

$$M^+ + B \rightarrow M + B^+ \quad (6)$$

Collision energy also affects the appearance of spectra and the type of information obtained in MS/MS experiments. Low-energy collisions tend to produce spectra with higher fragment ion abundances than those resulting from high-energy collisions. The low-energy experiment produces high fragment yields, suggesting that the E_{CM} is readily converted into internal energy upon collision.[27] An rf-only quadrupole effectively confines the ions that are widely scattered after collision, whereas sectors can only collect fragment ions over a narrow scattering angle near 0°_{LAB}.[1] Spectra resulting from high-energy collisions exhibit more extensive fragmentations.[28] Peak broadening due to kinetic energy release upon dissociation can be diminished by increasing the ion accelerating voltage, thereby improving resolution.[13]

High-E_{LAB} CAD is less susceptible to the skeletal rearrangements found following activation in the low-energy experiment,[1] so that the spectra better represent the selected ion's initial structure. Additionally, bond cleavage is a more important process at high E_{LAB}.[29] The relative abundances of fragments from high-energy collisions are determined primarily by the precursor ion structure rather than its prior internal energy.[30] Internal energy has a more significant effect on relative fragment ion abundances in the low-energy experiment.[1] A major underlying difference may be that high-energy CA occurs by electronic excitation whereas vibrational excitation is the mechanism for low-energy CA.[1]

In summary, the low-E_{LAB} experiment provides functional group information, can identify isomers because of varying reactivity, complements ion beam studies, and provides total cross section data.[1] The high-E_{LAB} experiment furnishes highly reproducible spectra. More importantly, it provides structurally characteristic fragments, largely independent of the precursor's prior history.

2.6. Energy-Resolved Mass Spectrometry

Energy-resolved mass spectrometry (ERMS) is a feature of QQQ and hybrid instruments and can be used to investigate the variation of fragment ion abundances with internal energy.[31] The amount of energy deposited in the internal modes of an ion can often be regulated by altering collision energy.[32,33] When a CAD experiment is performed at different collision energies, the resulting relative fragment ion abundances may be plotted as a function of ion kinetic energy. This produces an energy-resolved mass spectrum. Structural information obtained in this manner will, therefore, possess more detail and provide insight into the overall reactivity of an ion. Isomers of ions with low rearrangement barriers, such as unsaturated hydrocarbons, require sensitive methods of structure determination to be differentiated. Techniques such as charge-stripping CAD,[34-36] photodissociation,[37] and ERMS have been used successfully. The combination of ERMS and CAD enables differentiation of isomers which cannot be distinguished by CAD alone.[38,39] The counterpart of ERMS in the sector instrument field is angle-resolved MS.[1]

3. APPLICATIONS OF MS/MS IN METAL ION CHEMISTRY

3.1. Ion Beam Studies

Structures of adducts of metal ions and organic molecules can be determined by using CAD, but other properties of ions and reactions can also be

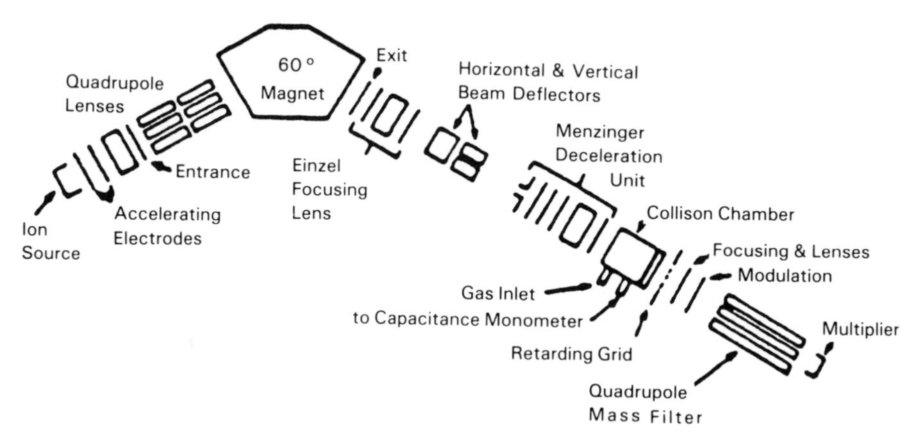

Figure 5. Schematic of an ion beam apparatus. (Reprinted with permission from *Chemical Physics*, copyright 1980.)

deduced from tandem mass spectrometry experiments. Beauchamp and co-workers have used an ion beam apparatus,[40] consisting of two mass spectrometers in tandem (Figure 5), to determine reaction cross sections and bond dissociation energies for many metals ranging from uranium[41] and uranium oxides[40] to the first- and second-row group 8 transition metals.[42] These thermodynamic data are important for understanding transition metal catalysts. The ion beam apparatus has been used to study the reactions of metal ions with hydrocarbons such as alkanes, alkenes, and other compounds over a range of ion kinetic energies. Metal ions are initially separated by a magnetic sector and then directed into a collision cell. The collision cell is filled to a low pressure (10^{-3} torr) with a reactant gas. Products of collisions of the metal ion with the reactant gas are mass analyzed by a quadrupole. By scanning kinetic energy while keeping the reactant gas pressure constant, it is possible to determine reaction thresholds and calculate cross sections. Bond dissociation energies are calculated from the energy-dependent cross sections.[43]

Elkind and Armentrout[44] assembled an improved ion beam apparatus at Berkeley and studied kinetic and electronic effects on reactions of metal ions such as Cu^+, Co^+, and Ni^+ with simple gases (H_2, HD, and D_2). They concluded that a transition metal's reactivity is controlled by the metal ion's electronic configuration rather than whether the ion is in a ground or excited state. The emphasis of the studies by Beauchamp and co-workers and by Elkind and Armentrout is thermodynamic and kinetic rather than structural, although structures can be inferred from the experimental results.

3.2. Triple-Quadrupole Studies

Tsarbopoulos and Allison[45] researched the structures of Co^+ adducts by using CAD and a triple-quadrupole mass spectrometer. $Co(CO)_x$ ($x = 0$–2) and $Co(CO)_xNO$ ($x = 0$–3) ions were formed in the source by electron ionization of $Co(CO)_3NO$ and allowed to react with selected *n*-butyl halides,

1-butanol, 4-halo-1-butanols, and 1,4-dihalobutanes. Specific product ions were selected by the first quadrupole and directed into the second quadrupole where collisions with argon, the target gas, took place, inducing dissociation. Fragment ions were mass analyzed by the third quadrupole. When $CoCO^+$ reacts with n-butanol, as shown in Equation 7, $CoC_4H_{10}O^+$ is formed.

$$CoCO^+ + n\text{-}C_4H_9OH \rightarrow CoC_4H_{10}O^+ \ (m/z\ 133) + CO \tag{7}$$

Collisional activation of the product ions points to a mixture of two structures, $Co(C_4H_8)(H_2O)^+$ and $Co(C_2H_4)(C_2H_5OH)^+$. The CAD spectrum of $CoC_4H_{10}O^+$ shows formation of fragment ions CoH_2O^+ at m/z 77, $CoC_4H_8^+$ at m/z 115, and $CoC_4H_6^+$ at m/z 113. These ions were postulated to be dissociation products of $Co(C_4H_8)(H_2O)^+$. Additional fragment ions are $CoC_2H_4^+$ at m/z 87 and either $CoCOH_2O^+$ or $CoC_2H_5OH^+$ at m/z 105. These ions were considered to be dissociation products of $Co(C_2H_4)(C_2H_5OH)^+$.

$CoCO^+$ also forms $CoCOC_4H_8O^+$ by reacting with n-butanol:

$$CoCO^+ + n\text{-}C_4H_9OH \rightarrow CoCOC_4H_8O^+ \ (m/z\ 159) + H_2 \tag{8}$$

Collisional activation of $CoCOC_4H_8O^+$ yields $CoC_4H_8O^+$ as the primary fragment ion. The implied ease of H_2 elimination in Equation 8 suggests that the molecular skeleton of the precursor ion is not altered. Structure **1** represents the structure of the precursor ion that conforms to the data obtained from collisional activation.

Co^+ OH

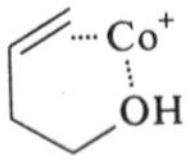

1

Similarly, collisional activation of $CoNOC_4H_9OH^+$ yields the fragment ions $CoNOC_4H_8^+$, from loss of H_2O, and $CoNOH_2O^+$, from loss of C_4H_8. The spectral data imply that the $C_4H_{10}O$ entity exists as C_4H_8 and H_2O bound to the metal ion. Collisional activation of the initial reaction products provides structural information about these products, which aids in determining the mechanism of the original reaction. This study demonstrates that careful use of the triple quadrupole permits organometallic reaction mechanisms to be deciphered. Other studies of transition metals with alcohols will be discussed in Section 3.4.1e.

3.3. Flowing Afterglow Studies

Squires and co-workers[46] have investigated nucleophilic addition reactions of metal anions by using a flowing afterglow (FAG) apparatus. By attaching a triple quadrupole to a flow reactor, the researchers are able to perform low-energy CAD experiments. The first quadrupole selects the ion of interest from the flow reactor. CAD occurs in the pressurized middle quadrupole, and the third quadrupole is scanned to produce the mass spectrum of the CAD fragments. Flowing afterglow mass spectrometry is primarily concerned with obtaining kinetic and mechanistic information concerning the reactivity of thermalized ions.

The Squires group studied[2c] the CAD of an iron carboxylic anion, $Fe(CO)_4COOH^-$, to determine its role as a catalyst in the water–gas shift reaction (WGSR):

$$H_2O + CO \rightarrow H_2 + CO_2 \tag{9}$$

It had been suggested that $Fe(CO)_4COOH^-$, thought to be a WGSR intermediate, either loses CO_2 to form $(CO)_4FeH^-$[47] or loses a proton in basic solution to form the dianion, $(CO)_4FeCO^{2-}$, which then expels CO_2.[48] The $Fe(CO)_4COOH^-$ CAD spectrum showed exclusive loss of CO, and no CO_2 loss. Squires *et al.* concluded that the barrier for decarboxylation was greater than that for decarbonylation in the gas phase and that decarboxylation required base catalysis. This combination of FAG and MS/MS appears to be uniquely powerful for studying ion structures without the complication of internal energy effects. A potential drawback is the sensitivity to tuning that characterizes spectra from QQQ tandem instruments.

3.4. High-Energy Collisional Activation

3.4.1. Activation of C—C and C—H Bonds by First-Row Group 8 Metal Ions

3.4.1a. Reactions with Alkanes and Cycloalkanes. The first instance of high-energy collisional activation of adducts of transition metal atomic ions and organic molecules was reported in 1980 by Freas and Ridge,[6] who used a reverse-geometry mass spectrometer to characterize complexes of metal ions with alkanes. These workers observed differences in the CAD spectra of complexes of a metal ion and isomeric alkanes as well as of complexes of different metal ions and the same alkane. Differences in the structures of the metal ion/organic adducts account for the variations in the spectra. A mechanism was postulated for the reactions of Fe^+, Ni^+, and Co^+ with butanes (Scheme 1) that is analogous to those proposed by Remick *et al.*[49] for Ta reactions with butanes and by Armentrout and Beauchamp[50] for Co^+ reactions with alkanes. The reaction is initiated by metal insertion into a C—C bond followed by a β-hydrogen atom shift and reductive elimination of an alkane.

Collisional activation of metal$^+$-*n*-butane complexes yields $M(C_2H_4)^+$ as the primary fragment ion. The appearance of this fragment is consistent with metal insertion into the weaker central C—C bond followed by a β-hydrogen shift. Although the $Fe(C_2H_4)^+$ fragment is formed upon CA of the *n*-butane complex, it does not appear in the CAD spectrum of the isobutane complex. This result is also in accord with the insertion/β-shift mechanism. $Fe(C_2H_4)^+$ is an expected product of the mechanism with the *n*-alkane, but not with isobutane.

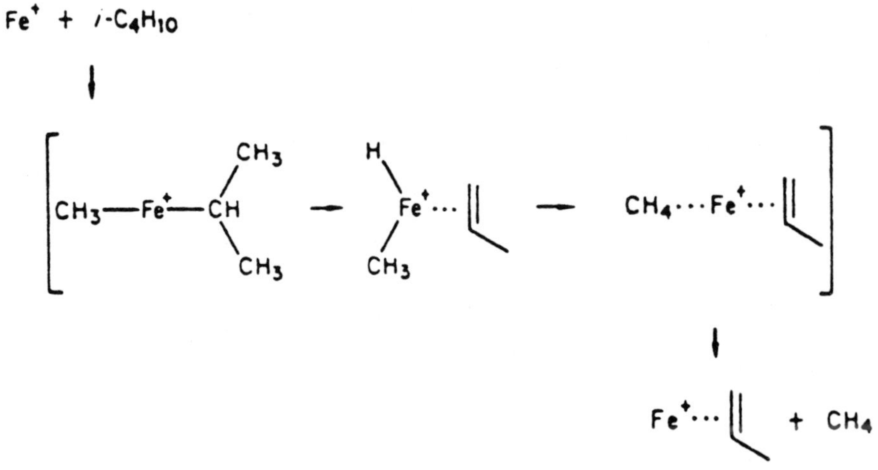

$Fe^+ + i\text{-}C_4H_{10}$

↓

[$CH_3—Fe^+—CH(CH_3)_2$ → $H—Fe^+(CH_3)\cdots$(propene) → $CH_4\cdots Fe^+\cdots$(propene)]

↓

$Fe^+\cdots$(propene) + CH_4

Scheme 1. (Reprinted with permission from *J. Am. Chem. Soc.*, copyright 1980, American Chemical Society.)

Freas and Ridge[6] also investigated the reactions of Cr^+ with *n*-butane and isobutane and found that the only C—C bond into which Cr^+ inserts is the central bond in *n*-butane. Cr^+ has a preference for C—H bond insertion, unlike the C—C bond insertion preference of Co^+, Ni^+, and Fe^+. Release of Cr^+ occurs most readily upon CA, suggesting a weak Cr^+-butane bond. The remaining fragments are primarily Cr^+-alkyl radical ions rather than Cr^+-olefin complexes. These fragments imply that bond formation between Cr^+ and butane occcurs, but that the complex does not dissociate via the same mechanism as the first-row group 8 transition metal ion complexes.

In a subsequent investigation by Larsen and Ridge,[51] on the triple-sector instrument at the University of Nebraska (see Section 2.1.3), the Fe^+ complexes with isobutane, propane, ethane, and methane were characterized. Electron ionization of $Fe(CO)_5$ forms the reactant, $Fe(CO)^+$, which reacts with an alkane, yielding a stable $Fe(\text{alkane})^+$ complex:

$$Fe(CO)^+ + \text{alkane} \rightarrow Fe(\text{alkane})^+ + CO \tag{10}$$

The $Fe(\text{alkane})^+$ of interest is selected by the double-focusing specctrometer and is transmitted to a collision chamber in the instrument's third field-free region where high-energy collisions with helium take place. CAD fragments are mass analyzed by scanning the third sector, an ESA. The EB–E instrument possesses the medium resolution needed to separate $M(CO)_x^+$ from $M(CH_2)_{2x}^+$.

The CAD spectra of $Fe(C_4H_{10})^+$ formed from Fe^+ and isobutane or *n*-butane agreed well with those from the previous study as discussed above.[6] Skeletal rearrangements do not occur either before or after collisional activation of the complex because the fragments in the spectra correspond to the expected products of Fe^+ with each butane. The abundance of the Fe^+ fragment, only

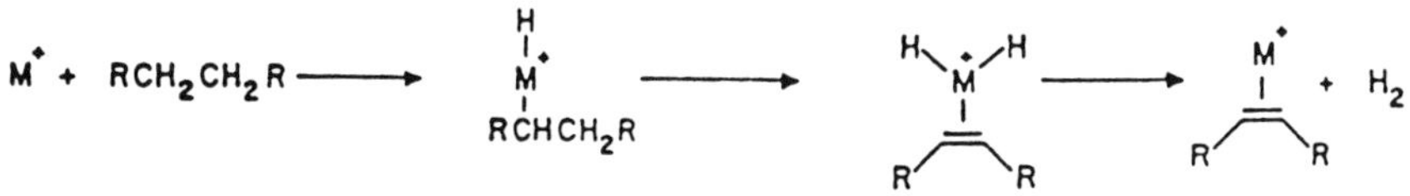

$M^{\bullet} + RCH_2CH_2R \longrightarrow H\text{–}M^{\bullet}\text{–}RCHCH_2R \longrightarrow (H)_2M^{\bullet}(RCH{=}CHR) \longrightarrow M^{\bullet}(RCH{=}CHR) + H_2$

Scheme 2. (Reprinted with permission from *J. Am. Chem. Soc.*, copyright 1983, American Chemical Society.)

about 25% of the total ion abundance, evinces the strong covalent bond formation in the complexes.

Insertion of Fe^+ into C—H bonds as well as C—C bonds was also noted in this study. The mechanisms suggested for bond insertion are outlined in Schemes 2 and 3. Fe^+, Co^+, and Ni^+ all dehydrogenate propane through the 1,2-elimination mechanism in Scheme 2.

CAD spectra of the dissociation products provide structural information about the products and evidence for their modes of formation. $Fe(C_4H_8)^+$ is a fragment of collisional activation of both Fe—*n*-butane$^+$ and Fe—isobutane$^+$. $Fe(C_4H_8)^+$ produced from the latter complex is an adduct of Fe^+ and 2-methylpropene, whereas it is produced from the former complex as a 2:1 mixture of two structures. The more abundant fragment in the CAD spectrum of $Fe(C_4H_8)^+$ is a dihydride of Fe^+ complexed with butadiene; the less abundant fragment is a Fe^+-*bis*(ethylene) complex. Fe^+ insertion into a C—H bond leads to a Fe(2-methylpropene)$^+$ adduct and the Fe(butadiene)$^+$ adduct, but the *bis*(ethylene) complex is due to C—C bond insertion.

Collisional activation of complexes of Fe^+ and alkanes smaller than butane[(51)] shows that the degree of metal-hydrocarbon bonding is less than for the butane complexes. Fe^+ inserts into C—C and C—H bonds of ethane, but a loosely bound Fe(ethane)$^+$ complex is of lower energy, and, therefore, preferred. Fe(methane)$^+$ also forms a weak complex, but insertion of the metal into a C—H bond does not occur. Although oxidative addition of C—C and C—H bonds does occur for Fe(propane)$^+$ complexes, the dominant fragment in the CAD spectrum of $Fe(C_3H_8)^+$ is Fe^+, so the complex is loosely bound, as for complexes of Fe^+ and the smaller alkanes.

Larsen and Ridge[(51)] concluded that the reactivity of various C—C and C—H bonds with Fe^+ is related to their bond strength. Reactivity was measured by determining how well cleavage of a particular bond competes with other

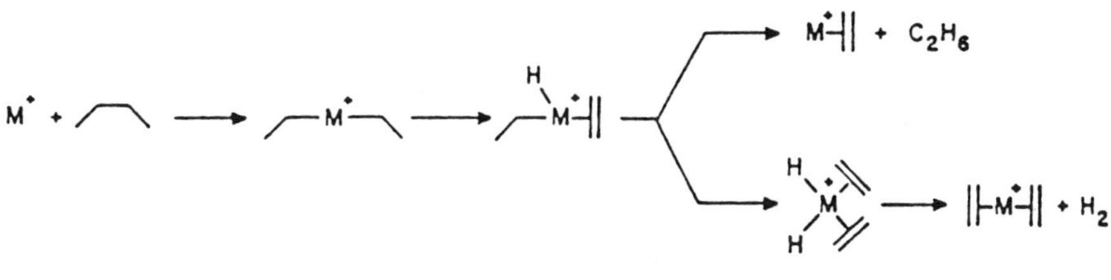

$M^{\bullet}$ + n-butane → M• inserted into the central C–C bond → H–M•(ethylene)(ethyl) → $M^{\bullet}(C_2H_4) + C_2H_6$, or → $(H)_2M^{\bullet}(C_2H_4)_2$ → $(C_2H_4)M^{\bullet}(C_2H_4) + H_2$

Scheme 3. (Reprinted with permission from *J. Am. Chem. Soc.*, copyright 1983, American Chemical Society.)

fragmentation processes. Bond reactivity was found to decrease with increasing bond strength.

The work of Ridge and co-workers[6,51] with high-energy CA and tandem mass spectrometry is the first for metal ion chemistry and shows quite conclusively the power of MS/MS in mechanistic studies.

The reactions of simple hydrocarbons (C_3-C_7) with Fe^+, Co^+, and Ni^+ were also studied by Jacobson and Freiser[20] using FTMS. Their results agree with and expand upon their earlier work[4a,52] and that of Ridge and co-workers[6,51] and Beauchamp and co-workers.[43] Fe^+, Co^+, and Ni^+ were found to insert selectively into C—C bonds. Co^+ and Ni^+ inserted into terminal C—C bonds much less frequently than Fe^+. Terminal bonds are stronger than internal C—C bonds by approximately 3 kcal/mol, making cleavage of terminal bonds more difficult. Also, the metal ion-alkyl bond energy increases with the length of the alkyl chain, promoting the selectivity of internal bonds for the metal ion.

Radecki and Allison[53] recently studied the reactions of linear alkanes (C_4-C_{10}) with Fe^+, Co^+, and Ni^+ by using an ICR mass spectrometer and reported a correlation between insertion into a particular C—C bond and the ionization potentials (IP) of the alkyl radicals created by cleavage of that bond. They suggested that the 3 kcal/mol difference in bond energies between internal and terminal bonds is insufficient to explain the observed metal ion insertion selectivity. The small degree of terminal bond insertion is better explained by the high ionization potential of the CH_3 radical, making the M^+—CH_3 bond weak compared to other M^+—alkyl bonds. Because the formation of strong bonds is favored, M^+ prefers to insert into a C—C bond that would form two alkyl radicals having a low IP sum.

High-energy collisional activation of metal-cycloalkane complexes was performed by Peake *et al.*[54] on the triple-sector instrument mentioned above. Dehydrogenation (Equations 11 and 12) dominates the CAD spectrum of Fe(cyclopentane)$^+$.

$$Fe(C_5H_{10})^+ \rightarrow Fe(C_5H_8)^+ + H_2 \quad (11)$$

$$Fe(C_5H_8)^+ \rightarrow Fe(C_5H_6)^+ + H_2 \quad (12)$$

The structures of the dehydrogenation products were identified by matching their spectra to the spectra of known ions. $Fe(C_5H_8)^+$ and $Fe(C_5H_6)^+$ correspond to Fe(cyclopentene)$^+$ and Fe(cyclopentadiene)$^+$, respectively. Dehydrogenation is also the primary process in the dissociation of Fe(cyclohexane)$^+$ (Equations 13 and 14).

$$Fe(C_6H_{12})^+ \rightarrow Fe(C_6H_{10})^+ + H_2 \quad (13)$$

$$Fe(C_6H_{12})^+ \rightarrow Fe(C_6H_6)^+ + 3H_2 \quad (14)$$

The $Fe(C_6H_6)^+$ fragment was identified as Fe(benzene)$^+$ by comparing its spectrum with that of the adduct of benzene and Fe^+ and with those of the

$Fe(C_6H_6^+)$ ions formed by electron ionization of cyclohexadiene iron tricarbonyl and of the iron-1,3,5,7-cyclooctatetraene complex.

Fe^+ insertion into C—H bonds is favored in five- and six-membered cycloalkanes. The ring is not cleaved. The lack of C—C bond insertion in cycloalkanes contrasts with the prevalence of C—C insertions in linear alkanes. These results also contrast with the results of ion beam studies of Co^+ with cycloalkanes.[55] C—C bond cleavage by Co^+ occurs in the ring series from cyclopropane to cyclohexane. Dehydrogenation (C—H insertion) was observed for all rings except cyclopropane.

An explanation emerges from another study in which Ridge and Kalmbach[56] also observed Fe^+ insertion only into C—H bonds of five- and six-membered rings. However, larger rings are cleaved by Fe^+ to produce alkene fragments. It was postulated that Fe^+ only forms a low-energy product by inserting into a C—C bond to form a nearly linear C—Fe^+—C moiety. Fe^+ insertion into cyclooctane or cyclononane C—C bonds would, therefore, occur more easily than insertion into cyclopentane or cyclohexane.

The CAD spectrum of $Fe(C_3H_6)^+$ formed by reaction of Fe^+ and cyclopropane indicates that the ion exists as an Fe^+ complex with ethylene and carbene ligands. This contrasts with the reactions observed for Fe^+, Co^+, and Ni^+ with linear propane[20] (Equations 15 and 16; M = Fe, Co, or Ni).

$$M^+ + C_3H_8 \longrightarrow M(C_2H_4)^+ + CH_4 \quad (15)$$

$$ \longrightarrow M(C_3H_6)^+ + H_2 \quad (16)$$

Low-energy collisional activation of $M(C_2H_4)^+$ and $M(C_3H_6)^+$ produces cleavage of the metal-organic bond in the ions. The data suggest that $M(C_2H_4)^+$ exists as a metal ion-ethene complex and $M(C_3H_6)^+$ exists as a metal ion-propene complex.

3.4.1b. Reactions with Alkenes. The reactions of Fe^+, Co^+, and Ni^+ with alkenes were investigated and characterized by using high-energy collisional activation. Peake *et al.*[54] demonstrated oxidative addition of allylic C—C bonds to Fe^+. Metal ion insertion is followed by a β-hydrogen shift to the metal and subsequently to the allyl fragment to produce a *bis*(olefin) complex. The smaller olefin is lost as a neutral fragment upon activation. The mechanism, shown in Scheme 4 for Fe(1-pentene)$^+$, is consistent with that proposed by Armentrout *et al.*[60,95] for the reaction between Co^+ and 1-pentene.

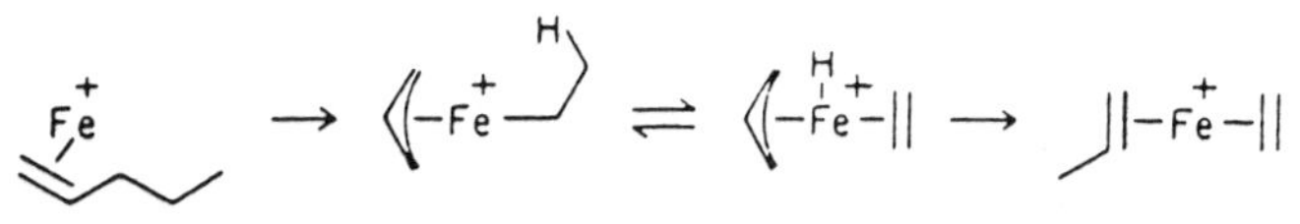

Scheme 4. (Reprinted with permission from *J. Am. Chem. Soc.*, copyright 1984, American Chemical Society.)

The mechanism was tested with the octene isomers shown as reactants in Equations 17–20. If Fe^+ inserts preferentially into an allylic C—C bond, then the expected products of reactions 17–20 should appear in the CAD spectra of the isomeric adducts (see Figure 6). The product expected from allylic insertion is dominant for all octenes.

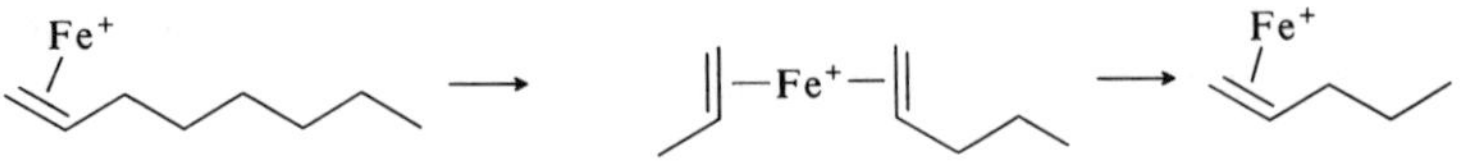

Fe^+ → —Fe^+— → Fe^+ (17)

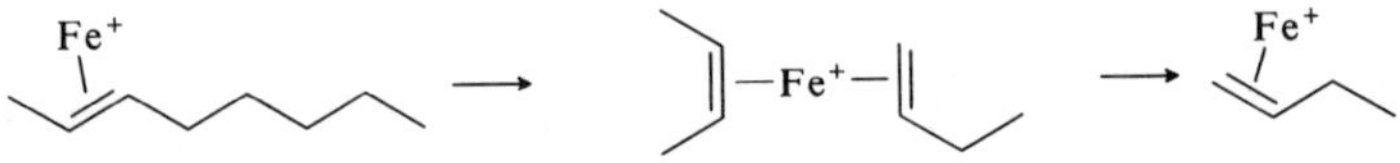

Fe^+ → —Fe^+— → Fe^+ (18)

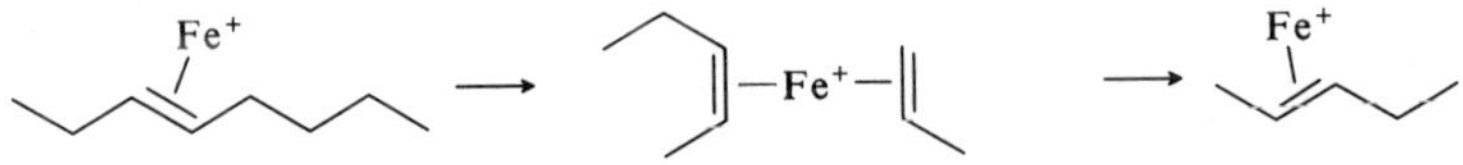

Fe^+ → —Fe^+— → Fe^+ (19)

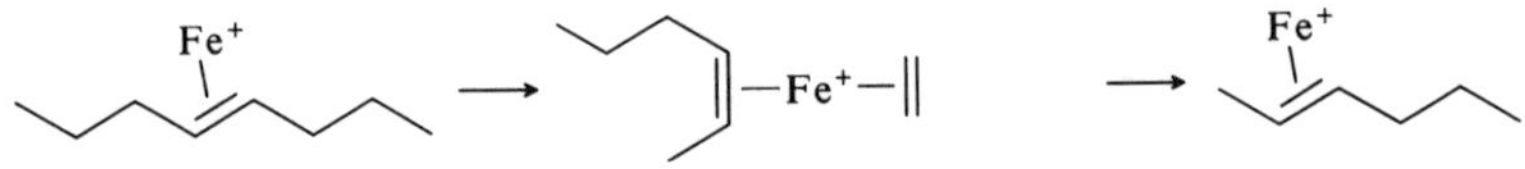

Fe^+ → —Fe^+— → Fe^+ (20)

Alkenes with less than five carbons are prohibited from following the mechanism in Scheme 4 because a β-hydrogen is not available in an adduct formed by allylic C—C bond activation. Two alternative mechanisms were proposed for alkenes containing five or more carbons but lacking β-hydrogens. After Fe^+ insertion into the allylic C—C bond, a hydrogen atom may undergo a 1,3-shift, coincidentally causing a 1,2-shift of the double bond. Alternatively, a six-membered cyclic transition state may form, expelling methane.

Jacobson and Freiser[57] employed low-energy CA and FTMS to study also the primary reactions of olefins with Fe^+. Cleavage of C—C bonds

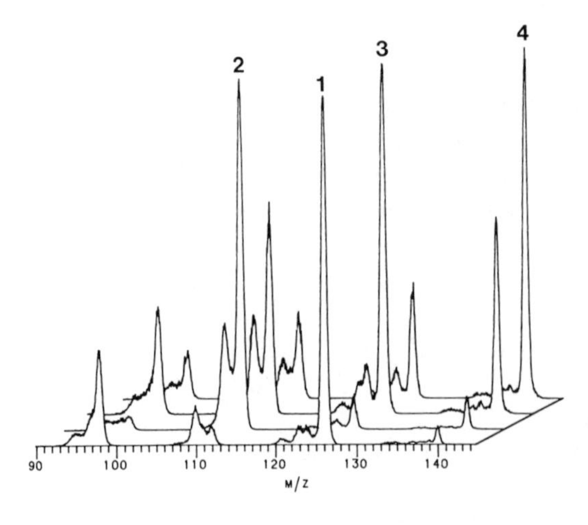

Figure 6. Partial CAD spectra of $FeC_8H_{10}^+$, m/z 168: (1) $FeC_5H_{10}^+$, m/z 126, from 1-octene; (2) $FeC_4H_8^+$, m/z 112, from *trans*-2-octene; (3) $FeC_5H_{10}^+$, m/z 126, from *trans*-3-octene; and (4) $FeC_6H_{12}^+$, m/z 140, from *trans*-4-octene. (Reprinted with permission from *J. Am. Chem. Soc.*, copyright 1984, American Chemical Society.)

followed by dehydrogenation was found to be the dominant process for reactions with linear alkenes. Branched olefins do not produce the same fragments as linear olefins. A metallocyclobutane intermediate was postulated to explain the fragmentation processes of branched alkene complexes. Fragment ions are also formed via the allylic insertion mechanism.

3.4.1c. Reactions with Alkynes and Dienes. The mechanism of reaction of Fe^+ with alkynes and dienes was investigated by Peake and Gross[58] with the triple-sector instrument. The dissociation of a series of Fe^+-alkyne and Fe^+-diene complexes and of their fragments provides new insight into the allylic and propargylic insertion mechanisms. The high dynamic range and precision of the triple-sector instrument permit conclusions to be drawn concerning subtleties of the reaction mechanisms. The conclusions could not have been drawn from low-energy CA data.

A more complete mechanism of Fe^+ insertion (shown in Scheme 5) was obtained as follows. The CAD spectra of a series of dienes (C_3-C_6) and alkynes (C_3-C_7) were obtained and used as reference spectra to determine the terminus of the migrating H atom in the Fe^+ insertion mechanism. Identification of the carbon to which the H atom is transferred after leaving the metal center had not been previously established. The H atom could transfer to C-3 of the propargylic entity along pathway a of Scheme 5 or to C-1 along pathway b. The structures of product ions from Fe^+-octyne complexes were identified by matching their CAD spectra to the reference spectra. The data demonstrate that the various $Fe(octyne)^+$ complexes decompose to a mixture of Fe(2-alkyne)$^+$ and Fe(1,2-diene)$^+$ fragment ions. Transfer of an H atom to the alkyne carbon distal to the union center creates a 1,2-diene complex, as in pathway a of Scheme 5. H atom transfer to the alkyne carbon proximal to the union center following pathway b in Scheme 5. Reaction along pathway b occurs to a lesser extent than along pathway a. Additional evidence for the mechanism was obtained from the CAD spectrum of $Fe(C_5H_8)^+$, a decomposition product of an Fe(3-nonyne)$^+$ complex in which 34% of the $FeC_5H_8^+$ fragment was identified as the pathway b product, Fe(2-pentyne)$^+$, whereas 66% was found to be Fe(1,2,-pentadiene), the pathway a product. The conclusions drawn in this work are based on small but highly reproducible differences in CAD spectra. The relative standard deviation (rsd) in the

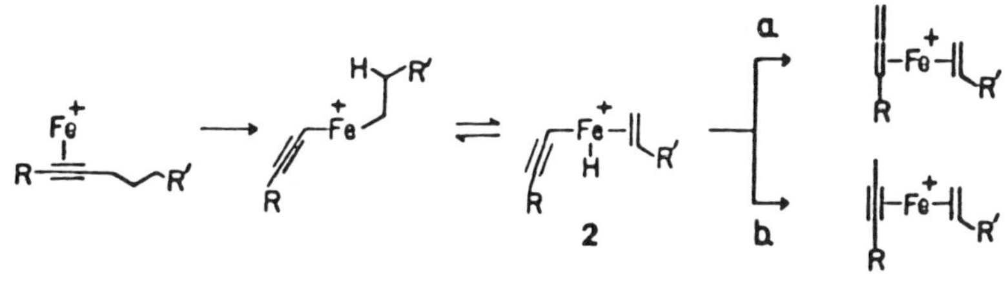

Scheme 5. (Reprinted with permission from *Organometallics*, copyright 1980, American Chemical Society.)

precision of CAD spectra acquired several months apart was 10%. The peak heights of small peaks in a limited mass range had only 3% rsd between scans. Reproducibility is a noteworthy and unique trait of high-energy CAD.

Collisional activation of Fe^+-octadiene complexes produces fragment ions as expected also from the allylic insertion mechanism:

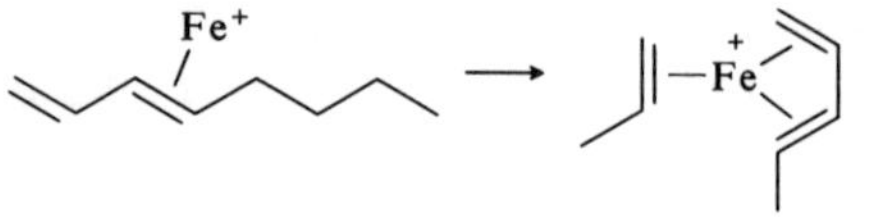

(21)

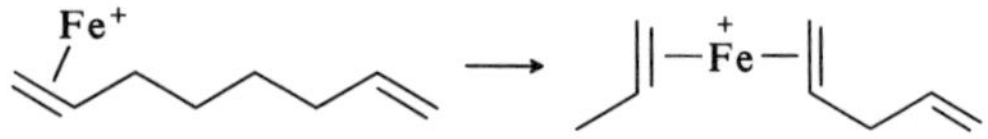

(22)

As shown in Equations 21 and 22, pentadiene-propene complexes are the expected products from decomposition of $Fe(1,3\text{-octadiene})^+$ and $Fe(1,7\text{-octadiene})^+$.

One example of isomerization successfully competing with allylic insertion was observed. Allylic Fe^+ insertion into $FeC_8H_{14}^+$ should produce a hexadiene-ethylene complex that would fragment to give $FeC_6H_{10}^+$. The $FeC_6H_{10}^+$ ion does appear in the CAD spectrum, but is only half as abundant as $Fe(C_5H_8)^+$, a fragment expected from allylic insertion of Fe^+ into 1,3-octadiene. The data suggest that $Fe(2,4\text{-octadiene})^+$ isomerizes readily to $Fe(1,3\text{-octadiene})^+$.

$Fe(1,3\text{-diene})^+$ or $Fe(1,4\text{-diene})^+$ complexes appear as fragment ions in the CAD spectra of Fe^+-octadiene and Fe^+-cyclooctene complexes, as expected. $Fe(2,4\text{-octadiene})^+$ fragments to $Fe(C_5H_8)^+$ upon collisional activation. The CAD spectrum of the $Fe(C_5H_8)^+$ ion demonstrates that a mixture of $Fe(1,3\text{-pentadiene})^+$ and $Fe(1,4\text{-pentadiene})^+$ is formed. The mixture components can be distinguished by comparing abundance ratios of the lower-abundance fragments in the spectra; these distinctions would have been more difficult under low-energy activation conditions.

A recent study of deuterium-labeled octynes furnishes additional evidence for the mechanism of reaction between Fe^+ and alkynes.[59] Collisional activation of $Fe(4\text{-octyne})^+$ produces predominantly $Fe(C_6H_{10})^+$ by the elimination of C_2H_4. Collisional activation of the $Fe(C_6H_{10})^+$ ion confirms the earlier conclusions of Peake and Gross.[58] The $Fe(C_6H_{10})^+$ ion exists more frequently as the diene complex (the pathway a product of Scheme 5) than as the alkyne complex (the pathway b product of Scheme 5). Data from the CA of the 1- and 3-octynes are also consistent with the interpretations of Peake and Gross.[58]

Unimolecular dissociations of labeled $Fe(4\text{-octyne})^+$ complexes were monitored to determine another subtle feature of the Fe^+ insertion mechanism, namely, the reversibility of β-hydrogen transfer and the rate-determining step of the mechanism.[59] Because no scrambling products were observed, hydrogen transfer to the metal center is concluded to be irreversible. β-Hydrogen transfer was compared for two differently deuterated $Fe(4\text{-octyne})^+$ complexes. A kinetic isotope effect, biased against D transfer, was not observed. It is concluded, therefore, that β-hydrogen transfer is not the rate-determining step of the mechanism. Lastly, a secondary kinetic isotope effect was observed for elimination of ethylene from C_6H_{10}—Fe^+—C_2H_4. Loss of C_2H_4 was favored over loss of deuterated C_2 neutrals.

These results are not those expected from the traditional propargylic insertion mechanism. According to this mechanism, β-hydrogen transfer should be rate-determining and irreversible if it requires more energy than competitive ligand dissociation of ethylene. The binding energy of ethylene to the group 8 metals is estimated to be 40–70 kcal.[60] Conversely, β-hydrogen transfer should be reversible and non-rate-determining if it is a lower-energy process than the bond dissociation.[59]

An alternate mechanism proposed in Reference 59 resolves the experimental contradictions. An analogous mechanism (see Scheme 6) was postulated for the reactivity of Fe^+ with nitriles (see Section 3.4.1d). This mechanism is initiated by Fe^+ insertion into a C—H bond to form a metallocyclic intermediate.

3.4.1d. Reactions with Nitriles. The reactions of Fe^+ with linear nitriles demonstrate a different mechanism than the one proposed for Fe^+ reactions with alkynes. Lebrilla *et al.*[61] collisionally activated $Fe(RCN)^+$ ions in a BE-B triple-sector instrument and observed the losses of 2- and 28-mass-unit neutrals to give the predominant products with C_5–C_7 nitriles. The 2-mass-unit loss is attributed to H_2. ^{13}C labeling of the cyano carbon proved the loss of 28 can be attributed to C_2H_4 (presumably ethylene) rather than initial loss of H_2 and subsequent loss of CN. The familiar propargylic insertion–cleavage mechanism for alkynes does not explain the results observed for CA of Fe^+-nitriles. Apparently, the nitrile triple bond does not activate C—C and C—H bonds in the same manner that the alkyne triple bond does.

Deuterium labeling of specific carbons along the chain of the nitrile provides key information for determining the fragmentation mechanism. No deuterium loss was observed when the C-2 or C-3 position was labeled, whereas both deuterated and nondeuterated losses occur when the C-4 position was tagged. Loss from the C-4 position depends upon the overall nitrile chain length. Hexanitrile does not lose a deuterium from the labeled C-4 position, though pentanitrile loses both HD and $C_2H_2D_2$. Thus, Fe^+ activates a bond remote to itself in the complex. H_2 loss takes place according to a 1,2-elimination mechanism at the C-4 and C-5 positions. Ethylene loss is a competing process involving the same positions.

Coordination of Fe^+ to the nitrogen end of the nitrile may explain the observed reactivity. Fe^+ coordinates to alkenes and alkynes through a side-on interaction. The C-2 and C-3 hydrogens are located sufficiently close to the metal center (2.8 and 2.0 Å, respectively) to allow interaction,[62] specifically hydrogen transfer. The "end-on" geometry of the Fe^+-nitrile interaction places the C-2 and C-3 hydrogens too far from the metal center (ca. 4.5 and 5.2 Å, respectively) for interaction to be possible. The C-5 and C-6 hydrogens,

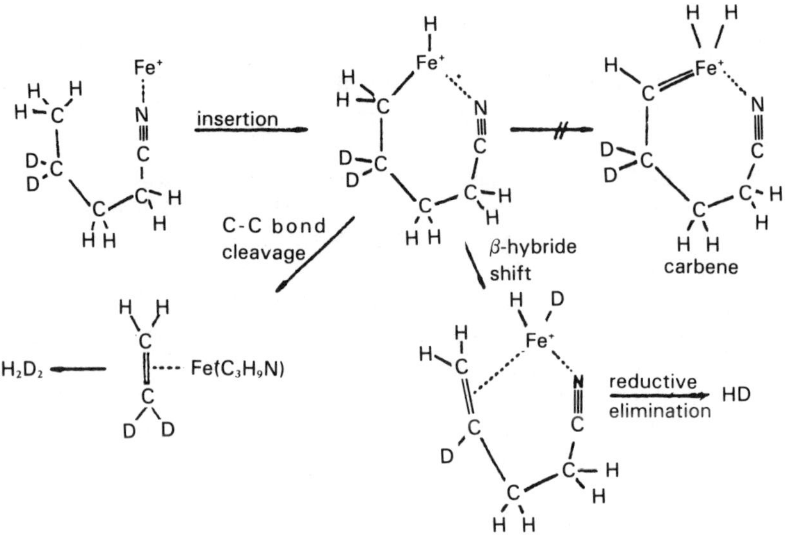

Scheme 6. (Reprinted with permission from *J. Am. Chem. Soc.*, copyright 1987, American Chemical Society.)

however, are closer to the metal (2.8 and 2.5 Å, respectively) so that hydrogen transfer occurs readily.

Scheme 6 illustrates the mechanism for remote activation of C—C and C—H bonds by Fe^+ in nitrile complexes. After end-on coordination with the nitrile, Fe^+ inserts into the C-5—H bond. A metallocycle is created by this step. If a β-hydrogen shift occurs next, a π-bonded metal-olefin complex forms. This step is followed by reductive elimination of HD from the labeled complexes. Otherwise, the C-3—C-4 bond is cleaved after the insertion step and $C_2H_2D_2$ is subsequently eliminated.

3.4.1e. Reactions with Alcohols. The gas phase chemistry of transition metals with alcohols has been a subject of recent interest. Weil and Wilkins[(22)] reported reactions of gold with alcohols and compared them to the reactions of silver and copper with alcohols. Their experiments utilized laser desorption/FTMS with metal oxides as precursors. Cu^+-alcohol reactions were also investigated earlier by Jones and Staley.[(63)] Ion–molecule reactions of iron, cobalt, and chromium anions with alcohols, cloro-alcohols, and alkyl halides were investigated by McElvaney and Allison[(64)] by using an ICR mass spectrometer. Huang *et al.*[(23)] used FTMS to study the reactivity of Fe^+, Cr^+, and Mo^+ with aliphatic alcohols.

Insertion of gas phase transition metals into R—OH bonds to eliminate H_2O was first proposed by Allison and Ridge.[(65)] Fe^+ reacts with methanol to produce $FeOH^+$, but not $Fe(CH_3)^+$. Metal insertion into the C—OH bond also occurs for $FeCO^+$ and CH_3OH. In a later study, the reactions of Fe^+, Co^+, and Ni^+ with alcohols were compared to those of Li^+ and Na^+.[(66)] Li^+ and

Na^+ also dehydrogenate alcohols, but by means of a mechanism that involves positive charge generation at the carbon bonded to the hydroxide.

Huang and Holman[67] recently studied the reactions of Fe^+ and Cr^+ with aliphatic alcohols (C_2-C_6) and compared high-energy CA reactions with unimolecular and low-energy CA reactions. Formation of $FeOH^+$ and $CrOH^+$ from $M(alcohol)^+$ (M = Fe or Cr) complexes dominates upon high-energy CA. CAD in an FTMS yields little $FeOH^+$ or $CrOH^+$. Unimolecular or metastable ion dissociation, which occurs wiithout target gas collisions, yields no $FeOH^+$ or $CrOH^+$. The high-energy CA produces other high-energy fragments that do not occur in metastable ion or low-energy CA fragmentations. The energy state differences account for the variations observed between spectra. The metal ion–alcohol complexes probably emerge from the spectrometer source in a ground electronic state. High-energy collisions with the target induce transitions to excited electronic states that undergo fragmentation. These excited states are not accessible upon low-energy excitations nor do they apply to ions undergoing metastable decompositions.

Comparison of the high-energy CAD spectra of Fe^+ and Cr^+ alcohol complexes suggests a difference in the reactivity of Fe^+ and Cr^+. Fe^+ forms a stronger covalent bond with alcohols as demonstrated by the greater abundance of fragments in the $Fe(alcohol)^+$ spectra. The high Cr^+ abundances in $Cr(alcohol)^+$ CAD spectra underscore the idea that the complex is more loosely bound, whereas the low Fe^+ abundances in the $Fe(alcohol)^+$ CAD spectra emphasize the formation of stronger Fe–alcohol covalent bonds.

3.4.1f. Analytical Applications. The study of gas phase metal ion chemistry has not only provided valuable information concerning ion–molecule reactions but has also enabled the development of a method for locating double and triple bonds in alkenes, dienes, alkynes, fatty acids, fatty acid esters, and alkenyl acetates.[68] This new method may be preferable to traditional derivatization techniques, which are more time-consuming, require larger amounts of sample to offset loss (and therefore may not be feasible with limited amounts of sample), and are not applicable to all olefins. Chemical ionization using transition metals as reagents was initially proposed by Burnier *et al.*[69] The new method discussed here finds its basis in the Fe^+-allylic insertion mechanism discussed earlier. After β-hydrogen transfer and formation of a *bis*(olefin) complex, collisional activation of the complex induces dissociation with release of the smaller olefin to yield information for identifying the point of unsaturation (see Equations 17–20).

Fe^+ reacts with double and triple bonds in this manner, even if other functional groups are present in the compound. Figure 7 illustrates the type of information obtained with a series of hexadecenyl acetate complexes of m/z 338. Ion C (m/z 295), a characteristic product in the spectra of these isomeric alkenyl acetates (C_{12}-C_{16}), results from the loss of the acyl radical. Fe-acetate$^+$ (m/z 115) is produced by cleavage of the alkenyl C—O bond. Ions A and B are products of allylic insertion of Fe^+. Other isomers such as ethyl 9-hexadecenoate are also readily distinguished from the acetate. Thus,

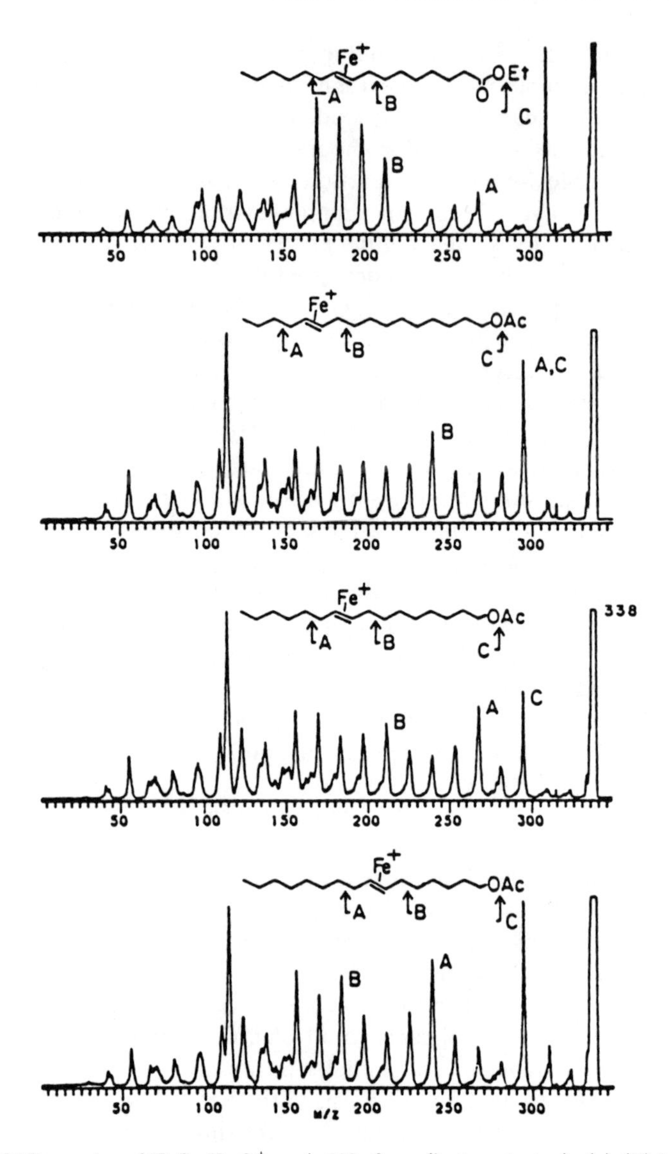

Figure 7. CAD spectra of $FeC_{18}H_{34}O_2^+$, m/z 338, from (bottom to top): (a) (Z)-hexadecenyl acetate, (b) (Z)-9-hexadecenyl acetate, (c) (Z)-11-hexadecenyl acetate, and (d) ethyl 9-hexadecenoate. (Reprinted with permission from *Analytical Chemistry*, copyright 1985, American Chemical Society.)

Fe^+ reactions provide information on both the nature and location of certain functional groups.

Another analytical application of CA in a tandem instrument is the analysis of simple mixtures of alkenes or alkynes.[70] Fe^+ is used as a chemical ionization (CI) reagent to locate multiple bonds. CI can be performed in a double-focusing mass spectrometer, but an MS/MS instrument is required to distinguish isomers in a mixture. The relative abundances of the characteristic fragment ions from specific isomers can be monitored to determine their mole fractions in the mixture. Standards of the mixture components are necessary. This approach is applicable to mixtures possessing a narrow dynamic range of concentrations but is susceptible to error due to collision cell pressure variations.

3.4.2. Bond Activation by Copper(I)

Research into the reactivity of copper(I) has revealed a surprisingly rich and varied chemistry. For example, a dissociative attachment mechanism (Scheme 7) was suggested to explain the reactivity of Cu^+ when used as a CI reagent for the analysis of esters and ketones[69] and to explain dehydrochlorination of alkyl halides by Cu^+.[71]

Another phenomenon, Cu^+ association with propene in an ICR cell, was observed by Burnier *et al.*[72] Association complexes between Cu^+ and alkenes and alkynes are formed as part of a CI protocol.

A third mode of reactivity, oxidative addition, was cited to explain Cu^+ activation of O—H bonds of small alcohols[63] and the O—NO bond of methyl nitrite.[73] In the following, the use of high-energy CA for understanding the reactivity of Cu^+ and alkanes, alkenes, and nitriles is established.

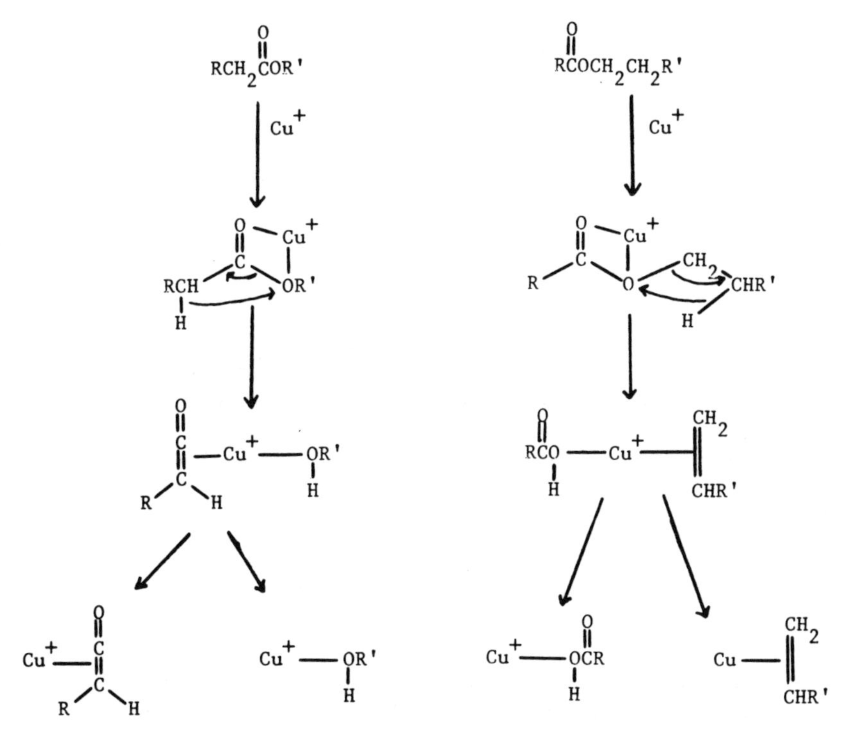

Scheme 7. (Reprinted with permission from *Analytical Chemistry*, copyright 1980, American Chemical Society.)

3.4.2a. Reactions with Alkanes. A study involving alkanes demonstrated that Cu^+ may insert into C—C and C—H bonds of isobutane.[74] Collisional activation of Cu^+-isobutane cluster ions leads primarily to loss of copper atoms, isobutane molecules, or copper–isobutane neutrals from the precursor. These fragmentations suggest that the copper ion and isobutane are weakly bound. Low abundances of metal insertion products, such as $CuCH_3^+$, $CuC_3H_6^+$, and $CuC_4H_8^+$, are also formed. These fragments are evidence for both C—C and C—H bond insertion by Cu^+; however, the reacting Cu^+ may be in an excited state. Collisional activation of the $CuC_4H_8{}^+$ fragment induces loss of mostly H_2 and C_4H_8. The C_4H_8 group, then, appears to be bound intact to the metal center rather than split into two pieces to create a *bis*(ethylene) complex. This evidence contrasts with that for *bis*(olefin) complex formation by the group 8, 9, and 10 transition metal ions reacting with linear hydrocarbons.[75]

3.4.2b. Reactions with Alkenes. A high-energy CA study presents interesting new information concerning the reactivity of Cu^+ with alkenes.[76] Data from two methods, FTMS to monitor reactivity coupled with pulsed laser ionization to produce Cu^+ and tandem sector mass spectrometry to permit collisional activation of Cu(olefin)$^+$ complexes, indicate that Cu^+ activates the allylic C—C bonds of alkenes. The highly similar CAD spectra of Cu(1-pentene)$^+$ and the 1-pentene complexes of Fe^+ and $TiCl^+$, both of which are known to activate C—C bonds, establish the analogies (see Figure 8). A highly abundant fragment in each spectrum is $MC_3H_6^+$, which arises by loss of C_2H_4 from the adduct.

Relative rate constants for disappearance of M^+, where M^+ refers to any first-row transition metal except Sc^+, which was not studied, were determined by using FTMS. The rate constants for Fe^+, Co^+, Ni^+, and Cu^+ reacting with 1-pentene are within the range of 0.96 (Co^+) and 0.65 (Cu^+). Cr^+ and Mn^+ fail to activate a C—C bond of 1-pentene.

Although the reactivity of Cu^+ with alkenes closely resembles that of Fe^+, allylic bond activation by Cu^+ is unexpected. Given the very stable electronic configuration of $3d^{10}$, one would expect Cu^+ to be similar to Cr^+, which has a stable $3d^5$ configuration and is unreactive with 1-pentene.

An oxidative addition mechanism may not be the means of bond activation by Cu^+. Oxidative addition is thought to require that two σ bonds be formed by using two $4s$ electrons on the metal.[77a,b] Depending on the ground state configuration of the metal, one or two electrons will need to be promoted from the $3d$ orbitals so that the metal will have a $3d^n\,4s^2$ configuration. Ions with high first or second promotion energies should, therefore, not react as easily as ions with low promotion energy. Fe^+, which is quite reactive, has a $3d^64s^1$ ground state configuration. The ion requires 65.3 kcal/mol to achieve the $3d^54s^2$ state.[77b] The promotion energies for Cu^+ are ~63 kcal/mol for the first electron and ~195 kcal/mol for the second electron.[77b] The promotion energies for Cr^+ are ~34 kcal/mol and ~148 kcal/mol for the first and second electrons, respectively. Accordingly, the reactivity of Cu^+ is expected to be

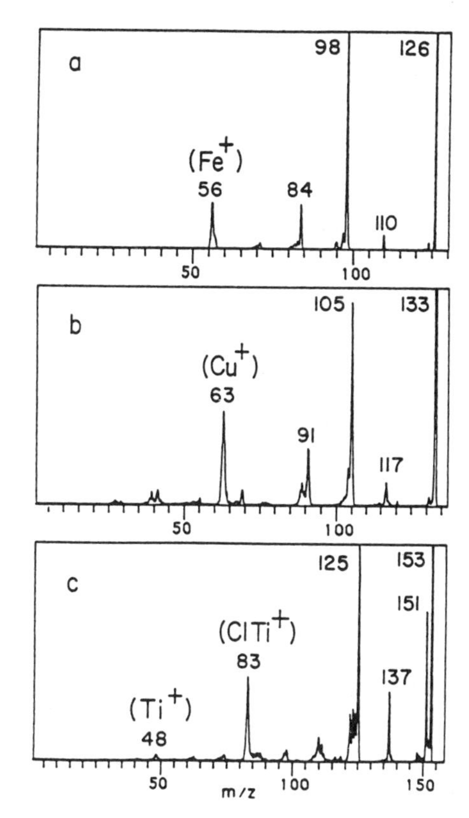

Figure 8. CAD spectra of 1-pentene complexes: (a) $FeC_5H_{10}^+$ m/z 126; (b) $CuC_5H_{10}^+$, m/z 133; $CuC_3H_6^+$, m/z 105; and, (c) Cl—$TiC_5H_{10}^+$, m/z 153; Cl—$TiC_3H_6^+$, m/z 125. (Reprinted with permission from *J. Am. Chem. Soc.*, copyright 1987, American Chemical Society.)

more similar to that of Cr^+, but it actually shows greater similarity to the reactivity of Fe^+.

Formation of a C—Cu^+ bond does not free enough energy to outweigh the energy required to cleave an allylic C—C bond (>70 kcal/mol). $(Cu—alkyl)^+$ bond energies are not known, but should be slightly greater than that of the $(Cu—H)^+$ bond, which has a calculated value of 21 kcal/mol.[42c] Mandich *et al.*[42c] observed periodic trends in the bond energies of first- and second-row transition metal cations and alkyl and hydrogen ligands. The $M^+—CH_3$ bonds were 2 to 20 kcal/mol stronger than the $M^+—H$ bonds for every case studied.

The dissociative attachment mechanism (Scheme 7) may be responsible for Cu^+ reactivity with alkenes, but it does not explain the inertness of Cr^+. The mechanism is initiated by Cu^+ association with a functional group, usually a heteroatom, to form an activated complex. The complex then rearranges or undergoes a cleavage reaction. The bond dissociation energy of $Cr^+—H$ is greater than that of $Cu^+—H$,[42c] so the Cr^+-alkyl bond should be stronger than the Cu^+-alkyl bond. Thus, Cr^+ would be expected to provide a greater driving force for the reaction with 1-pentene. More research is needed to clarify the reactivity of Cu^+ with alkenes.

3.4.2c. Reactions with Nitriles. A study by Lebrilla *et al.*[78] indicates that the reactivity of Cu^+ with nitriles differs from the mechanism of end-on coordination by other first-row transition metals such as Fe^+[61] and Co^+.[61] End-on coordination of Cu^+ with the nitrogen lone pair of electrons is favored thermodynamically but is thought to create unreactive $Cu(nitrile)^+$ complexes. CA of $Cu(nitrile)^+$ complexes, where the nitrile ranged from propyl to hexyl cyanide, generates three major fragments:

$$(N{\equiv}C{-}Cu{-}H)^+$$

2

$$(N{\equiv}C{-}CH_2{-}Cu)^+$$

3

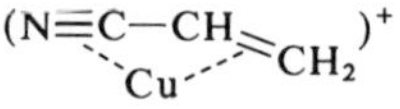

$$(N{\equiv}C{-}CH{=}CH_2)^+ \text{ (N≡C and CH=CH}_2\text{ coordinated to Cu)}$$

4

The nitriles were labeled to aid in the determination of the mechanism of reaction and structure of the complex. The results indicate that Cu^+ participates in a complicated mechanism which does not resemble those of Fe^+ or Co^+ with nitriles. Cu^+ inserts between the β- and the γ-carbon to form the ion with structure **3**. Ion **4** could be formed through either of two possible processes: a mechanism involving allylic C—H insertion, followed by β-alkyl transfer and subsequent reductive elimination, or cleavage of the β-γ carbon–carbon bond followed by hydrogen atom transfer. The dissociative attachment mechanism, thought to be important for the reactions of Cu^+ with alcohols[79,63] and ketones[69] may occur with nitriles also. This mechanism is considered a factor in the reactivity of alkali metal ions,[80] but fragments similar to **2**, **3**, and **4** are not observed upon reacting Na^+ with pentyl cyanide.

3.4.3. Cluster Ions

Considerable effort has been applied in recent years to understand the chemistry of metal cluster ions. Ion beam techniques,[82] ICR mass spectrometry,[83] and FTMS[84,19] are among the methods used to study cluster ion reactivity. High-energy collisional activation has also provided information about cluster ions.

High-energy collisional activation of cobalt/oxygen cluster ions was recently performed on a reverse-geometry instrument by Freas *et al.*[85] Three types of cobalt/oxygen ions were observed: oxygen-equivalent clusters $[(CoO)_x]^+$, oxygen-deficient clusters $[(Co(CoO)_x]^+$, and metal-deficient clusters $[(CoO)_xO]^+$. Collisional activation of these clusters provides insight into their structures.

Collisional activation of the oxygen-equivalent clusters produces $[(CoO)_{x-y}]^+$ ($x = 1$ to 5) fragments, reflecting the loss of $(CoO)_y$.[82] The most abundant fragment from $[(CoO)_5]^+$ is $[(CoO)_4]^+$ from loss of CoO. $[(CoO)_3]^+$ and $[(CoO)_2]^+$ are also abundant. Loss of a single cobalt ion from any of the precursor ions was not observed. The fragmentation data suggest that Co and O are paired together in the cluster ion, interacting ionically. The $[(CoO)_x]^+$ formula for the ions is a better representation of the ionic structure than $[Co_xO_x]^+$.

$(CoO)_yO$ groups also are liberated from the oxygen-equivalent clusters, but these fragmentations are more prevalent for small clusters. Larger clusters lose $(CoO)_y$ primarily. The dependence of fragment type on cluster size supports the contention that the oxygen-equivalent cluster ions interact ionically. The overall positive charge on a large cluster ion affects fragmentation to a lesser extent than does the charge on a small cluster ion. Large clusters, therefore, lose paired Co—O fragments. The positive charge on small clusters causes them to dissociate to products that can best accommodate the charge.

Freas and Campana[85b] reported C—H bond activation of isobutane complexed with oxygen-deficient clusters. CA of $[Co_2OC_4H_{10}]^+$ yields primarily $[Co_2O]^+$ and $[Co_2OC_4H_8]^+$. $[Co_2O]^+$ forms by loss of C_4H_{10}, presumably isobutane, and $[Co_2OC_4H_8]^+$ is the product of dehydrogenation. The adduct involving an oxygen-equivalent cluster ion only loses isobutane upon collisional activation; dehydrogenation does not occur. The fragments are indicative of an isobutane weakly bound to the oxygen-equivalent cluster ion. Insertion occurs with oxygen-deficient clusters.

Structural differences between the two types of cluster ions may account for their different reactivities. The low-energy forms of oxygen-equivalent cluster ions are rings, ladders, and cages.[86] These closed structures render the clusters unreactive. The low-energy structures for the reactive oxygen-deficient clusters are chains and other structures possessing terminal cobalt atoms. With Co^+ in a sterically accessible position, these structures permit Co^+ bond insertion.

Freas and Campana[74] investigated CA of copper cluster ions. Metal insertion into both C—C and C—H bonds occurs for $[Cu_x$—isobutane$]^+$ ($x = 1, 2$), but the fragmentation products do not support C—C bond insertion by Cu_3^+. C—H bond activation is indicated for the latter cluster, however. The clusters of isobutane and three copper atoms lose Cu and Cu_2, suggesting that the interaction between the metal and isobutane is limited to a single metal atom. Loss of Cu and Cu_2 also suggests that the theoretical triangular structure[87] proposed for Cu_3^+ may be incorrect. Instead, Cu_3^+ may be linear or bent in form, thereby allowing a single copper to associate with the organic moiety.

3.5. Metastable Ion Studies

Metastable decompositions constitute a second category of fragmentations that provide the metal ion chemist with structural and mechanistic information.

A metastable decomposition is a unimolecular fragmentation occurring in a field-free region of a mass spectrometer. It is indicated by a wide, diffuse spectral peak when analyzed by using an ESA. When an ion, m_1, decomposes to form m_2, a peak appears in the normal mass spectrum at $m/z\, m^*$ such that

$$m^* = (m_2^2)/(m_1) \tag{23}$$

where m_2 is the mass of the fragment ion and m_1 is the mass of the precursor ion.

The residence time of an ion in the source and the length of the field-free region are factors that determine the extent of metastable ion reactions. The metastable ion reaction will occur within 1 to 10 μs after an ion is formed, and thus the rate constant is in the narrow range of 10^5 to $10^6\ s^{-1}$. The time range depends upon the ion acceleration and physical makeup of the mass spectrometer. Ions that remain stable for 100 μs are able to reach the detector intact. Those highly unstable ions that decompose immediately after ionization but prior to acceleration out of the source (within less than 10^{-7} s) are not detected as metastable ions.

Metastable transitions are unequivocally studied by using the same scanning methods as for collisionally activated processes.[88,89] Metastable decompositions occurring prior to the electric sector in a forward-geometry (EB) sector instrument will not be seen using normal scan modes. The electric sector voltage and accelerating voltage are normally coupled so that only ions with the full amount of kinetic energy (or accelerating voltage) can pass through the electric sector. Because metastable ions have reduced kinetic energy owing to energy partitioning upon decomposition, they will not be transmitted. Reducing the electric sector voltage from E_1 to E_2 (Equation 24) while holding the accelerating voltage constant will allow m_2 to pass.

$$m_2 = (E_1/E_2)m_1 \tag{24}$$

A scan of the electric sector voltage (E scan) will produce an ion kinetic energy (IKE) spectrum.[90] The ions are only separated by kinetic energy and not mass so multiple combinations of ions can satisfy Equation 24.

A reverse-geometry instrument eliminates ambiguities concerning M_1 and M_2. The magnetic sector can be set to select M_1 prior to metastable decomposition. The electric sector is then used to analyze the fragment ions.[91]

A V scan on a forward-geometry instrument will also specifically identify m_1 and m_2.[92] For this method, the m_2 ion is brought to focus and the electric sector voltage is held constant while the accelerating voltage, V, is raised to V_1 such that

$$V_1 = (m_1/m_2)V \tag{25}$$

When the accelerating voltage equals V_1, m_2, formed via a metastable process, is able to pass through the electric sector.

The study of metastable reactions helps determine ion structure. Metastable decompositions are generally low activation energy processes, such as rearrangements as opposed to simple cleavages.[93] Metastable peak shapes provide additional structure information and help differentiate isomers. The kinetic energy release (KER) upon decomposition determines the peak shape and, for large energy release, flat-topped peaks are observed because of slit discrimination. The degree of flatness is directly related to the amount of energy freed.

Beauchamp, Bowers, and co-workers[94] recently investigated metastable and collisionally activated decompositions to gain insight into the potential energy surfaces of the reactions of Co^+ with alkanes and pentenes. Results from these studies complement the findings from earlier work.[95,60]

The amount of kinetic energy released during a metastable decomposition can provide details of the potential energy surface for the reaction. An activated adduct may dissociate along one of two general potential energy surfaces. A surface without a reverse activation barrier is indicative of simple bond cleavage reactions. A more complex reaction, such as one incorporating the concurrent breakage and formation of several bonds, will usually possess a significant reverse activation barrier. Simple cleavages release less kinetic energy whereas more complex reactions show kinetic energy release originating from the reverse activation energy.

Hanratty *et al.*[94] used kinetic energy release distributions to verify mechanisms of reaction of Co^+ with alkanes. KER distributions for dehydrogenation of a deuterated Co^+-butane adduct were found to be similar to that of a labeled Ni^+-butane adduct known to eliminate D_2 via a 1,4-elimination. These results are additional evidence for 1,4-elimination of H_2 from the cobalt complex and are in accord with Jacobson and Freiser's[4a,20] earlier conclusions. CAD of a Co^+-*n*-butane adduct demonstrates that the ion exists as a *bis*(olefin)-metal complex. The CAD results, thus, also support the 1,4-elimination mechanism.

A theoretical KER distribution for dehydrogenation was modeled by using phase space theory,[96,97] assuming no reverse activation barrier. The calculated distribution[94] is broader than the observed one. The discrepancy between these results indicates that the energy release does not occur in a wholly statistical manner, as is assumed in phase space theory.

Elimination of alkanes from the Co^+-butane adducts produces KER distributions that agree with statistical models. The match in distributions implies that the metal-alkane interaction is strong and the transition state is loose.

A study of Co^+ reactions with pentene isomers and cyclopentene[94b] also employed CAD and potential energy surface determinations to obtain information about ion structure and reaction mechanisms. Losses of H_2 and CH_4 from the adducts appear to occur with nearly equal activation barriers. Interconversion between isomeric adducts was observed. Also, H_2 loss from the Co^+-cyclopentane adduct appears to occur without ring cleavage. This observation corroborates the results of Jacobson and Freiser.[98]

4. CONCLUSIONS

Collisional activation is an especially versatile and useful technique for studying metal ion reactivity and determining structure of adducts of metal ions and organic molecules. Inelastic collisions with a target gas increase the internal energy of an ion and induce different reactions depending on the amount of energy transferred. High-energy collisions induce decompositions into structurally characteristic fragment ions. The many MS/MS instruments available enable the metal ion chemist to select an instrument with the energy, resolution, mass range, and transmission capabilities needed to study the metal ion chemistry of interest.

High-energy collisional activation has presented new information and has also corroborated and extended research performed with other methods. The allylic mechanism of alkene bond activation by Fe^+ has been verified and elaborated by high-energy CAD.[54,58] A new mechanism involving a metallocycle intermediate has been proposed for the collisionally activated reactions of Fe^+ with nitriles.[61] Bond activation by Cu^+ was first observed using high-energy CAD.[76] Cluster ion studies incorporating the technique have suggested a correlation between reactivity and cluster geometry.[74,85]

There are several reasons why high-energy collisional activation is an effective method for elucidating the structure and reactivity of transition metal ions and their complexes. Spectra obtained via high-energy CAD are highly reproducible and contain fragments that are characteristic of ion structure. Ions activated via the high-energy experiment are less susceptible to skeletal rearrangements than those activated in the low-energy experiment. Additionally, fragment ion abundances from high-energy CAD show little sensitivity to small changes in precursor ion internal energy. The same instruments used for high-energy activation are ideal for determining kinetic energy release in decomposition reactions to provide details of the reaction coordinate. High-energy collisional activation and metastable ion decompositions are already a powerful and sensitive means to understand the chemistry of metal ions and promise to continue to be so in the future. Given the ever increasing availability of tandem mass spectrometers, studies of gas phase metal ions should expand to many new laboratories.

ACKNOWLEDGMENT

This chapter was prepared with the support of the NSF Midwest Center for Mass Spectrometry (Grant No. CHE-8620177).

REFERENCES

1. McLafferty, F.W., Ed. *Tandem Mass Spectrometry*; John Wiley: New York, 1983.
2. (a) Gross, M.L.; Chess, E.K.: Lyon, P.A.; Crow, F.W.; Evans, S.; Tudge, H. *Int. J. Mass Spectrom Ion Phys.* 1982, *42*, 243; (b) Glish, G.L.; McLuckey, S.A.; Ridley, T.Y.; Cooks, R.G.

Int. J. Mass Spectrom. Ion Phys. 1982, *41*, 157; (c) Lane, K.R.; Lee, R.E.; Sallans, L.; Squires, R.R. *J. Am. Chem. Soc.* 1984, *106*, 5767.

3. Louris, J.N.; Wright, L.G.; Cooks, R.G.; Schoen, A.E. *Anal. Chem.* 1985, *57*, 2918.
4. (a) Jacobson, D.B.; Freiser, B.S. *J. Am. Chem. Soc.* 1983, *105*, 736; (b) Jacobson, D.B.; Freiser, B.S. *J. Am. Chem. Soc.* 1984, *106*, 4623; (c) Jacobson, D.B.; Freiser, B.S. *Organometallics* 1985, *4*, 1048.
5. Beynon, J.H.; Cooks, R.G.; Amy, J.W.; Baitinger, W.E.; Ridley, T.Y. *Anal. Chem.* 1973, *45*, 1023A.
6. Freas, R.B.; Ridge, D.P. *J. Am. Chem. Soc.* 1980, *102*, 7129.
7. (a) Russell, D.H.; Smith, D.H.; Warmack, R.J.; Bertram, L.K. *Int. J. Mass Spectrom. Ion Phys.* 1980, *35*, 381; (c) Burinsky, D.J.; Cooks, R.G.; Chess, E.K.; Gross, M.L. *Anal. Chem.* 1982, *54*, 295.
8. Bursey, M.M.; Hass, J.R. *J. Am. Chem. Soc.* 1985, *107*, 115.
9. Biemann, K.; Gibson, B.W.; Mathews, W.R.; Pang, H. In *Mass Spectrometry in the Health and Life Sciences*; Burlingame, A.L.; Castagnoli, N., Jr., Eds.; Elsevier: Amsterdam, 1985, 239.
10. (a) Yost, R.A.; Enke, C.G. *J. Am. Chem. Soc.* 1978, *100*, 2274; (b) Yost, R.A.; Enke, C.G. *Anal. Chem.* 1979, *51*, 1251A.
11. McGilvery, D.C., Morrison, J.D. *Int. J. Mass Spectrom. Ion Phys.* 1978, *28*, 81.
12. McLafferty, F.W.; Todd, P.J.; McGilvery, D.C.; Baldwin, M.A. *J. Am. Chem. Soc.* 1980, *102*, 3360.
13. Cottrell, J.S.: Greathead, R.J. *Mass Spec. Rev.* 1986, *5*, 215.
14. McLuckey, S.A.; Glish, G.L. *Anal. Chem.* 1986, *58*, 1887.
15. Ciupek, J.D.; O'Lear, J.R.; Cooks, R.G.; Dobberstein, P.; Schoen, A.E. Extended Abstracts of the 32nd Annual Meeting on Mass Spectrometry and Allied Topics, 1984, 378; American Society for Mass Spectrometry.
16. Comisarow, M.B.; Marshall, A.G. *Chem. Phys. Lett.* 1974, *25*, 282.
17. Lehman, T.A.; Bursey, M.M. *Ion Cyclotron Resonance Spectrometry*; John Wiley: New York, 1976.
18. Kaplan, F. *J. Am. Chem. Soc.* 1968, *90*, 4483.
19. Tews, E.C.; Freiser, B.S. *J. Am. Chem. Soc.* 1987, *109*, 4433.
20. Jacobson, D.B.; Freiser, B.S. *J. Am. Chem. Soc.* 1983, *105*, 5197.
21. (a) Lombarski, M.; Allison, J. *Int. J. Mass Spectrom. Ion Phys.* 1985, *65*, 31; (b) Radecki, B.D.; Allison, J. *J. Am. Chem. Soc.* 1984, *106*, 946; (c) Huang, S.K.; Allison, J. *Organometallics* 1983, *2*, 883.
22. Weil, D.A.; Wilkins, C.L. *J. Am. Chem. Soc.* 1985, *107*, 7136.
23. Huang, S.; Holman, R.W.; Gross, M.L. *Organometallics* 1986, *5*, 1857.
24. Fredeen, D.A.; Russell, D.H. *J. Am. Chem. Soc.* 1985, *107*, 3762.
25. Bricker, D.L.; Adams, T.A., Jr.; Russell, D.H. *Anal. Chem.* 1983, *55*, 2417.
26. Berstein, R.B.; Levine, R.D. *Molecular Reaction Dynamics*; Clarendon Press: Oxford, 1974.
27. Yost, R.A.; Enke, C.G.; McGilvery, D.C.; Smith, D.; Morrison, J.D. *Int. J. Mass Spectrom. Ion Phys.* 1979, *30*, 127.
28. McLafferty, F.W. *Acts. Chem. Res.* 1980, *13*, 33.
29. McLafferty, F.W.; Todd, P.J.; McGilvery, D.C.; Baldwin, M.A. *J. Am. Chem. Soc.* 1980, *102*, 3360.
30. McLafferty, F.W.; Bente, P.F., III; Kornfeld, R.; Tsai, S.C.; Have, J. *J. Am. Chem. Soc.* 1973, *95*, 2120.
31. Porter, C.J.; Proctor, C.J.; Beynon, J.H. *Org. Mass Spectrom.* 1981, *16*, 62.
32. Fetterolf, D.D.; Yost, R.A. *Int. J. Mass Spectrom. Ion Phys.* 1982, *44*, 37.
33. McLuckey, S.A.; Glish, G.L.; Cooks, R.G. *Int. J. Mass Spectrom. Ion Phys.* 1981, *39*, 219.
34. Holmes, J.L.; Terlouw, J.K.; Burgers, P.C.; Rye, R.T.B. *Org. Mass Spectrom.* 1980, *15*, 149.
35. Dass, C.; Sack, T.M.; Gross, M.L. *J. Am. Chem. Soc.* 1984, *106*, 5780.
36. Dass, C.; Peake, D.A.; Gross, M.L. *Org. Mass Spectrom.* 1986, *21*, 741.
37. Wagner-Redekar, W.; Levsen, K. *Org. Mass Spectrom.* 1981, *16*, 538.
38. Mason, R.S.; Jennings, K.R.; Verma, S.; Cooks, R.G. *Org. Mass Spectrom.* 1985, *20*, 727.
39. Verma, S.; Ciupek, J.D.; Cooks, R.G. *Int. J. Mass Spectrom. Ion Proc.* 1984, *62*, 219.

40. Armentrout, P.B.; Beauchamp, J.L. *Chem. Phys.* 1980, *50*, 21.
41. Armentrout, P.B.; Beauchamp, J.L. *Chem. Phys.* 1980, *50*, 27.
42. (a) Halle, L.F.; Klein, F.S.; Beauchamp, J.L. *J. Am. Chem. Soc.* 1984, *106*, 2543; (b) Houriet, R.; Halle, L.F.; Beauchamp, J.L. *Organometallics* 1983, *2*, 1818; (c) Mandich, M.L.; Halle, L.F.; Beauchamp, J.L. *J. Am. Chem. Soc.* 1984, *106*, 4403.
43. Halle, L.F.; Armentrout, P.B.; Beauchamp, J.L. *Organometallics* 1982, *1*, 963.
44. Elkind, J.L.; Armentrout, P.B. *J. Phys. Chem.* 1987, *91*, 2037.
45. Tsarbopoulos, A.; Allison, J. *Organometallics* 1984, *3*, 86.
46. Lane, K.R.; Sallans, L.; Squires, R.R. *J. Am. Chem. Soc.* 1986, *108*, 4368.
47. Kang, H.; Mauldin, C.H.; Cole, T.; Slegeir, W.; Cann, K.; Pettit, R. *J. Am. Chem. Soc.* 1977, *99*, 8323.
48. Pearson, R.G.; Mauermann, H. *J. Am. Chem. Soc.* 1982, *104*, 500.
49. Remick, R.J.; Asunta, T.A.; Skell, P.S. *J. Am. Chem. Soc.* 1979, *101*, 1320.
50. (a) Armentrout, P.B.; Beauchamp, J.L. *J. Am. Chem. Soc.* 1980, *102*, 1736; (b) Armentrout, P.B.; Beauchamp, J.L. *J. Am. Chem. Soc.* 1981, *103*, 784.
51. Larsen, B.S.; Ridge, D.P. *J. Am. Chem. Soc.* 1984, *106*, 1912.
52. Byrd, G.D.; Burnier, R.C.; Freiser, B.S. *J. Am. Chem. Soc.* 1982, *104*, 3565.
53. Radecki, B.D.; Allison, J. *Organometallics* 1986, *5*, 411.
54. Peake, D.A.; Gross, M.L.; Ridge, D.P. *J. Am. Chem. Soc.* 1984, *106*, 4307.
55. Armentrout, P.B.; Beauchamp, J.L. *J. Am. Chem. Soc.* 1981, *103*, 6628.
56. Ridge, D.P.; Kalmbach, K.A. *Adv. Mass Spec.* 1985, 813.
57. Jacobson, D.B.; Freiser, B.S. *J. Am. Chem. Soc.* 1983, *105*, 5197.
58. Peake, D.A.; Gross, M.L. *Organometallics* 1985, *5*, 1236.
59. Schulze, C.; Schwarz, H.; Peake, D.A.; Gross, M.L. *J. Am. Chem. Soc.* 1987, *109*, 2368.
60. Armentrout, P.B.; Halle, L.F.; Beauchamp, J.L. *J. Am. Chem. Soc.* 1981, *103*, 6624.
61. Lebrilla, C.B.; Schulze, C.; Schwarz, H. *J. Am. Chem. Soc.* 1987, *109*, 98.
62. Brookhart, M.; Green, M.L.H. *J. Organomet. Chem.* 1983, *250*, 395.
63. Jones, R.W.; Staley, R.H. *J. Phys. Chem.* 1982, *86*, 1669.
64. McElvaney, S.W.; Allison, J. *Organometallics* 1986, *5*, 416.
65. Allison, J.; Ridge, D.P. *J. Am. Chem. Soc.* 1976, *98*, 7445.
66. Allison, J.; Ridge, D.P. *J. Am. Chem. Soc.* 1979, *101*, 4998.
67. Huang, S.K.; Holman, R.W. Extended Abstracts of the 34th Annual Conference on Mass Spectrometry and Allied Topics, 1986, 614; American Society for Mass Spectrometry.
68. Peake, D.A.; Gross, M.L. *Anal. Chem.* 1985, *57*, 115.
69. Burnier, R.C.; Byrd, G.D.; Freiser, B.S. *Anal. Chem.* 1980, *52*, 1641.
70. Peake, D.A.; Huang, S.K.; Gross, M.L. *Anal. Chem.* 1987, *59*, 1557.
71. Jones, R.W.; Staley, R.H. *J. Am. Chem. Soc.* 1980, *102*, 3794.
72. Burnier, R.C.; Carlin, T.J.; Reents, W.D., Jr.; Cody, R.B.; Lengel, R.K.; Freiser, B.S. *J. Am. Chem. Soc.* 1979, *101*, 7127.
73. Cassady, C.J.; Freiser, B.S. *J. Am. Chem. Soc.* 1985, *107*, 1566.
74. Freas, R.B.; Campana, J.E. *J. Am. Chem. Soc.* 1985, *107*, 6202.
75. (a) Jacobson, D.B.; Freiser, B.S. *J. Am. Chem. Soc.* 1983, *105*, 5197; (b) Larsen, B.S.; Ridge, D.P. *J. Am. Chem. Soc.* 1984, *106*, 1912.
76. Peake, D.A.; Gross, M.L. *J. Am. Chem. Soc.* 1987, *109*, 600.
77. (a) Armentrout, P.B.; Halle, L.F.; Beauchamp, J.L. *J. Am. Chem. Soc.* 1981, *103*, 6501; (b) Babinec, S.J.; Allison, J. *J. Am. Chem. Soc.* 1984, *106*, 7718.
78. Lebrilla, C.B.; Drewello, T.; Schwarz, H. *Organometallics Comm.* 1987, *6*, 2450.
79. Weil, D.A.; Wilkins, C.L. *J. Am. Chem. Soc.* 1985, *107*, 7316.
80. Wysocki, V.H.; Kenttamaa, H.I.; Cooks, R.G. *Int. J. Mass. Spectrom. Ion Proc.* 1987, *75*, 181.
81. Armentrout, P.B.; Loh, S.K.; Ervin, K.M. *J. Am. Chem. Soc.* 1984, *106*, 1161.
82. Ervin, K.; Loh, S.K.; Aristov, N.; Armentrout, P.B. *J. Phys. Chem.* 1983, *87*, 3593.
83. (a) Jacobson, D.B.; Freiser, B.S. *J. Am. Chem. Soc.* 1985, *107*, 1581; (b) Larsen, B.S.; Freas, R.B.; Ridge, D.P. *J. Phys. Chem.* 1984, *88*, 6014.
84. Meckstroth, W.K.; Ridge, D.P.; Reents, W.D., Jr. *J. Phys. Chem.* 1985, *89*, 612.

85. (a) Freas, R.B.; Dunlap, B.I.; Waite, B.A.; Campana, J.E. Preprint of 191st ACS meeting presentation; (b) Freas, R.B.; Campana, J.E. *J. Am. Chem. Soc.* 1986, *108*, 4659.
86. Campana, J.E.; Freas, R.B. *J. Chem. Soc. Chem. Commun.* 1984, 1414.
87. (a) Post, D.; Baerends, E.J. *Chem. Phys. Lett.* 1982, *86*, 176; (b) Basch, H. *J. Am. Chem. Soc.* 1981, *103*, 4657.
88. Cooks, R.G.; Beynon, J.H.; Caprioli, R.M.; Lester, G.R. *Metastable Ions*; Elsevier: Amsterdam, 1973.
89. Beynon, J.H. *Anal. Chem.* 1970, *42*, 97A.
90. Beynon, J.H.; Caprioli, R.M.; Baitinger, W.W.; Amy, J.W. *Int. J. Mass Spectrom. Ion Phys.* 1969, *3*, 313.
91. Smith, D.H.; Djerassi, C.; Maurer, K.H.; Rapp, U. *J. Am. Chem. Soc.* 1974, *96*, 3482.
92. Barber, M.; Wolsteholme, W.A.; Jennings, K.R. *Nature* 1967, *214*, 664.
93. McLafferty, F.W.; Fairweather, R.B. *J. Am. Chem. Soc.* 1968, *90*, 5915.
94. Hanratty, M.A.; Beauchamp, J.L.; Illies, A.J.; Bowers, M.T. *J. Am. Chem. Soc.* (submitted).
95. Armentrout, P.B.; Halle, L.F.; Beauchamp, J.L. *J. Am. Chem. Soc.* 1981, *103*, 6628.
96. Pechukas, P.; Light, J.C.; Rankin, C.J. *Chem. Phys.* 1966, *44*, 794.
97. Chesnavich, W.J.; Bowers, M.T. *J. Am. Chem. Soc.* 1976, *98*, 8301.
98. (a) Jacobson, D.B.; Freiser, B.S. *Organometallics* 1984, *3*, 513; (b) Jacobson, D.B.; Freiser, B.S. *J. Am. Chem. Soc.* 1983, *105*, 7492.
99. (a) Elkind, J.L.; Alford, J.M.; Weiss, F.D.; Laaksonen, R.T.; Smalley, R.E. *J. Phys. Chem.* 1987, *87*, 2397; (b) Alford, J.M.; Weiss, F.D.; Laaksonen, R.T.; Smalley, R.E. *J. Phys. Chem.* 1986, *90*, 4480.

Index